Life Histories of
North American Blackbirds,
Orioles, Tanagers, and Allies

by Arthur Cleveland Bent

Dover Publications, Inc., New York

Published in the United Kingdom by Constable and Company Limited, 10 Orange Street, London W. C. 2.

This Dover edition, first published in 1965, is an unabridged and unaltered republication of the work first published in 1958 by the United States Government Printing Office, as Smithsonian Institution United States National Museum *Bulletin 211.*

International Standard Book Number: 0-486-21093-6
Library of Congress Catalog Card Number: 65-12256

Manufactured in the United States of America

Dover Publications, Inc.
180 Varick Street
New York, N. Y. 10014

CONTENTS

Introduction

This is the twentieth in a series of bulletins of the United States National Museum on the life histories of North American birds. Previous numbers have been issued as follows:

107. Life Histories of North American Diving Birds, August 1, 1919.
113. Life Histories of North American Gulls and Terns, August 27, 1921.
121. Life Histories of North American Petrels and Pelicans and Their Allies, October 19, 1922.
126. Life Histories of North American Wild Fowl (part), May 25, 1923.
130. Life Histories of North American Wild Fowl (part), June 27, 1925.
135. Life Histories of North American Marsh Birds, March 11, 1927.
142. Life Histories of North American Shore Birds (pt. 1), December 31, 1927.
146. Life Histories of North American Shore Birds (pt. 2), March 24, 1929.
162. Life Histories of North American Gallinaceous Birds, May 25, 1932.
167. Life Histories of North American Birds of Prey (pt. 1), May 3, 1937.
170. Life Histories of North American Birds of Prey (pt. 2), August 8, 1938.
174. Life Histories of North American Woodpeckers, May 23, 1939.
176. Life Histories of North American Cuckoos, Goatsuckers, Hummingbirds, and Their Allies, July 20, 1940.
179. Life Histories of North American Flycatchers, Larks, Swallows, and Their Allies, May 8, 1942.
191. Life Histories of North American Jays, Crows, and Titmice, January 27, 1947.
195. Life Histories of North American Nuthatches, Wrens, Thrashers, and Their Allies, July 7, 1948.
196. Life Histories of North American Thrushes, Kinglets, and Their Allies, June 28, 1949.
197. Life Histories of North American Wagtails, Shrikes, Vireos, and Their Allies, June 21, 1950.
203. Life Histories of North American Wood Warblers, June 15, 1953.

The same general plan has been followed, as explained in previous bulletins and need not be repeated here. The nomenclature of the 1931 Check-List of the American Ornithologists' Union, with its latest supplements, has been followed.

Many who have contributed material for previous volumes have continued to cooperate. Receipt of material from over 530 contributors has been acknowledged previously. In addition to these, our thanks are due to the following new contributors: Hildegarde C. Allen, F. S. Barkalow, Jr., Ralph Beebe, H. E. Bennett, A. J. Berger, Virgilio Biaggi, Jr., C. H. Blake, Don Bleitz, B. J. Blincoe, L. C. Brecher, Maurice Broun, J. H. Buckalew, H. L. Crockett, Ruby Curry, J. V. Dennis, M. S. Dunlap, J. J. Elliott, A. H. Fast, Edith

K. Frey, J. H. Gerard, H. B. Goldstein, L. I. Grinnell, Horace Groskin, G. W. Gullion, R. H. Hansman, W. R. Hecht, J. W. Hopkins, R. F. James, Verna R. Johnston, Malcolm Jollie, R. S. Judd, Louise de K. Lawrence, G. H. Lowery, J. M. Markle, D. L. McKinley, Lyle Miller, A. H. Morgan, R. A. O'Reilly, K. C. Parkes, G. H. Parks, O. M. Root, Doris H. Speirs, E. A. Stoner, and R. B. Williams. If any contributor fails to find his or her name in this or in some previous volume, the author would be glad to be advised.

As the demand for these bulletins is much greater than the supply, the names of those who have not contributed to the work within recent years will be dropped from the author's mailing list.

Winsor M. Tyler rendered valuable assistance by reading and indexing four of the leading ornithological journals for references. He and Alfred O. Gross each contributed two complete life histories. Alexander F. Skutch, Alexander Sprunt, Jr., Laidlaw Williams, and Robert S. Woods have each contributed one complete life history.

The greater part of the egg measurements were taken from the register sheets of the United States National Museum by William George F. Harris, who also relieved the author of a vast amount of detail work by collecting and figuring hundreds of egg measurements and by collecting, sorting, and arranging several thousand nesting records to make up the "egg dates" paragraphs.

Through the courtesy of the Fish and Wildlife Service, Mr. Chandler S. Robbins has compiled the migration paragraphs. The distribution data have been taken from advance sheets of the fifth edition of the A. O. U. "Check-List of North American birds." The author claims no credit and assumes no responsibility for these data, which are taken from the great mass of records on file in Washington.

The manuscript for this bulletin was completed in 1949. Contributions received since then will be acknowledged later. Only information of great importance could be added. The reader is reminded again that this is a cooperative work; if he fails to find in these volumes anything that he knows about the birds, he can only blame himself for not having sent the information to—

THE AUTHOR.

Publications of the U. S. National Museum

The scientific publications of the National Museum include two series known, respectively, as *Proceedings* and *Bulletin*.

The *Proceedings* series, begun in 1878, is intended primarily as a medium for the publication of original papers based on the collections of the National Museum, that set forth newly acquired facts in biology, anthropology, and geology, with descriptions of new forms and revisions of limited groups. Copies of each paper, in pamphlet form, are distributed as published to libraries and scientific organizations and to specialists and others interested in the different subjects. The dates at which these separate papers are published are recorded in the table of contents of each of the volumes.

The series of *Bulletins*, the first of which was issued in 1875, contains separate publications comprising monographs of large zoological groups and other general systematic treatises (occasionally in several volumes), faunal works, reports of expeditions, catalogs of type specimens, special collections, and other material of similar nature. The majority of the volumes are octavo in size, but a quarto size has been adopted in a few instances. In the *Bulletin* series appear volumes under the heading *Contributions from the United States National Herbarium*, in octavo form, published by the National Museum since 1902, which contain papers relating to the botanical collections of the Museum.

The present work forms No. 211 of the *Bulletin* series.

REMINGTON KELLOGG,
Director, United States National Museum.

Arthur Cleveland Bent

Arthur Cleveland Bent died in his home in Taunton, Massachusetts, in his eighty-ninth year on December 30, 1954. In keeping with his ability, as a successful business man, to plan for eventualities he had long foreseen the improbability of his living to complete the Life Histories of North American Birds, and had taken steps to assure completion of the work. Quite naturally, he turned to the Nuttall Ornithological Club, which he had originally joined in 1888, later becoming one of the four Honorary Members in the history of that organization. He arranged through James Lee Peters for the Club to take ultimate charge. Subsequently, he chose me to head up a Committee of national distribution, largely ornithologists he designated.

In addition to Mrs. Bent, whose cooperation has been invaluable, members of the Committee are Messrs. Arthur W. Argue in charge of photographs, Charles H. Blake, Alfred O. Gross, William George F. Harris, so well known already in this series for his work on eggs, Frederick C. Lincoln, who has handled so much of the detail in Washington, Robert A. Norris, Christopher M. Packard, and Lawrence H. Walkinshaw.

Mr. Bent could anticipate completion. He could not foresee the fervor with which ornithologists throughout North and Central America have rallied to ensure fulfillment of his great undertaking. Than this there can be no higher praise.

WENDELL TABER,
Chairman, Arthur Cleveland Bent Life History Committee,
Nuttall Ornithological Club.

LIFE HISTORIES OF NORTH AMERICAN WEAVER FINCHES, BLACKBIRDS, ORIOLES, AND TANAGERS

By Arthur Cleveland Bent

Order Passeriformes: Family Ploceidae, Weaver Finches

PASSER DOMESTICUS DOMESTICUS (Linnaeus)

English Sparrow

HABITS

The common name English sparrow is a misnomer, but it has stuck to this bird for some hundred years and is likely to survive indefinitely. It was quite natural to call it the English sparrow, as most of the birds were imported from England, but the species is widely distributed in Europe and Asia, with closely related forms in North Africa. For a full account of its distribution and geographical variations the reader is referred to an excellent paper on the subject by Dr. John C. Phillips (1915). And, after calling it a sparrow for these many years, and our commonest and best known sparrow at that, we must recognize it as a weaver finch and separate it widely from our sparrows in the A. O. U. Check-List. Who wants to call it the European weaver finch? The scientific name has not been changed, for which we may be truly thankful!

Many years ago, when I was a small boy, probably in the late 1860's or early 1870's, my uncle, who lived next door to us in Taunton, was the first to introduce English sparrows into that immediate vicinity. He built a large flying cage in his garden that was roofed over, covered with netting on four sides, and well supplied with perches and nesting boxes. Here the sparrows were so well fed and cared for that they soon began to breed. It was not long before the cage became overcrowded, and he ordered his coachman to put up

1

numerous nesting boxes all over the place and to liberate the sparrows. They soon filled all the new boxes, and also drove away the purple martins, tree swallows, and house wrens from all the older boxes. When the neighbors' cats killed a few of the precious sparrows, which were the newest pets and were zealously guarded, my uncle became so angered that he ordered his coachman to "kill every cat in the neighborhood." My uncle drove in that night to find the coachman with nine of the neighbors' cats laid out on the stable floor, a cause for some profanity. It was not long, however, before my uncle began to miss the martins, swallows, and wrens and to realize that the sparrows were not as desirable as expected; so he ordered the coachman to reduce them. This he did effectively by digging a trench and filling it with grain, so that he could kill large numbers with a single raking shot. But the martins, swallows, and wrens never returned. This incident is typical of what happened in many other places before we realized that we had made a great mistake in importing this undesirable alien.

Walter B. Barrows (1889), in his Bulletin on this species, quotes the following account by Nicolas Pike of his efforts to get the English sparrow established in this country:

It was not till 1850 that the first eight pairs were brought from England to the Brooklyn Institute, of which I was then a director. We built a large cage for them, and cared for them during the winter months. Early in the spring of 1851 they were liberated, but they did not thrive.

In 1852 a committee of members of the Institute was chosen for the re-introduction of these birds, of which I was chairman [sic].

Over $200 was subscribed for expenses. I went to England in 1852, on my way to the consul-generalship of Portugal. On my arrival in Liverpool I gave the order for a large lot of Sparrows and song birds to be purchased at once. They were shipped on board the steam-ship *Europa*, if I am not mistaken, in charge of an officer of the ship. Fifty Sparrows were let loose at the Narrows, according to instructions, and the rest on arrival were placed in the tower of Greenwood Cemetery chapel. They did not do well, so were removed to the house of Mr. John Hooper, one of the committee, who offered to take care of them during the winter.

In the spring of 1853 they were all let loose in the grounds of Greenwood Cemetery, and a man hired to watch them. They did well and multiplied, and I have original notes taken from time to time of their increase and colonization over our great country.

Barrows lists the following other places in which the sparrows were introduced directly from Europe: Portland, Maine, in 1854 and 1858; Peacedale, R. I., in 1858; Boston, Mass., 1858–60; New York, 1860–66; Rochester, N. Y., between 1865 and 1869; New Haven, Conn., 1867; Galveston, Tex., 1867; Charlestown, Mass., 1869; Cleveland, Ohio, 1869; Philadelphia, Pa., 1869 or earlier; Salt Lake City, Utah, 1873–74; Akron, Ohio, 1875; Fort Howard, Wis., 1875; Sheboygan, Wis., 1875;

and Iowa City, Iowa, 1881. He gives a long list of places in which the sparrows were introduced, probably by transplanting from other places in the United States, and adds: "A study of these tables shows that even before 1875 there were many large sparrow colonies throughout the United States, east of the Mississippi, as well as several in Canada, one or more in Utah, one at Galveston, Tex., and probably another in San Francisco, Cal. There were small colonies also in eastern Iowa and in Missouri, Kansas, and Nebraska."

With all these importations and transplantings, it is no wonder that the English sparrow was soon able to overrun the whole country. Barrows (1889) estimated that by 1886 the sparrow was "found to have established itself in thirty-five States and five Territories." It spread and increased very rapidly. Between 1870 and 1875 it spread over 500 square miles; from 1875 to 1880 it spread over 15,640 square miles; between 1880 and 1885 it spread over 500,760 square miles; and in the year 1886 alone it added 516,500 squares to its range. This range was naturally spotty, there being many portions of each State that had not been invaded, and the centers of abundance were near the points of introductions. For example, although it made its first appearance in California, in the San Francisco Bay region, in 1871 or 1872, it extended its range very slowly during the next 20 years into adjacent regions; it apparently did not become established in Los Angeles County until about 1906 and in San Diego County about 1913, according to Grinnell and Miller (1944), but by 1915 it "had spread to virtually all sections of the State, at least locally in towns and about ranches, inclusive of desert areas and the larger islands offshore."

It evidently invaded Arizona in 1903 and 1904, and New Mexico about 1909. Finley (1907) reported it at Portland, Oreg., in 1889, but Rathbun tells me that it did not appear in Seattle until 1897 and Bellingham, Wash., in 1900. It spread through Colorado between 1895 and 1906.

E. R. Kalmbach (1940) writes: "At present the range of the English sparrow in North America covers the entire continental United States except Alaska, all thickly settled parts of the contiguous Canadian Provinces, and similar areas in Mexico south at least as far as San Luis Potosi and Guadalajara in Jalisco. * * * The most northerly point of occurrence of which the writer has record is Two Islands Indian Village on the Mackenzie River, 30 miles below Fort Simpson, Mackenzie, latitude 62 N. * * * The bird is known also at Athabaska Landing in northern Alberta and is present in most of the settlements in the coastal region of British Columbia."

Leonard Wing (1943), in comparing the spread of the English sparrow with that of the starling, states that "the English sparrow spread much faster than the starling, and its occupation was substantially completed forty years after its introduction. The starling spread is still far from finished fifty years after its introduction." The rapid spread of the sparrow is largely due to the fact that it was much more widely introduced and artificially transplanted. Being largely a grain-eating bird, it traveled extensively from place to place along the highways, where it could pick up waste grain dropped by passing vehicles and find some semidigested grain in the droppings of horses. Undoubtedly many found their way to more distant places by securing unintentional transportation in grain cars or cattle cars, as shown by the fact that towns and cities along the principal railway lines, especially where there has been heavy traffic in grain, have been the first communities occupied. When these places have become over-crowded, the birds have spread out into the surrounding rural or suburban regions.

At the peak of its abundance, during the early part of this century, the English sparrow was undoubtedly the most abundant bird in the United States, except in heavily forested, alpine, and desert regions. Within its favorite haunts one could easily see twice as many sparrows as all other birds combined. Mrs. Nice (1931) says: "The English or House Sparrow appears to be the most abundant breeding bird in Oklahoma. On 1,166 miles of 'roadside censuses' taken in May, June, and early July, 1920–1923, in all sections of the state, we counted 2,055 of these birds; this was 26 percent of all the birds seen and twice as many as the most common native bird—the Dickcissel."

Tilford Moore tells me that in his counts in St. Paul, Minn., he recorded 42 English sparrows to 20 other birds. On the other hand, Wing (1943) estimates that in the Eastern States these sparrows constitute about 3 percent of the breeding bird population and about 4.5 percent of the wintering bird population. If this is true today, there has been a marked decrease in the East during the past two decades; and this is quite evident to the casual observer.

The decrease is most marked in the Eastern States, especially in the cities and towns, though the sparrows are still common in the rural districts, about the farmyards and poultry farms, where there is still plenty of grain being fed to livestock. The vast hordes that formerly roosted in the trees of the King's Chapel and Granary burying grounds, in the center of the city of Boston, are no more, though a few may still be found in the public parks, where the pigeons seem to find some food. Warren F. Eaton (1924), then of Weston, Mass., published some interesting data showing the decline in the numbers of these sparrows in eastern Massachusetts, between 1914 and 1922. The number of days on which the sparrows were seen declined from 232 in 1916 to 101 in 1922; and the total number of sparrows seen declined from 2,705, an average of 13.7 per day, between November 20, 1914, and January 1, 1916, to only 570, an average of 5.6 per day, between January 1, 1921, and June 28, 1922. He claims to have kept a careful record and "accurate account of every bird seen at any time."

According to W. H. Bergtold (1921), there was a marked decrease in the number of English sparrows in Denver, Colo., during the 15 years previous to 1921. His observations were made in "a well grassed and timbered area surrounding the Court House," opposite his office; he estimated conservatively that 15 years ago "the sparrow population of this area * * * could not have been less than one thousand birds." Careful counts made by him on 7 days in October 1919, in the same area, varied from 5 to less than 20 birds seen each day.

Considerable falling off in numbers has been noted in the cities and towns of eastern Canada, but in the rural districts and in some of the towns in the Prairie regions the decline in numbers is less marked and the population has become more static.

Natural causes do not seem to have been sufficient to account for the decrease in numbers; English sparrows have no more serious natural enemies than other birds; there is no evidence of any widespread epidemics or diseases; the elements have caused some wholesale destruction in a few places, but other birds have recovered from the results of such disasters. There remains a generally accepted cause, the diminution in the food supply, especially in the cities. It is significant that the decrease in the sparrow population in urban and suburban areas coincides very closely with the increased use of motor vehicles and the decrease in the number of horses that formerly spread a bountiful food supply along our streets and highways. Even in the farming districts, the tractor and other mechanical agricultural machines have largely replaced the horse; and in the cities and towns a horse-drawn vehicle is a rare sight today. Bergtold (1921) gives us some figures to illustrate the passing of the horse; official statistics show that, in Denver, the number of horses declined from 5,904 in 1907 to 3,832 in 1917, a reduction of about 33 percent. He says further: "There can be, however, little question concerning the reality of the 'vanishing horse', for it has been shown (Saturday Evening Post, Sept. 13, 1919) that the number of horses in New York City recently declined from 108,036 to 75,740, and it is probable that what amounts to decrease (by displacement or substitution) has occurred also in suburban areas, since statistics seem to show a decrease or displacement of 33% of the horses in one of the Dakotas. Finally in this connection it can be said that early in November 1919 there were enough tractors in use in Colorado to displace 16,000 horses."

The above is largely ancient history, but the figures show the trend, which has been going on ever since at an increasing rate until the horse, as a tractive force, has almost reached the vanishing point. The motor vehicle has driven the English sparrow out of our cities not only by removing its principal, almost its only, food supply, but by making its street life so hazardous among the swiftly moving vehicles as to cause it to seek safer surroundings, where food is more easily obtained. Studies of stomach contents show that no very large proportion of the food of this bird consists of semidigested oats, from which it may be contended that the passing of the horse was not a primary factor in the decline of the sparrow; nevertheless the passing of the horse certainly resulted in driving the great concentrations of sparrows out of our large cities.

Courtship.—The courtship of the English sparrow is more spectacular and strenuous than elegant. It used to be a common experience to see a group of these dirty, soot-begrimed street gamins struggling and fighting almost under our feet in our streets and gutters, oblivious to their surroundings. Charles W. Townsend (1909) thus describes the actions of the ardent male:

With flattened back, head held up and tail down [up?], wings out from the body, the tips of the primaries touching or nearly touching the ground, he hops back and forth before the coy female as if on springs. Not one but several dance thus before a lady who barely deigns to look at them, and then only to peck in feigned disgust at the love-lorn suitors. These pecks are often far from love pats. At times she stands in the middle of a ring of males at whom she pecks viciously in turn as they fly by, all chirping excitedly at the top of their lungs. The casual observer might think the lady was being tormented by a crowd of ungallant males, but the opposite is in reality the case for the lady is well pleased and is showing her pretended feminine contempt for the male sex, who on their part are trying their best to attract and charm her. At other times she plants her bill firmly on the head of the suitor, and pecks at him violently from time to time without letting go her hold. I have seen several such one-sided fights, for the oppressed rarely fights back, where the male seemed to be on the verge of exhaustion, lying panting on the ground, but on being disturbed both birds flew off apparently none the worse. * * * About a year ago I watched two males in fierce encounter on a small grass plot in front of my house. One had the other by the bill and held him back downwards on the grass. They were both using their claws vigorously and bracing with their wings. Occasionally they would roll over, or go head over heels. Breaking apart they would fly up at each other like enraged barn-yard cocks. Although I stood within two feet of them, so intent were they that they did not notice me until I made an incautious movement and they fled to fight elsewhere.

A disgraceful fight between two female English Sparrows occurred in front of my house one April day. Catching each other by the bills they pulled and tugged and rolled over on the grass. When they broke away the fight was renewed a few inches above the ground in fighting cock style. Three males appeared, and watched the fight. One, evidently scandalized, endeavored to separate the Amazons by pecking at them, but they paid no attention to him and only after some time flew away, one chasing the other.

Claude T. Barnes has sent me the following interesting account of the mating of this strenuous species: "The incredible English sparrow is the best illustration of *furor amatorius*. The male suffers from satyriasis, the female from nymphomania. In the several years that we have observed them breeding, in two instances copulation took place fourteen times in succession, with a stopwatch record of five seconds for the act and five seconds for the interval. In each instance it was the soft *tee tee tee tee tee tee* of the female, sitting with outstretched wings, that attracted our attention, and our count one was perhaps in reality two or three. Since other males within 20 feet took no interest, we believe that despite its reputation for promiscuity the domestic sparrow, after earlier imbroglios are settled, actually does mate with at least a short period of fidelity. Once mated, however, the female seems willing to continue the venery beyond the capacity of the male, for in every instance we have observed she continued her fluttering chant until he ceased to respond."

Nesting.—The resourceful and adaptable English sparrow will build its bulky, unkempt, and loosely constructed nest in almost any conceivable spot that will give it support, some security, and a reasonable degree of concealment, though some of the locations seem to lack even these requirements.

Their favored site appears to be a nesting box, from which other box-nesting birds are often excluded or sometimes even evicted. But sometimes, even where boxes are available, natural sites have been occupied. Richard L. Weaver (1939) found in his studies at Ithaca, N. Y., that "boxes were not preferred to natural sites if the natural ones were well hidden. This was shown at the sanctuary pavilion where boxes were placed besides the natural sites. Only five of the twenty-five broods raised there were in boxes, the others being in rafters under the overhanging porch."

In eastern Massachusetts, favorite nesting sites are found in the dense growths of Boston ivy which climb luxuriantly over many of our large buildings and offer good support, security, and some concealment. Similar ivy and other vines are favored in different parts of the country. These vines sometimes harbor so many nests that they become a nuisance; the slovenly nests disfigure the walls, while the vines and the ground beneath are defiled by the droppings of the birds, and the noisy chattering of so many birds disturbs the occupants of the buildings. Attempts to drive away the birds by pulling down the nests have not always been successful; some of the nests are 40 or 50 feet from the ground and difficult to reach; and the birds are so persistent that they return to build again. But repeated efforts will eventually succeed and the sparrows will learn to build their nests in less conspicuous places, where they are less likely to be disturbed.

Charles R. Stockard (1905) writes: "There was a church in Columbus [Miss.] the walls of which were completely covered with ivy and the ivy was almost as completely filled with sparrows' nests. Permission was obtained to raid this colony and in one day four hundred and fifty-nine eggs were taken and about seven hundred young sparrows were killed. * * * Several compound nests were found, one a large ball of hay with three small openings each leading to a separate feather-lined chamber containing a set of eggs."

In the days when we had arc lights over the streets, sparrows built their nests on the supports under the hoods, where they had shelter and warmth at night. Occasionally I have seen sparrows occupying one compartment in a large martin box while the purple martins lived in the others. The bulky nests of large hawks often have sparrow nests in their lower portions; many times I have seen one in an osprey's huge nest, and have even found one occupying a crevice in the nest of a Swainson's hawk. The sparrows do not seem to bother the martins, nor are they afraid of the large hawks. Sometimes sparrows drive out cliff swallows and occupy their bottleneck nests; occasionally they use deserted nests of cliff or even barn swallows.

A. Dawes Du Bois tells me that he found "a great many English sparrows" occupying the lower holes in a large colony of bank swallows. He also saw one building a nest in a woodpecker's hole.

Natural cavities in trees, especially apple trees, cavities such as are used by starlings, bluebirds and tree swallows, offer convenient nesting sites. When the sparrows were more abundant in our cities they nested in large numbers, sometimes as many as half a dozen nests in a tree, in the shade trees along the streets and in the parks; their large nests were very conspicuous before the trees were in full leaf. Most of these nests were in deciduous trees at heights ranging from 10 up to 50 feet. Some nests were in spruces; and, in California, nests are often seen in the tallest eucalyptus trees, and even palms.

In or about buildings the sparrows will build their nests wherever they can find lodgement for them, on a rafter or a brace, on the corbel of a pillar, on a rain spout under the eaves, behind blinds or shutters, or in the pocket of a drawn-up awning; in the last case, if the awning is lowered, the nest is destroyed, but the sparrows will build there again, if the awning remains drawn-up long enough. Hervey Brackbill has sent me the following description of an awning nest: "A nest built in a deep fold of a drawn-up awning in Baltimore city was a great mass of loosely-placed and loosely-woven material, in the middle of which was a comparatively small and rather neat pocket for the eggs. In its extreme dimensions, the whole unkempt thing was 20 inches tall and 12 by 5 inches in breadth. The actual nest pocket had an extreme depth of 7 inches, but from the point where the walls became solid the depth was only 5; the inner diameter was 3 inches. The nest was made chiefly of very long, coarse grass stems with the heads still on, but it also contained some leaves, a few small feathers, a small wad of cotton, several pieces of string, a piece of cloth, and a piece of waxed chewing gum paper."

Nests in other cavities vary greatly in size, the space, whether large or small, being filled with the material. Nests in open situations in trees are usually large, more or less globular in shape, with the entrance on the side.

William L. Finley (1907) published a photograph of a nest in an unusual location, of which he says: "Down near the end of sparrow row some hornets built a nest up under the projecting eaves of the front porch of a cottage, just beside the bracket. I can understand how a pair of sparrows will fight for a bird-box and drive other birds away, but I never dreamed they would be envious of the hornets. But a sparrow must have a place to nest. Whether the hornets left voluntarily or with the aid of the sparrows I do not know, but the next time I passed I found the birds in possession—actually making a home in the hornet's nest. They had gone in through the bracket and pulled out a large part of the comb, and were replacing it with grass and feathers."

Weaver (1939) observed that—

Nest sites were chosen both before and after mating had occurred. If before, the male selected the site and performed his courtship from there, but if afterward, the female helped with or probably did most of the choosing. * * *

"The variations in nest structure resulted mostly from the presence or absence of certain nesting materials. The commonest form of nest was one with an outer structure of coarse hay or dried weeds, and a lining of finer materials such as feathers, cord, hair, and frayed rope. Hay and dried weeds were preferable to straw. Feathers were preferred to other lining materials and the birds often traveled several hundred yards to the chicken yards to obtain them.

* * * "Coarse materials were brought to the site selected and layed down rather loosely for a foundation. When a strong support was necessary many stiff stems of hay or weeds were forced into small crevices around the sides and bottom of the nest. As the bulk increased upwards, the female formed the cup by turning round and round in the center. This movement caused the long strands to bend into a 'U' shape. The ends were, therefore, forced up along the sides and helped to support the roof, which was added next. After the outer shell was constructed, the lining was added."

The English sparrow is an early and a late nester, as well as a prolific breeder, raising several broods a year. Clarence Cottam (1929) holds the record for an early nesting date in Utah. On January 1, 1929, a boy showed him a nest containing five naked young, recently hatched. The temperature was near the zero point at the time and went down to 14 below during the month. "During the first 18 days one or both parents were almost constantly on the nest. During the night both parents remained within the bird house. Contrary to the usual custom of these birds, the young were practically grown when they left the nest and began to fly. One of these juveniles, collected late in February, disclosed a body that was fat and in perfect physical condition."

This was, of course, a very unusual date for nesting, as the eggs were laid in December. J. J. Murray writes to me that he has seen these birds carrying nesting material as early as February 1, and that a neighbor took a practically complete nest out of a wren box on February 21. He has also seen them carry material into a hole as late as November 2. These dates indicate possible nesting activities in every month of the year. Weaver (1943), on the other hand says: "The season of nesting is from April to September for most of the United States but may start as early as March 2 in the South and may be delayed until near the first of May in parts of Canada and in Europe."

Eggs.—The number of eggs laid by the English sparrow varies from three to seven; five seems to be the commonest number, though sets of six are not very rare; as many as nine have been recorded, and four seems to be the normal minimum. The eggs are mainly ovate in shape, with a tendency toward elongate ovate, and they have very little gloss. Niethammer (1937) gives a very good description of them, of which the following is a translation: "Eggs—very variable with basic color almost pure white, greenish or bluish, less often green-gray or brownish. Marks limited to a few gray or brown dots, usually consisting of closely packed, clouded or sharply limited spots, which vary from deep black-brown through all tone ranges to bright ashy gray and can crowd in toward the blunt end, however, without forming a genuine wreath structure. Usually the last egg is abnormally colored; basic color brighter, spotting more pronounced and not so frequent. Likewise the next to the last egg has a darker basic color and very dense marking."

Weaver (1943) gives the measurements of 54 eggs, of which the average is 22.8 by 15.4 millimeters; the eggs showing the four extremes measure 25.0 by 16.0, 22.0 by 16.8, 20.2 by 15.0, and 20.4 by 14.5 millimeters.

Young.—Weaver (1943) found that incubation was performed wholly by the female, 12 days being the commonest incubation period: "Three of the twenty-two sets required thirteen days, nine required twelve days, one required eleven days, and three sets required ten days." He figured that incubation began with the laying of the third egg. Of the hatching operation he writes:

A clicking sound usually announces the readiness of the young to start hatching. It is made by contact of the egg tooth with the shell and possibly also by a clicking together of the mandibles. The egg tooth presses against the shell and makes an upraised crease around the larger end of the egg about one-fourth of the way from the end. The young bird may break the shell with the egg tooth before the crease is noticeable. In either case, a slit now appears, starting at a point where the egg tooth first pushed through the shell. The slit is made in a circular direction around the egg and meets the point where it started. The young is able to turn itself or its head in the egg making a complete circular slit possible. The head is located in the larger end of the egg and as the slit nears completion the piece of shell around the head is broken off and the head is freed. The larger piece of shell is now kicked free and the young forces itself out. The feet are crowded into the depressions on either side of the neck while in the shell, and after hatching they have a tendency to remain doubled up for several hours. Often the shell does not come free from the young immediately and the female will help to remove it, and when doing so may often carry the young and the shell out of the nest causing early death to the unfortunate young.

In another paper (1942) he describes the development of the young in great detail, with illustrations, but I quote only from his summary:

English Sparrows are hatched without natal down. * * *

The egg tooth disappears and the edges of the bill change from white to lemon-yellow by the fourth day after hatching. * * *

The greatest development in the plumage of young sparrows is delayed until the latter part of the period in the nest. The greatest change in appearance of young English Sparrows occurs between the age of six and seven days, when most of the feathers emerge and many of them lose their sheaths.

By the tenth day after hatching the color pattern is evident, showing a wing bar, and in some males a black bib.

Practically all of the sheaths have disappeared from the contour feathers and all but one-fourth of the flight feathers are unsheathed by the fifteenth day. These sheaths may remain one to two weeks after the young depart from the nest. The greatest amount of sheath is present in the flight feathers on the eleventh day. The amount of sheathing present gives an accurate criterion of the age of young birds in the nest.

Most of the young left the nest at about the fifteenth day, but English Sparrows may remain in the nest for seventeen days if entirely undisturbed.

Males and females share about equally in the feeding of the young at the nest, but the females do the greater portion of the nest sanitation. Both birds may brood the young, although the female does the greater part of it, and always stays in the nest during the night. The young are fed by regurgitation during the first part of the period after hatching.

There was 70.5 per cent success of survival in thirty-eight nests which produced 127 young from 180 eggs laid. This corresponds closely to that reported for other hole-nesting species.

The older young are able to command the most advantageous positions in the nest and thus receive relatively more food and often are able to leave the nest several days before the other young. The young can fly rather well upon leaving the nest, considerably better than do the young of most species that nest in the open.

The young are fed by the adults for a period of two weeks, and probably more, after leaving the nest. The young have a strong bond for one another.

The young, out of the nest, may be fed entirely by one adult or by both.

A complete post-juvenal molt begins about five weeks after the young leave the nest. It began in early August and ended in mid-September at Ithaca in 1937.

Probably the English sparrow raises regularly two broods of young in a season and often three, but it is doubtful if more than three are often raised. Reports of four or five broods have not been definitely proven, so far as we know. Weaver (1943) suggests: "Since one individual nest may be used by three or more different females in one season, the actual number of broods raised by one female in a season is questionable. Two banded females are known to have raised but two complete broods in one season while one nest site is known to have been used four times with three successful broods. Therefore, it is suspected that the large number of broods claimed by some writers may refer to clutches per nest site rather than broods per female."

When a nest is robbed or destroyed, however, the sparrows lose no time in making another attempt to raise a brood. Murray writes to me: "I tore a nest out of a hole on April 11, and 26 days later, on May 7, the pair had two young and two eggs ready to hatch."

Barrows (1889) quotes Otto Widmann as saying: "A Sparrow never deserts its brood. If one of the parents is killed, the other will do all the work alone. If a young one happens to fall down from the lofty nest, it is not lost; the parents feed it, shelter, and defend it. If a young Sparrow is taken from the nest and placed in a cage, the mother feeds it for days and weeks, even if she has to enter a room to get to it."

Plumages.—Both sexes are alike in the juvenal plumage, which Dwight (1900) describes as follows: "Above, hair-brown somewhat buffy, wings and tail slightly darker, and streaked broadly with clove-brown on the back; secondaries, tertiaries and wing coverts edged with wood-brown. Below, mouse-gray, darkest across jugulum and on the sides, the chin and mid-abdomen nearly white. A dusky postocular stripe."

A complete postjuvenal molt takes place about 5 weeks after the young bird leaves the nest, at which the male acquires the black throat and becomes practically indistinguishable from the adult. Dwight (1900) describes this handsome plumage as follows:

Pileum, rump and upper tail coverts smoke-gray, the feathers brownish edged and dusky basally. The back streaked with black each feather partly Mar's-brown and edged with buff. Below, dull white tinged with French-gray on throat and sides, the feather tips with buffy wash, the shafts faintly grayish; the chin and throat, loral and postocular stripe, black veiled with grayish or buffy edgings; sides of chin and throat and mid-abdomen nearly white; auriculars olive-gray; posterior part of superciliary line, postauricular and nuchal regions chestnut veiled with buff edgings. Wings and tail dull black edged with pale cinnamon, rich chestnut on the greater and lesser coverts, the median coverts white, buff edged forming a wing band.

The first and subsequent nuptial plumages are acquired by wear, which brings out the contrasts in and the brilliancy of the colors. A complete postnuptial molt for both yound and old birds begins late in August. Adult males in winter plumage are not very different from the first winter males; the black of the throat is usually more extensive, the buff less evident, the crown grayer and the median coverts whiter. After the postjuvenal molt, females resemble the males above, but lack the black throat and the chestnut patches; the molts are the same.

Stimulated by Wetmore's (1936) feather counting, Arthur E. Staebler (1941) took the trouble to count the contour feathers on eight English sparrows of different ages and sexes and at different seasons, from which he made the expected discovery that the sparrows wear more feathers in winter than in summer; only such feathers as formed parts of the outer covering were counted. He found that an adult male taken in January had 3,615 feathers, while an adult male taken in July had 3,138 feathers.

Food.—The latest and most comperehensive study of the food of the English sparrow was made by E. R. Kalmbach (1940), based on the examination of 8,004 stomachs of adults and nestlings, 337 of which "were found to be too nearly empty or otherwise unfit for use in the computation of bulk percentages." The stomachs of the 4,848 adults were found to contain 3.39 percent of animal matter and 96.61 percent vegetable. The proportions were largely reversed for the nestlings, 68.13 percent animal and 31.87 percent vegetable matter. Grouped to show whether the consumption of the various items is beneficial, neutral or harmful to the interests of man, his tables give the following figures: Adults' animal food is 2.67 percent beneficial, 0.64 percent neutral, and 0.08 percent harmful. Adults' vegetable food is 16.97 percent beneficial, 24.14 percent neutral, and 55.50 percent harmful. Nestlings' animal food is 59.21 percent beneficial, 4.48 percent neutral, and 4.44 percent harmful. Nestlings' vegetable food is 0.17 percent beneficial, 7.85 percent neutral, and 23.85 percent harmful. By adding the totals it will be seen that the feeding habits of the adults are 55.58 percent harmful while those of the nestlings are 59.38 beneficial; but, unfortunately, the feeding time of nestlings is very limited.

He gives the sparrow credit for destroying many weevils, particularly the very destructive alfalfa weevil, scarabaeid beetles, click beetles, leaf beetles, grasshoppers, locusts, crickets, caterpillars, moths, and some flies; but blames it for eating the useful predaceous ground beetles and spiders. He gives it some credit for destroying some weed seeds, but condemns it for the large amounts of feed grains such as oats, wheat, and corn that it eats. These grains alone made up over 55 percent of the food of the adults.

Forbush (1929) gives the following general account of the food of this bird: "The food of the House Sparrow includes many substances, chiefly vegetal, and ranging from fruit and grain to garbage, and undigested grain and seeds in horse droppings. It eats greedily all the small grains and bird seeds, crumbs of bread, cake and other foods of mankind, small fruits and succulent garden plants in their tender stages. It destroys young peas, turnips, cabbage and nearly all young vegetables, and it often eats the undeveloped seeds of vegetables. When numerous it attacks apples, peaches, plums, pears, strawberries, currants and all other common small fruits."

It has repeatedly been seen eating apple blossoms and those of peas and beans. Tilford Moore (MS.) saw one trying to get at the seeds in a large sunflower head, "but, as the bloom was face downward and as he was unable to hang upside down, he was unsuccessful. The above method, sometimes successful, is used on seeds at the edge of the bloom only. For seeds toward the center, they hover beneath the bloom and reach up to draw one out. Then they fly to the ground where they remove the husks to get at the kernels. Thus the birds of this species have harvested almost all the seeds."

Irving W. Burr writes to me: "One of the bird's commonest food in late summer and fall is the seed from crab grass (*Digitaria* sp.). That is the explanation of the foraging flocks on the lawns. Binoculars reveal that the bobbing heads are busy shoveling in the seeds, just as a boy will strip a weed stalk. The number of crab grass seeds which a flock of forty sparrows will eat in a day must be enormous. Surely everyone would regard this as a commendable trait in the bird." I have seen the birds doing this on my lawns, but cannot see that the crab grass is materially reduced.

Judd (1896) also refers to the sparrows as eating the seed of crab grass, chickweed, and dandelion, but none of these lawn pests have been exterminated anywhere, though they may be somewhat controlled. He says that "more than half of the dandelions that bloomed in April on the lawns of the U. S. Department of Agriculture were damaged by Sparrows." Kalmbach (1940) mentions ragweed seeds as dominant in the food of this sparrow, but says that crab grass seeds are "taken in greater bulk and numbers but found in fewer stomachs. * * * As many as 1,274 were taken from the crop of a single English sparrow from Alabama; more than 900 each from 2 others; and 150 or more each from fully 40 others." He then gives a long list of other weed seeds and grass seeds eaten.

Judd (1901) said of the vegetable food, as then known and not very different from our present knowledge: "Of the 98 percent constituting the vegetable food, 7 percent consisted of grass seed, largely of plants of the genera *Zizania* (wild rice), *Panicum*, and *Chaetocloa*, and notably crab-grass and pigeon-grass, and 17 percent of various weeds not belonging to the grass family. The grass and weed seeds taken are not noticeably different from those usually eaten by native sparrows. But what especially differentiates the vegetable food from that of all other sparrows is the large proportion of grain consumed, which formed 74 per cent of the entire food of the year and 90 percent of that of the period from June to August."

In late summer, when the numbers of these sparrows are augmented by the addition of two or three broods of young, the sparrows swoop down on the grain fields to raid the standing crops; they alight on the stalks to pluck the grain from the fruiting heads or to shake the kernels down to the ground to be picked up at their leisure, all of which results in heavy damage to the crops.

Among the few redeeming features in the food of English sparrows is the small percentage of harmful and annoying insects that it eats. Hervey Brackbill writes in his notes: "During one October when aphids heavily infested the silver and Norway maples that line several blocks in northwest Baltimore, English sparrows were among the most persistent of 13 species of birds that fed upon them. The sparrows appeared daily, foraged throughout the 35- to 50-foot trees, and used many different methods. In a heavy vertical fork, a type of place in which the aphids sometimes collected in particular numbers, one bird once clung for some seconds head downward, much like a nuthatch, while snatching up the insects on all sides. Another clung to a silver maple trunk practically like a woodpecker and foraged over and beneath the flaky bark. The English sparrows also often picked the aphids off the under sides of leaves. The English sparrow is one of the heaviest bird feeders on the Japanese beetle, which has become such a pest in parts of the East. It is the most versatile bird in its hunting of them, too. It flies to commanding perches on rose bushes and trellises, scans the leaves and flowers thoroughly, and upon locating a beetle makes its capture with a swoop. It searches the bushes from below, hopping along the flower beds, peering intently and then darting upward to seize its prey. I have seen it make catches as high as 18 feet up in trees. It also pursues low-flying beetles through the air and captures them on the wing."

I can remember that many years ago, before the sparrows became abundant, we were greatly annoyed by inch worms spinning down upon us from the trees which they had partially defoliated. Then the sparrows came and began foraging in our shade trees for these little caterpillars or canker worms, as well as for the elm-leaf beetles.

We frequently saw several sparrows at work in a single tree. At the height of the sparrow abundance these little worms were nearly exterminated about my home, or at least very materially reduced in numbers. But, since the decline in the sparrow population, the inch worms have increased decidedly. The scarcity of vireos and other insectivorous birds may have partially accounted for the increase in the inch worms.

I remember once, when there was a plague of army worms here, the sparrows gathered in large numbers and fed greedily upon them. In the West, outbreaks of Mormon crickets have been at least checked by English sparrows.

According to William J. Howard (1937), English sparrows were the most numerous and most active of all the birds in attacking an emergence of 17-year locusts in Indiana. "Although there were multitudes of dead and dying insects upon trees and the ground, the sparrows were very active in pursuing flying locusts. As many as three sparrows were seen to chase a single insect, and the squabble and fight characteristic of this bird usually ensued when one of the birds caught an insect."

Sparrows are often seen picking insects off the radiators of automobiles where they have been caught and killed. But they can also catch many flying insects in the air—wasps, bees, flying ants, and other Hymenoptera. They are, in fact, very resourceful in their varied feeding habits. They eat the snow-white linden moths when they appear in July, and feed on the tent caterpillars and brown-tailed moths.

Behavior.—The English sparrow is a noisy, boisterous, and aggressive bully in its relations with other species. Generally cordially hated by both birds and men, its record is almost wholly black. It drives bluebirds, swallows, and wrens from their nesting boxes by force, or by preempting them in advance. Some of these rightful tenants of the boxes can resist eviction by an aggressive pair of sparrows, but they cannot withstand mob violence when the sparrows attack in superior numbers, as they sometimes do; then the gentler birds give up the fight and retire to find more peaceful quarters elsewhere. But the box-dwellers are not the only sufferers; the sparrows seize and occupy the bottleneck nests of cliff swallows, the open nests of barn swallows, and even the burrows of bank swallows. It was thought that the English sparrow might meet its match in the house finch in Colorado, but such was not to be. Bergtold (1913) writes:

The loss of nests, eggs and young of the House Finch through direct destruction by the English Sparrow is very large. It was 16% in some of the nests studied by the writer, and, moreover, this 16% loss of eggs does not include the very large potential loss of House Finch eggs and young brought about by destruction of nests by English Sparrows before the House Finch eggs are laid in them. * * * The writer has personally witnessed English Sparrows going into the House Finches' nests, and has seen them throw out the young, these nestlings having the heads pecked open by the Sparrows before they were thrown out. The House Finch will often put up a mild fight against the invaders, giving at the same time a very characteristic squeak but the Finch is almost invariably beaten in these battles. In many years' observations on this phase of the Finch question, the writer has but once seen a Finch whip a Sparrow.

The sparrows destroy the eggs and young of the birds that nest in boxes and throw out the nesting material, also those of the other birds mentioned above. Nests of birds as large as robins have been robbed of their eggs and young. On the other hand, an English sparrow has been seen by reliable observers to defend the nest and feed the young of a pair of red-eyed vireos, together with the parent birds (see The Cardinal, vol. 2, pp. 191–192). We often see one or more sparrows trailing a robin or a starling on the lawn, seeming to know that the larger birds are more successful than they would be in digging out worms or grubs, and they look for a chance to steal their food. They are such aggressive and persistent bluffers that they sometimes succeed, by force of numbers or by strategy.

They show their intelligence and ingenuity in other ways. John Burroughs (1879), with the remark that "it is too good not to be true" tells the following story:

A male bird brought to his box a large, fine goose feather, which is a great find for a sparrow and much coveted. After he had deposited his prize and chattered his gratulations over it, he went away in quest of his mate. His next-door neighbor, a female bird, seeing her chance, quickly slipped in and seized the feather; and here the wit of the bird came out, for instead of carrying it into her own box she flew with it to a near tree and hid it in a fork of the branches, then went home, and when her neighbor returned with his mate, was innocently employed about her own affairs. The proud male, finding his feather gone, came out of his box in a high state of excitement, and, with wrath in his manner and accusation on his tongue, rushed into the cote of the female. Not finding his goods and chattels there as he expected, he stormed around a while, abusing everybody in general and his neighbor in particular, and then went away as if to repair the loss. As soon as he was out of sight, the shrewd thief went and brought the feather home and lined her own domicile with it.

Mr. Brackbill says in his notes: "An instance of real ingenuity was witnessed at a fountain and pool in Mount Vernon Place, in downtown Baltimore. On a flat rim of this pool, covered by water to an ideal depth, sparrows were accustomed to gather and bathe. One day the pool was drained, leaving the birds only some steep-sided bowls on a surrounding wall as watering places. From the rims of these they could lean forward and bathe their heads and shoulders, but the water's depth quite precluded complete baths in normal fashion. The birds got their baths nonetheless. To wash their hind parts they turned around and tipped slightly backward. And to wash their breasts and underparts generally they flew very low across the water, dipping down into it one to several times on the way. Some contrived even better baths by swimming flutteringly the whole way across the bowls.

"An instance of drinking was noted at a building where an exhaust pipe projected through the wall a yard or so above the ground. The pipe was about an inch in diameter and it projected from the wall just about an inch. Water was dripping from it, and although the drip was not fast enough to form even a small puddle on the ground, the sparrows of the neighborhood had solved the problem of getting a drink. Every now and then one would come flying, alight on the bit of exposed pipe, bend downward—sometimes in a very awkward position—and drink the drops as they collected on the lip of the pipe. Four birds drank in this way during ten minutes that I watched."

English sparrows will roost for the night wherever they can find a little shelter in, under, or about buildings or other human structures, under electric light hoods, or in dense evergreen trees. But they often huddle together for mutual protection where there is no shelter whatever, just as they did in the big city roosts, when the sparrows lived in the cities.

On Chestnut Street, the principal business street in Philadelphia, writes J. P. Norris (1891), stood an old-fashioned dwelling, the only one on the street: "On the lower side of the house, just inside the brick wall that encloses the garden, stands a tree about forty feet high, with many branches; and every afternoon the English Sparrows roost here literally by thousands. Every branch is covered with them, and they are huddled together as close as they can sit. To count them all would be impossible, but I have seen over fifty on one branch. A long wall of an adjoining store is covered with ivy and Virginia Creeper, and this forms a convenient roosting place for those birds that cannot find places on the tree."

The spectacular sparrow roost that formerly existed in the King's Chapel burying ground in the center of Boston is well described by Dr. Townsend (1909). The birds—

frequent the place throughout the year but are decidedly less numerous in the spring months and most numerous during the fall and winter. Thus on November 25, 1905, between 4 and 5 p. m., I estimated that about 3,000 were in this place in five trees. The other two trees were empty. On February 20, 1906, on a mild pleasant day, when the sun set 5:24 p.m., the roost was studied from the near-by City Hall. The roosting trees seen from above looked as if their limbs had been whitewashed and the ground and grass beneath were similarly affected. The first arrivals appear at 3:45 p.m., about a dozen in all. At 4 the birds are coming singly and in small groups alighting in the trees but frequently changing from place to place, chirping continuously and fighting for positions. At 4:05 a flock of 12 fly swiftly and directly to one tree; 4:10 p.m.: there are now about 150 sparrows present, but new ones are coming sailing in with wings wide spread from over or between the surrounding high buildings. They fly with astonishing swiftness and directness, projected as it were from space directly into the roost— is it the city rush and scramble for position? At 4:15 p.m. It is now raining birds. I have seen only one alight on a building before entering the roost; they are in too much of a hurry to get there. The trees are a scene of great activity and the noise rises above the roar of the city's streets. The birds are crowding together in the trees, constantly fighting and flying about as they are forced from their perches. At 4:30 the birds are still coming, but by 4:45 there is a noticeable diminution in the numbers of the coming birds and by 5 o'clock the movement has ceased with the exception of a few stragglers. Many are now spreading their wings and tails and composing themselves for sleep. At 5:30 the roost is still noisy but many are fast asleep, and before long all is quiet.

He describes the morning awakening as follows:

On November 26, 1905, I watched the King's Chapel roost wake up and depart about its day's business. All were asleep and quiet until 6 o'clock when the first chirp was heard, while the stars were still shining, and the first movement took place at 6:05, when a sparrow flew from one branch to another. The sleeping ones had their heads depressed in front, or the head turned around with the bill concealed in the feathers of the back. A sudden general chirping begins at 6:07 and a few buzz about from branch to branch. The chirping swells into a continuous volume of sound, not the chorus of the spring, but a confused conversational chirping noise as if all were talking at once. Birds buzz about with rapid wing vibrations, suggestive of hummingbirds. The first one flies off in an unsteady way as if still half asleep at 6:12. The sound grows louder, although the majority still appear to be asleep. Some are stretching their wings and preening their feathers. The stars are nearly gone. At 6:20 no. 2 flies off uncertainly. 6:25. Now there is greater noise and activity. Many are flying about and a dozen or more have left. All awake seem to enjoy spreading their tails. A considerable proportion sleep on through the hubbub. There is very little fighting compared with the evening. 6:26. Now the birds are leaving constantly. 6:27. They are leaving in bands of 15 or 20 at a time. 6:30 a. m. The stream of outgoers, mostly down Tremont Street to the north, is now continuous and too great to count. The remaining birds are noisy in the extreme, flying about vigorously and filling up the empty trees. 6:35 a.m. It is now broad daylight and the birds are flying off like bees, but more or less in waves. A few still sleep on undisturbed. The sun rose about 6:50 and by that time doubtless all or nearly all of the birds had gone.

Voice.—Little can be said in favor of the English sparrow's voice, except that it expresses cheerfulness under adverse weather conditions, indicates abundant energy and aggressiveness, but the incessant chirping and chattering that one hears on spring mornings often seems monotonous and soon becomes tiresome.

Dr. Townsend (1909) describes it very well as follows:

The "chorus" begins from twenty to thirty minutes before sunrise in April, May and June on bright days—fifteen or twenty minutes later on cloudy days— and lasts in full volume nearly an hour. A few scattering chirps are first heard from the early ones, but the multitudes on vines and trees and house-tops soon take up the theme, and the din is almost deafening. The chief note is *chis-ick* or *tsee-up* monotonously repeated, with various modifications, for the most part high pitched and ear racking, but occasionally deeper and almost melodious.   Certain individuals repeat notes or even series of notes that are not unattractive, and may even be called musical.   These are not common but may be heard every spring, and, on mild days, even as early as January.   At the height of the morning chorus, for such it must be called, there is at times a distinct rhythm, caused by some of the birds keeping time.   This chirping rhythm I have frequently tried to count but generally without success, for each bird appears to chirp manfully on his own hook without regard to time.   I have, however, sometimes found its rate to be 60 or 70 times a minute, slowing down to 40 on hot days.   In this respect the Sparrow differs directly from the cold blooded insect that sings faster the hotter the weather.

Ernest Thompson Seton (1901) tells a remarkable story of an English sparrow that was hatched in a cage by canaries and learned to sing.

It escaped and was frequently heard to sing "a loud sweet song, much like that of a Canary."   I wrote to Thompson-Seton, to learn if the story was just pure fiction.   He replied that he met the bird many years ago in Toronto, and that "in the main the story is founded on fact," but he "expounded and developed the details."   The story, "A Street Troubadour," is attractively written and well worth reading as a character study.

We have other evidences of singing ability.   Dr. Dayton Stoner (1942) writes of a versatile captive English sparrow: "This unique sparrow possessed various types of vocal ability which he utilized to express insistence concerning certain kinds of food, absence of the cage cover at night, general well-being, disgust and the like.   Moreover, he acquired a remarkable proficiency in singing ability through the medium of two canaries which were his companions—in separate cages—for about six years.   His imitations of the 'rolling' notes of the one and the 'chopping' notes of the other were sometimes so well done as to deceive even his mistress."   Furthermore, Tilford Moore has sent me the following note: "Sept. 15, 1941.   I heard a male in the honeysuckle beside my bed actually sing today.   He was uttering the usual harsh chatterings of his kind, but about once a minute he'd substitute a short song for a squawk.   The song was a thin and squeaky but rather pretty one, with *tut-tut* at beginning and end.   Not a warble, it varied over several notes, was rather like an incomplete song which I have heard from a white-throat."

Field marks.—This impudent and aggressive little pest is easily recognized by its behavior, its familiarity, and its noisy chirps and chatter.   The male is a handsome fellow in fresh, clean plumage, with his black bib, gray crown, and conspicuous markings of chestnut, black, and white about the head.   The female is more soberly colored, lacking the conspicuous markings about the head, but similar to the male above and below.

Enemies.—The resourceful English sparrow is more than a match for its small-bird enemies, of which there are plenty. Shrikes, grackles, and small hawks and owls take their toll. Many are killed on the highways by speeding automobiles; motor vehicles are increasing in numbers and in speed faster than the birds can learn to avoid them, and nearly as many English sparrows are killed on the roads as all other species of birds combined; most of those killed are young and inexperienced birds.

The natural elements take heavy toll on rare occasions. Sparrows are hardy birds and generally can stand extreme cold and ordinary winter storms, but those that roost in city trees, as they formerly did in large numbers, are sometimes killed by heavy sleet, hail or rain storms. Ruthven Deane (1908) mentions such a catastrophe that occurred in Chicago in August. A "torrent of rain which is seldom exceeded in force or quantity" lasted most of the night, while the birds were roosting in the trees. He quotes Luther E. Wyman as saying: "My own observations were confined to Garfield Park, where they roost in great numbers. Here I found them dotting the grass under the trees, but massed around the trunks of the larger trees, though many lay even under such dense-growing shrubs as the lilac. * * * The area I examined would cover probably less than a third of a city block, yet I found upward of a thousand birds, all sparrows but one, a young robin."

But the worst enemy of the English sparrow is man. Repenting of his folly in introducing this alien species, he has tried his best to exterminate it or control it, but with indifferent results. Various types of traps have been used with temporary success, but the sparrow soon learns to avoid them. Grain poisoned with strychnine will kill a few sparrows, especially if they have been baited to some chosen spot with wholesome grain, but after a few have been seen to die, the survivors will avoid it; furthermore, it may kill other birds.

The most effective method of driving sparrows away from premises where they are not wanted is by persistent shooting and by repeatedly destroying their nests, both of which methods will eventually discourage them. A long trench can be baited with grain, and, after the birds have learned to feed there, a large number can be killed by a raking shot. The use of light charges anywhere about the premises will not frighten away other birds. Nests should be removed regularly from the boxes, and most of those on trees and in vines can be reached by a long pole with a hook at the end. Where sparrows roost in large numbers in vines on buildings, they can be driven away by heavy spraying with a hose for several nights in succession; this also cleans their filth from the vines. This method is also effective on their nests, as it makes the nests uncomfortable and is likely to kill small young. But any of these methods, to be successful, must be followed up persistently. Several important papers on this subject have been published by the U. S. Department of Agriculture. The latest and perhaps the best of these is by E. R. Kalmbach, "English Sparrow Control" (U. S. Dep. Agric., leaflet 61, 1930).

In spite of its many enemies, however, this sparrow is probably a fairly longlived bird; Stoner (1942) tells of one in captivity that lived to be 12 years old.

Economic status.—A study of what has been written above on the food and the behavior of the English sparrow will throw much light on this subject. Almost everyone who has written anything about the bird has had something to say about its faults and virtues, with a decided emphasis on the former. Barrows (1889) received 1,048 original reports on the relation of this sparrow to other birds, of which 168 were favorable to the sparrow, 837 unfavorable, and 43 indeterminate. He gives a list of 70 kinds of wild birds that are known to be molested in one way or another by English sparrows. And he remarks: "For our own part, after careful consideration of each bit of testimony presented, we believe that the proportion of one hundred to one against the Sparrow is the most favorable estimate which any unprejudiced person is likely to make."

About all that can be said in favor of the sparrow's food habits is that it destroys a few noxious insects and weed seeds; only 2.67 percent of the food of adults consists of harmful insects, to which should be added 0.64 percent of neutral insects and a few (0.08 percent) that are beneficial. Among harmful insects destroyed, in addition to those mentioned in its food (pp. 13–16), we should include the cotton boll weevil, the San José scale, other scale insects, and the caterpillars that attack cotton and tobacco plants.

On the other side of the ledger we may charge up against the sparrow the great damage it does to our agricultural interests. Dr. B. H. Warren (1890) expresses this very well by saying: "In the spring it feeds largely on the fruit buds of trees, bushes and vines, chief among which may be mentioned pear, apple, peach, plum, cherry, currant and grape. Different garden products, such as lettuce, beans, peas, cabbage, berries, pears, apples and grapes are greedily fed upon. The Sparrow greatly damages the corn crop, tearing open the husks, devouring the tender part of the ear and exposing the remainder to the ravages of insects and to atmospheric changes. It alights on fields of wheat, oats and barley, consuming a large quantity, and, by swaying to and fro on the slender stalks and flapping its wings, showers the remainder on the ground."

The sparrow is also accused of spreading the germs of blackhead, so fatal to turkeys, and the germs of hog cholera; as it feeds regularly in all poultry yards and pig pens, these germs could easily be picked up and carried to other yards. It has been proven by H. E. Ewing (1911) that it "frequently harbors and is the host of one of the worst, if not the worst, of poultry pests, the chicken louse or chicken mite, *Dermanyssus gallinæ* Redi. * * * The English Sparrow likewise harbors and is the host of perhaps the most important of all the external parasites of our native song birds, and likewise our tamed cage birds, the bird mite, *Dermanyssus avium* De Geer."

A sparrow that he picked up in a weak and sickly condition was found to "possess scores, if not hundreds," of the chicken mites. A recently deserted sparrow nest was found to be heavily infested; by counting the number of mites on a moderately infested feather, he esimated that there were some 18,000 chicken mites in the nest.

By summing up the evidence regarding its food, its behavior, and the damage that it does, it can be plainly seen that the English sparrow is one of the worst avian pests ever introduced into this or any other country. Barrows was probably not far wrong in estimating that the evidence is 100 to 1 against it.

Winter.—The English sparrow is not a migratory bird, except that it may be driven by very severe weather to leave the northern extremities of its range. Even in much of Canada it is a permanent resident and able to stand ordinary winter weather, provided it can find food and shelter, but when the temperature remains as much as 30 or 40 degrees below zero for a week or more at a time, they must huddle together in buildings for protection and many may die for the lack of food or succumb to the cold.

Farther south there is much less winter mortality; there they cluster about barns and farmyards, find a ready food supply, often under cover, and can roost at night inside the buildings. Except in the severest weather, they find sufficient shelter in hedges, vines, brush piles, or even huddled together in city trees. Formerly very abundant in our cities all winter, they can no longer find sufficient food there; but in rural districts they are our commonest winter birds, and we must admit that, no matter how much we dislike them, they add a little cheer to the bleak winter landscape.

DISTRIBUTION

Range.—The English sparrow was introduced in North America, and is now a permanent resident from central and northeastern British Columbia, central-southern Mackenzie, northwestern and central Saskatchewan (Emma Lake), northern Manitoba (Churchill), central-western, central, and northeastern Ontario, southwestern and central-southern Quebec (Blue Sea Lake, Anticosti Island), and Newfound-

land; south to northern and central-eastern Baja California, Guerrero,
Coahuila, Michoacán, southern Tamaulipas, the shores of the Gulf of
Mexico, southern Florida (to Key West), Cuba, Jamaica, and
Bermuda.

Egg dates.—Ontario, Canada: 9 records, April 4 to July 5.
California: 50 records, March 18 to July 19.
Illinois: 34 records, May 10 to July 5.
North Dakota: 12 records, June 10 to 20.

PASSER MONTANUS MONTANUS (Linnaeus)

European Tree Sparrow

HABITS

This pretty little weaver finch is widely distributed throughout the
Eurasian continent, but less common and more local in England.
Witherby's Handbook (1919) gives its distribution as "Europe gener-
ally and Siberia. Replaced by closely allied forms in east Siberia,
Japan, Turkestan, and Persia, India and China, Greater Sunda
Islands, Hainan and Formosa."

Although we have only comparatively recently decided to place our
two species of *Passer* in the family Ploceidae, it is interesting to note
that this was being discussed as long ago as when Yarrell (1876) wrote
his "History of British Birds."

As an American bird, the European tree sparrow is known only in the
vicinity of St. Louis, Mo., where it was originally introduced and
whence it apparently has not spread very far.   Otto Widmann gave
Barrows (1889) the following interesting account of its introduction
and its struggle with the more aggressive house sparrow:

Early in 1870 a Saint Louis bird dealer imported, among other birds, twenty
Tree Sparrows (*Passer montanus*) direct from Germany.   Mr. Kleinschmidt,
hearing of it, persuaded Mr. Daenzer, of the *Anzeiger des Westens*, who was at
that time experimenting with the introduction of European singing birds, to contribute to the purchase of these birds.   Accordingly they were bought and taken
to Lafayette Park, in the then southwestern part of the city, and liberated April
25, 1870.   All left the park immediately, and none were seen again until April 24
of the following year, when a single bird was seen one mile east of the park.
This discovery was considered worthy of mention in the public press, since at
that time the introduction of the European Sparrow at Saint Louis was thought
to be a failure.

During the next few years dealers had pairs of House Sparrows sent from
New York, and well-meaning citizens bought them for liberation, but the exact
number can not be learned, since the principal parties have died.   Both species
increased amazingly, and as early as 1875 *Passer* had spread over the entire
64 square miles which make up the city of Saint Louis.   In the Southern part
the Tree Sparrow prodominated [sic], and as late as 1877 no House Sparrow

was seen on my premises, one mile south of the arsenal, which latter point they had then occupied in large numbers. Also during the winter of 1877–'78 all of my twelve boxes set up for Sparrows were in undisputed possession of the Tree Sparrows.

On March 28, 1878, the first House Sparrow appeared on the scene, and trouble began. One pair of Tree Sparrows was dislodged and a pair of House Sparrows began nest-building. That summer no increase in House Sparrows took place in my colony, and the Tree Sparrows reared their broods in peace, but when the first cold October nights forced the Sparrows to change their roost from the now nearly leafless trees to some warm shelter, a whole flock of House Sparrows took possession of the boxes and the Tree Sparrows had to leave. Thereafter the weaker Tree Sparrow had little chance to gain a suitable nesting site around its old home. Only one pair continued breeding for a few years longer, in a box which, besides hanging lower than the rest, had an entrance which the bigger House Sparrow found uncomfortably small. It appeared to me that the Tree Sparrow would be much more of a house-sparrow if his stronger cousin did not force him to be a tree sparrow by robbing him of every suitable nesting and roosting place about human habitations.

With the increase of the House Sparrow the Tree Sparrow had to yield the city almost entirely to him and betake himself to the country, spreading in all directions and resorting to tree-holes and out-of-the-way places, while the other took the cities and towns.

Nesting.—The European tree sparrows in Widmann's colony evidently preferred to nest in the birdboxes until they were driven out by the English house sparrows and were forced to nest in holes in trees in the suburbs, as they do in Europe. Witherby (1919) says of its breeding habits: "More retiring than House-Sparrow but locally common in suburbs of some large towns, breeding in holes of ivy-covered trees, pollarded willows, haystacks, thatched roofs, quarries, and old nests of larger birds; also in Woodpecker's holes and Sand-Martin's burrows." The nest is similar to that of the house sparrow, "though smaller, but never built in open among branches; often shows little trace of roof." The nesting materials are also similar, mainly grass, hay, and feathers, with a mixture of various bits of rubbish and trash.

Eggs.—Witherby (1919) describes the eggs as "4–6, rarely 7 or 8, much smaller than House-Sparrow's, darker, with finer stippling, browner in general tint, and more glossy. One light egg also commonly found in each clutch. Average of 103 eggs, 19.5 x 14 mm."

J. P. Norris (1890) describes and gives the measurements of five sets of from 3 to 6 eggs collected for him by Widmann, between June 10 and July 2, near Saint Louis, some of which were colored much like the eggs of the long-billed marsh wren.

Young.—Incubation is shared by both sexes and lasts for 13 or 14 days.

In some notes sent to Bendire, Widmann says: "They begin to lay eggs soon after the first of April (6 eggs April 10) and the young leave

the nest about May 18.   A second brood is commenced soon afterward (7 eggs, June 6) and it takes nearly all July to raise them.   I do
not think that they make a third brood, but they succeed in raising 4
or 5 young of each brood.   The young ones are sweet little creatures,
showing the same pattern of coloration as the old ones from the
beginning.   They gather into flocks as soon as they can fly well and
retire to out of the way places, where they enjoy themselves in hedges,
brush-heaps, and similar retreats, spending hours in frolicking and
twittering.   Other members of the sparrow family often associate
with them and seem to be perfectly satisfied with their behavior."

Plumages.—The young European tree sparrow is hatched naked,
with no natal down appearing.   Witherby (1919) describes the juvenal
plumage as much resembling that of the adult, "but crown mostly
smoky-brown, feathers with small blackish tips, sides of crown and
back of neck dull chestnut-brown; mantle less rufous; throat, lores,
and ear-coverts greyish-black; greater wing-coverts and outer webs
of wing-feathers brown, not chestnut-brown; tips of greater and
median coverts buff; lesser coverts browner, not so chestnut."   The
sexes are practically alike in this plumage and nearly alike in subse
quent plumages.

A complete postjuvenal molt occurs in late summer and early fall to
produce a first winter plumage indistinguishable from that of the
adult, the crown becoming uniform magenta-chocolate and the black-
and-white areas on the head and throat becoming clearer.   The
nuptial plumage is acquired by wear, producing little change except
brighter colors.   Adults have a complete postnuptial molt between
August and October.

Food.—These sparrows sometimes visit the grain fields and eat
some wheat, oats, and corn, but probably not enough to do any great
amount of damage.   A large part of their food consists of weed seeds
and various insects, but no thorough study of their food seems to have
been made, at least not in North America.

Behavior.—This tree sparrow is evidently a very different bird
from its pugnacious and aggressive relative, the house sparrow.
Widmann says in his notes: "The St. Louis tree sparrow is a gentle,
sociable bird, seldom seen quarreling among themselves or with other
birds.   They like to live in large flocks of 50 to 100 birds.   Some tree
sparrows remain paired, or pair, during the winter, build warm nests
and spend much time in each other's company, away from the flock,
and in anticipation of the joys of approaching spring.   When sitting
together in a tree it is one of their peculiarities to sit so close as to
touch each other's side.   This affords a very pleasing picture of peace
and good companionship, showing how much they are attached to
each other."

Voice.—The same observer says that "though they have no real song, a medley of their various tinkling notes answers well the same purpose. There is a slight resemblance between the twittering of a flock of this species and that of *Spizella monticola,* and this has probably led the first European settlers to give the name of tree sparrow to the latter species, an appellation which would otherwise be difficult to account for."

Yarrell (1876–82) writes: "The common call-note of the Tree-Sparrow is a chirp, not unlike though shriller than that of the House-Sparrow, but, as Blyth remarks (Mag. Nat. Hist. vii, p. 488), it has others in great variety. The cock has also a proper song, which the same observant naturalist describes as "consisting of a number of these chirps, intermixed with some pleasing notes, delivered in a continuous unbroken strain, sometimes for many minutes together; very loudly, and having a characteristic sparrow tone throughout."

Field marks.—The European tree sparrow bears a remote, superficial resemblance to our well-known English sparrow, but Witherby (1919) gives the following field characters: "Both sexes are alike and differ from male House-Sparrow in smaller size, trimmer build, black patch on ear-coverts, and chocolate-coloured, not grey, crown. Double white wing-bar is another, though less obvious, distinction. Notes bear general resemblance to House-Sparrow's, but are perhaps shriller, and *chee-ip, chup* is distinctive. Birds flying across open fields—often singly—may be detected by their sharp *teck, teck.*"

DISTRIBUTION

Range.—Introduced and now resident in central-eastern Missouri (Creve Coeur Lake, St. Charles, St. Louis), southwestern Illinois (Alton, Grafton, Belleville), and Bermuda (no recent records).

Family ICTERIDAE
Meadowlarks, Blackbirds, and Troupials

DOLICHONYX ORYZIVORUS (Linnaeus)

Bobolink

Plates 1, 2, and 3

HABITS

Our familiar bobolink is known by various names in different parts of its seasonal wanderings. We know it in the north by the above common name, which has stood for many years and is evidently an abbreviation of "Robert of Lincoln" in the classic poem of that name by William Cullen Bryant. In New England it is sometimes called by the pretty name, "meadow-wink," and the less complimentary name, "skunk blackbird," owing to its fancied resemblance in color pattern to that unpopular animal, On its fall migration it is recognized as "ortolan," "reed bird," and "rice bird" on account of its haunts and habits, and, in Jamaica, where it has grown exceedingly fat, they call it "butter bird." On its spring migration through the southern States, it is often called the "May bird."

Unfortunately for us New Englanders our beloved bobolink has largely disappeared, or at least has been greatly reduced in numbers in most of its former haunts, during the past 50 years. In my youthful days nearly every mowing field of long, waving grass, many of the damper meadows near our streams, and some of the drier portions of the brackish marshes furnished attractive homes for one or more pairs, often many pairs, of bobolinks. In driving through the open country past such places we could always count on seeing some of these showy birds hovering in ecstatic flight just above the tall grasses, the waving white daisies, and the bright yellow buttercups, pouring out a flood of bubbling, erratic song. They were always conspicuous to both eye and ear, forming one of the delights of a springtime ramble. But this is now mainly a happy memory, for there are so few places where they can now be found that it is an event of importance if we see one.

The partial disappearance of the bobolink from the Northeastern States has been due to several very evident causes. The heavy slaughter of the migrating hordes, both spring and fall, as will be discussed later, has perhaps killed off a large proportion of the birds that formerly nested in New England. Fortunately, due to the reduction in the cultivation of rice in the Southern States, this slaughter has been largely stopped and the birds are more rigidly protected everywhere. Another cause of less importance was the wholesale killing of "reed birds" for the market, but this is now prohibited by law. But the New England population of bobolinks has not been built up to its former proportions. A local cause here that has also had its effect in driving away our breeding birds is a decided change in the time and in the methods of harvesting our hay crops. Former- ly, the grass in our mowing fields, the favorite nesting places for bobolinks, was cut by hand and rarely before the first or middle of July. By that time the young bobolinks were out of the nest and safely on the wing. Now the mowing is done earlier, usually before the end of June, the grass is cut close with mowing machines, and the hay is scraped off by machine rakes. Many young birds would thus be killed while still in the nests or before they were able to escape by flight. This naturally drove the birds away to seek safer nesting grounds. Furthermore, with the passing of the horse much less hay has been needed, and there are fewer fields of the tall grass so much preferred by the bobolinks. The haying fields in Massachusetts are largely a thing of the past.

Southern New England is not the only place in the east where the bobolink has decreased in numbers. Robie W. Tufts writes to me from Nova Scotia: "My notes indicate a marked scarcity of these birds during the summer of 1919 and again in 1920. They were noticeably scarce again during the summer of 1930, and during the past summer of 1945 seemed alarmingly scarce." Ludlow Griscom (1923) wrote referring to the New York city region: "This distinguished songster was formerly a common summer resident throughout our territory, but is now found only in the outlying and more rural dis- tricts. Its great decrease started fifty years ago when trapping the males for cage-bird purposes was a profession on large scale." Todd (1940) remarks, for Pennsylvania: "Observers from various parts of the state agree that since the early twenties there has been a marked falling off in the numbers of this species." And even as far west as Minnesota the bobolink is yielding ground, but not for the same reasons. Thomas S. Roberts (1932) writes: "There is some indication that the Bobolink has been decreasing in numbers in recent years and that, locally, it has almost disappeared from lowlands where it was formerly abundant. Its place has been taken by the Brewer's Blackbird, which has swept eastward across the state and is now abundant even in the southeastern counties. It lives and nests here under exactly the same conditions as the Bobolink and, being a larger and more aggressive bird, there is reason to fear that it is driving the Bobolink from its former domain."

While the bobolink has been discouraged and its numbers have been depleted in many of its eastern breeding resorts, it has been encouraged to extend its range and to increase in abundance farther west until it is now a common breeding bird across the entire continent in the northern States and the southern Provinces of Canada. It apparently never liked to nest on the virgin prairies but it followed civilization westward, and with the settlement of the country it found congenial nesting sites in cultivated grasslands and clover fields. The westward movement evidently began many years ago, for Ridgway (1877) wrote: "The Bobolink seems to be spreading over all the districts of the 'Far West' wherever the cultivation of cereals has extended. We found it common in August in the wheat-fields at the Overland Ranch, in Ruby Valley [Nevada]." W. L. McAtee (1919) says: "The trend of the bird's breeding range to the northwest is unmistakable; for instance in the first edition of the A. O. U. Check-List, the Western limit of the breeding range was given as the Great Plains; in the second edition, 1895, as Nevada, Idaho and Alberta, and in the third edition, 1910, as British Columbia."

The bobolink began to be common and well distributed in Montana during the first decade of this century; Aretas A. Saunders (1921) recorded it as "a common summer resident of all except extreme eastern Montana, breeding in the wet meadows and irrigated fields of the prairie region, and in the valleys of the mountain region. * * * In most parts of the state the Bobolink is increasing with the extension of irrigation."

It apparently first appeared in Oregon about 1903. Gabrielson and Jewett (1940) say: "The Bobolink seems to be a comparatively new arrival in this State, as so good an observer as Bendire failed to find it in the Harney Valley during his stay, through it is now a regular resident of that area."

And for California, Dawson (1923) writes: "It was the chief surprise of a visit paid in 1912 to the Surprise Valley in Modoc County to find the Bobolink common and, apparently, breeding." According to Grinnell and Miller (1944), it is now a "summer resident in extreme northeastern part of State, where there is at least one colony. Rare straggler to other sections, chiefly in the autumn."

Spring.—From its winter home in South America, as far south as south Brazil, northern Argentina, and Paraguay, the bobolink makes a very long and somewhat hazardous flight to its summer home, which extends from Nova Scotia to British Columbia. It enters the United States on a broad front, from Florida to Louisiana, with possibly a few migrating along the coast of Texas.

Just when the bobolinks leave their winter home or by what route they reach the north coast of South America does not seem to be known. Thence the main flight is almost directly northward. Only a few, perhaps only stragglers, follow an eastern route, through the Lesser Antilles and the Bahamas to Florida.

The species is rarely mentioned by observers in the West Indies, but Baird, Brewer, and Ridgway (1874) say:

Dr. Bryant, in his visit to the Bahamas, was eye-witness to the migrations northward of these birds, as they passed through those islands. He first noted them on the 6th of May, towards sunset. A number of flocks—he counted nine— were flying westward. On the following day the country was filled with these birds, and men and boys turned out in large numbers to shoot them. He examined a quantity of them, and all were males in full plumage. Numerous flocks continued to arrive that day and the following, which was Sunday. On Monday, among those that were shot were many females. On Tuesday but few were to be seen, and on Wednesday they had entirely disappeared.

The main flight passes farther westward, where thousands make the 500-mile flight directly across the Caribbean Sea to Jamaica, then 90 miles more to Cuba, and another oversea flight of 150 miles to Florida. Jamaica is passed in April, it does not linger long in Cuba, and reaches northern Florida before the end of April. While we were cruising south of the Florida Keys, on April 24, 1903, a steady stream of bobolinks, water-thrushes, and other small land birds passed our boat, flying northward from Cuba to Florida against strong northerly winds; they seemed much exhausted; a bobolink attempted to alight on our boat but missed it and fell into the water, from which we did not see it rise; another alighted on the cabin and was so tired that it allowed us to pick it up. Many birds must perish on these long flights over open water against adverse winds, but some are probably able to rise from the surface after resting there for a while. Vincent E. Shainin (1940) saw this happen off the coast of Florida, during the spring migration from the Bahamas. "Using my eight-power binocular I was amazed to see a male Bobolink (*Dolichonyx oryzivorus*) riding the swells with both its head and tail held at right angles to the surface. Occasionally its back would appear above the water. * * * For a few seconds it remained very still, then it began to struggle vigorously for several seconds, finally leaving the water directly without pattering along in coot fashion."

Bobolinks also reach the United States by a trans-Gulf migration, from Yucatán to Louisiana, another long overwater flight. George H. Lowery, Jr. (MS.), reports that one came aboard his ship for a few minutes and then disappeared, on May 2, 1945, while the ship was 210 miles from Yucatán and 328 miles from the coast of Louisiana. Audubon (1842) says:

In Louisiana, small detached flocks of males or of females appear about the middle of March and beginning of April, alighting in the meadows and grainfields, where they pick up the grubs and insects found about the roots of the blades. * * *

During their sojourn in Louisiana, in spring, their song, which is extremely interesting, and emitted with a volubility bordering on the burlesque, is heard from a whole party at the same time; when, as each individual is, of course, possessed of the same musical powers as is his neighbours, it becomes amusing to listen to thirty or forty of them beginning one after another, as if ordered to follow in quick succession, after the first notes are given by a leader, and producing such a medley as it is impossible to describe, although it is extremely pleasant to hear it. While you are listening, the whole flock simultaneously ceases, which appears equally extraordinary. This curious exhibition takes place every time that the flock has alighted on a tree, after feeding for awhile on the ground, and is renewed at intervals during the day.

Bobolinks are, apparently, somewhat irregular in their appearance and never very abundant in Louisiana in spring, whence they migrate northward through the Mississippi Valley in moderate numbers; they are generally regarded as only fairly common, or even rare, in the southern half of this Valley. This would seem to indicate that the large numbers of these birds that nest in the more western States and Provinces must reach their breeding grounds by a westward migration from some of the Atlantic States.

The flood tide of the great spring migration flows rapidly in a nearly northward direction through the Atlantic States, mainly east of the Alleghenies, and reaches the northern breeding grounds during the last of April or early in May. Some observers say that it migrates by day and others regard it as a night traveler; perhaps circumstances vary and both are partially correct. Bendire (1895) quotes the following from W. M. Hazzard, of Annandale, S. C.: "The Bobolinks make their appearance here during the latter part of April. At that season their plumage is white and black, and they sing merrily when at rest. Their flight is always at night. In the evening there are none. In the morning their appearance is heralded by the popping of whips and firing of musketry by the bird minders in their efforts to keep the birds from pulling up the young rice. This warfare is kept up incessantly until about the 25th of May, when they suddenly disappear at night."

In Massachusetts, we eagerly await their arrival around the 10th of May and are seldom disappointed, as their jovial, rollicking songs bring life to the fresh, green meadows.

Courtship.—The males arrive a few days or a week in advance of the females, to select their nesting territories and to indulge in a few days of jolly frolic and exuberant song; the fields and meadows are now being clothed with fresh green grass and the trees are bursting into new foliage. The carefree birds are singing in little groups in the trees, or chasing each other about over their chosen homes. When the females come, courting begins in earnest; this largely consists of rivalry in song, as the handsome male in full nuptial dress pours out his joyous melody while perched on some tall, waving weed stalk, low tree, or fence. Often a "game of tag" ensues, as the female flies across the field, with two males in hot pursuit, as if she were saying "catch me if you can." More often the female seems coy and indifferent, hiding in the long grass, until the rival males find her and display their charms before her. As Townsend (1920) says: "One may see a male courting on the ground. He spreads his tail and forcibly drags it like a Pigeon. He erects his buff nape feathers, points his bill downward and partly open his wings, gurgling meanwhile a few of his song notes. The female indifferently walks away."

The following attractive account is written by Miss Ruth Trimble (Todd, 1940):

On a morning in early May, in one of their favored haunts, a tinkle of fairy music, like the strains of an old Greek harp, seems to come from the sky and may be traced to a company of male bobolinks, circling on fluttering wings high above. While you watch, the tinkling notes descend earthward, and an exuberant male sinks to a swaying weed stalk; with tail spread, wings partly opened, and feathers of his nape ruffled, he concludes his song with a few enchanting notes addressed to the mate he is wooing. Up she darts from the grasses to engage him in a lively chase, and in a flash he is off again in pursuit—an ardent troubadour, serenading his lady as he follows her; at times, seemingly forgetting her, he mounts skyward, his throat fairly bursting with the ecstatic melody that bespeaks his joie de vivre. No other bird courtship exhibits such reckless abandon. None is attended by such a flood of joyous music bubbling forth irrepressibly, with never a plaintive strain. No other wooing seems so delightfully spontaneous and gay. This wanton frolic may continue for a week or more before nest-building is actually begun and the female assumes responsibility for her family.

Dr. Kendeigh (1941) made some interesting observations on the family relations of the bobolink on a restored prairie in Iowa:

There were ten females here, but evidence for no more than six males, with polygamy strongly indicated. The male at nest No. 1 was frequently present also at nest No. 7 about 200 feet away, although he was only seen to feed the young at No. 1. He was recognized by the characteristically clipped tail given him when caught at nest No. 1; no other male was seen around nest No. 7. Nests number 9 and 10 were separated by only 44 feet and the male appeared equally concerned for both nests, although he was observed feeding young only at number 9. No other male was seen here. * * *

Notable in this species was the lack of territorial defense by either the adult male or female. If these birds establish a territory at all, it must be only for the mating and early nesting period. A fairly good spacing of the nests over the area would indicate that they may establish territories during the period when nests are started, but certainly after the young are hatched there is very little evidence for their continued maintenance. * * *

Lack of territory was also manifested by the tolerance of other males close to the nest. This was often noticed; once two foreign males were observed near the nest with the male who owned it disregarding them.

P. L. Buttrick (1909) gives further evidence of polygamy among bobolinks. One male and two females, the only bobolinks in the vicinity, raised four broods in two adjacent fields.

Nesting.—The nest of the bobolink is a very simple affair—a hollow, either scraped in the ground or selected for the purpose, loosely surrounded with coarse grasses or weed stems and thinly lined with finer grasses. The nest is sometimes placed in an old wagon rut or in a depression made by a horse's hoof. A. D. Du Bois mentions in his notes a nest that was in a hollow 3 inches in diameter by 1 inch deep; the measurements of the nest were: "Diameter 2.12 to 2.25 inches; depth 1.62 inch." This was evidently quite typical as to size. He mentions another nest that was "sunk among the bases of the standing tall grasses; but there was no hollow in the ground; the bottom of the nest was approximately at the ground surface."

What the flimsy nest of the bobolink lacks in construction it makes up for in concealment; it is almost invariably placed in a dense stand of tall vegetation, in the long grass of some luxuriant mowing field or damp meadow, in a field of clover, alfalfa, or in a thick growth of weeds or other wild plants.

Elisha Slade (1881a), who formerly lived in a town near me and was well known locally, describes two most remarkable nests. The first was—

occupying the space between four stalks of a growing narrow dock (*Rumex crispus*). This nest was suspended from four points of its circumference, 90° apart, to the four stalks of the plant which grew from the same root. The bottom of the nest was about six inches above the ground. It was constructed entirely of vegetable material and consisted of two distinctly separate parts. A hemispherical cup, in one piece of coarse but neatly woven cloth, very strong and very light, was fastened to the living, growing supports by strong fibres passing around each stalk above and below a joint firmly woven into the rim of the cup with some of the longer strings interlacing the sides. * * *

In this hanging basket was an elaborate lining of very soft blades of grass between which and the cup was an elastic padding. The woven cup was about five inches in diameter and five inches deep, the padding about half an inch thick, and the lining about the same thickness. The whole structure, dock and nest, swayed in every passing breeze but the nest was so strongly fastened to the stalks and the plant so securely held by the nest that it would have required a hurricane or tornado to have blown it away.

He claims to have found a similar nest, 22 years later, at the same place and in a similar plant. This all sounds like a fairy tale, but is printed here for what it is worth, as an interesting suggestion that the cloth cup may have been placed there by human hands. It seems incredible that a bobolink could have built such a nest, or even been tempted to occupy it.

The evidence indicates that the male selects the general locality for the nesting, which he occupies until the female arrives and is persuaded to remain there; she, then, probably selects the exact spot in which the nest is to be placed and does all the simple construction. Alexander F. Skutch, in his notes from Ithaca, N. Y., says: "In a field of alfalfa and grass where many males are singing, I have watched long to see a female building, but all in vain. I think that they must work under cover to avoid molestation by the too ardent males. May 28, 1931: Today I found a nest that already contained two eggs—a sparse and shallow cup of dried grass stems placed on the ground in the center of a clump of alfalfa, in a rather bare part of the generally lush meadow."

The nest of the bobolink is one of the most difficult to find. The female can almost never be traced to it during the simple process of building, for the small amount of building material can generally be picked up in the immediate vicinity, without having to bring in anything from a distance. The female can seldom be flushed directly from the nest, as she runs for some distance through the grass before flying. I have tried dragging a rope over a field where the birds were nesting, but there was never any nest where any of the birds flushed. The only method I have used with any degree of success is to run wildly back and forth over the field until all the females were flushed, then conceal myself and watch for their return; after marking down the exact spot at which a female alighted, I then might, if I ran quickly to the spot, flush her near enough to me to be able to find the nest by going over the ground carefully on hands and knees.

Dawson (1903) says: "If you care to spend an hour or so hunting for the treasures, the safest way is to mark the spot where the bird rose, and then hunt toward your original position along the line of approach." Skutch (MS.) tells of his method, which worked successfully: "Whenever I came close to their nest, the bobolinks made no cries nor demonstrations of alarm, but withdrew to a very respectful distance and eyed me quietly—only the male at times letting a few melodious tinkles escape his muffled bell. The parent bobolinks trusted implicitly in their nest's concealment; any demonstration would be superfluous or foolhardy. * * * Finally I set up some branches in the ground in the general region of the nest. Returning with food, the bobolinks rested on these before dropping down out of sight amid the grass. Noting the direction they took when leaving the first branch to go to the nest, I set up another on that side, on which the parents alighted when they next returned. And so, by giving myself closer and closer points of reference, I at length discovered the frail cup of grasses, on the ground between the stems of a daisy plant."

Lyle Miller, of Youngstown, Ohio, writes to me: "Twice I have flushed the female directly from the nest. On several occasions I have found it necessary to touch the brooding bird before she would leave the nest."

Eggs.—The bobolink lays from four to seven eggs to a set, usually five or six, and only one brood is raised in a season. Bendire (1895) describes the eggs as follows:

The eggs are ovate or short ovate in shape. The shell is close grained and somewhat glossy. The ground color varies from pearl gray or pale ecru drab to a pale reddish brown or pale cinnamon rufous. They are irregularly blotched and spotted with different shades of claret brown, chocolate, heliotrope purple, and lavender markings, intermingled with each other, and varying greatly in size and intensity. Almost every set is differently marked, and it is extremely difficult to give a fair average description. In some specimens the ground color is almost hidden, the markings being nearly evenly distributed in the shape of large blotches over the entire surface of the egg. In the majority, however, the darker markings are mainly confined to the larger end of the egg, while the paler ones are more noticeable in the middle and about the smaller end.

The average measurement of seventy-seven specimens in the United States National Museum collection is 21.08 by 15.71 millimeters, or 0.83 by 0.62 inch. The largest egg in this series measures 22.35 by 16.26 millimeters, or 0.88 by 0.64 inch; the smallest, 17.53 by 15.24 millimeters, or 0.69 by 0.60 inch.

William George F. Harris has in his collection a set of eggs larger than any in the National Museum; these measure 23.9 by 16.2, 24.1 by 16.1, 23.9 by 16.4, 23.8 by 16.2, and 24.0 by 16.1 millimeters.

Young.—The period of incubation for the bobolink is given by F. L. Burns (1915) as 10 days, but Mrs. Wheelock (1904) says: "The mother bird broods alone for thirteen days, while Robert frolics gayly over the fields with others of his sex, always within call, but seldom or never feeding her. When the young are hatched, however, he takes charge of them, and I have found him alone with a brood of seven nestlings huddled in a fence corner in Michigan." E. H. Eaton (1914) says: "The young are hatched in about 11 days and develop very rapidly so that they are able to take wing in from 10 to 14 days." He probably means that they leave the nest at this age, for it is well known that the young leave the nest and wander around in the grass for several days before they learn to fly; at this stage many would be killed by early mowing and raking. While still in the nest, the young are practically invisible; packed in as closely as sardines in a box, they show no form or shape, remaining absolutely immovable and with eyes closed; and their colors match the surrounding earth so closely that one could step on them without seeing them. Only when one of the parents comes with food do they wake up and give the buzzing food call.

A. D. Du Bois has sent me some very full notes on the behavior of a pair of bobolinks and their young, from which I can quote only a few parts: "June 14 (midmorning): All the eggs appear to have hatched. The female jumps over the grass for a distance of three or four feet, then hobbles along in the grass; and, if I follow her, she repeats this—and continues to repeat until we are perhaps a hundred feet from the nest, when she flies for a short distance. This is the pattern of her ruse.

As I return toward her nest she sits on a small shrub, and now and then utters a note which sounds like *quick*. In the afternoon, while I was sitting on the ground adjusting a small observation blind five or six yards from the nest, the female came rather near, calling *quick*; the male came up and perched on a tall spray of wild asparagus, calling rather anxiously a note different from hers.

"June 16 (between 8 and 9 a. m.): While hidden in the blind trying to photograph the male on the nearby asparagus, I saw the female go to the nest two or three times to feed the young. Both parents were agitated and continued chirping for some time after I was hidden. Upon examining the nest I found one almost naked nestling outside (two or three inches away) with its head down in the grass. It kicked when I touched it. As I was leaving, the male used the jumping retreat, similar to that of the female.

"June 18: I can see only three young in the nest; they have grown rapidly; and there is one unhatched egg, which had hitherto been hidden. The male is very communicative. When I was thinning the onion seedlings in the garden he sat on a nearby bean pole berating me and singing to me repeatedly. This evening he followed me across the garden.

"June 21: There was a great deal of alarm-calling on the part of both parents when I went out this morning. After I had become well settled in the blind both parents came to the tall asparagus, and I thought I saw a little billing. Soon I saw one of the youngsters scrambling out of the nest; and before 8:30 a. m. they all had left and were crawling away in the tall grass. * * * During my stay in the blind, the male bobolink did the watching and guarding while the female did the feeding; then he brought food to the young who were hidden in the grass at some distance from the nest. He alighted on the asparagus before taking food to them. He brought green larvae, a dark miller, and an insect which looked like a brown wasp. The female once brought a white object which had the appearance of being hard and about the size of a small seed of sweet corn. All the commotion was repeated when I made another visit, near noon, to try to locate the young. After I was hidden and all had become quiet again, I could hear a youngster in the grass only a few feet from the nest. Occasionally it uttered a note resembling the syllable *chib*, and it moved the grass so that I got its location and went out and found it. But before I reached the spot it had half buried itself, head downward, in the thick mat about the grass roots. Thus all its forward parts were hidden, and only its legs and the posterior extremity of its body were visible. While I was trying to part the overhanging grass sufficiently to photograph it in this position, it was overcome by some great discomfort and quickly unburied itself, squirmed, shook itself, and sat right-side-up. * * * The last three nestlings which left the nest in the forenoon of June 21, were not less than 7 nor more than 8½ days old when they left."

Plumages.—The striking plumage changes of the male bobolink are well, and I believe correctly, described by Jonathan Dwight, Jr. (1900), one of the best authorities on the plummages of passerine birds. He calls the natal down buff, and describes the juvenal plumage as follows: "Above, dull brownish black, median crown stripe, superciliary line, nuchal band and edgings of the other feathers of back and wings buff deepest on nape; primaries, their coverts, secondaries and alulae tipped with grayish white. Below, rich buff paler on chin and faintly flecked on sides of throat with clove-brown. A dusky postocular streak. * * * This plumage is worn but a short time and the postjuvenal moult is well advanced by the end of July as shown by four specimens in my collection."

The first winter plumage is "acquired by a partial postjuvenal moult in July which involves the body plumage, tertiaries and wing coverts, but not the rest of the wings nor the tail." This is "similar to the previous plumage, but darker above and yellower below, a rich ochre or maize-yellow prevailing, paler on chin and abdomen, the sides of the breast and flanks and under tail coverts conspicuously streaked with dull black veiled by the overlapping feather edges."

The first nuptial plumage is acquired by a complete prenuptial molt. Dr. Dwight describes this plumage as "almost wholly black, the body plumage veiled by long maize-yellow feather tips. The nape is rich ochre and the scapulars white, the inner plumbeous, both edged with olive-gray. The outer primary is edged with white, the two adjacent with maize-yellow, the tertiaries, greater coverts and interscapularies with wood-brown. Rump plumbeous, upper tail coverts white, both areas veiled with olive-gray or olive-buff. Tail tipped with olive-gray."

This is the plumage in which the birds arrive in the United States, and in many cases the maize yellow tips have not entirely worn away by the time that the birds reach their breeding grounds, the tips persisting longest on the abdomen, flanks, and under tail coverts.

The adult winter plumage is acquired "by a complete postnuptial molt beginning the end of July. Similar to first winter plumage, usually whiter below especially on the chin and middle of the abdomen, and above with rich-brown edgings expecially of the tertiaries. The bill becomes clay colored or purplish [it was black in the spring]. The chief differential character is however the presence of a few black feathers, usually yellow tipped, irregularly scattered on the chin and breast."

The adult nuptial plumage is "acquired by a complete prenuptial moult in midwinter. Differs inappreciably from first nuptial dress, but it is probable that (as in other species) the yellow edgings diminish with age."

The fact that certain male bobolinks in captivity have assumed the black spring plumage without any apparent signs of molting has led to some discussion of the old, threadbare theory of color change without molt. Dwight (1900, p. 123) has discussed fully this and the evidence offered by others, and concludes that there is not the slightest evidence to support the theory. "Nowhere among living organisms do restorative changes in tissue take place without destruction or casting off of the old. Consequently belief that a feather which regularly develops, dies and is cast off, can possibly violate such a universal law is not only contrary to common sense but contrary as well to every established fact regarding the moulting of birds."

The sequence of molts and plumages of the female is apparently similar to that of the male, but not so conspicuous.

Food.—As we know the pretty bobolink on its northern breeding grounds we can find little to complain of in its feeding habits, which are mainly beneficial to our interests, or at their worst only neutral, but when it becomes the "rice bird" in the Southern States the planters have a strong case against it. While with us in the north it feeds on insects and the seeds of useless plants, and the young are fed almost exclusively on insects, mostly harmful species. After the young are on the wing, the flocks wander about, living mainly on weed seeds, a little waste grain, and the seeds of the wild rice which grows along the borders of our streams and marshes.

Of the 291 stomachs examined by Beal (1900), 231 were collected in the Northern States from May to September, inclusive. The food was found to consist of 57.1 percent animal matter and 42.9 percent vegetable. His table lists the following average percentages for the five months: Predaceous beetles 0.6; May-beetle family 2.7; snout-beetles 9.0; other beetles 6.7; wasps, ants, etc., 7.6; caterpillars 13.0; grasshoppers 11.5; other insects 4.6; spiders and myriapods 1.4; oats 8.3; other grain 4.1; weed seeds 16.2; and other vegetable food 14.3 percent. Aside from the trace of useful predaceous beetles and a few parasitic Hymenoptera, all the other insects are more or less harmful; the large percentages of caterpillars and grasshoppers may be placed to the credit of the bobolink. The small amount of grain eaten is of little account when compared with the large amount of noxious weed seeds such as barn-grass, panic-grass, smartweed, and ragweed.

Beal evidently found no corn in the stomachs he examined, but Warren (1890) says that, in Pennsylvania, "they visit the cornfields, and in company with the English Sparrow, prey to a more or less extent on the corn; like the sparrow they tear open the tops of the husk and eat the milky grain."

Forbush (1927) says of its food in New England:

The food of the Bobolink on its breeding grounds consists chiefly of insects, which comprise from about 70 per cent to over 90 per cent of its sustenance in May, June and July. Of the vast quantity of insects consumed by the bird less than 3 per cent, on the average, are beneficial species. In May it takes a small and fast dwindling per cent of grain, and in July an increasing amount, which rises in August to about 35 percent, but decreases rapidly until at the end of September it is only about 3 percent. Its consumption of weed seeds averages about 8 percent in May, June and July, but increases rapidly in August until at the end of September it reaches over 90 percent; by that time most of the Bobolinks have left New England.

On their southward migration they feed almost exclusively on the seeds of wild rice and other useless plants, together with some grain, until they reach the cultivated rice fields in the south, where they do enormous damage; this will be discussed in a later paragraph.

E. R. Kalmbach (1914) writes: "The bobolink does exceptionally good work as a weevil destroyer, for wherever it lives near infested alfalfa fields the insect forms its most important animal food. * * * Seven bobolinks collected in June [in Utah] had taken the weevil at an average of about 8 adults and 42 larvae per bird, to the extent of 68 per cent of the stomach contents.

In the stomach of one, 6 adults and 90 larvae formed the entire food.

Another had eaten no less than 28 adults and 77 larvae, amounting to 86 per cent of the stomach contents, while a third had eaten 3 adults and 61 larvae."

Arthur H. Howell (1932) says: "Every one of 15 Bobolinks collected in celery fields near Sanford [Florida] had fed on the destructive celery leaf-tyer (*Phlyctaenia rubigalis*), the remains of this insect forming 67 per cent of the total food in their stomachs."

Behavior.—Except when on their nesting grounds, bobolinks live largely in open flocks, congregating in favorable feeding grounds. The males are still in small flocks when they first arrive in their summer haunts, perching on trees or fences and indulging in frequent outbursts of glorious song. While migrating they fly high in open formation, appearing much like other small blackbirds. The hovering flight of males over their nesting grounds, or when courting their mates, is very characteristic; they proceed rather slowly on rapidly vibrating wings, a short distance above the tops of the tall grass, singing rapturously; their striking color pattern makes a pretty picture above the buttercups and daisies. On the ground, they seldom walk with the dignified gait of other blackbirds, but usually proceed by hopping or running.

Frederick C. Lincoln (1925) describes a flight behavior which I have never seen: "There was a small colony nesting near North Napoleon Lake [North Dakota] in a rank growth of milkweed (*Ascle-*

pias), and while watching them on July 10 I observed a curious performance. On several occasions the males would flock together as at a prearranged signal, fly rapidly from the field in close formation for a considerable distance, and then scatter like the fragments of a bursting shell, each bird turning about and returning in a leisurely fashion to his own part of the cover."

Skutch says in his notes: "The brown female bobolinks remain hidden in the tall grasses and weeds, where it takes sharp eyes to pick them out. As soon as a female makes her appearance, even when she has just been driven from her cover by my passage over the meadow, one or as often two males dash after her, twisting and turning to follow every quick maneuver she makes in her effort to escape their attentions, and not relaxing their hot pursuit until she dives again into the vegetation, all unmindful of the approaching man."

Du Bois speaks several times in his notes of the male bobolink following him about, perhaps through solicitude for his young or perhaps as an evidence of curiosity. "June 19: This morning when I went over into the Wilkins lot to photograph a yellow warbler's nest, the male bobolink followed me, and 'hung around' to supervise the job, though this place is probably more than a hundred yards from his nest. Later he went away; but when I had made a second photograph (after first returning to the house on an errand), and was sitting on a box writing notes, here came the bobolink again. He was carrying something white in his bill (excrement no doubt) which he dropped as he alighted on a small sapling about eighteen feet from me."

Evidently the male is a good "watchdog." The female is also very alert for approaching danger while she is brooding, which makes it very difficult to flush her directly from the nest. While he was in his blind, close to the nest, and she was brooding the young, "she was very alert, continually looking about, and often stretching up her neck to see over the matted grass which surrounded the nest. In this upstretched position the streaking of her head matched wonderfully well the mixture of dead and green grass blades and stems through which, and against which, I saw her—so well in fact that she was rendered invisible except when she was moving, or when her eye was in plain sight. The male remained on the nearby asparagus almost all the time that his mate was on the nest."

Forbush (1927) writes of the slaughter in the South:

Early in September 1912, I left Boston for Georgetown, South Carolina, and remained until after the fifteenth in the coastal region of the state. I found that the negroes used two methods of taking the birds: (1) hunting with a gun by daylight, (2) hunting at night with torches, when the men poled skiffs along the irrigation ditches and picked the dazzled birds off the reeds where they roost, or else threshed them off into the boat with branches cut for the purpose. This could be done on dark nights. The following quotation from my report of the trip will give an idea of the conditions there at that time regarding the Bobolink.

On my arrival at the rice fields, colored gunners were seen in all directions, and the popping of guns was continual. All the shooting appeared to be done by creeping up to birds when they were sitting on stubble or on heaped-up rice, selecting a time when a large number flocked together. One of the negroes said that he often frightened up the birds in the rice fields and shot into the flocks as they flew, but I saw nothing like this. One man with a full bag told me that he had 8 dozen birds at noon and that he killed 16 dozen the day before. Another stated that he had six dozen so far, and shot about 12 or 13 dozen daily on an average, but that formerly he used to get 14 or 15 dozen, or even more, when the birds were numerous. He said it was not unusual formerly to kill 20 to 30 dozen at night and sometimes even 40 dozen, but all the negroes that I talked with agreed that they were getting very few at night now. Some said that nights must be dark for successful hunting. They said they received 20 cents a dozen now for "shoot" birds and 25 to 30 cents for "ketch" birds. One gunner said that when he could not get 25 cents a dozen he would knock off. * * *

Mr. James H. Rice, then chief game warden of South Carolina, wrote to me that he had checked up the game shipped from Georgetown, South Carolina, in one year. The result was 60,000 dozen of rice-birds and about 20,000 dozen of Carolina Rails and Virginia Rails, but at the time of writing (1912) the number shipped had fallen off greatly on account of the reduced number of the birds. He also stated that these birds were not shot to protect the rice, as they were not killed until they had grown fat on rice and until they would bring a good price in the market.

All this, happily, is now ancient history. The killing of songbirds and their sale in the market for food is prohibited by law, and the cultivation of rice in the Southeastern States in the main migration path of the bobolink has been greatly reduced. But the bobolink still has its natural enemies. Birds of prey still take their toll and prowling quadrupeds still rifle the nests of these ground-nesting birds. In low meadows the nests are sometimes flooded by heavy rains. The wily cowbird finds the well-hidden nests and deposits one or two of its unwelcome eggs; Friedmann (1929) gives only a few records, but so few bobolinks nests are ever found that the scarcity of records does not mean much.

Voice.—Aretas A. Saunders contributes the following comprehensive study of this subject: "The song of the bobolink is a loud, clear series of short notes, no two consecutive notes on the same pitch. The song begins on a comparatively low pitch. The pitch rises higher and higher, and the notes follow each other more and more rapidly as the song progresses. It is most commonly sung in flight, the bird flying horizontally above the meadow as it sings. It is occasionally sung from a perch in a tree or on the top of a tall weed. The song from a perch is frequently curtailed and not full length, but the flight song is usually complete.

"The song is difficult to record and I have only 15 records of complete songs, and 5 other records of the beginnings of songs. From these records I find the number of notes in a song varies from 18 to 43, averaging 27. The length of the songs varies from 2⅘ to 4⅗ seconds. The range in pitch is from E flat '' to B ''', or ten tones. Single songs range from 3½ to 8 tones, averaging about 6 tones.

"In 9 records the first note of the song is the lowest in pitch, and in 11 the first note is next to the lowest, the second note lowest. In 9 songs the last note is the highest in pitch, and in 6 songs the last note is next to the highest. Often, in the middle of the song a group of 2 to 4 notes is repeated two or three times in rapid succession.

"The beginning notes of the song are loud and rich in quality, but this richness seems to decrease as the song progresses, probably because as the notes become higher, the number of overtones that are low enough to affect the human ear decrease. To the bird it is quite possible that the quality is just as rich at the end as at the beginning.

"Consonant sounds, such as liquids like the letter l, and explosives like K or T, are common throughout the song.

"The season of song lasts from the arrival of the birds in early May (in Connecticut) to the early days of July. They evidently sing on the spring migration. On May 5, 1944, the day I saw the first bobolinks of that year, 10 male birds flew north over a woodland, and several were in full song as they flew. Only in the last 5 years have I been where I could observe the cessation of song in this bird. The date when the species as a whole had ceased singing averaged July 4, the earliest July 2, 1943, and the latest July 8, 1942. The date on which the last individual was heard to sing averaged July 9, the earliest July 2, 1943, and the latest July 18, 1942.

"Bobolinks have number of short call notes. I have written some of these, in the field, as *tschick* and *tchow* and *pink*. A three-syllable call is *tcheteeta* and, when repeated several times, it suggests the flight notes of a goldfinch. The *pink* note is commonly heard in late summer and fall, and is used by birds flying southward in fall migration."

Francis H. Allen writes to me: "In August a continuous warbling song may sometimes be heard from a flock feeding in grain fields. It seems to be formless, though at times it is suggestive of the regular song of the breeding season. On August 5, 1917, I heard snatches of the full song from a bird in Vermont."

Du Bois (MS.) writes of notes that he heard about the nest: "The female, agitated, utters her *quick, quick, quick*. The male, also much concerned, says *chow*, or *chaup*, in a pitch lower than the female's *quick*. As I moved away from the nest, or stood still at a little distance, he exclaimed: *Gee,whiz-ic*!, repeating it several times in his excitement (the *gee* higher than the *whiz-ic*).   This was followed by his *chow* notes; and sometimes he flew near to me, alighting on a weed and adding a portion of excited song to his entreaties or complaints."

Albert R. Brand (1938), in recording the vibration frequencies in the songs of passerine birds, gave as the approximate mean for the bobolink 3,000 and as the highest note 6,950 vibrations per second.

No description of the song of the bobolink is adequate to convey to the reader who has not heard it any appreciation of its beauty and vivacity.   It is unique among bird songs, the despair of the recorder or the imitator; even the famed mockingbird cannot reproduce it.   It is a bubbling delirium of ecstatic music that flows from the gifted throat of the bird like sparkling champagne.

F. Schuyler Mathews (1921) calls it "a mad, reckless song-fantasia, an outbreak of pent-up, irrepressible glee.   The difficulty in either describing or putting upon paper such music is insurmountable.   One can follow the singer through the first few whistled bars, and then, figuratively speaking, he lets down the bars and stampedes.   I have never been able to 'sort out' the tones as they passed at this breakneck speed."

The song has often been rendered in human words; these attempts give a good impression of the vivacity of the song, but no idea of its musical quality.   Among the best of these are those oft-quoted words from the classic poem of William Cullen Bryant (Robert of Lincoln): "Bob-o'-link, bob-o'-link, Spink, spank, spink."   Henry D. Minot (1877) suggests the following:

"Tom Noodle, Tom Noodle, you owe me, you owe me, ten shillings and sixpence!"

"I paid you, I paid you!"

"You didn't, you didn't!"

"You lie, you lie; you cheat!"

Field marks.—The male bobolink, in spring plumage, is so conspicuously marked that it cannot be mistaken for anything else; there is no other bird in its summer haunts that is at all like it.   It is the only one of our small land birds that reverses the almost universal law of concealing coloration by being wholly black below and mainly light-colored above.   The female is a shy, retiring bird, never much in evidence and generally out of sight in the long grass, where its yellowish-brown colors and its stripes help to conceal it among the grass stems. In the fall, all ages and sexes look much alike and can be recognized only by where they are and what they are doing.

Enemies.—The bobolinks were largely driven out of New England by early mowing and raking of the hayfields.   They were slaughtered in enormous numbers, for the market as "reed-birds," on their fall migration.   And thousands, probably millions, were killed as "ricebirds" in the Southern States, where they did great damage in the ripening ricefield in the fall and some harm to the sprouting grain in the spring; many were also shot there for food.

Coues and Prentiss (1883) comment on the reed-bird market: "The familiar 'clink' of the Reed-bird begins to be heard over the tracts of wild oats along the river banks about the 20th of August, and from that time until October the restaurants are all supplied with 'Reed-birds'—luscious morsels when genuine; but a great many Blackbirds and English Sparrows are devoured by accomplished gourmands, who nevertheless do not know the difference when the bill of fare is printed correctly and the charges are sufficiently exorbitant."

Economic status.—Much information on the economic status of the bobolink will be found in the foregoing paragraphs and need not be repeated here. While it is with us on its breeding grounds there is no doubt that it is a beneficial species. Most of the insects that it eats are of harmful species, or those of no value. The greater part of its vegetable food consists of weed seeds, or the seeds of useless plants; much of the very little grain it takes is waste. On its migration southward it feeds mainly on the seeds of wild plants, such as wild rice or wild oats, of no economic value. But, in the cultivated ricefields of the Southern States, it does, or has done, immense damage to the ripening grain in the fall and the sprouting grain in the spring. A few quotations from Beal (1900) will serve to illustrate the damage formerly done in the ricefields. Mr. J. A. Hayes, Jr., of Savannah, Ga., reported that a field—

which consisted of 125 acres of rice that matured when birds were most plentiful, and which, in spite of 18 bird-minders and 11 half kegs of gunpowder, yielded only 18 bushels per acre of inferior rice, although it had been estimated to yield 45 bushels. * * *

As a sample of actual loss, the following statement, furnished by Colonel Screven, gives his account with the bobolink at Savannah, Ga., for the year 1885:

Cost of ammunition_____	$245. 50
Wages of bird-minders_____	300. 00
Rice destroyed, say 400 bushels_____	500. 00

1, 045. 50

Colonel Screven cultivated in that year 465 acres of tidal land, so that he has estimated a loss of less than 1 bushel of rice to the acre, while most of the rice growers estimate the loss at from 4 to 5 bushels.

Captain Hazzard states that in cultivating from 1,200 to 1,400 acres of rice, he has paid as much as $1,000 for bird-minding in one spring.

He wrote to Major Bendire (1895):

The Bobolinks make their appearance here during the latter part of April. * * * Their next appearance is in a dark yellow plumage, as the Ricebird. There is no song at this time, but instead a chirp which means ruin to any rice found in the milk. My plantation record will show that for the past ten years, except when prevented by stormy south or southwest winds, the Ricebirds have come punctually on the night of the 21st of August, apparently coming from seaward. All night their chirp can be heard passing over our summer homes on South Island, which is situated 6 miles to the east of our rice plantations, in full view of the ocean. Curious to say, we have never seen this flight during the day. During the nights of August 21, 22, 23, and 24, millions of these birds make their appearance and settle in the rice fields. From the 21st of August to the 25th of September our every effort is to save the crop. Men, boys, and women, with guns and ammunition, are posted on every 4 or 5 acres, and shoot daily an average of about 1 quart of powder to the gun. This firing commences at first dawn of day and is kept up until sunset. After all this expense and trouble our loss of rice per acre seldom falls under 5 bushels, and if from any cause there is a check to the crop during its growth which prevents the grain from being hard, but in milky condition, the destruction of such fields is complete, it not paying to cut and bring the rice out of the field. * * *

Fall.—As soon as the young are on the wing, in July, and the males have ceased to sing, the bobolinks, old and young, disappear from their nesting grounds and retire to more secluded haunts in the marshes and along the banks of sluggish streams, where they feed on the seeds of wild rice, wild oats, and various weeds, and become quite inconspicuous during the molting season. Migration from the northern part of the bird's range begins in July, and by August it is in full swing all through the United States. The fall flight is mainly if not wholly by night; we often hear the distinctive *clink* note coming to us out of the darkness, as the scattered flocks pass over us. They stop to feed during the day, but probably do not move on every night, for they are known to roost in enormous numbers in the ricefields. Where food is plentiful and attractive, they probably stop over for a night or two and become excessively fat.

The fall migration of the bobolink, a long one and a remarkable one, in the main is a reversal of the spring route, but more concentrated. From the far western extension of the breeding range the birds retrace the steps by which they extended their range westward, flying almost east to the Atlantic coast. Only a scattering few take the shorter route southward through the Western States and Central America, and a comparatively few migrate through the Mississippi Valley, mostly east of the river. As Dr. Wetmore (1926) puts it: "When southward flight begins, it comes with a rush that distributes the flocks far southward, so that on the east coast the birds arrive at suitable points in the region from Maryland south to Georgia and Florida almost simultaneously at some date between the middle of August and the first of September." Meantime, the heavy flight from the eastern Provinces and States has poured down along the Atlantic coast, to join the western birds, where the main stream of migrants converges into a narrow funnel on the southeastern coast and overflows the whole peninsula of Florida. From there, three routes are available: the least popular of these, over which comparatively few birds travel, is the eastern route through Puerto Rico and the Antilles to British Guiana; the main trunkline, followed by a majority of the species, is an overseas route to Cuba and Jamaica and a long flight across the Caribbean Sea to South America; the third route, also well patronized, leads to Yucatan and thence along the east coast of Central America. After reaching South America, its route to its winter quarters is not as well known, but Chapman (1890) has this to say about it: "Salvin gives the bird from British Guiana and this, with the Cayenne record, seems to form the eastern limit of its range, there being, as far as I know, no records for eastern Brazil or the lower Amazon, while Darwin's record, already referred to, of a specimen taken in October, 1835, on James Island in the Galapagoes, is the only one with which I am familiar from west of the Andes. Indeed our bird's further wanderings seem now to be largely confined to the eastern slope of this range of mountains and the head waters of the Amazon, until it reaches what may be its true winter quarters in southern or southwestern Brazil."

Skutch tells me that "the bobolink appears only exceptionally to migrate through Central America. On October 12, 1930, I saw a few birds which I took to be bobolinks in winter plumage among the swamp grasses around the Toloa Lagoon in northern Honduras." But Todd and Carriker (1922) record it in Colombia as "a common visitor in September and October in the lowlands, from Santa Marta around to Fundación and all along the shores of the Cienaga Grande."

Examining the migration in more detail, a few published remarks are worth quoting. In Manitoba, according to Seton (1891), they gather into large flocks toward the end of July, and "then leave the prairie and attack the oat fields, doing, with the assistance of the Grackles and Redwing Blackbirds, an immense amount of mischief. After the oats are cut they resort to the marshes, feeding on wild rice, etc., until the cool nights inform them it is time to leave."

Milton B. Trautman (1940) describes an unusually heavy migration in Ohio as follows:

While in a boat near Sellars Point, between 5:30 a.m. and 6:00 a.m. on September 3, 1931, I heard flight notes of Bobolinks, and looking into the cloudless sky I saw a flock of approximately 50 flying in a southerly direction. The roughly rectangular flock was about one-fourth as deep as long and was advancing with the long side in front. At approximately 200-yard intervals behind this group came 31 other such flocks. No flock in this long irregular column contained less than 35 individuals nor more than 75, and the distance between each was remarkably constant. The birds appeared to be about 200 feet above the water, and could barely be seen with the naked eye. This migration was unusual because of its large size and its regularity and uniformity.

Of the normal flight, he says: "In the southward migration between 100 and 500 individuals could usually be seen daily as they migrated overhead or fed in marshes and fallow fields. Upon a few occasions from 600 to 2,000 were observed in a day."

After converging in Florida, the migrating hosts make two long, oversea flights, to Cuba and then to Jamaica. In Jamaica, the now overfat bobolink is called the "butter-bird" and is shot in large number for food. Gosse (1847) writes: "In ordinary seasons this well-known bird arrived in vast numbers from the United States, in the month of October, and scattering over the lowland plans, and slopes of the seaside hills, assembles in the guinea-grass fields, in flocks amounting to five hundred or more. The seed is then ripe, and the black throngs settle down upon it, so densely, that numbers may be killed at a random discharge. * * * Early in November they depart for the southern continent, but during their brief stay they are in great request for the table."

A long flight from Jamaica across the Caribbean Sea lands the birds on the northern coast of Venezuela. Dr. Wetmore (1939) says: "Shortly after sunrise on October 16 as our ship entered the harbor at La Guaira a flock of about 75 small birds swept in along the shore in close formation and rose to pass over the docks. At a casual glance I took them for sandpipers, but as I obtained a better look I saw that they were bobolinks. I supposed that they had just arrived in migration and were making a landfall as there was no place here for them to feed. At Ocumare de la Costa before seven on the morning of October 28, one flew with a low call from a large sea-grape tree on the beach and went uncertainly toward the marsh beyond. It seemed to be newly arrived. The following day I flushed half a dozen from rushes growing in the lagoon."

Winter.—Although there are a few scattering late fall and early winter records for even the northern States, practically all the bobolinks have left the United States before November, and nearly all have reached their winter home in central South America. Dr. Wetmore (1926) says: "During winter it continues to frequent swamps and grass-grown marshes, and seems to have its centre of abundance in the Chaco, a vast area of poorly drained, swampy land, with broad grass-grown savannas, that extends west of the Paraná and Paraguay rivers, from northern Santa Fé in north central Argentina, north into Bolivia and Brazil." Although the bobolink has had a safe haven here for many years, he remarks that the country is being settled, rice is being cultivated, and the birds are being killed for food by the foreign settlers. This prospect does not look favorable for the bobolink, which is also popular as a cage bird there.

DISTRIBUTION

Range.—Canada to Argentina.

Breeding range.—The bobolink breeds from central-southern and southeastern British Columbia (Vernon, Waldo), southern Alberta, southern Saskatchewan (Eastend, Quill Lake), southern Manitoba (Brandon, Winnipeg), central and southern Ontario (north sporadically to Chapleau and Bigwood), southwestern and central-southern Quebec (Blue Sea Lake, Newport), New Brunswick, Prince Edward Island, and northern Nova Scotia (Cape Breton Island); south through eastern Washington (rarely) and eastern Oregon (Blue Mountains) to northeastern California (Eagleville), northern Nevada (Ruby Valley), northern Utah (Springville), central and southeastern Colorado (Gunnison, Fort Lyon), central Nebraska (North Platte), northeastern Kansas (Manhattan), northern Missouri, central Illinois (Peoria, Urbana), south-central Indiana (Worthington, Columbus), southwestern and central-eastern Ohio (Hillsboro, Scio), northern West Virginia (south in the mountains to Greenbrier County), western Maryland (Red House), Pennsylvania, and central New Jersey. There are summer records from southwestern British Columbia (Chilliwack), central Alberta (Glenevis, Edmonton, Camrose), central Saskatchewan (Ladder Lake), western and northern Ontario (Emo, Missanabie, Strickland), the north shore of the St. Lawrence River in Quebec (Godbout), central Nevada (Toyabe Mountains), central-eastern Arizona (Showlow), central-northern New Mexico (between Park View and Chama), and north-central Kansas (Rooks County).

Winter range.—Winters in eastern Bolivia, central-southern Brazil, Paraguay, and northern Argentina, migrating from North America mainly through the Mississippi Basin, the Atlantic coastal States, Florida, and across the Gulf of Mexico and the Caribbean Sea; casually through eastern México and Central America, south to Ecuador, the Galápagos Islands, and Perú; east to the Bahamas, Hispaniola, Puerto Rico, the Lesser Antilles, Trinidad, French Guiana, and southeastern Brasil.

Casual records.—Casual in western Arizona (Wikieup). Accidental in Greenland (Godthaab, Arsuck), Labrador (Gready Island), southeastern Quebec (Bradore Bay), and northern Ontario (Moose Factory).

Migration.—Early dates of spring arrival are: Cuba—Havana, March 29. Bahamas—Cay Sal, March 28. Florida—St. Marks, April 9. Alabama—Dadeville, April 15. Georgia—Milledgeville, April 3; Athens, April 14. South Carolina—Charleston, April 7. North Carolina—Chapel Hill, April 13; Raleigh, April 19 (average

of 24 years, May 2). Virginia—Naruna, April 18. West Virginia—
White Sulphur Springs, April 27. District of Columbia—April 25
(average of 33 years, May 3). Maryland—Baltimore County, April
16. Delaware—Delaware City to Rehoboth, April 13. Pennsyl-
vania—Berwyn, April 25; Philadelphia, average of 12 years, May 4.
New Jersey—Cape May, April 21. New York—Rochester and
Bronx County, April 19; Ithaca, May 4. Connecticut—Glastonbury,
April 24. Rhode Island—Jerusalem, April 21. Massachusetts—
Norwich, April 19; Bernardston and East Longmeadow, April 22.
Vermont—St. Johnsbury, April 26; Wells River, average, May 12.
New Hampshire—Charlestown, April 30. Maine—Bar Harbor, April
27; Saco, May 2. Quebec—Montreal, May 7; Kamouraska, May 8.
New Brunswick—Scotch Lake and St. Andrews, May 12. Nova
Scotia—Bridgetown, May 7. Prince Edward Island—North River,
May 23. Louisiana—Grand Isle, April 1. Mississippi—Oak Vale,
April 24. Arkansas—Rogers, April 15. Tennessee—Nashville, April
19 (average of 12 years, April 27). Kentucky—Bardstown, April 22.
Missouri—Bolton and Corning, April 15. Illinois—Murphysboro,
April 19; Freeport, April 20; Chicago region, April 25 (average, May
5). Indiana—Brookville and Gary, April 6. Ohio—Bowling Green
and Oberlin, April 8 (average of 19 years at Oberlin, April 27).
Michigan—Ann Arbor, April 19; Blaney Park, May 2. Ontario—
Hamilton, April 15; Ottawa, May 3 (average of 31 years, May 16).
Iowa—Hudson, April 25. Wisconsin—Madison and Oshkosh, April
25 (average of 15 years in Dane County, April 30). Minnesota—
Minneapolis, April 23 (average of 7 years, May 7); Duluth, April 26
(average of 23 years for northern Minnesota, May 8). Texas—
Houston, April 24. Oklahoma—Tulsa County, May 2. Kansas—
Clearwater, April 17. Nebraska—Whitman, April 28. South Da-
kota—Faulkton, April 29; Sioux Falls, 4-year average, May 9. North
Dakota—Marstonmoor, May 1; Cass County, average, May 12.
Manitoba—Margaret, April 28; Treesbank, average of 22 years, May
15. Saskatchewan—McLean, May 7. Colorado—Weldona, April
28. Utah—San Juan River, May 19. Wyoming—Wheatland, May
1; Laramie, average of 9 years, May 20. Idaho—Meridian, May 18.
Montana—Jackson, May 7. Oregon—Harney County, May 18.
British Columbia—Vaseaux Lake, May 24.

Late dates of spring departure are: Brasil—Marabitanas, April 13.
Venezuela—Aruba Island, April 25. Honduras—Northern Two Cays,
May 18. Yucatán—Celesta, May 12. Cayman Islands—Grand
Cayman, May 1. Haiti—Tortue Island, May 16. Cuba—Remedios,
June 1; Isle of Pines, May 29. Bahamas—New Providence, May 12.
Florida—St. Augustine, June 7; Franklin County, June 5; Key West,
May 30. Alabama—Birmingham, June 1. Georgia—Athens, June

10. South Carolina—South Carolina coast, June 5. North Carolina—Raleigh, May 27 (average of 5 years, May 23). Virginia—Rosslyn and Charlottesville, May 30. West Virginia—Cranberry Glades, May 27. District of Columbia—June 6 (average of 21 years, May 22). Maryland—Baltimore and Anne Arundel Counties, June 12. Pennsylvania—Butler, Limerick, and Jeffersonville, June 10. New Jersey—South Orange, June 9. New York—Bronx County, June 9. Louisiana—Grand Isle, June 16; New Orleans, May 29. Mississippi—Vicksburg, May 19. Arkansas—Monticello, May 28. Tennessee—Clarksville, June 9. Kentucky—Bardstown, May 27. Missouri—Corning, June 2. Illinois—Chicago, May 31 (average of 16 years, May 20). Texas—Dallas County, June 8; Cove, May 26. Oklahoma—Oklahoma City, May 29. Nebraska—Omaha and Lincoln, May 24.

Early dates of fall arrival are: Kansas—Cimarron, August 15. Oklahoma—Payne County, August 2. Texas—Edinburg, August 25. Illinois—Chicago, August 10 (average of 6 years, August 14). Kentucky—Eubank, August 15. Tennessee—Elizabethton, August 17. Louisiana—Kaplan, July 13; Abbeville, August 4. Rhode Island—South Auburn, July 23. Connecticut—Hartford, August 2. New Jersey—Camden, July 8. Pennsylvania—Philadelphia, July 18. Delaware—Odessa, July 19. Maryland—Laurel, July 18. District of Columbia, July 26 (average of 23 years, August 17). Virginia—Warwick, July 24. North Carolina—Raleigh, August 15 (average of 11 years, August 29). South Carolina—Frogmore, July 13. Georgia—Savannah, July 27. Florida—Pensacola, July 10; Key West, August 4. Bahamas—Cay Lobos Light, September 1. Cuba—Trinidad, September 1. Jamaica—September 25. Dominican Republic—San Juan, September 21. Panamá—Obaldia, September 30. Colombia—Santa Marta region, September 11. Venezuela—Merida, September 20. Bolivia—Alto Paraguay, October 15. Paraguay—Trinidad, November 9. Argentina—Ocampo, November.

Late dates of fall departure are: British Columbia—Okanagan Landing, August 27. Oregon—Harney County, September 18. Nevada—Montello, September 20. Montana—Fortine, September 28. Wyoming—Fort Laramie, October 2. Colorado—Boulder County, September 9. Saskatchewan—Indian Head, September 20. Manitoba—Treesbank, September 22 (average of 11 years, September 14). North Dakota—Cass County, September 22 (average, September 12). South Dakota—Sioux Falls, October 9 (average of 5 years, September 12). Kansas—Osawatomie, October 13. Oklahoma—Oklahoma City, November 27. Texas—Comanche County, November 28. Minnesota—St. Paul, October 15; Bradford, October 12. Wisconsin—Burlington and Prairie du Sac, October 16. Iowa—National, October

18.   Ontario—Point Pelee, October 3; Ottawa, September 29 (average of 11 years, September 10).   Michigan—Vicksburg, September 25. Ohio—Buckeye Lake, October 27 (median, October 12).   Indiana— Bicknell, October 17.   Illinois—Chicago region, October 9 (average, September 10).   Missouri—Bolivar, October 1.   Kentucky—Bardstown, September 28.   Tennessee—Knoxville, October 7.   Mississippi— Biloxi, October 8.   Louisiana—Diamond, September 27.   Nova Scotia—Yarmouth, September 5.   New Brunswick—Scotch Lake, September 25.   Quebec—Kamouraska, September 25;   Montreal, September 16.   Maine—Jefferson, October 10.   New Hampshire— Tilton, September 19.   Vermont—Rutland, October 1.   Massachusetts—Northampton, November 8; Scituate, October 23.   Rhode Island—Quonochontaug, October 9.   Connecticut—Portland, October 15.   New York—Flushing, November 2; Guilderland Center, October 16.   New Jersey—Newark, October 22.   Pennsylvania—Renovo, November 1; Shermansville, October 10.   Maryland—Patapsco River, November 8.   District of Columbia—October 21 (average of 14 years, September 29).   West Virginia—Morgantown, October 9.   Virginia— Naruna, October 5.   North Carolina—Dare County, October 24; Raleigh, October 7 (average of 7 years, September 29).   South Carolina—Dillon County, December 6; Mount Pleasant, November 26. Georgia—Savannah, October 21.   Alabama—Dauphin Island, September 21.   Florida—Alligator Reef Light, November 28; Defuniak Springs, November 15.   Bahamas—Watlings Island, October 12. Cuba—Havana, October 5.   Jamaica—Spanish Town, October 10. Dominican Republic—San Juan, September 28.   Barbados—October 26.   Nicaragua—Rio Escondido, October 10.   Panamá—Permé, October 18.   Colombia—Santa Marta region, October 14.   Venezuela— Ocumare de la Costa, October 28.

Egg dates.—Connecticut: 12 records, May 27 to June 4.

Illinois: 25 records, May 25 to July 11; 15 records, May 25 to May 31.

Massachusetts: 25 records, June 1 to June 25; 20 records, June 1 to June 9.

Minnesota: 7 records, June 2 to June 15.

New York: 24 records, May 18 to June 11; 12 records, May 29 to June 4 (Harris).

STURNELLA MAGNA MAGNA (Linnaeus)

Eastern Meadowlark

Plates 3 and 4

Contributed by ALFRED O. GROSS

HABITS

The meadowlark is the outstanding and the most characteristic bird of the American farm. It is revered by the farmer not only because of its charming simplicity and its cheerful, spirited song, but also for its usefulness as a destroyer of harmful insects and the seeds of obnoxious weeds. The coming of the meadowlark in the early spring, while the fields are still brown, is a thrilling event. His arrival is made known by his plaintive but not complaining or melancholy song as he stands mounted atop some tall tree in a grassy meadow, with his bright yellow breast surmounted by a black crescent gleaming in the morning sun.

The meadowlark has the build and the walk, as well as the flight, of the quail; and since it frequents the marshes, especially in its winter quarters, it has sometimes been called the marsh quail. This name has probably lead many a hunter to think of it as a game bird. Fortunately in recent years fewer meadowlarks are killed for food, and this may be at least one factor responsible for the increasing numbers as well as the extension of its nesting range.

When I first came to Maine 35 years ago the meadowlark was a comparatively rare bird in the southern part of the State. Since that time it has steadily increased in numbers, until today almost every suitable meadow and grass field has its quota of meadowlarks. Similar increases in the number of meadowlarks have been reported from other sections of its range. Milton B. Trautman states in a letter that he counted 400 pairs of meadowlarks while walking through suitable fields, during the course of a few days in the Buckeye Lake region, Ohio. He estimated the amazing number of 1,400 pairs as nesting in the area, an average of 1 meadowlark for every 7 acres, or about 91 to the square mile.

In 1906–1908 I conducted the fieldwork of a statistical survey of the birds of Illinois for the Illinois Natural History Survey. In making the census counts, I walked many times through fields and woods over the length and breadth of the State. An assistant traveled at 30 yards distant and parallel to my line of march and was responsible for measuring the distance of each field traversed in terms of paces, which later were translated into feet. The species and the numbers of birds flushed in a strip 50 yards in width, including those flying across the strip within a hundred yards to our front, were recorded. Thus we covered all types of crops and vegetation during all conditions of weather and at all seasons of the year to obtain a comparative sample of the birdlife. During the summer months alone an area equivalent to 7,793 acres was covered, on which 85 species of birds were recorded. The meadowlark proved to be the most abundant of the native Illinois birds, being represented by 1,025 individuals, or 13.2 percent, of the total bird population. There was an average of 85 meadowlarks to the square mile for the whole area traversed. As the birds were unequally distributed, never occurring, for example, in woodlands or among shrubbery, their numbers rose to 266 to the square mile in stubble, 205 in meadows, 160 on untilled lands, 143 in pastures, and 131 on wastelands, but fell to 10 per square mile in fields of corn.

The meadowlark population varied in numbers from the northern to the southern part of the State, 100 in northern Illinois being represented by 175 in the central and by 215 in the southern part. The center of density of the summer meadowlark population at that time was in the southern section, and during the winter months the concentration of meadowlarks in southern Illinois reached an average of 373 per square mile. Many of the birds which nest further north winter in that section of the State.

From various reports I have recently received from the Middle West, it is probable that if the census were repeated today the average meadowlark population would exceed the average of 85 to the square mile obtained during the summer months of 40 years ago.

Spring.—The migration of the meadowlark is a comparatively limited movement, and the bird retires completely from only the most northern sections of its breeding range. It is a regular winter resident as far north as Maine, southern Ontario, and Michigan; and the southern summer residents do not go beyond the Carolinas, Alabama, Louisiana, and southeastern Texas. In spring the migrants reach Missouri and southern Illinois by the middle of March, arriving in the north central States during the first weeks of April, and in Minnesota and the Dakotas usually during the latter part of the month. The first arrivals in Manitoba, Saskatchewan, and Alberta appear in the last of April or the first week of May. The vanguard of the migrants reaches southern New England about the middle of March, and a marked movement extends well into April.

According to William Brewster (1886b), the meadowlarks are among the birds which migrate exclusively by night. He states: "Species which migrate exclusively by night habitually feed in or near the shelter of trees, bushes, rank herbage or grass, and when not migrating are birds of limited powers of flight and sedentary habits, restricting their daily excursions to the immediate vicinity of their chosen haunts. As a rule they are timid, or at least retiring disposition, and when alarmed or pursued seek safety in concealment rather than extended flights."

Meadowlarks migrate by night because they are either afraid to venture on long exposed journeys by daylight, or unable to continue these journeys day after day without losing much time in stopping to search for food. By taking the nights for traveling they can devote the days entirely to feeding and resting in their favorite haunts.

Milton B. Trautman (1940) in his observations at Buckeye Lake, Ohio, differs from the conclusions reached by Brewster. He writes:

Few Eastern meadowlarks were seen or heard migrating in very late evenings and early mornings, but many more were observed in the daylight hours. In late March and April individuals and loose flocks of as many as 60 flew northward at a low elevation across the lake. Loose flocks of 5 to 100 birds were often observed flying during spring and fall. The flocks generally flew a short distance and began to feed. Presently, those in the rear rose into the air and, flying over the flock, alighted in front to feed again. This maneuver was many times repeated. When the flock reached an obstruction, such as woods, cattail swamp, or lake, it flew over in a long loose column. The flocks traveled in this leisurely manner 2 to 6 miles an hour. Sometimes the flocks stopped feeding and flew 1 to 3 miles at a low elevation before dropping to the earth to feed again.

At Ithaca, N. Y., G. B. Saunders (MS., see p. 56) has found that the first meadowlarks to be seen early in the year are males. As early as January young birds which may be classed as vagrants are reported. Upon the advent of warmer weather more vagrants which have wintered only a short distance to the south wander in, feeding in manured fields and about farm buildings. Stormy weather often covers the fields with snow and sends them into barnyards where they may pick their food along with domesticated animals. Not infrequently many of these early meadowlarks perish in the long blizzards which put an end to their food supply.

Later, usually about the middle of March, the first migrants appear. These are old males, few in number and quiet in manner, which have wintered far to the south. By the end of March the migrant males become abundant. Song is less common among these early flocks of migrants than among the first resident males, which come a week or so later. When these arrive in the latter part of March they are active in the mornings and late afternoon, but during midday they often retire to a common feeding ground which the birds from different territories share without any apparent hostility. Their early song, although sweet and full of spirit, is not of the brilliance which characterizes it in April when the first resident females arrive.

Groups of migrants and resident males continue to arrive until the latter part of April. The last resident males establish themselves in areas left by the first wave of migrants, or carve their territories from the domains of earlier males unable to defend their original holdings. The migrants remaining in flocks resume their journey northward.

The vanguard of migrant females arrives 2 weeks or more after the first resident males and are followed closely by the first resident females. The coming of the resident females stimulates the first song peak of the males, whose songs become longer and more brilliant and animated. They engage in territorial combats over females, in defense displays and sexual flights, as well as posturing and sexual displays for the benefit of the female. The female is at a much lower sexual pitch at this time and responds only by preliminary stages of posturing such as the erection of the body to a vertical position, pointing her bill upward and twitching her wings. Late in April she reaches the necessary sexual level and begins building the nest and laying the eggs.

Other migrant and resident females, which are young birds, continue to come in April and May and even in June and July. These birds, which are late in maturing, become mates of polygamous or late arriving males, or remain unmated.

The following accounts of territory and courtship are based primarily on an exhaustive treatment of the subject contained in an unpublished thesis, submitted at Cornell University in 1932 by George Bradford Saunders ("A Taxonomic Revision of the Meadowlarks of the Genus *Sturnella* Vieillot and the Natural History of the Eastern Meadowlark *Sturnella Magna Magna* Linnaeus").

Territory.—Soon after his arrival during the latter part of March, the resident male leaves his companions and selects a territory, preferably a grassland or meadow, because of the great abundance of food as well as his decided liking for this type of habitat. The size and shape of the territory depend chiefly on the area of suitable land available, the local abundance and strength of competing males, the relative concentration of food supply, and certain barriers and individual range requirements of the male. The size of the territory may be increased as a result of polygamous relations, particularly if the females choose widely separated nesting sites, but the average size of 15 territories at Ithaca, N. Y., was found to be 7 acres. There is a decided difference between the total area of the territory and that which is regularly used. The more concentrated the food supply the less need there is for foraging, and the smaller the area frequented. Of two territories studied by G. B. Saunders (MS., see above) throughout the breeding season of 1931, one contained 9 and the other 20 acres; but due to the abundance of food in a meadow separating the two families, one monogamous and the other with three females, this common feeding ground was shared. In each territory, however, the area of land used regularly was the same, 6 acres.

Important to the male are the various commanding perches from which he can survey his territory. Telephone wires or electric power lines with unobstructed views often furnish favorite song and lookout posts. Mounds of earth, farm implements, and fence posts provide perches near the ground. During the first few days of his occupancy he visits them from time to time and selects one for his primary headquarters. Here he sings and watches during the day, usually roosting nearby at night. This territorial center is frequented faithfully during the entire season, unless his routine is changed by polygamy, in which case secondary headquarters are often established nearer his mates. As he may have as many as three females, each having two broods, the chief center of interest in the territory may change as each female reaches the peak of sexual responsiveness.

By intimidating songs and alarms, displays and disputes, the male meadowlark defends his domain against the encroachment of covetous rivals. It is clear that from the beginning of his tenure the male has a definite conception of his territorial acreage and chases all resident males of his species beyond these boundaries. The competitors may come to blows, but it is usually a matter of vehement displays or competitive singing, ending when the vanquished bird takes wing. The loser may be pursued rapidly as far as the boundary line; the victorious male then returns to the sentinel station, singing a spirited flight song as he flies. G. B. Saunders (MS., see p. 56) describes a territorial flight he observed on March 25, 1929: "My attention was drawn to two birds fighting savagely in the grass. From a distance I could see the flashing white rectrices and was able to identify them as meadowlarks. One male was on top of the other jabbing him fiercely with his long bill. Then they rolled about for a moment wrestling and stabbing with their feet locked together. Instead of taking wing they hopped at each other, grappled, and again fell on their sides. Wings were held loosely and white tail feathers flashed repeatedly as their tails opened and closed spasmodically. After more than a minute of jabbing, one bird arose and flew, pursued hotly by the victorious contestant who gave a jubilant flight song during the chase."

Courtship.—The arrival of the females on the breeding territory stimulates the resident males, who by this time are well prepared for an animated and lively courtship, to a frenzied rivalry that often becomes furious. Two rival males have been seen tumbling about on the ground on their backs with their feet firmly locked together, striking at each other with their bills in mortal combat.

The courtship is featured by elaborate displays, spectacular flights, and intensive singing. G. B. Saunders (MS., see p. 56) describes a performance he observed from a blind on April 16, 1931: "Today instead of witnessing the usual routine I observed the first resident female seen since the preceding fall serve as the center of attraction for three competing males. As I reached the blind all four birds took wing and began a most exciting and spectacular chase in which they zig-zagged back and forth, describing circles 200 yards in diameter and maintaining for the most part a steady flight, but occasionally sailing on set wings or giving pulsating strokes. Throughout most of the exhibition the female was pursued by all the males, but now and then two of the latter would engage in a private chase after each other (rarely striking in midair) only to return quickly to the magnetic female. Finally all four came down to the spot from which they had flown. The female began walking about, feeding in the short grass; occasionally she paused to give a conversational chatter that impressed me as being softer, finer, and more modulated than the alarm chatter. The males vied in following her, first one then another arching his body, pointing his bill up, and flying jerkily toward her at an elevation of from 3 to 6 feet. At times they would walk near her with quick, short steps, their bodies held vertically, bills pointed to the zenith, wings twitching so rapidly that the remiges (particularly the tertiaries) described a blurred arc above their backs, and tails convulsively spreading and flashing the white areas. Then they would spring into the air, fanning their wings powerfully but jerkily for six or eight strokes. This 'jump flight' apparently serves two purposes, that of displaying to the female, and of observing and intimidating other males.

"In this way the four birds proceeded for some 50 yards, the female for the most part apparently uninterested but occasionally pointing her bill, twitching her wings and tail, and revealing her excitement. The males at such times would attempt to intimidate each other with violent displays. The female would chatter her approval. After several minutes all four birds flew out of sight, but very soon one male returned followed by the female. Now and then a new note which sounded like the *beert* of the nighthawk was given. She resumed her feeding, while he continued to post himself nearby. Instead of making himself tall and slim however, he fluffed out his body feathers until the yellow breast he presented to her gaze was a broad, flat golden shield set with a shining black gorget. He continued to make advances, pointed his bill upward now and then, flirted his wings, etc. She chattered or gave the *weet, weet, weet* call in answer to almost every song. It is noteworthy that while he alone was displaying, he did not sing. Then, resuming his perch at headquarters, he sang brilliantly for 19 minutes, averaging 11 songs per minute. During the next 4 days he continued to spend much of his time near her, frequently displaying and posturing, but she seldom displayed, usually continuing to feed quietly near him."

The usual courtship display of the male is summarized by Saunders as follows: "Taking a direct stance near the female he raises his body to its full height, stretches his neck to its full length, and points his bill to the zenith. The tail is fanned, showing all the white, and is also jerked up and down; the wings are flirted rapidly over the back, either simultaneously or alternately; and the breast feathers are fluffed out to form a lovely shield of contrasting yellow and black. The *beert* note may be given. He may spring from the ground as has been described, even flashing his tail in midair.

"The female's reaction to this performance is to raise her body to its full height, stretch her neck, and point her bill; flashing her wings and tail in answer to his song and chattering or giving a *beert*.

"Throughout this period they spend much of their time together. When he is aloft singing she is usually feeding or perching nearby. As a rule, however, she frequents an elevated perch much less often than the male and seems less sure of herself while doing it. If he is singing, she often answers each song with soft conversational chatter, "*dzert-tet-tet-tet-tet*." They seem to enjoy their companionship very much, remaining together in long flights across the territory to feeding grounds and maintaining this proximity while feeding; and occasionally indulging in a sexual flight, the male singing a beautiful flight song. After such a flight he often repeats his sexual advances in a more or less obvious manner, but she responds either weakly or not at all. While feeding, they pass hours in which little if any show of sexual interest is witnessed."

Nesting.—The meadowlark is primarily a bird of the grasslands, meadows, and pastures; and it is in such places that we usually find its nest. I have also found them in corn, alfalfa, and clover fields and weedy orchards, as well as in grassed islands among plowed fields. The nest is made of dried grasses lined with finer materials. In Illinois I have found nests lined with small amounts of horsehair; in Maine, wiry grasses and even pine needles are sometimes employed for this purpose. Most of the nests have a dome-shaped roof constructed of grass more or less interwoven with the attached and growing parts of the clump of grass or weeds against which it is built. The interior of the nest is open to view from only one side, and this opening may be more or less obscured by overhanging grasses. Sometimes there is a covered passageway to the nest especially to those built in a field where the tall grass was not cut during the previous season. Some of the nests are so well hidden that they are difficult to find, and are discovered only when the bird is flushed by the accidental encroachment of someone walking through the field. The colors and markings of the plumage blend so perfectly with the surroundings that if a nesting bird could restrain its fear, a person might pass within inches of a nest and never be aware of its presence. Most nests that I have found were those I nearly trampled under my feet. Most nests are built in a small depression of the ground, the depth of which may be augmented by some excavation by the bird until it is about 1 to 2½ inches deep. Skutch writes of a nest he found in an alfalfa field near Ithaca, N. Y., on May 26, 1931; it was a sparse structure of grasses only half covered over and set in a depression of the ground so that the upper side of the five eggs were about level with the surface of the field.

The nest varies in size and bulk according to the situation in which it is found. Of five nests measured, the average total height was 7 inches, the outside diameter 6½ inches and the inside measurements of the nesting cavity approximately 4 by 5 inches. The average opening is 3¼ inches wide and 4 inches high.

A. C. Bent found an unusual nest containing three eggs in an unusual site at Sea Isle City, N. J., on June 23, 1928. Located in short grass on Black Rail Marsh, it was made of dried coarse grasses, completely arched over with a thick, dense canopy of coarse dry grasses and weeds, and was much like the nest of the black rail in appearance. He found another nest containing four eggs among the beach grass at Chatham, Mass., on May 28, 1904, sunk into the sand just back of the crest of the beach.

F. W. Rapp reports a meadowlark's nest, containing four eggs, which was located within 9 feet of the track of the Grand Trunk Western Railroad. Trains running at a high rate of speed, making much noise and jarring the ground, apparently did not disturb the birds. The nest, in short grass, was completely covered over with dried grasses and the entrance was away from the tracks.

Robert L. Denig (1913) reports unusual conditions under which a meadowlark nested at Wakefield, Mass., where the U. S. Marine Corps conducted rifle practice during the summer of 1909. Mounds of earth about 3 feet high were built to elevate the firing points at 100-yard intervals. The meadowlark built its nest on the far side of the 400-yard mound directly in line with the target, so that the muzzle of the rifle of the man lying on the mound was directly over the nest and not more than 2 feet above it. At first when the firing skirmish line was about 400 yards distant, the birds would fly away; but as the practice continued they became more and more accustomed to the noise; they would allow the men to approach nearer and nearer before leaving the nest and would return at once when the firing ceased at that point. As the time came for the eggs to hatch, one of the birds would remain on the nest throughout the firing, even when the gun was being discharged directly over its head, not more than 2 feet away. Finally, the eggs hatched, and the young birds were brought up, so to speak, "under fire."

G. B. Saunders (MS., see p. 56) provides the following notes on nest building: At Ithaca, N. Y., the meadowlark begins nesting in late April or early May. The time required for building the nest varies from 3 to 18 days. One pair began carrying nesting material on April 20 and completed the nest on May 8, another began building on April 21 and finished on May 8. Still another began carrying grass on June 3 and finished her nest and laid her first egg on June 6. In every case except the last one mentioned, several nests were started and worked upon. Usually the first beakfulls of nesting materials are deposited at different places and at first there seems to be no concentration on any particular location. During these first days copulation with the male takes place. When the female is responsive to the advances of the male she crouches close to the ground, shortens her neck, and points her bill upward at an angle. She flutters and flirts her wings and lifts and spreads her tail. The male displays a few feet distant, and the copulatory act is finally achieved with no sounds being uttered by either bird.

Trips with nesting material are most frequent early and late in the day, but may sometimes continue during midday as well. The details of nest building are presented in the following typical case. On May 12, 1931, a nest which Saunders had under observation was scarcely begun, but both cup and roof had been started. A natural depression 2 inches deep had been slightly modified by the female, who used her bill for the retouching. Into this cavity a thin layer of last year's grass blades had been laid. Much of the tuft of grass surrounding the cup had been arched over it and woven together, being secured by a few long dry grass stems woven among the growing blades. In half a morning's work she had both the cup and roof well started. The male gave no assistance in the enterprise and offered none later.

On May 13 as well as on the two following days, the female made regular trips with material every 5 to 10 minutes. She usually remained at the nest from 35 seconds to 2 minutes placing the material that she had brought. On this day the nest was half completed; the lining of the bowl was much deeper but was still flimsy. On May 15 the nest was complete except for occasional additions of material. The first egg was laid on May 17. Occasionally the building continues for several days after egg laying is begun.

Saunders' observation of the nest-building process is thus summarized:

Following the choice of a nesting site, the customary first step is to prepare the earthen foundation for its cup of withered grass. There may be a natural depression, a hoofprint or similar hollow already present, in which case the female remodels it by using her bill as a combination pick and forceps tool. In some soil the marks of her beak remain after the young have departed. The use of the bill for digging the soil is not surprising, for the habit is often shown in feeding, when the meadowlark employs it to probe for insects and grain and to dislodge clods of earth.

Once the hollow is satisfactory, the adjacent grasses or other growing plants are pulled over the pit and interlaced, or secured by the addition of long stems or blades of dead grass until they form a more or less complete dome which later conceals the eggs from view and protects them from the sun and rain. Other nests are not below the general surface level but are built entirely above it, there being a front step as a result of this variation in architecture.

The cup and nest lining are usually fashioned while the dome is in the process of construction, first one part and then another receiving the attention of the female. Many more than a hundred loads of dried grass going to the making of the finished home. Although many authors credit both sexes of the eastern meadowlark with the job of building, I have never observed a male sharing in this activity. Perhaps he does, but such a male would be an exceptional individual, and a far more helpful mate than any of the dozen males which had their intimate lives scrutinized daily during my study at Ithaca, N. Y.

G. B. Saunders (MS., see p. 56), the first to discover the common practice of polygamy among meadowlarks, reports that the secretive nature of the females and the inconspicuousness of their nests are two of the principal reasons why the eastern meadowlarks have been able to keep their polygamous habits a secret for so long. Although meadowlarks breed in every one of the 48 States and are abundant in most of them, no mention has appeared in the voluminous literature on *Sturnella* regarding the frequent bigamy of the males of this subspecies. His intensive field work in more than 20 territories at Ithaca, N. Y., in 1931 revealed that about 50 percent of the males were polygamous. One of them was found to have three females, all of which were nesting at the same time. He adds that among the several reasons why polygamy is common among meadowlarks is the fact that the females are not hostile to one another as they are in many other species; they feed together, associate with the male together, and often nest within 50 feet of each other. Another is that the males are repeatedly attracted by desirable females.

Eggs.—The number of eggs of a set of the meadowlark varies from three to seven, but sets of five eggs are most common. Sets of four are more usual in the second brood nests of the season. Birds breeding in the southern part of the nesting range on the average lay smaller sets.

According to Bendire (1895):

The eggs of the Meadowlark vary considerably both in shape and size; the majority are ovate, while others are short, elliptical, and elongate ovate. The shell is strong, closely granulated, and moderately glossy. The ground color is usually pure white; this is occasionally covered with a pale pinkish suffusion, and it is very rarely pale greenish white. The eggs are more or less profusely spotted, blotched, and speckled over the entire surface with different shades of brown, ferruginous, pale heliotrope purple, and lavender; these markings generally predominate about the larger end of the egg, and are rarely heavy enough to hide the ground color.

In some sets the markings consist mainly of a profusion of fine dots; in others the spots are well rounded and fewer in number; and again they occur in the shape of irregular and coarse blotches, mixed and finer specks and dots; in fact, there is an endless variation in the style of markings.

The average measurement of a series of two hundred and one specimens in the United States National Museum collection is 27.75 by 20.35 millimeters, or 1.09 by 0.80 inches. The largest egg measures 30.78 by 22.61 millimeters, or 1.21 by 0.89 inches; the smallest, 21.59 by 18.29 millimeters, or 0.85 by 0.72 inch.

Ninety-five eggs weighed by G. B. Saunders (MS., see p. 56) had an average weight of 6.6 grams, a minimum weight of 5.4 grams, and a maximum weight of 7.7 grams. Eggs of any one clutch are usually similar in size, coloration, and weight. He found that of 85 eggs in 20 nests found at Ithaca, N. Y., 14 eggs, or 15.5 percent, were sterile.

Incubation.—When an average clutch of five eggs is laid, incubation may begin with the deposition of the third, the fourth, or the fifth egg, but more frequently begins with the laying of the fourth egg. The incubation period is usually 14 days, but under certain unusual conditions it may be only 13 or as many as 15 days.

Although the male has been credited with a share in incubation, Saunders (MS., see p. 56) has never witnessed any such cooperation in the many nests that he has closely observed. Once incubation has begun the female remains on the eggs most of the day, leaving only long enough to feed. The nest is never left at night unless she is frightened by an intruder. During 12 hours and 40 minutes of daylight, one female spent 9 hours and 40 minutes, or 76 percent of the time, on the nest. The longest absences from the eggs were during the middle of the day when the temperature was highest. On cool or rainy days less time is spent away from the nest.

While incubating, the female is continually active. These activities include listening to songs and sounds of approaching danger, "humming" when the male sings, turning the eggs, feeding on insects which come within reach of the nest, probing in the nest, rearranging nest materials, preening, etc. The female on the nest often responds to the flight song of the male by voicing low, sweet chuckling notes that are unlike any others uttered by the meadowlarks. The softness of these sounds precludes their detection by man at distances greater than 15 or 20 feet. The eggs are turned many times during the day, and in the course of 1 hour in the late afternoon a female was observed to turn them five times. The eggs do not hatch in the order that they are laid; for example, the fourth egg may hatch several hours in advance of the third. Individual variation in development causes differences of several hours or even a day in the hatching time.

Young.—The young at the time of hatching, according to G. B. Saunders (MS., see p. 56) have a smooth orange-red skin; the bill and nails are flesh color; and the natal down is pearl gray. The down is longest on the capital and spinal tracts and shorter on the humeral and femoral tracts. When the down is dry it fluffs out and appears quite abundant, particularly on the spinal tract. On the head the down is localized chiefly above the eyes and on the occiput. The dark sheaths of juvenal feathers are visible on the dorsal surface of the head and on the spinal and dorsal regions, and less easily discernible in the humeral, alar, femoral, and crural tracts.

Shortly after hatching, the nestling reacts to the food call of the female and holds up its mouth in a wobbly and uncertain manner, at the same time uttering weak notes, *see see* or *seep seep*. The female during the first few days spends long hours in the nest brooding the nearly naked young. The young are fed by both the male and female but the male feeds them much less often. During the first 2 days the young evacuate in the nest, being too weak and lacking the instinct to void their droppings outside the doorway. Later, by the third or fourth day, each youngster may be observed to turn about and expel the mucous-covered sac beyond the rim of the nest. These sacs are removed by the adults but in the early stages of the development of the young they may be eaten. Egg shells are removed to a considerable distance by the female immediately after the young are hatched. Infertile eggs are usually left in the nest during the entire period of occupation.

By the third or fourth day a slitlike opening appears in the eyelids, so that the youngster can see whenever it is fed or disturbed. At other times the eyes remained closed. The position of the young birds is now more upright and alert, and the wings have grown enough to be useful as props for maintaining balance. The legs are still almost useless; there is little muscular coordination and they are well sprawled out at the sides.

By the fifth day the eyes are fully opened and the voice is stronger. The nestlings now face the opening of the nest, expectantly waiting for food. Growth is rapid, and the juvenile plumage is rapidly acquired. When disturbed, the young now exhibit signs of fear. Wing exercises and stretching of the legs and neck are indulged in frequently.

By the eighth day the young are very alert and receptive to sounds coming from outside the nest. They may be seen frequently preening their feathers, apparently to facilitate the unsheathing process. During the remaining days in the nest the young become so active the nest is wrecked and the roof worn away, exposing the nestlings to view and to the hot, direct rays of the sun. When thus exposed to the sun they pant violently in order to control their temperature. Sometimes they leave the nest, but return after being fed.

By the eleventh or twelfth day the birds normally take their final leave of the nest, although if molested they may desert it as early as the eighth day. Feather growth of the wing tracts has proceeded sufficiently by the eleventh day to allow the nestling to fly, in case flight is necessary. However, the newly departed young meadowlark seldom takes to its wings during the first few days except to make short jumps in the grass. The young are fed by the adults for a period of at least 2 weeks or longer after they leave the nest. Their food call is a loud bisyllabic *tseup, tseup,* and it is by these notes that they are located and fed by the adults. The second nest may be started within 2 or 3 days of the desertion of the first one. While the female is building and laying she continues to feed the first brood, but when the second incubation is begun the male assumes the major part of the work of caring for the young of the first brood, which are about 3 weeks old at this time.

Gradually they learn to catch insects for themselves and become more and more independent. When they are able to shift for themselves, they are apparently chased out of the territory by the male. They probably do not travel far before September, when they acquire their first winter plumage.

Four birds taken from a nest when 8 days old were raised in captivity by G. B. Saunders. Since they were given long hours of freedom in their native fields, their development and habits were similar to those of wild juveniles. On the fifteenth day they all took dust baths, fluffing and shaking their plumage as adults would, Following this exertion they drank heartily from a basin of water. On the sixteenth day they began prying into the soil with their bills, which marked the inception of the boring habit which is so typical of the adults. On the seventeenth day they began to stand high on their legs and to hold bodies erect whenever they heard a sound which startled them. On the twentieth day one was observed to take a thorough bath in the water, after which he spent several minutes in a systematic dressing of his plumage, during which he apparently used his oil gland frequently. On the twenty-second day, two of the four began feeding for themselves; before that Saunders had been feeding each bird about 175 grasshopper nymphs daily. On the same day one of the two that had begun feeding themselves gave a rolling chatter very similar to that uttered by the adults.

Plumages.—Jonathan Dwight, Jr. (1900) gives the following description of the plumages and molts:

Juvenal plumage acquired by a complete moult. Above, clove-brown, the feathers broadly edged with buff palest on the nape, those of the back having double subapical spots of russet. Median crown stripe, and superciliary line cream-buff. Wings sepia-brown, the primaries and secondaries obscurely barred on the outer web with darker brown and edged with pale vinaceous cinnamon shading to white on the first primary, the tertials clove-brown broadly edged with buff and having a row of partly confluent vinaceous cinnamon spots on either side of their shafts producing a barred effect, * * * the rest of the wing converts obscurely mottled with light and dark browns and edged with buff, the alulae with white. The three outer pairs of rectrices are white with a faint dusky subapical shaft-streak, the next pair largely white and the others hair-brown confluently barred with clove-brown, and whitish edged. Below, including "edge of wing" pale canary-yellow, nearly white on the chin, the sides of the throat, breast, flanks, crissum and tibiae washed with pinkish buff, streaked and spotted with brownish black which forms a pectoral band. Bill and feet pinkish buff, the former becoming slaty, the latter dull clay color.

First Winter Plumage acquired by a complete post-juvenal moult beginning about September first after the juvenal dress has been worn a long time, young birds and old becoming practically indistinguishable.

Above, similar to the previous plumage, but all the browns even to the wing and tail quills much darker, often black, and distinct barring rather than mottling, the rule. The feathers of the back have large single subapical spots of rich Mar's-brown crossed by two faint dusky bars, and the primary edgings are usually grayer. Below, a rich lemon-yellow (including the chin and supraorbital dash) veiled with buff edgings and a black pectoral crescent is acquired completely veiled with deep buff and ashy edgings. The streakings below are heavier and darker, many of the feathers with subapical russet spots and the wash on the sides is deeper and pinker.

First Nuptial Plumage acquired by wear which is excessive by the end of the breeding season producing a dingy brown and white appearance above with yellow and black below. The subapical spots of the feathers of the back are almost entirely lost by abrasion and the same force scallops out the light portions of the tertiaries, wing coverts, and tail. Neither the yellow nor the black below fades very appreciably, but the shining denuded shafts of the feathers project far beyond the abraided barbs. The yellow seems even to be intensified by the loss of paler barbules.

Adult Winter Plumage acquired by a complete postnuptial moult in September. Usually indistinguishable from first winter dress.

Adult Nuptial Plumage acquired by wear as in the young bird.

Female.—In natal down and juvenal plumage the sexes are indistinguishable. Later the female differs only in slightly duller colors and a more restricted black area on the throat. The moults are exactly the same as in the male.

Abnormal plumages involving albinism and melanism are known to occur in the meadowlark. The majority of the cases of albinism which have been reported are actually only partially albinistic; in most the brown of the upperparts is white or whitish, whereas the yellow of the underparts seems to be retained in varying degrees of intensity.

James Savage (1895) collected an albino meadowlark near Buffalo, N. Y., in which "The usual brown of the upper parts was of a pale buff color with the pattern of the feather markings indistinctly discernible, while the yellow on the breast was as pure as in an ordinary Lark."

Louis S. Kohler (1915b) gives an account of a partial albino meadowlark he observed near Bloomfield, N. J.: "On October 7th during the afternoon while strolling over the fields I came upon a partly albino bird. This bird was of normal plumage except the tail and wings in which parts, more than half the feathers were devoid of color. This bird during its association with others of its kind was continually being attacked and presented a very bedraggled appearance from their frequent onslaughts and was forced into solitude by them at close intervals. But in spite of their pugnacity it always returned to the vicinity of its tormentors and was immediately set upon and driven off."

G. B. Saunders (MS., see p. 56) states: "There is an albino eastern meadowlark in the Cornell University museum which has upperparts and wings whitish, the bill pale brown, the jugular crescent buffy brown, but the yellow underparts nearly normal." There are many other similar cases of partial albinism in the meadowlark but I have discovered no report of a pure albino eastern meadowlark.

Chas. H. Townsend (1883) describes a melanistic specimen collected in New Jersey as follows: "The upper plumage is of the normal color, while the whole head, neck and under parts are perfectly black. There is the faintest possible trace of yellow along the sides, and no white feathers in the tail, which is very dark above and below."

Food.—Few birds of the agricultural areas can claim a higher rank in its economic relations to man than does the meadowlark. During the summer months most of its food consists of insects and closely allied forms. It eats practically all of the principal pests of the fields and is particularly destructive to the dreaded cutworms, caterpillars, beetles, and grasshoppers. In the autumn, and especially in winter, when insect life is scarce, it resorts in a large measure to seeds. It does feed on certain grains useful to man, such as corn, wheat, rye, and oats; but most of these are waste left behind at harvesttime. It seldom disturbs these cereals when growing or before being harvested. I have seen flocks of them in weedy cornfields where apparently they were feeding exclusively on seeds of smartweeds and ragweeds. Meadowlarks have been known to eat certain fruits such as wild cherries, strawberries, and blackberries; but in general these constitute but a very small part of their subsistence.

An account of the food habits of the meadowlark among the sand hills of North Carolina in winter is given by M. P. Skinner (1928):

During the winter the number of Meadowlarks remained quite constant, although there were temporary variations each day. But in February it became noticeable that some of the winter birds were leaving. They seem to stand the cold weather, but snows cover their usual food and then these birds may be found in very unusual places, on any little patch of bare ground they can find, and about barns and stock-yards.

During the winter in the Sandhills the Meadowlarks depend largely on seeds and waste oats for food, but also catch caterpillars, cutworms, earthworms, and as many kinds of insects as they can. These foods are secured on the ground and in the short stubble and grasses. At times these birds seem to give preference to seeds and at other times to feed almost entirely on insects even during the depth of winter when insects might be supposed to be scarce. For securing the two different kinds of food, the Meadowlarks use quite different methods. When after seeds they hunt through the grass and weeds, stopping occasionally to gather seeds from the standing or fallen stalks. When they find places where the seeds are numerous on the ground, they both scratch with their feet and dig with their bills. If there is a wind blowing, they usually fly to the lee side of the field and then advance on foot across it and against the wind. This is apt to scatter the flock, especially as one individual often has better luck than another, and the unsuccessful ones usually hunt up new places for themselves rather than share the first ones' success. Even when scattered over a large field the flock retains its organization, and when one bird leaves, the others usually follow one by one at short intervals until all have left. When they are feeding on insects the Meadowlarks move more rapidly, and perhaps separate more. Then, they do not search the ground or dig with their bills, but they look very closely at the bases of the bunches of grass as they pass by. At times they appear to find insect-catching very profitable at the stock-yards and near barns.

Occasionally a Meadowlark takes both insects and seeds indiscriminately. Such a bird came walking through the rough at the edge of a golf links; like a Flicker, it thrust its bill into the soil experimentally every step or two. At the foot of a tuft of grass it dug out two white grubs and ate them, then it walked over to a spray of dried everlasting, pulled it down and ate several seeds while holding the stalk down under one foot.

In Florida and sections of southern United States more of the food during the winter months consists of insects, chiefly beetles but also cutworms. caterpillars, and grasshoppers. Howell (1924) has found the meadowlark to be an important enemy of the cotton-boll weevil in the south. Since it feeds regularly upon this insect during the winter months, it very materially reduces the number which might otherwise descend on the cotton crop the following season.

Investigations in South Carolina and other Southern States as far west as Texas, according to Beal, McAtee, and Kalmbach (1927), have substantiated accusations that the meadowlark is guilty of destroying sprouting corn.

This habit seems to be confined to the migrating or wintering flocks before they have broken up for the breeding season and is probably occasioned by a scarcity of other available food. North Carolina seems to be the most northerly State in which this objectionable trait of the meadowlark manifests itself. Corn planted in March is most susceptible to attack and cases may be frequently encountered where whole fields must be replanted, resulting in a delayed and less profitable crop. In attacking the sprouts the birds usually drill a small conical hole down to the germinated kernel which they eat, leaving the tender sprout exposed to the withering effect of sun and air.

F. E. L. Beal (1926), in a report of a detailed analysis of the contents of 1,514 stomachs of meadowlarks, found that 74 percent consisted of animal food and 26 percent of vegetable matter. The animal food consisted of practically all insects, chiefly "ground" species such as beetles, bugs, grasshoppers, and caterpillars, with a few flies, wasps, and spiders. Of the various insects eaten, crickets and grasshoppers are the most important, constituting 26 percent of the food of the year and 72 percent of the food in August. Of the 1,514 stomachs collected at all seasons of the year, 778, or more than half, contained remains of grasshoppers, and one was filled with fragments of 37 of these insects. Next to grasshoppers, beetles are the most important food item of the meadowlark, food amounting to about 25 percent. Forty-two adult May beetles and numerous white grubs of this beetle, a most destructive insect, notably to grasses and grain, were found. Among the weevils the cotton-boll and alfalfa weevil were the most important economically. Caterpillars, including many cutworms, form a constant element of the food and in May constitute over 24 percent of the entire food. Adult moths and butterflies are seldom eaten. The remainder of the insect food is made up of ants, wasps, and spiders, with some bugs, including chinch bugs, and a few scales.

The vegetable food, according to Beal (1926), consists of grain and weed and other hard seeds. Grain was found chiefly in stomachs collected in winter and early spring; hence it represented waste material. Clover seed was found in only six stomachs and but little in each. Seeds of ragweed, barnyard grass, and smartweed are eaten from November to April, inclusive, but during the rest of the year are replaced by insects.

As for the food and behavior of meadowlark young, G. B. Saunders (MS., see p. 56) says that within an hour after the meadowlark is hatched it receives its first meal of cutworms, other small insects, and spiders. Young grasshopper nymphs which the female has mashed between her mandibles may be included in these early meals. When the adult arrives at the nest, insects can be seen projecting from her bill, and these she feeds to the young by squatting on her tarsi in front of the entrance and putting morsels well down the throat of each youngster. When the young are satisfied she resumes her brooding. During brooding, a nestling may get hungry, in which case the female raises her breast and reaching down into the nestling's open mouth, gives it either some insects which were left over, or a meal of regurgitated food. That she regurgitates is clearly shown by the pumping action of her neck and head. During the first few days there is no pronounced change in the routine of the female, for she continues to spend long hours in the nest brooding the nearly naked young.

When the young later become stronger, hardier, and somewhat insulated by feathers, the female spends much less time at the nest and feeds the young no regurgitated food. By the time the young are about 6 days old they are receiving the usual fare of grasshoppers and larvae, plus ground beetles, crickets, and other heavily chitinized insects. During later nest life their hunger must be appeased about every 5 to 10 minutes early and late in the day, and at intervals of about every 15 minutes during the hotter hours. Most of these trips are made by the female, whereas the male makes few visits and is much less solicitous in his attentions to the young. The female averages nearly a hundred trips a day to the nest during the 12 days the young are in the nest. The food daily given each nestling weighs 8 to 20 grams, a weight equivalent to that of about 100 to 300 small grasshopper nymphs. Saunders estimates, on the basis of various methods of determination, that a 10-days' supply for 10 nestlings, when the chief food is grasshoppers, would be 5,000 to 7,000 grasshoppers. These figures again emphasize the great economic importance of the meadowlark.

Voice.—The plaintive and very pleasing whistled notes of the meadowlark, heard on its arrival, stand out among my most delightful memories of early spring on an Illinois farm. There, where a tall Osage orange tree stood at the edge of a rolling meadow, a meadowlark came each year to announce his arrival. This song may be rendered by the words *Ah-tick-seel-yah* or *Heetar-see-e-oo*, but others have translated it variously such as *Spring-o'-the-yeear; Peek-you can't see me; Toodle-te, to-on*, etc. There is an infinite number of variations of the territory song, but all have much the same quality. This song is not only the first heard from the meadowlark in spring, but is the one repeated from the singing posts throughout the season.

The meadowlark is known to alternate the versions of its song. Frances H. Allen (1922) writes of a bird he observed on an April morning:

He had four or more songs in his repertoire. The first, which was repeated a number of times in succession, resembled the opening notes of the white-crowned sparrow's song, but had three high notes on the same pitch, instead of two, before the lower one—*ee-ee-ee-hew*. It was a beautiful song and so different from anything we commonly hear from the meadowlark that I did not suspect its author at first. * * * then the bird began to alternate this song with another which seemed a good musical complement to it. This second song began low and ended high. It was something like *hew-hew-he-hee*, the third note shorter than the others. After a few alternations of these two songs the bird dropped the first and sang only the second a number of times, but dropped that in turn and finally took up two or three simpler and more normal songs, of which one, at least, was sweeter than most meadowlark songs.

The peak of singing activity, when the most beautiful songs may be heard, occurs during the first part of the breeding season, prior to incubation. During incubation there is a distinct lull in singing which lasts until the return of sexual activities in preparation for the second brood. Another lull occurs during the rearing of the second brood and lasts until fall, when singing is again renewed. In sections of the country where the meadowlark is represented by individuals during all seasons, its characteristic territory song may be heard throughout the year, even during the winter months.

The versatile meadowlark has also a flight song, a truly ecstatic performance. Prefacing the flight song with a few notes from a perch, it flies swiftly upward, sometimes spirally into the air. It vibrates its wings rapidly and utters penetrating and chattering notes in rapid concert not unlike that of the bobolink. After flying more or less in a circle, it slowly descends to the ground. This song too is variable but is very different and not at all suggestive of the ordinary song.

The songs of the eastern and western meadowlark have frequently been compared. Albert Brand (1938) who has made a study of vibration frequencies of passerine bird song, found that for the eastern meadowlark the highest note had 6,025, the lowest 3,150, and the approximate mean 4,400 vibrations per second. Those of the western meadowlark are much lower in pitch—3,475 for the highest, 1,475 for the lowest, and 3,475 for the approximate mean.

When the meadowlark is alarmed or excited it nervously flits and twitches its tail, exposing the white tail feathers. This behavior is accompanied by a sharp nasal call note, which changes to a rolling chatter followed by a plaintive but pleasing whistle. G. B. Saunders (MS., see p. 56) describes the call notes of the meadowlark in detail, as follows:

The day-old nestling first voices his calls for food with a faint *tseep, tssep, seep, seep* or *tsp, tsp*. As he gains strength this utternace is a lisping *sweet, sweet, sweet*. By the seventh or eighth day the note becomes a bysyllabic *tscheep, tscheep, tscheep* or *tschip', tschip'*. All of these notes are of the same general type. When out of the nest, the juvenile's call is a loud peeping *tseup', tseup'* or *sweet, sweet*, similar to the *weet, weet* notes of the adults.

The adult call notes may be expressed phonetically as *weet, weet, weet*. Those of the female are usually softer and more modulated than those of the male. There is an infinite variation in the expression of these notes. Other conversational calls of the adults are the low pitched and modified alarm notes *dzert, dzert* and the *tet-tet-tet-tet* notes of the chatter. The female often joins them, i. e., *dzert, tet-tet-tet-tet-tet-tet* in answering the male's song.

The common alarm chatter, *dzert-tet-tet-tet-tet*, seems to be a modification of the call notes just mentioned. The speeding up due to excitement gives the notes a much harsher quality. The notes *dzert-dzert* are usually given when a pre—liminary alarm is uttered. Another note fairly common during the breeding season, but one not heard except at that time, is the queer *beert* or "nighthawk" note. It may be given as an alarm when the birds are greatly excited, or it may be given during sexual displays and competitions. It is uttered by both sexes.

Aretas A. Saunders (MS.) has written a very excellent analysis of the song and notes of the meadowlark: "The song of the Eastern meadowlark is a short series of sweet, clear, very high pitched whistled notes. It is loud, carries a long distance, and, when one is near the bird, is rather shrill. The notes are few, compared to those of other birds, and downward slurs from a high to a lower note are frequent. In spite of the few notes, it is exceedingly variable.

"In pitch and time the song is remarkably like human music. The notes are usually on the same eight notes of the octave as in the simpler kinds of human music. The shorter notes are commonly half or a third the length of the longer notes, so that the songs could be recorded on the musical scale, as human music is written, with considerable accuracy. The different songs are easily and quickly recorded by the graphic method. My earliest experiments in recording bird songs were with the meadowlark, and although many of the records have proved to be duplicates, I have at the present time more than a thousand different songs of this species on record. The following data are based on a study of 962 of these records that I have filed and catalogued, the remaining records being still only in my field note books.

"These records show that the songs vary from 2 to 8 notes each, the great majority 3 to 6 notes. There are 4 songs of 2 notes; 65 of 3 notes; 352 of 4 notes; 391 of 5 notes; 132 of 6 notes; 15 of 7 notes and 3 or 8 notes. In spite of the great variation, many records prove to be duplicates, and it is a common experience to hear two or three birds singing the same song, one after the other, and also common to record songs from widely separated localities that are exact duplicates. While the majority of my records are from southwestern Connecticut, I have a good many from various localities in New York, and scattering records from other States. Songs that are common in Connecticut are often equally common in southwestern New York, approximately 400 miles distant. I have also recorded duplicates of Connecticut songs from the vicinity of Dover, Del.

"The pitch of songs varies from C''' to D#'''', a range of 1½ tones more than an octave, the highest notes being a little higher than the highest on the piano. The range of individual songs varies from 1 tone to an octave; 12 songs have a range of only 1 tone, and only one has a range of an octave. Nearly half of the records, 446, have a range of 2½ tones, and the average of pitch of all of them is 2.7 tones.

"The duration of meadowlark songs varies from about ⅖ second to nearly 3 seconds, averaging about 1⅘ seconds. It is difficult to measure short songs accurately with a stopwatch. The time factor of greater interest is the perfect rhythm of the notes and the great number of variations in time arrangement that, with the variations in pitch, go to make the great number of different songs that this species possesses.

"Not only does the meadowlark, as a species, sing a great variety of songs, but each individual has many variations. I once recorded 53 different songs from one individual in less than an hour, and recorded altogether 96 different songs of birds singing in that location in that season.

"Consonant sounds are not prominent in meadowlark songs. In some songs notes are linked together with a liquid consonant sound, like the letter *l* that occurs in about 10 percent of the songs I have recorded. Another consonant sound, which occurs at the beginning of certain notes, most commonly at the beginning of downward slurs, is sibilant and sounds like the letters *ts*, making a slur sound like '*tseeyah*', or something similar. The sound is rather faint however. I have recorded it in less than 5 percent of the songs, but it may be commoner than this indicates, for it is not easily audible from more distant singers.

"In early spring, usually late March and early April, the meadowlark frequently sings two different songs in alternation, usually with a pause of about one second between them. I have eight records of these alternated songs, all different. In most of them one song ends on a high-pitched note and the other on a low pitch, so that they sound something like a question and an answer, and form a pleasing musical combination. All my records but one, recorded at Cross Lake, N. Y., in July, are dated between March 7 and April 11.

"In addition to this form of song, the meadowlark has a flight song, very different in character, that is rather rarely heard. In a good many years I have not heard it at all, whereas in others I have heard it several times, most commonly in late April. The performance begins from a perch, the bird calling at intervals on a rather harsh, nasal, downward slurred note. After several of these notes the bird rises into the air and flies across the meadow singing a song made up of groups of 4 or 5 notes, separated by short pauses. These notes are fricative and not especially musical, nor are they so loud as the common song. Such a song takes 10 to 12 seconds from the beginning notes on the perch until the bird is silent.

"I have heard songs of the meadowlark in every month of the year. The regular period of singing, however, begins in March and lasts until late August. Songs in January are rare, and in 32 years of records I have heard the song in that month only 4 times. In February there is often quite a bit of singing, and in 16 of these years the first

song of the year was heard in February, the average date of the first
song being February 19. Regular singing, however, does not begin
until March, and in 6 years it did not begin until April. The average
date of its beginning is March 26.

"I have less full data on the cessation of song, as I have frequently
been in places where I could not hear it at the proper season. Five
years in Cattaraugus County, N. Y., give an average of August 11
for the last song, whereas 5 years in Connecticut average August 18.

"The song is revived in September or October, and is to be heard
quite frequently through the fall until November. In Connecticut
20 years of observations give an average of September 30 for the
beginning and November 13 for the end of the fall singing, but such
singing is much more erratic than spring singing. Songs in December
are rare, though more frequent than in January."

Enemies.—In most sections of its range the eastern meadowlark is
not commonly imposed upon by the cowbird. I have never found a
nest in New England that contained an egg of the cowbird, and
G. B. Saunders (MS., p. 56) states that of over 50 nests studied in
Oklahoma and New York, none contained other than meadowlark
eggs. However, during the course of a statistical survey of the birds
of Illinois in 1906–1908 I found four cases of cowbird parasitism: One
nest in northern Illinois near Rockford contained three eggs of the
meadowlark and one cowbird's egg; of two nests in Champaign
County, central Illinois, one contained two meadowlark and three
cowbird eggs and the other, three meadowlark and two cowbird eggs,
with a broken meadowlark's egg outside of the nest; and a nest near
Benton, Franklin County, in southern Illinois, contained two eggs of
the meadowlark and two young, one of which, judging from its size
and appearance, was a freshly hatched cowbird.

G. Eifrig (1915, 1919) writing on the birds of the Chicago area
states that he has repeatedly found nests of the meadowlark with one
or more eggs of the cowbird. He also states that one or more or all
the eggs of the rightful owner were apparently rolled out. It would
seem that the meadowlark is a common victim of the cowbird in the
State of Illinois. Milton B. Trautman (1940) found two nests of the
meadowlark containing cowbirds eggs at Buckeye Lake, Ohio. Bendire (1895) reports an instance where a second nest was built over one
containing the parasitic egg. This is a common habit of certain birds
such as the warblers but presumably it is rare in the case of the
meadowlark. Herbert Friedmann (1929) has obtained records of
cowbird parasitism of the eastern meadowlark from New England,
New York, Pennsylvania, Michigan, Illinois, and Iowa, but states
that the meadowlark is not a common host.

It is of passing interest to note that eggs of the bob-white quail and bobolink have been found in the nests of meadowlarks, although these instances are not to be classed as parasitism but merely unusual accidents. J. B. Lackey (1913) reports finding eggs of the bob-white in two meadowlark's nests near Clinton, Miss., and Edward R. Ford of Chicago found a meadowlark's nest with four eggs of the meadowlark, one of the cowbird, and one egg of the bobolink.

G. B. Saunders (MS., see p. 56) in an examination of 45 adult meadowlarks found 8 contained internal parasites. The tapeworm *Anonchotaenia* sp. was found in 3 and the parasite *Mediorhynchus grandis* in 6 birds. The roundworm *Diplotriaenoides* sp. was found in both Oklahoma and New York birds. Of 5 young in a nest at Ithaca, N. Y., 3 were found to have dipterous larvae, probably of the genus *Chrysomyia*, in their nasal passages. The meadowlark like most other birds is host to a number of external parasites including lice, ticks, and mites, among which Harold S. Peters (1936) has found the three lice *Degeeriella picturata* (Osborn), *Menacanthus chrysophaeum* (Kellogg) and *Philopterus subflavescens* (Geof.), the three ticks *Haemaphysalis leporis-palustris* Packard, *Ixodes* sp., and *Amblyomma tuberculatum* Marx; and the mite *Trombicula hominis* Ewing. Occasionally nests of the meadowlark are heavily infested with mites, and G. B. Saunders cites one case where a nest was deserted because of an unusually heavy infestation.

Because the eastern meadowlark has two broods of four or five young during each season, we need not be alarmed at the large number of enemies and of its great mortality. Man, directly or indirectly, is responsible for the loss of a great many meadowlarks and probably he is the most important factor in the control of the species and thus preventing overpopulation. Perhaps the most disastrous but unwitting acts of man is the mowing of alfalfa, clover, and timothy fields in which the meadowlarks nest. In Illinois, while traversing the various sections of the State on foot for hundreds of miles in connection with the statistical bird survey in 1906–1908, the loss I noted from this source was appalling. In June and July I saw nest after nest that had been destroyed by mowing machines and it is probably safe to state that more meadowlarks are destroyed by this means, which is repeated year after year, than by any other.

In autumn, when meadowlarks congregated in large flocks in southern Illinois, it was a common experience to see groups of a dozen or more gunners out killing meadowlarks in large numbers, to be carried home for food for themselves and their neighbors. Such practices have been common in some of the Southern States in the past, but I am convinced that in recent years there has been less of this kind of destruction because of the more rigid enforcement of protective laws and the general education of the public to the economic value of this bird.

Automobiles, which constitute a menace to certain of our birds, are not such a menace to the meadowlark; however, when the birds frequent dirt roads in autumn to dust their plumage and possibly to pick up stray bits of food, such as grasshoppers, a considerable number have been reported killed.

Since the meadowlark nests on the ground, predatory mammals and birds and probably snakes are responsible for a number of deaths. The domestic cat ranks high as a destroyer of meadowlarks, especially those that nest in fields adjacent to farm homes. Farm dogs, which also roam the fields and which are able to locate the nesting birds through the sense of smell, probably destroy a number of nests. Saunders (MS., see p. 56) states that he has seen Bonaparte's weasel attack juvenile meadowlarks.

The examination of the stomach contents of owls and hawks has revealed that the horned and snowy owls, the goshawk, duck hawk, sparrow hawk, red-tailed hawk, red-shouldered hawk, and Cooper's hawk have taken meadowlarks, chiefly during the winter months.

Since the meadowlark is one of the earliest spring migrants, snowstorms frequently cut off its food supply and, the accompanying cold, cause the death of many of the birds. Frederick C. Lincoln (1939) states that during the early part of June 1927 a hailstorm of exceptional violence in and around Denver, Colo., killed a large number of meadowlarks and other birds. The ground was strewn with dead birds and many lay dead in their nests where they were incubating eggs or brooding young when the storm broke.

Fall and winter.—In fall the meadowlarks leave their nesting grounds in Quebec and Ontario during September and October, and by the middle of October the bulk of them have departed. A few individuals may linger on until well into November. Since the meadowlark normally winters in northern United States, the time of departure of migrants is difficult to ascertain.

In southern Illinois during the month of October I have seen immense flocks made up of hundreds of individuals concentrated in the lowlands above Cairo, at the junction of the Mississippi and Ohio Rivers. These flocks were made up largely of birds that had migrated from points farther north. Also, in going through cornfields and stubblelands of this part of the State, I frequently saw smaller companies of them waddling about the clustered stalks. As they paused to inspect me they would hold their bodies in a vertical position with their bills pointed skyward. At the same time they would flick their tails displaying the conspicuous white markings as they opened and closed the fan of feathers.

On the New Jersey coast the meadowlarks start flocking about the middle of August, when it is common to see parties of 20 to 25 individuals. In October the birds band together in large flocks of 200 to 300. Many of these birds pass on farther south, but flocks of 50 to 75 are to be seen throughout the winter. They become much tamer in winter, especially when food is scarce and it is then that they frequent the habitations of man and even enter the towns, where they may be seen in vacant lots feeding in company with English sparrows and starlings. They have also been reported as seen feeding on garbage in alley ways during times of severe blizzards. During the winter months meadowlarks have been flushed from the tall grass of marshes, where the great accumulation of droppings indicated that they had roosted during the night. Meadowlarks have also been known to accompany grackles to their roosts in trees, but this is not common practice.

In recent years meadowlarks have been wintering in increasing numbers in the salt marshes of southeastern Maine in the region of Scarboro, Pine Point, and southward. During October and November as many as 100 to 200 meadowlarks may be started from a single marsh. These birds are probably individuals which had nested in the interior of the State and concentrated on the coast in the autumn.

Fred S. Walker (1910) reports that he has seen meadowlarks at Pine Point throughout the winter. A flock of 30 to 40 were frequently seen in the adjacent marshes.

In very cold weather, when the grasses and weeds of the marsh were buried beneath the snow, they would venture up to the railway station and pick up grain which had fallen from freight cars. * * * In February, when the marsh was deeply covered with snow, I frequently walked out near the river, scraped off snow from small patches of grass and fed the larks with grain—cracked corn, oats, and barley. They evidently relished this, for it was eagerly devoured. On warm days in January and February they often alighted on the telegraph wires and sang.

In South Carolina the meadowlarks arrive in large numbers in October to take up their winter residence in stubble, corn and cotton fields, and in old fields grown up in weeds and brown sedge. These birds, like those that winter along the Maine and New Jersey coasts, spend the nights in the salt marshes. In various parts of the State they swarm about the rice plantations, where they are often killed by hunters who know the meadowlark as the "marsh quail."

At Buckeye Lake, Ohio, M. B. Trautman (1940) writes of wintering meadowlarks as follows: "In an average winter 10 to 30 birds could be found during a day's field trip, but when the species was most numerous as many as 210 were seen in a day. The wintering birds were found in fields and meadows whenever these were largely free of snow. When there was deep snow the birds congregated about manure piles, straw stacks, and in barnyards and adjacent fields where stock was fed."

DISTRIBUTION

Range.—Central United States and eastern Canada to the Gulf coast.

Breeding range.—The eastern meadowlark breeds from southwestern South Dakota (Martin), northwestern Iowa (Sioux City, Ashton), central-northern and northeastern Minnesota (Itasca County, Two Harbors), northern Wisconsin (Lake Owen), northern Michigan (Baraga, Whitefish Point), southeastern Ontario (Sault Ste. Marie, North Bay), southwestern and central-southern Quebec (Blue Sea Lake, Kamouraska) and, rarely, southern New Brunswick (Sussex, Grand Manan), and central Nova Scotia; south to southern Nebraska (Stapleton, Hastings), through east-central Kansas and central Oklahoma (Woods County, Stillwater) to central Texas (Hamilton, Waco), northwestern Arkansas, central-eastern Missouri (St. Louis), central Illinois, southern Indiana (Wheatland), northern and eastern Kentucky (Corydon, Monticello), northeastern Tennessee (Shady Valley), central-northern North Carolina (Chapel Hill), and southeastern Virginia (Cobbs Island).

Winter range.—Winters rarely north to Nebraska, central Wisconsin, central Michigan, southeastern Ontario, central Vermont, southern Maine and central Nova Scotia; south to eastern Texas, southern Louisiana, central Alabama, northwestern Florida, central Georgia, central South Carolina, and northeastern North Carolina.

Casual records.—Rare in northwestern Minnesota (eastern Red River Valley), and east-central Ontario (Englehart). Casual in northeastern Colorado (Wray). Accidental in northwestern Quebec (east Maine), and Newfoundland (St. Shotts).

Migration.—The data deal with the species as a whole. Early dates of spring arrival are: South Carolina—Greenwood, February 15. North Carolina—Asheville, February 17. West Virginia—Wheeling, February 20. District of Columbia—February 21. Maryland—Laurel, February 28. Pennsylvania—State College, February 13 (average, February 27). New Jersey—Elizabeth, February 12. New York—Rochester, February 21; Watertown, March 2. Connecticut—Meriden, February 24. Rhode Island—Westerly, March 1. Massachusetts—Groton, February 15. Vermont—Bennington, March 5. New Hampshire—Sanbornton, March 3. Maine—Cumberland County. March 5. Quebec—Montreal, March 5. New Brunswick—Kent Island and Woodstock, March 26. Arkansas—Helena, February 16. Tennessee—Knoxville, February 16. Kentucky—Versailles and Carrollton, February 17. Missouri—Kansas City, Columbia, and St. Louis, February 17. Illinois—Urbana, February 14 (median of 20 years, February 26); Chicago, February 26 (average of 16 years,

March 10). Indiana—DeKalb County, February 12. Ohio—Buckeye Lake, February 9 (median, February 23). Michigan—Newberry, February 17; Blaney Park, March 15. Ontario—Port Dover, February 24; Ottowa, average of 34 years, April 2. Iowa—Delaware County, February 19. Wisconsin—Dane County, February 19. Minnesota—Fairbault, February 26 (average of 21 years in southeastern Minnesota, March 12); St. Louis County, April 1. Oklahoma—Oklahoma City, February 6. Kansas—Topeka, February 12. Nebraska—Red Cloud, February 16.

Late dates of spring departure are: South Carolina—Spartanburg, April 24. North Carolina—Raleigh, April 29 (average of 7 years, April 20). Maryland—Laurel, April 29. Louisiana—Avery Island, March 16. Mississippi—Cat Island, March 21. Illinois—Chicago, June 1 (average of 16 years, May 17). Ohio—Lucas County, April 25; Buckeye Lake, median, April 15. Texas—Dallas, March 31.

Early dates of fall arrival are: Texas—Dallas, September 15. Michigan—Charity Islands, September 26. Ohio—Buckeye Lake, median, September 1. Illinois—Chicago, September 19 (average of 11 years, October 7). Arkansas—Delight, September 29. Mississippi—Deer Island, October 21. Louisiana—Slidell, October 28. Maryland—Laurel, September 16. North Carolina—Raleigh, September 28 (average of 10 years, October 7). South Carolina—Frogmore, September 17.

Egg dates.—Connecticut: 8 records, May 10 to Aug. 6; 4 records, May 18 to June 6.

Illinois: 91 records, April 6 to July 1; 47 records, May 10 to May 22.

Massachusetts: 50 records, May 5 to July 4, 26 records, May 28 to June 10.

Ontario: 2 records, June 3 and July 6.

Florida: 13 records, April 11 to May 27; 7 records, May 1 to May 12.

Louisiana: 5 records, May 3 to June 4.

Arizona: 1 record, April 27.

Texas: 6 records, April 10 to May 29, 4 records, May 3 to May 17.

México: 3 records, April 13 to May 13 (Harris).

Late dates of fall departure are: Nebraska—Badger, November 24. Kansas—Onaga, November 30. Oklahoma—Oklahoma City, November 14. Minnesota—Hutchinson, November 29; Minneapolis, October 31 (average of 20 years in southeastern Minnesota, October 18). Wisconsin—Greenbush, November 13. Iowa—Buchanan County, December 4; Sioux City, November 30. Ontario—Wellington County, December 3; Ottawa, November 13 (average of 22 years, October 15). Michigan—Detroit, December 17; Mackinac County, November 9. Ohio—Lucas County, December 19; Buckeye Lake, December 15 (median, November 14). Indiana—North Manchester,

November 29; Hobart, November 21. Illinois—Deerfield, December 3; Chicago, November 20 (average of 11 years, October 25). Missouri—Columbia, December 8. Kentucky—Danville, November 25. Arkansas—Helena, November 23. New Brunswick—Grand Manan, October 23. Quebec—Hatley, November 14. Maine—Portland, November 10. New Hampshire—West Littleton, November 15. Vermont—Rutland, December 5. Massachusetts, Essex County, November 26. Connecticut—Fairfield, November 28. New York— New York City, November 27; Watertown, November 18. New Jersey—Kirkwood, December 5. Pennsylvania—Erie, November 29. Maryland—Laurel, November 27. West Virginia—Bluefield, December 2.

STURNELLA MAGNA ARGUTULA (Bangs)

Southern Meadowlark

Plate 5

HABITS

In naming and describing this southern race, Outram Bangs (1899) gives it the following subspecific characters: "Size much less than in true *S. magna*, though the proportions remain about the same; yellow of under parts more intense; upper parts much darker in color, the dark central areas of the feathers being much greater in extent and the light edges much less; tail and wings darker, the barring on middle rectrices, and on secondaries, tertials and wing coverts, much wider and more pronounced. The general effects produced by these differences are, in *S. magna magna*, a large bird with paler yellow under parts and a lighter brown back; in *S. magna argutula* a small bird with deeper yellow under parts and a very dark brown back."

He says that its range, "though reaching its extreme differentiation in peninsular Florida, extends along the Gulf coast to Louisiana, and thence up the Mississippi Valley to Indiana and Illinois." The A. O. U. Check-List extends its range to South Carolina and to northeastern Oklahoma and northern Arkansas.

The southern meadowlark is widely distributed and fairly common throughout Florida in all suitable localities, the prairies, the grassy plains, and the more open places in the flat pine woods, where the ground is not covered with saw palmettos. A few miles west from Melbourne, in 1902, we drove through a fine stand of tall pines, widely scattered, with large areas of open grasslands between them and an occasional slough or shallow pond. Here, and on the broad expanse of open prairie which extended for miles toward the St. Johns marshes, we found the meadowlarks really abundant.

Nesting.—The nesting habits of the southern meadowlark do not differ materially from those of its northern relative. The only nest I have ever seen was found on Merritt's Island, Fla., on April 26, 1902. While tramping across a broad grassy plain, I flushed a meadowlark from her nest almost at my feet. The nest was sunken into the ground between two small tussocks of short grass in a rather open place; it was made of dry grass and weed stems, and arched over with dead and green grass; it was rather poorly hidden.

Maynard (1896), referring to certain plains in southern Florida, says: "The growth of grass on the margins of these plains is low, seldom exceeding 6 inches in height, and consequently forms the homes of countless Meadow Larks, for these birds always exhibit a decided preference for low herbage." Howell (1932), on the other hand, says that the nests "are well concealed in thick grass."

Donald J. Nicholson (1929) states that this meadowlark is very sensitive to any examination of its nest, and will usually desert it if it is discovered before the set is complete; he tells of a pair that built three nests before they felt safe in laying their eggs in the third, the first two having been examined before the eggs were laid.

Eggs.—The set for the southern meadowlark usually consists of three or four eggs, very rarely five. These are practically indistinguishable from those of the northern bird. The measurements of 40 eggs average 27.5 by 20.4 millimeters; the eggs showing the four extremes measure 30.5 by 21.3, and 23.8 by 17.5 millimeters.

In the inhabited regions of Florida, especially near towns and villages, where the birds are sometimes hunted as game, the southern meadowlark is as wild and shy as is its relative in the north. But, in the more remote, unsettled regions, it is often very tame and unsuspicious of man.

Maynard (1896) says that, in such a wilderness in 1871—

The birds which occurred there were seldom if ever disturbed so that I found them exceedingly tame; in fact they would start up at my feet, fly a few yards, and either settle down again in the grass or alight on a low limb of a pine, where they would quietly gaze at me, even allowing me to pass directly beneath them without attempting to move. Then as if satisfied that I intended doing them no harm, would sound a loud, strange note which was so utterly at variance with the song of the same species in New England, that when I first heard it could scarcely believe it was a Meadow Lark. This lay even in the North has a peculiar intonation which is quite suggestive of freedom, but that given by the birds which inhabit the trackless piney woods and widespread plains of Florida is, although very melodious and pleasing, so wild, clear and ringing, that it is in perfect harmony with surroundings where Nature reigns supreme.

Others have noticed a difference between the songs of the northern and southern birds. I wrote in my journal in 1902, when my hearing was good, that it was similar to the song of the eastern meadowlark, "but rather more musical and richer in tone, slightly suggestive of the song of the western meadow lark."

DISTRIBUTION

Range.—The southern meadowlark is resident from central southern and northeastern Oklahoma (Love County, Vinita), northern Arkansas, (Fort Smith), southeastern Missouri (Portageville), southern Illinois (Wabash, Richland, and Lawrence Counties), southwestern Indiana (Knox County), southwestern Kentucky (Fulton County, Rockport), Tennessee (except northeastern), Georgia (except extreme northern), South Carolina and central southern and northeastern North Carolina (Rockingham, South Mills); south to southeastern Texas (Pierce, Galveston), the Gulf Coast and southern Florida, south to Cape Sable.

Casual records.—Casual in winter to southern Texas (Corpus Christi, Cameron County).

STURNELLA MAGNA HOOPESI Stone

Rio Grande Meadowlark

HABITS

The name Rio Grande meadowlark was formerly applied to the meadowlarks of this species that live along our southern borders, from Brownsville, Tex., to southern Arizona. But when Oberholser (1930) described *S. m. lilianae*, the name *S. m. hoopesi* was restricted to the birds of central-southern Texas. This meadowlark was described and named by Dr. Witmer Stone (1897) from a specimen in the collection of Josiah Hoopes, from Brownsville, Tex. He gives as its characters: "Color below as in *magna*, but rather lighter and less buff on the sides and under tail coverts; upper surface much grayer and generally lighter. The brown tints of *magna* are very largely replaced by gray, especially on the wings. Sides of the face whiter than in *magna;* tail bars almost always distinct, i. e., not confluent along the shaft of the feather.

"This bird is the lightest of all the Meadow Larks, averaging a little lighter than *neglecta*, the tail bars are also more distinct than in any of the other races."

I can find nothing recorded on its habits to indicate that they are in any way different from those of the other southern races.

DISTRIBUTION

The Rio Grande meadowlark is resident from southeastern Texas (Eagle Pass, Port Lavaca) to northern Coahuila, Nuevo León, and northern Tamaulipas.

STURNELLA MAGNA LILIANAE Oberholser

Arizona Meadowlark

HABITS

In naming this western subspecies, Dr. Oberholser (1930) says that it is "similar to *Sturnella magna hoopesi,* of central southern Texas, but wing longer; other dimensions smaller, particularly the feet; upper parts much paler, more grayish; the dark bars on wings and tail still narrower, and even more disconnected; under parts averaging still more deeply golden yellow."

He gives as its range, "central western Texas and southern New Mexico, west to central and southern Arizona, and south to Sonora and Chihuahua."

And adds that "this new bird is most closely allied to *Sturnella magna hoopesi,* described from Brownsville, Tex., and, in fact, is its western representative.

"It is strikingly similar to *Sturnella neglecta,* more so, indeed, than is any of the other subspecies of *Sturnella magna.* Meadowlarks from Arizona and New Mexico have commonly been referred to *Sturnella magna hoopesi,* but comparison of a series shows at once that they are different."

The race was named in honor of Mrs. Lillian Hanna Baldwin (Mrs. S. Prentiss Baldwin), who presented the Cleveland Museum with a collection of birds, including the type of this subspecies.

We found meadowlarks fairly common on the grassy plains and low foothills of the mountains in southeastern Arizona, but found no nests. They could easily be distinguished from the western meadowlarks by their songs, and were undoubtedly of this race. Nothing peculiar was noted as to their haunts and habits, which apparently resembled those of the other southern races.

DISTRIBUTION

The Rio Grande meadowlark breeds from northwestern and central Arizona (Juniper Mountains, Springerville) east to southern New Mexico (Gila River, Hachita), and western Texas (El Paso, Chisos Mountains), and south to northeastern Sonora and northern Chihuahua. It winters north to central Arizona.

STURNELLA NEGLECTA NEGLECTA Audubon

Western Meadowlark

HABITS

I shall never forget the day I first heard the glorious song of the western meadowlark; the impression of it is still clear in my mind, though it was May 30, 1901! It was my first day in North Dakota, and we were driving from Lakota to Stump Lake when we heard the song. I could hardly believe it was a meadowlark singing, so different were the notes from those we were accustomed to in the east, until I saw the plump bird perched on a telegraph pole, facing the sun, his yellow breast and black cravat gleaming in the clear prairie sunlight. His sweet voice fairly thrilled us and seemed to combine the flutelike quality of the wood thrush with the rich melody of the Baltimore oriole. I have heard it many times since but have never ceased to marvel at it. It seems to be the very spirit of the boundless prairie.

Audubon (1844) gave this bird the above scientific name, but called it the Missouri meadowlark. He says of its discovery:

> Although the existence of this species was known to the celebrated explorers of the west, Lewis and Clark, during their memorable journey across the Rocky Mountains and to the Pacific; no one has since taken the least notice of it. * * *
> We found this species quite abundant on our voyage up the Missouri, above Fort Croghan, and its curious notes were first noticed by Mr. J. G. Bell, without which in all probability it would have been mistaken for our common species (*Sturnella Ludoviciana*). When I first saw them, they were among a number of Yellow-headed Troupials, and their notes so much resembled the cries of these birds, that I took them for the notes of the Troupial, and paid no further attention to them, until I found some of them by themselves, when I was struck with the difference actually existing between the two nearly allied species.

During the latter part of the last century considerable discussion arose among leading ornithologists as to its status as a full species, a subject fully covered by Widmann (1907). As a result, this bird in the first two editions of the A. O. U. Check-List stood as a subspecies of *S. magna*, and it was not until the third edition (1910) that it was restored to full specific status. There is a striking resemblance in the general appearance of the two species; intergradation has been claimed, but probably no more than might be accounted for by hybridizing. But the songs of the two are strikingly different; and, where the ranges of the two come together and even overlap, typical birds, with typical songs are sometimes found breeding in the same region, a condition not supposed to occur with subspecies.

The western meadowlark is widely distributed in all suitable regions throughout western North America, from southern Canada to northern

México and from the eastern borders of the prairies and plains to the Pacific. There is some evidence to suggest that it may be extending its range eastward.

The favorite haunts of this meadowlark are the prairies and the grassy plains and valleys, but it also ranges well up into the mountain parks and foothills, as high as 5,600 feet even in Washington, from sea level to 7,000 in California, 8,000 feet in Utah, 10,000 feet in Arizona, and 12,000 feet in the mountains of Colorado. Dawson and Bowles (1909) say of its haunts in Washington: "It is found not only on all grassy lowlands and in cultivated sections but in the open sage as well and upon the half-open pine-clad foothills up to an altitude of four thousand feet." In other parts of the west, where it is common, it is likely to be seen wherever there is a thick growth of weeds and grasses, along country roads and even in vacant lots in the thinly settled parts of towns and villages.

The specific characters of the western meadowlark do not appear to be very conspicuous to the casual observer. Ridgway (1902) says that it is "similar to *S. magna hoopesi*, but different in proportions, the wing averaging longer, the tail, tarsi, and toes shorter; coloration much grayer and more 'broken' above, the broad lateral crown stripes never uniform black, but always (except in excessively worn plumage) more or less conspicuously streaked with pale grayish brown; malar region always largely yellow, usually including both anterior and extreme posterior portions; blackish streaks on sides and flanks varied with spots of pale grayish brown, the ground color of these parts paler buffy (often white, scarcely if at all tinged with buff); black jugular crescent averaging decidedly narrower."

Territory.—Kendeigh (1941), in his study of the birds of a prairie community, has this to say:

Territorial behavior is well established in this species, although only the male defends the territory. At least two variations of song were given from singing posts, and a song was given occasionally while flying. Flight songs were not so frequent as one might expect. Possibly they were given more often during the earlier mating season. Most of the singing was from fence or telephone poles or from tall weeds or small trees. The song served as an advertisement to other males that the area was occupied. When another meadowlark encroached on the area or simply flew high over it, the male met the challenge and gave chase until the intruder passed the limits of the owner's jurisdiction. The females, on the other hand, were at no time observed to be concerned about territorial boundaries. * * *

In computing the bird population only three pairs of meadowlarks were counted for the area although four territories were represented. Three of the four territories extended well outside the area under study. The male at nest No. 1 had the smallest territory of approximately 10 acres. The male at nest No. 2 at various times maintained right over about 24 acres. The other two territories were about 21 and 32 acres, respectively, as near as could be estimated.

Courtship.—I can find no published account of the courtship of the western meadowlark. It probably consists of song and plumage display. I have some interesting notes, sent to me by J. W. Slipp, who watched a bird that displayed before its reflected image in the "shiny chromium hub caps of three parked cars," on the campus of a college at Tacoma, Wash., on May 8, 1940. "Visiting seven of these hub caps in succession, it spent an average of about a minute, at each, apparently fascinated by its own reflected image. Walking up to the first wheel the bird stretched itself nearly erect, then began to strut excitedly back and forth, turning first one side and then the other to the mirroring of the hub cap, and repeatedly flirting its wings and tail in such a way as to flash the white outer rectrices. All this was accompanied by frequent short ejaculatory notes, interspersed occasionally with full-throated snatches of the beautiful song characteristic of the species." On another occasion a similar performance was given on the running board of a Plymouth sedan, with his image reflected in the lustrous surface of the car. This all may have been only "shadow-boxing," but it suggests what the courtship display might be like.

Nesting.—The nesting habits of the western meadowlark are not very different from those of other meadowlarks, due allowance being made for any difference in environment. Samuel F. Rathbun tells me that in western Washington this bird begins nesting as early as the first week in April; he describes in his notes a very fine nest: "This nest was beautifully built, and placed in a growth of low grass, a small tuft, on rather rocky land. It was finely arched over with strips of fine, dry, fibrous bark taken from a nearby small dead tree. The body of the nest was made of dry, fine grasses, it being lined with very fine, dry grass. It was placed in a shallow depression of the ground. This nest, if removed from the ground would be nearly round, with an entrance on the side." The site was on the shore of a lake in eastern Washington, in a nesting colony of about 150 pairs of ring-billed gulls. A nest that I found on the shore of Many Island Lake, in Alberta, was similarly located, but well concealed in long grass.

E. S. Cameron (1907) says that, in Montana, "Meadowlarks make their nests entirely of grass under the sage-brush or in tussocks of grass, and roof them over with the same material. * * * On June 30, 1906, I noticed a bird sitting in a flowering cactus patch which was the prettiest nest I have seen."

Jean M. Linsdale (1938) describes a nest found in Smoky Valley, Nev., as follows: "The nest was in an open part of a meadow, and was built in a depression in the ground, fully 3 inches deep and 8 inches in diameter. It was well covered with a dome-shaped roof composed of fibers of bark and plant stems woven in with the growing vegetation.

The top of the roof was about 5 inches above the surface of the ground. The inside of the nest was globular and 4½ or 5 inches in diameter. The round entrance on the south side was 2½ inches in diameter. The lower margin of the entrance was about an inch below the surface of the ground. The lining was of small grass stems."

Kendeigh (1941) mentions two Iowa nests: "The first nest with six eggs was well concealed in *Poa pratensis* under a clump of *Solidago rigida.* * * * The second nest was under a tuft of *Andropogon* and had a tunnel a foot long, slightly curved, leading to it."

John G. Tyler (1913), of Fresno, Calif., writes:

Other nests have been seen in alfalfa fields and among thick growths of weeds; but what I consider the most unusual site was located April 23, 1908 when a Meadowlark was plainly seen sitting on her nest while I was yet over one hundred feet distant. The nest was found near a berry patch, the ground having been plowed early in the winter, later a sparse, stunted growth of oats springing up. At the time the nest was found the oats were not over six inches in height, and so thin and scattering as to afford almost no protection or concealment. In a slight hollow, not over three-quarters of an inch in depth, were four eggs resting on the bare, damp ground, without a semblance of nesting material either over, under, or around them.

Bendire (1895) mentions a nest "placed in a hole in the ground fully 8 inches deep." Dr. Harold C. Bryant (1914) found that, in California, "a preference for pasture land for nesting sites was shown, at least eighty per cent of the nests found being so situated. * * * A canopy of dry grass stems usually arches the top of the nest and a runway two to five feet long leads to the nest. Ofttimes this runway is the only clue to the location of the nest."

Eggs.—From three to seven eggs constitute the set for the western meadowlark, five being the commonest number. They are practically indistinguishable from those of the eastern bird. According to Bendire (1895), "the average measurement of 206 specimens in the United States National Museum collection is 28.33 by 20.60 millimeters, or about 1.12 by 0.81 inches. The largest egg in the series measures 30.78 by 21.84 millimeters, or 1.21 by 0.86 inches; the smallest, 25.65 by 20.07 millimeters, or 1.01 by 0.79 inches."

Young.—Bendire (1895) writes: "Both sexes assist in the construction of the nest and also in incubation, which lasts about 15 days. An egg is deposited daily until the set is completed. The young leave the nest before they are able to fly, depending for safety on hiding themselves in the grass, and they are cared for by the parents until they can provide for themselves. When they are able to do this they gather into small companies and roam over the surrounding country. I do not believe that any of the young of the year remain in our Northwestern States through the winter; they probably move slowly southward in the late fall."

Mrs. Irene G. Wheelock (1904) says that incubation—

Lasts thirteen days, and the young remain in the nursery twelve days longer, leaving it before they are able either to fly or to perch. Yet so protective is their coloring and so jealously does the long grass guard its secret that, search as you may within a circle where you know they are hidden, you will not find one of them. For two weeks longer they remain with their parents, learning to hunt grasshoppers, beetles, and crickets, to hide in the shadow of a green tuft, to bathe in the shallows at the brook's edge, and last of all, to perch in low bushes at night with others of their kind. As soon as they have mastered these things, they are able to provide for themselves and are abandoned by the parents.

Several observers have reported that two broods are raised in a season, even in the more northern parts of the bird's range. This seems likely, for the nesting season begins early and continues well into the summer. Dawson and Bowles (1909) say that in Washington, "one brood is usually brought off by May 1st and another by the middle of June"; they add that the young are "very precocious and scatter from the nest four or five days after hatching, even before they are able to fairly stand erect." Bryant (1914) says that, in California, "the first nesting usually occurs in April and May and the second in July and August." His figures show the rapid increase in the weights of young birds; the egg ready to hatch weighed 0.135 ounce, the young 1-day-old 0.25, the 8-day-old 2.50, and an adult 4.00 ounces.

Plumages.—The molts and plumages of the western meadowlark are the same as those of the eastern bird, which are fully explained under that species and need not be repeated here.

Food.—A great mass of information has been published on the food of the western meadowlark, mainly from investigations made in California. The most concise account, though based on the study of comparatively few specimens, is given by F. E. L. Beal (1910). In an examination of 91 stomachs, distributed throughout the year, he reported that "the food consists of 70 percent of animal matter to 30 of vegetable. Broadly speaking, the animal matter is made up of insects and the vegetable of seeds." Beetles constitute the largest item in the animal matter; the amount for the year is almost 27 percent, practically half of which consists of predatory ground beetles, an argument against the meadowlark, as the beetles prey on other insects.

Lepidoptera, largely caterpillars, amount to about 15 percent, wasps and ants nearly 6 percent, bugs (Hemiptera) a little more than 4 percent, and grasshoppers only 12 percent for the whole year, but 42 percent in August. Other items included crickets, craneflies, spiders, sowbugs, and a few snails. Of the vegetable food, only one stomach contained anything "doubtfully identified as fruit pulp." And only 2 percent of the yearly food was weed seeds, a surprisingly small amount for a ground-feeding bird. The remainder consisted of grain, the average monthly consumption amounting to 27.5 percent; this consisted of oats, wheat, barley, and a little corn eaten in various amounts at different seasons.

A much more elaborate report, based on the examination of nearly 2,000 stomachs, is made by Bryant (1914), from which only a few extracts can be included here. "Stomach examination has shown that sixty-three and three-tenths percent of the total volume of food of the western meadowlark for the year is made up of animal matter and thirty-six and seven-tenths percent of vegetable matter. The animal matter is made up mostly of ground beetles, grasshoppers, crickets, cutworms, caterpillars, wireworms, stink-bugs, and ants, insects most of which are injurious to crops. The vegetable matter is made up of grain and seeds. Grain as food reaches a maximum in November, December, and January, insects in the spring and summer months, and weed seeds in September and October." These foods break down into the following percentages for the year: Grain 30.8 percent, weed seeds 5.3, miscellaneous vegetable food 0.6, Coleoptera 21.3, Orthoptera 20.3, Lepidoptera 12.2, Hemiptera 1.7, Hymenoptera 5.6, Diptera 0.1, Arachnida 0.2, and miscellaneous insects 1.9 percent. Of the food of the nestlings, he says:

Stomachs of nestling western meadowlarks examined contained as high as two grams of insect food. Maxima of seven large cutworms, of twelve grasshoppers (three-quarters of an inch in length), and of eight beetles have been found in the stomachs of nestlings. One stomach contained twenty-four ants and parts of a ground beetle. * * *

A nestling western meadowlark after obtaining no food for three hours was fed twenty-eight small grasshoppers (one-half inch in length) equal in volume to about three cubic centimeters. Another one was fed four grasshoppers (one inch in length), twelve small grasshoppers (one-half inch in length), one robberfly, one beetle, and five ants. A third one was fed thirty grains of wheat inside of ten minutes."

Bendire (1895) mentions seeing meadowlarks probing in the ground, probably for locust eggs deposited just below the surface of the ground. The alfalfa weevil, which does so much damage to the crop in Utah and other Western States, it's largely eaten by the western meadowlark where these insects are abundant. E. R. Kalmbach (1914) says: "In April, 27 of these birds were collected, and the weevil, which was found to comprise one-sixth of their food, was present in all but seven. The insects taken were adults, and the average was 14.4 weevils per bird. One bird had taken 75 of these insects, another 60, and three others 51, 48, and 33, respectively."

Ira La Rivers (1941) writes: "This species is by far the ablest avian predator of the Mormon cricket, for it specializes upon the eggs of the pest. Meadowlarks have been reported at various times as destroying entire, vast cricket egg-beds, and I have, on many occasions, seen them hard at work in such egg-beds, digging industriously for the palatable eggs, which are generally laid in clusters from a few to over fifty."

Behavior.—In a general way the habits of the western meadowlark are very similar to those of the well-known eastern species. Ridgway (1877), however, noted the following differences in its manners: "It is a much more familiar bird than its eastern relative, and we observed that the manner of its flight differed in an important respect, the bird flitting along with a comparatively steady, though trembling, flutter, instead of propelling itself by occasional spasmodic beatings of the wings, then extending them horizontally during the intervals between these beats, as is the well-known manner of flight of the eastern species."

In his notes from western Iowa, Dr. J. A. Allen (1868) writes: "At the little village of Denison, where I first noticed it in song, it was particularly common, and half domestic in its habits, preferring apparently the streets and grassy lanes, and the immediate vicinity of the village, to the remoter prairie. Here, wholly unmolested and unsuspicious, it collected its food; and the males, from their accustomed perches on the house-tops, daily warbled their wild songs for hours together."

Grinnell and Storer (1924) say: "In spring and early summer meadowlarks are seen chiefly in pairs; but throughout the fall and winter they forage in flocks numbering anywhere from 10 to 75 individuals. The flock organization is loose; in fleeing from danger each bird takes its own course, remaining with or leaving the flock at will. It usually happens that certain individual birds fail to take wing when a flock is first flushed, and these belated birds subsequently rise one after another as their field is invaded, to straggle off independently."

Kendeigh (1941), speaking of some birds he had under observation, states: "Through July, six to a dozen or more meadowlarks were seen frequently in the evenings as they went to roost in the grass within the former territory of the male of nest No. 1 or in other parts of the area. Male No. 1 was not a member of this group; his tail had been clipped for recognition purposes. These birds do not roost on any perch above the grass cover. Although they could not be observed at very close range, it appeared that they passed the night on the ground under some clump of grass, where they were relatively well protected."

Voice.—Much has been written in praise of the western meadowlark's sweetly beautiful song, but only a few of the many references to it in the literature can be quoted. Its song is the bird's greatest charm, which is bound to attract attention to it. My first impressions of it are mentioned at the beginning of this story. Aretas A. Saunders sends me his impressions of it as follows: "Probably all bird lovers who know the songs of both eastern and western meadowlarks will agree that the song of the western is far superior to that of the eastern. While I have no records of the western bird's songs, and cannot give detailed statistics, I have heard it many times and can compare the two songs in some of their details. The western meadowlark's song probably averages about the same in length, but contains more notes, and the notes are shorter and more rapidly repeated. The pitch is lower than that of the eastern species. Consonant sounds, both liquids like *l* and explosives like *k* or *t* are much more frequent, occurring in practically every note of the song. Individual birds sing a great number of variations, and it is probable that the variation in this species is as great as in the eastern bird. Finally, the quality of the song is richer and fuller, resembling that of thrushes or the Baltimore oriole. This matter of richer quality is what makes the song superior to our ears. It is undoubtedly due to the lower pitch. Physicists tell us that quality of musical sounds is caused by overtones, and a lower-pitched note will have more overtones that are audible to the human ear.

"The western meadowlark sings a flight song that is quite unlike the commoner song and very similar to the flight song of the eastern bird. The introductory notes, however, are not harsh or nasal, but clear and thrushlike, while the rest of the song is far inferior in quality to the commoner song."

A. D. Du Bois, who has heard the song in both Montana and Minnesota, says in his notes: "It seems to me this westerner is something of a yodeler. * * * To my ear, its song has a very pleasing alto quality which makes the eastern bird's song seem a rather thin falsetto by comparison. In the vicinity of my home in Minnesota we have both species; but in this locality I do not hear quite the same songs of the westerner that I heard in Montana."

Impressions of two of the earlier travelers in the west are worth quoting. J. A. Allen (1868) did not at first recognize it as the song of a meadowlark, saying: "It differs from that of the Meadow Lark in the Eastern States, in the notes being louder and wilder, and at the same time more liquid, mellower, and far sweeter. They have a pensiveness and a general character remarkably in harmony with the half-dreary wildness of the primitive prairie, as though the bird had received from its surroundings their peculiar impress; while if less loud their songs would hardly reach their mates above the strong winds that almost constantly sweep over the prairies in the hot months. It differs, too, in the less frequency of the harsh complaining chatter so conspicuous in the Eastern birds, so much so that at first I suspected this to be wholly wanting."

And Robert Ridgway (1877) writes:

We know of no two congeneric species, of any family of birds, more radically distinct in all their utterances than the eastern and western Meadow Larks, 2 years of almost daily association with the latter, and a much longer familiarity with the former, having thoroughly convinced us of this fact; indeed, as has been the experience of every naturalist whose remarks on the subject we have read or heard, we never even so much as suspected, upon hearing the song of the Western Lark for the first time, that the author of the clear, loud, ringing notes were [sic] those of a bird at all related to the Eastern Lark, whose song, though equally sweet, is far more subdued—half-timid—and altogether less powerful and varied. As to strength of voice, no eastern bird can be compared to this, while its notes possess a metallic resonance equalled only by those of the Wood Thrush.   The modulation of the song of the Western Lark we noted on several occasions, and found it to be most frequently nearly as expressed by the following syllables: *Tung-tung-tung ah, tillah'-tillah', tung*—the first three notes deliberate, full, and resonant, the next two finer and in a higher key, the final one like the first in accent and tone.   Sometimes this song is varied by a metallic trill, which renders it still more pleasing.   The ordinary note is a deep-toned *tuck*, much like the *chuck* of the Blackbirds (*Quiscalus*), but considerably louder and more metallic; another note is a prolonged rolling chatter, somewhat similar to that of the Baltimore Oriole (*Icterus baltimore*), but correspondingly louder, while the anxious call-note is a liquid *tyur*, which in its tone and expression calls to mind the spring-call (not the warble) of the Eastern Blue-bird (*Sialia sialis*), or the exceedingly similar complaining note of the Orchard Oriole (*Icterus spurius*).   In fact, all the notes of the Western Lark clearly indicate its position in the family *Icteridae*, which is conspicuously not the case in the eastern bird."

Charles N. Allen (1881), evidently an accomplished musician, has published an excellent study of the song of the western meadowlark, to which the reader is referred, as it is too long to quote from satisfactorily.   Twenty-seven distinct songs are illustrated in musical notation, in which the bird apparently sings from 120 to 200 notes per minute.   Referring to the quality of the song, he says: "I know of no musical instrument whose quality of tone—*timbre*—is like that of *Sturnella neglecta*.   I have thought that a combination of the tones of the Boehm flute and a good, glass dulcimer might represent it pretty accurately.   It has qualities heard in the notes of the Bobolink, and of the Baltimore Oriole."   He says that he cannot apply the syllables, quoted above from Ridgway, to any of the songs he has studied; and adds that, while the songs of many birds may be well represented in syllables, he has "as yet heard nothing of the kind in any of the songs of the bird under consideration."

While Allen's musical notations may convey some impressions to a trained musician, they are of no help to the average layman; nor, in my opinion, do the many attempts, which I have seen in print, to express the songs in syllables, give any adequate idea of them.   While attempts to express the songs in human words are entirely inadequate to show their quality, they at least indicate the rhythm and serve to recall the songs to one who has heard them. One of the best of these is written by Dawson and Bowles (1909): "One boisterous spirit in Chelan I shall never forget for he insisted on shouting, hour after hour, and day after day, '*Hip! Hip! Hurrah! boys; three cheers'!*." And Fred J. Pierce (1921) describes what he calls a one-sided imaginary conversation: "We see the Meadowlark standing on a post repeating, '*Oh, yes, I am a pretty-little-bird*' (the '*pretty-little-bird*' winds up with a trill). In a moment he says, '*I'm going to-eat pretty-soon.*' Then, suiting the action to the word, he drops out of sight into the grass, and presently we hear him say, '*I cut 'im clean off, I cut 'im clean off*' (this is often followed by '*Yup*'). He flies back to his perch with a bug in his bill, and when he has deliberately eaten it, he—in a fast, sing-song and unmusical voice—says, '*It makes me feel very good.*' " Fanciful as these renderings are, they *do* suggest the song.

Claude T. Barnes writes to me from Utah that, on April 10, 1925, he "heard a meadowlark give the song '*Tra la la traleek*'; the '*traleek*' was a jumble of sounds, short, emphatic. Rising into the air, it sang, while a-wing, a song quite like that of the bobolink, then alighted on a post and uttered occasionally the first song. After a while, it sang the common '*U-tah's a pretty place*'." He has heard the bird singing at midnight, and others have said that it sings at all hours of the day and night, though mainly in the early morning. Weather makes very little difference; it sings in sunshine, rain, wind or snow. It is also a very persistent singer. Linsdale (1938), writing of the birds of the Great Basin, says:

The songs of meadowlarks were conspicuous among the sounds in the inhabited areas. Usually they were given from some rather high perch. One, on the morning of May 25, 1932, sang 22 times in 3 minutes: 8, 7, and 7 times each minute. It then uttered 3 single whistles and moved about 75 yards to another perch where it resumed singing. Another on June 6, 1933, sang 10 times in 1½ minutes; 7 times the first minute.

One type of song was given regularly in flight. The singing bird would rise gradually in a straight line and then drop abruptly. One that was watched flew up 75 to 100 feet, at a 45° angle, singing on rapidly beating wings, and went down 50 to 75 yards away.

Albert R. Brand (1938) in his study of vibration frequencies of birds' songs gives the western meadowlark a low-pitch rating; he recorded an approximate mean of 2,500 vibrations per second for this bird, and only 3,475 for the highest note and 1,475 for the lowest; this latter figure is lower than for any other passerine bird tested, except the catbird, crow, starling, yellow-breasted chat, and eastern red-wing, and twice as low as for the eastern meadowlark.

Field marks.—This species can be easily recognized as a meadow-lark by its three well-known characters, white lateral tail feathers, yellow breast, and black crescent, but there is no visible character by which it can be distinguished from the eastern meadowlark. Its song is the most easily distinguished character, being very conspicuous and quite diagnostic. The differences in behavior referred to above are very slight and not very constant.

Enemies.—Cameron (1907) writes from Montana:

Meadowlarks have many enemies, more especially Golden Eagles, Prairie Falcons, Marsh Hawks, and Red-tailed Hawks. A pair of the latter, which nested for several years, close to my ranch in Custer County, fed their young almost entirely upon these birds. Whereas heaps of Meadowlark feathers lay on a log near the tree, other remains were scarcely ever found, although the hawks did occasionally procure snakes and cotton-tail rabbits. * * *

On June 15, 1898, I surprised the female hawk just after she had seized a newly flown Meadowlark which was immediately dropped. Mr. M. M. Archdale has seen a female Marsh Hawk standing by a Meadowlark's nest and devouring the young birds. I have several times found Meadowlarks impaled, or hanging, on a barbed wire fence, and a few perish from the buffeting of spring storms.

Dawson and Bowles (1909) say: "The Meadowlark is an assiduous nester. This is not because of any unusual amativeness but because young Meadowlarks are the *morceaux délicieux* of all the powers that prey, skunks, weasels, mink, raccoons, coyotes, snakes, magpies, crows. Hawks and owls otherwise blameless in the bird-world err here—the game is too easy." Even the little sparrow hawk will stoop for a young meadowlark. Only the fecundity of the meadowlark and its skill in concealing its nest serve to perpetuate the species.

Some nests are probably trodden upon by cattle or sheep grazing in the nesting fields. Many meadowlarks die from eating grain poisoned with thallium and spread on the ground to kill rodents; they eat this grain readily.

The cowbirds sometimes find the nests and lay one or more eggs in them, but Dr. Friedmann (1929) knew of only five definite records; it is doubtful if a young cowbird could compete with the larger young of the meadowlark.

Economic status.—A study of the food of the western meadow-lark, as outlined under that heading, above, will prove it a very useful and beneficial bird. The small amount of sprouting or mature grain it eats is of little consequence when compared with the enormous number of injurious insects it destroys, while the number of useful insects it eats is too small to have much effect on the balance in its favor. For an exhaustive study of the subject, the reader is referred to two of Bryant's important papers (1912 and 1914). The following paragraph from the latter paper is significant: "As a destroyer of cutworms, caterpillars, and grasshoppers, three of the worst insect pests in the State of California, the western meadowlark is probably unequaled by any other bird. The stomachs of meadowlarks examined have averaged as high as 6 cutworms and caterpillars and 16 grasshoppers apiece. Maximum numbers of 66 cutworms and of 32 grasshoppers have been taken from a single stomach. As the time of digestion is about four hours, three times the average must be consumed daily."

Fall.—After the last brood of young are strong on the wing, old and young gather into groups or larger flocks and begin their late summer wanderings, both regional and altitudinal. Fred M. Packard (1946), writing of such movements in Colorado, says that "in late summer, they increase in numbers through the mountain parks and may even be found then above timberline. They leave the mountains in September and early October."

John G. Tyler (1913) witnessed a heavy concentration of these birds in the Fresno district of California: "October 10, 1905, just at sundown I witnessed a flight of Meadowlarks unlike anything I had ever seen. A very large flock of these birds, estimated at about one hundred and twenty-five, came sweeping in from a half-section of stubble, and settled for just a moment in an adjoining vineyard; then the whole mass arose again and in a compact body flew back to the stubble. In every movement this flight was suggestive of ducks and the flight resembled a flock of Sprigs coming in from some irrigated wheat field, settling for an instant on a pond and then again taking wing."

The fall migration of the western meadowlark is not greatly extended or very conspicuous, for the bird is mainly resident over most of its breeding range. It amounts to a gradual withdrawal from the more northern summer haunts, or from regions where its feeding grounds are covered with snow. Even in California, according to Grinnell (1915), in the "highest localities, which are subject to snowfall, there is evidently an exodus of meadowlarks for the winter, and in complementary fashion many birds winter on suitable portions of the Colorado and Mohave deserts, where the species in unknown in summer."

Winter.—Even as far north as Montana, according to Cameron (1907), the western meadowlark sometimes stays for the whole winter, "During the last winter, 1906–1907, no less than seven Meadowlarks remained on Mr. Al. Johnson's property situated on the outskirts of Miles City."

Referring to Oregon, Gabrielson and Jewett (1940) say: "During the winter the birds withdraw somewhat from the State and those remaining gather into small wintering bands that seek the sheltered valleys during the worst weather. In late February or early March, they increase in numbers as the migrants move north."

The Point Lobos Reserve, on the coast of Monterey County, Calif., seems to be a favorite winter resort for this species. Grinnell and Linsdale (1936) write: "In the open portions of Point Lobos the western meadowlark was the most numerous kind of bird and the most persistently conspicuous one throughout the whole year. Repeated counts and estimates fixed the highest number present at one time, in winter, as around two hundred. The meadowlarks in winter were banded into two or three flocks varying from forty to one hundred individuals, with additional scattered individuals always present in the neighborhood. * * * Possibly not more than 50 pairs remained to nest."

DISTRIBUTION

Range.—Western North America from British Columbia and Ontario south to Mexico.

Breeding range.—The western meadowlark breeds from southeastern British Columbia, central Alberta, central Saskatchewan (Manitoba Lake, Hudson Bay Junction), southern Manitoba (Dauphin, Shoal Lake), western Ontario (Emo, Fort William), northeastern Minnesota, northern Wisconsin (Superior), northern Michigan (Marquette), southern Ontario (Sault Ste. Marie; rarely Hamilton), northwestern Ohio (casually); south through western Montana, eastern Idaho, Nevada, and southeastern California to northwestern Baja California (San Quintín), northwestern Sonora, central and southeastern Arizona (Chandler, Safford, rarely Tucson), eastern Sonora, Sinaloa, Jalisco, northwestern Durango, Guanajuato, southeastern Coahuila, central Texas (Eagle Pass, Austin), northwestern Louisiana (Gilliam), northwestern Arkansas, central-eastern Missouri, southwestern Tennessee (Memphis), southern Illinois, southern Michigan, and (casually) central Ohio.

Winter range.—Winters north to southern Alberta, southern Saskatchewan, southern Manitoba, and southern Wisconsin (Racine); south to southern Baja California, Michoacán, Mexico, Nuevo León, Tamaulipas, southern Texas (Brownsville, Cove), Louisiana, and southern Mississippi.

Casual records.—Casual in Alaska (Craig), northern British Columbia (Ispatseeza River), Mackenzie (30 miles below Fort Simpson), northern Alberta (Fort Chipewyan), and Kentucky (Louisville, Bowling Green). Accidental in northern Ontario (Moose Factory), New York (Rochester), and Georgia (St. Marys).

Migration.—Early dates of spring arrival are: Missouri—St. Louis and St. Joseph, March 21. Illinois—Port Byron, March 6. Indiana—Posey County, February 11; Newton County, March 31. Ohio—Salem, March 13. Michigan—Three Rivers, March 10. Iowa—Indianola, February 24. Wisconsin—Hammond, March 5; Superior, March 15. Minnesota—Red Wing, March 1 (average for southern Minnesota, March 12); Wilkins County, March 9 (average of 18 years in northern Minnesota, March 25). Texas—Dallas, February 12. Oklahoma—Skiatook, February 27. Kansas—Johnson County, February 23. Nebraska—Red Cloud, January 18 (average of 23 years, February 20). South Dakota—Yankton, February 20. North Dakota—Fargo, March 8 (average for Cass County, March 19). Manitoba—Rosser, March 4. Saskatchewan—Qu'Appelle, March 18. Colorado—Weldona, February 24. Utah—Ogden, February 18. Wyoming—Barnum, March 2 (average of 10 years, March 15); Yellowstone National Park, March 18. Idaho—Rupert, March 3. Montana—Kirby, February 20; Fortine, March 2; average of 18 years in Custer County, March 30. Alberta—Camrose, March 8. California—Tule Lake, February 26; Twentynine Palms, March 7. Nevada—Carson City, February 23. Oregon—Corvallis, February 28. Washington—Pullman, February 25; Richardson and Bellingham, February 27. British Columbia—Okanagan Landing, February 28.

Late dates of spring departure are: Sonora—Oposura, April 1. Baja California—Guadalupe Island, March 22. Alabama—Fort Morgan, March 19. Georgia—St. Marys, March 16. Mississippi—Bolivar County, April 26. Illinois—Port Byron, May 17. Texas—Atascosa County, April 15. Oklahoma—Oklahoma City, May 5. Kansas—Douglas County, May 7. California—Death Valley, April 29.

Early dates of fall arrival are: Washington—Blaine, September 3. Oregon—Prospect, September 28. Nevada—Charleston Mountains, September 11. Oklahoma—Norman and Oklahoma City, October 8. Texas—Atascosa County, October 8. Mississippi—Deer Island, October 13. Baja California—San José del Cabo, October 14. Sonora—Hermosillo, October 20.

Late dates of fall departure are: British Columbia—Okanagan Landing, November 30. Washington—Westport, December 5. Oregon—Weston, November 23. Nevada—Clark County, November 27. California—Twentynine Palms, November 27. Alberta—Morrin, November 16. Montana—Charlo, November 10. Idaho—Lewiston, November 2. Wyoming—Sundance, November 10; Laramie, November 7 (average of 9 years, October 29). Utah—Ogden, November 18. Colorado—Yuma, November 18. Saskatchewan—Eastend, November 15; Indian Head, November 14. Manitoba—Treesbank and Brandon, November 17. North Dakota—Stutsman County, November 27; Cass County, November 24 (average, October 26). South Dakota—Sioux Falls, December 3 (average of 7 years, October 30). Nebraska—Lincoln, November 30. Kansas—Douglas County, November 12. Texas—Denison, November 30. Minnesota—Hutchinson, November 19 (average for southern Minnesota,

October 18); Sherburne County, November 1 (average of 10 years in northern Minnesota, October 17). Wisconsin—Dunn County, November 19. Iowa—Newton and Emmetsburg, November 17. Michigan—McMillan, October 21. Illinois—Port Byron, October 17. Arkansas—Hot Springs National Park, November 13.

Egg dates.—Alberta: 2 records, June 15 and June 20.

Arizona: 2 records, April 22 and April 30.

California: 100 records, February 11 to July 2; 50 records, Apr. 20 to May 30.

North Dakota: 20 records, May 2 to June 10; 12 records, June 2 to June 6.

Utah: 6 records, April 20 to May 25; 3 records, May 7 to May 17 (Harris).

STURNELLA NEGLECTA CONFLUENTA Rathbun

Pacific Western Meadowlark

HABITS

The Pacific western meadowlark was described and named by S. F. Rathbun (1917) from a specimen taken at Seattle, Wash. He gives as its characters: "Similar to *Sturnella neglecta neglecta*, but the bars on tail and tertials broader and much more confluent; upper parts darker throughout, and their black areas more extensive; yellow of under parts averaging darker; spots and streaks on the sides of breast, body, and flanks larger and more conspicuous." Its range is the Pacific coast region of southwestern British Columbia and northwestern Washington, south to northwestern Oregon and east to the Cascade Mountains.

I can find nothing recorded on its habits to indicate that they are in any way different from those of the interior race.

In Washington it is both a migrant and a summer resident, also, especially in southwestern Washington, it is an irregular permanent resident. The breeding season near Seattle and Tacoma extends from April 21 to June 5.

DISTRIBUTION

Range.—British Columbia to Oregon, west of the Cascade Mountains.

Breeding range.—The Pacific western meadowlark breeds from southwestern and central British Columbia south through Washington, western Idaho (Payette), and Oregon to southern California, intergrading with the western meadowlark in central Idaho, Death Valley, and San Diego County, California.

Winter range.—Winters from Vancouver Island and the adjacent mainland southward, casually north to southern British Columbia. Migrant, in part, in the northeastern portion of its range.

XANTHOCEPHALUS XANTHOCEPHALUS (Bonaparte)

Yellow-headed Blackbird

Plates 6, 7, and 8

HABITS

Many years ago I wrote (Bent, 1903) of my first impressions of the showy yellow-headed blackbird in North Dakota:

Seated in a comfortable buckboard, with two congenial companions, and drawn by a lively pair of unshod bronchos, we had driven for many a mile across the wild, rolling wastes of the boundless prairies, with nothing to guide us but the narrow wagon ruts which marked the section lines and served as the only highways. It was a bright, warm day in June, and way off on the horizon we could see spread out before us what appeared to be a great, marshy lake; it seemed to fade still farther away as we drove on, and our guide explained to us that it was only a mirage, which is of common occurrence there, and that we should not see the slough we were heading for until we were right upon it.

We came at last to a depression in the prairie, marked by a steep embankment, and there, ten feet below the level of the prairie, lay the great slough spread out before us. Flocks of Ducks, Mallards, Pintails, and Shovellers, rose from the surface when we appeared, and in the open water in the center of the slough, we could, with the aid of a glass, identify Redheads, Canvasbacks and Ruddy Ducks, swimming about in scattered flocks, the white backs of the Canvasbacks glistening in the sunlight, and the sprightly upturned tails of the Ruddies serving to mark them well. A cloud of Blackbirds, Yellowheads and Redwings, arose from the reedy edges of the slough, hundreds of Coots were scurrying in and out among the reeds, a few Ring-billed Gulls and a lot of Black Terns were hovering overhead, and around the shores were numerous Killdeers, Wilson's Phalaropes and other shore birds. The scene was full of life and animation. * * *

But by far the most abundant birds in the slough were the Yellow-headed Blackbirds, the characteristic bird of every North Dakota slough; they fairly swarmed everywhere, and the constant din of their voices became almost tiresome. The old male birds are strikingly handsome with their bright yellow heads and jet black plumage, offset by the pure white patches in their wings, the duller colors of the females and young males making a pleasing variety. * * * The song most constantly heard, suggests the syllables *Oka wée wee*, the first a guttural croak, and the last two notes loud, clear whistles, falling off in tone and pitch, the whole song being given with a decided emphasis and swing.

Although it was some 50 years ago that I heard it, the rhythmic swing of that impressive chorus still seems to ring in my ears whenever I think of a North Dakota slough and its yellow-headed blackbirds.

Throughout its wide range in western North America, from Canada to México and from the eastern border of the prairie regions to the Pacific slope, small or very large colonies of yellow-headed blackbirds may be found wherever there are lakes bordered with suitable aquatic vegetation, or marshes or sloughs with permanent water of sufficient depth. Damp marshes are not suitable for them, neither are the shallow-water sloughs; they prefer to nest over water that is from two to four feet deep, or even much deeper.

Deep water serves to protect the nests and young from prowling predators, and a thick growth of tall vegetation, tules, reeds (*Scirpus* or *Phragmites*), or cattails (*Typhus*), serves to shield them from birds of prey.

In the Rocky Mountain region, the breeding range extends to somewhat higher levels. Fred M. Packard (1946) says that, in Colorado, these birds "nest commonly from the plains to about 5,500 feet in the foothills, rarely as high as 6,000 feet." There is some evidence that the bird is extending its range somewhat farther east than formerly. And Gordon W. Gullion writes to me (in 1948) that this blackbird is becoming widely distributed as a breeding bird in the Willamette Valley, in western Oregon.

Spring.—The yellow-headed blackbird winters as far north as some of the Southwestern States, not far north of the southern limits of its breeding range. The northward movement starts about the middle of March, continues through April and reaches the breeding grounds before the middle of May. Thomas S. Roberts (1932) writes:

In the northward spring movement the vanguard of the Yellowheads that are to breed in Minnesota arrives during the first half of April, the males preceding the females by a few days. Stragglers may enter the southern part of the state during the very first days of that month, but it is not until toward the last of April or early in May that they become numerous. While the females are busy building their nests in the sloughs, the males assemble in little parties and feed on the adjoining uplands. Should they select a grassy plot where dandelions are in full bloom, the bright yellow of the blossoms and the heads of the birds match so well that they are almost indistinguishable.

Arthur C. Twomey (1942) noticed the first migrants in Utah on May 2, when "from forty to two hundred males could be seen flying in compact flocks, but no females were in evidence. It was not until May 15 that females were noticed, and they likewise were in segregated flocks. * * * By May 20 there were females among the flocks of males, and soon after this the nesting season commenced."

At a colony studied by George A. Ammann, in northwestern Iowa, the adult males were first seen on April 8 and were numerous on April 23; the adult females came on May 2, but were not common until May 12; the first-year males arrived on May 11, and were numerous on May 22, the young females coming about the same time. These dates are taken from a manuscript copy of his thesis (submitted to the University of Michigan), which he has very kindly loaned me. I shall quote freely from parts of this excellent and extensive monograph on the yellow-headed blackbird.

Courtship.—While exploring in a canoe, on May 31, 1913, the extensive marshes surrounding Lake Winnipegosis, we found the yellow-heads fairly swarming in the tall bulrushes (*Scirpus*), growing in water 3 or 4 feet deep and extending higher than a man's head above the water along both sides of the Waterhen River. Courtship was in full swing. The males were chasing the females all over the marshes; the female usually returned to the place from which she started, after which the male alighted near her, as this was probably the chosen territory for the pair. Grasping a tall, upright cane, or perhaps two in a straddling attitude, he displayed his fine plumage by spreading his black tail and half opening his wings to show the white patches; he leaned forward, pointing his bright yellow head downward until it was almost parallel with his tail and poured out his grotesque love notes. The female seemed indifferent.

We must admit that the courtship is more spectacular than beautiful, but we should hardly condemn it in the following words of W. L. Dawson (1923): "Grasping a reed firmly in both fists, he leans forward, and, after premonitory gulps and gasps, he succeeds in pressing out a wail of despairing agony which would do credit to a dying catamount. When you have recovered from the first shock, you strain the eyes in astonishment that a mere bird, and a bird in love at that, should give rise to such a cataclysmic sound."

Alexander Wetmore (1920) gives the following account of the courtship of the yellow-headed blackbird, as observed at Lake Burford, N. Mex.:

The adult males were settled in large part on their breeding grounds on my arrival, though many of them were not yet mated. Each selected a stand in the tules at the border of the lake, and, unless away feeding, were certain to be found in the immediate vicinity constantly from that time on. * * * At this season the male seems fully conscious of his handsome coloring and in his displays makes every effort to attract attention. In the most common display the male started towards the female from a distance of 30 or 40 feet with a loud rattling of his wings as a preliminary. The head was bent down, the feet lowered and the tail dropped while he flew slowly toward his mate. The wings were brought down with a slow swinging motion and were not closed at all so that the white markings on the coverts were fully displayed, the whole performance being reminiscent of a similar wing display of the Mocking-bird. In flying from one perch to another males often dangled their feet, frequently breaking through small clumps of dead tules with considerable racket. Or they clambered stiffly along, hobbling over masses of bent-over rushes, with heads bent down, tails dropping and back humped, appearing like veritable clowns.

Jean M. Linsdale (1938) noticed a form of display which was apparently made in defense of territory. Two males which owned adjoining territories "were seen on the ground halfway between their respective singing posts which were in separate cattail patches about 20 feet apart. For 3 or 4 minutes they kept close to each other, walking back and forth along the boundary with fluffed feathers and arched necks. In turn, they made short flights, getting scarcely more than a foot above the ground and moving, altogether, only 3 or 4 feet. Once one went as far as 10 feet. In these flights the wings were flapped violently, but the bird moved slowly, and the body was held with the bill pointing upward 80° above the horizontal. Finally, each bird returned to its own singing post, having had no actual combat."

On another occasion he noticed severe fighting for about 30 seconds, one holding the other down and pecking at it. Referring to the territorial behavior, he says in part:

Judging from continued watching at this pond through the greater parts of 2 nesting seasons, territory for these yellow-headed blackbirds was a definitely recognized area for males only. Moreover, this area was a remarkably small one, when the size of the birds is considered. Each male established itself in 1 small patch of cattails or a portion of a patch. * * *

From the first establishment of the territories one of the chief concerns of each male was to keep other male yellow-headed blackbirds off his area. The enmity seemed aroused in inverse proportion to familiarity with the trespassing individual. When 2 males owned portions of the same cattail patch, they were much more tolerant of each other than of males from another part of the pond. Newly arrived, strange males arouse a quicker response than ones already settled in the same pond. * * *

Besides their vigorous defense against intrusion by other male yellow-headed blackbirds the males were especially active in driving red-winged blackbirds from their territories. The pursuit, however, was usually a short one. In observed instances male red-wings were pursued for only about 30 feet, or just to the limits of the yellow-head territory. If, in leaving, a red-wing crossed the territory of a second yellow-head, the latter would take up the chase and the first yellow-head would turn back.

Of the actual mating, he says:

Males were noticed flying to females especially when the latter uttered a certain type of screeching note. Sometimes these notes were given on the pursuit flight. The notes along with the posturing of the female seemed to be the signal that the female was near the mating stage.

Mating of yellow-headed blackbirds was noticed in late afternoon on May 26, 1932. In territory III a female flew to the top of a currant bush where it postured, and then male III flew there from the cattail patch and they copulated. The union lasted about 2 seconds during which the male flapped his wings rapidly. After perching a few inches away for a short time, the whole procedure was repeated until it had taken place 9 times in quick succession. Then the male flew back to the cattails where it perched on a dead stem and shook its plumage. The female may have been the same one noted earlier in territory II. Posturing, with bill and tail pointed upward, had been noticed there, but male II had made no response. The circumstances seemed to indicate that male III was the only one ready for mating, and the female hunted it out.

The behavior described above suggests promiscuity between the sexes. He noted that there were many more females than males engaged in nesting. No females were noticed that were not nesting, but many males less brightly colored and with less perfect songs were seen day after day half a mile or more away from the breeding colonies. He inferred from this that females mature and are ready to breed when 1 year old, but that the males require 2 years to mature, and that the less brightly colored males, seen away from the breeding colonies, were yearlings and would not breed until the following year.

Ammann (MS.) says: "these young males, with their distinctive plumage, were not welcome when they invaded the breeding grounds. Wherever they went they were immediately driven away by the adult males and thus became nomads by necessity. Breeding females did not resent their intrusion, however; once a female was seen to take a receptive position in front of a first-year male (thus evidently recognizing him as a male, in this plumage so similar to her own) to which he did not respond."

He gives a full account of the actual mating, as follows: "The female stops in the midst of nest building and selects a more or less solid stand low in the bulrush clump or on a mass of floating debris and assumes the mating posture, at the same time giving the low, soft mating call. If the male is anywhere in the vicinity, he responds immediately; it seems almost incredible sometimes how far distant he may be and yet hear this call.

"He proceeds toward the female in one or more short, jerky flights— thus causing the wings to beat very loudly, with bill pointing almost straight up. Then he draws in his head, erects the feathers of breast and back, droops his tail and approaches the female indirectly by short hops through the rushes or over floating debris, sometimes completing a half circle before reaching her. Then he may strut, twist, or turn in a foolish manner and rarely give vent to the buzzing song before mounting. Meanwhile, the female remains in the mating posture—body tilted slightly forward, tail spread and pointing straight up, bill raised high in the air. As the male comes closer, she watches him attentively with open bill and alternately quivering wings, and may repeat the mating call. She turns her head as he walks around her, or he may stop directly in front of her and both remain motionless, except for the quivering of her wings, for 15 or 20 seconds. Then the male mounts the female, placing first one foot on her back, then the other; at the same time he flaps his wings vigorously high above him and brings the bill close to his breast so that the neck is quite arched; his tail is pressed down between the two central tail feathers of the female, allowing the cloacae to come in contact. While the male is on her, probably a great deal of the weight is supported by the material on which her breast is resting. The male does not maintain his balance in this position for more than a second, and probably the first attempt at copulation has been unsuccessful. He jumps off, takes two or three short hops, and mounts the female again. This may be repeated five or six times until finally it appears that a successful copulation has taken place, because of the slightly longer time (about 1½ seconds) the male remains on the back of the female. As many as 16 consecutive attempts at copulation have been counted— all in rhythmical succession. The female remains in the same position during the whole performance."

Nesting.—Yellow-headed blackbirds nest in colonies, often of very large size. The colonies are not as densely packed with nests as are those of the tricolored redwings in parts of California, though in the most thickly populated colonies as many as 25 or 30 nests may be found in a space 15 feet square. The colonies are not always continous, and may be scattered in separate groups along the shores of a lake or slough where the vegetation is most suitable for nest construction. Red-winged blackbirds are usually more or less loosely associated with the yellow-heads on their breeding grounds, but generally the two species occupy different portions of the marsh. The nests of the yellow-heads are invariably built over water, preferably from 2 to 4 feet deep and rarely much deeper. Should the water recede during the process of nest building, unfinished nests found to be over dry land are likely to be abandoned.

The nest is built entirely by the female, without any help from the male. In his study of a nesting colony in Minnesota, Roberts (1909) gives the following good description of the construction of the nest:

The body of the nest was invariably constructed of water soaked dead grass blades picked out of the water of the marsh. This sort of material being soft and pliable was easily woven and wound around the reed stems to the smooth surface of which it closely adhered; and when the structure, which was at first very wet, soggy and dark colored, dried in the sun and wind, it contracted and drew the included reed stems nearer together thus forming a compact, firm, and securely attached basket-like nest. The lining consisted of pieces of broad, dry, reed leaves and the rim of the nest was well finished off with fine branches of the plume-like fruiting tops of the reeds. Occasionally the lining was not placed for a day or two until the nest had dried somewhat, but usually the coarse lining was added, in part at least, to the bottom and around the walls while the body of the nest was still in course of construction and soft and wet. The finishing touches to the nest consisted in adding the fine material about the upper walls and rim which, in the more perfect nests, partially closed and formed a sort of canopy over the entrance.

These nests were all built in quill-reeds (*Phragmites*), and were placed from 2 to 3 feet above the water. Of the 62 nests started in the colony, 28 were abandoned before completion, "due to faulty workmanship or poor judgment in selecting a site. * * * In one instance it was positively determined that the same bird built four imperfect nests before being able to construct one that was habitable. * * * A skillful, industrious bird would build one of these large beautifully woven and lined nests, all complete, in from two to four days. Of twenty well built nests, nine were finished in two days, nine in three days, and two in four days. * * * From one to five days was allowed to elapse after the completion of the nest before egg-laying began. Eggs were invariably deposited one each day."

In the North Dakota slough, referred to at the beginning of this story, red-winged blackbirds were nesting commonly around the edges of the marsh in the shorter vegetation growing in the shallow water, but all through the deeper parts of the slough, in the tall reeds (*Scirpus*) and flags (*Typhus*), the yellow-headed blackbirds fairly swarmed, with nests often close together.

The nests were firmly attached to the reeds or flags at height ranging from 6 inches to 3 feet above the water of varying depths. Four of these nests are now before me. They were evidently built, after the manner described above by Roberts, of wet, dead material picked up from the water, which dried and shrunk enough to hold the nest firmly to its support. This material consists of strips of dead leaves of flags, coarse grasses, items of dead reeds, roots of water plants, and general swamp rubbish. My nests are not decorated around the rim with the fruiting tops of the quill-reeds for the simple reason that there were no *Phragmites* growing in the vicinity. All the nests that I saw were neatly and smoothly lined with narrow strips of dry grass blades of a dull orange color, evidently carefully selected and probably brought from dry land; these formed a very distinct feature in all the nests.

The nests are all bulky and very firmly woven; all but one of them were somewhat crushed in packing, but one that is apparently in its original shape measures 5 by 6 inches in outside diameter, fully 4 inches in depth, and the inner cup is about 3 inches in diameter and 2½ inches deep. A nest figured by Roberts (1909) measured 11 inches from the rim of the nest to the long extension between the reeds below it; it was also partially canopied at the top.

In southwestern Saskatchewan, where Bear Creek enters Crane Lake, that wonderful bird paradise more fully described in my account of the western grebe (Bent, 1919), we found yellow-headed blackbirds' nests in abundance. The nests were firmly attached to the tall, waving bulrushes, from 10 to 30 inches above the water, which was in many places more than waist deep. They were much like those described above, but instead of the distinctive lining seen in the North Dakota nests they were lined with fine strips of dead flags or with fine grasses, and they were not decorated like those described by Roberts.

We noticed that many nests were abandoned because of unfortunate location in growing tules; the nests had been attached to several stalks which had grown unevenly, overturning the nests and rendering them useless.

In Nevada, Linsdale (1938) found these blackbirds nesting in willows. "In the early summer of 1932 water from streams in the Toyabe Mountains flooded parts of Smoky Valley. Within the flooded area was a patch of willows 5 to 7 feet high and approximately 100 by 50 yards in diameter. At this place the water was 1 to 1½ feet in depth. Yellow-headed blackbirds took over the willow patch and nested there." On June 3, he counted 30 nests there, all but 3 of which contained eggs or young.

Ammann (MS.) gives a very full account of the process of building the nest: "Once a nest is begun the female works feverishly, picking up long wet strands from the surface of the water and bringing several at a time, in her bill, to the nest site. These are suspended between conveniently arranged stems of vegetation several inches apart—sometimes as much as six inches. They are probably wound around the supporting stems singly or a few at a time and the loose ends attached to other supports.

"Soon a number of nearby stems are connected by a loose network of these coarse, wet fibers. At first 4 or 5 supporting stems are used but as the structure grows, more are included—sometimes as many as 25 or 30—if the nest is built in bulrushes or quill-reeds. * * * This frail network is reenforced by more fibers until a strong saucer-shaped base with a rather angular outline is formed. * * * As soon as this structure can support her body, the female begins adding material around the margin for the outer wall, the next stage in construction. It is that part of the nest which envelops the supports and forms its main bulk. After gathering suitable material in her bill from the surface of the water, the female flies straight to the edge of the nest, jumps into the cavity, drops her load on the edge, and immediately begins to arrange it. With quick, deft movements of her head, she snatches individual strands and winds them around the nest supports that have already been included in the construction of the base. Usually the strands are given a half twist around each support as follows: an end is pushed beyond the rim adjacent to the support, then the female reaches around and snatches this end from the other side, pulls it down and anchors it with a thrust of the bill to the inside of the nest. The other free end may likewise be anchored.

"Often the strands are given a complete turn around each of several nest supports in a row, or may be placed along the rim and woven in and out among the upright supports. Of course, there is great variation in placing each individual shred, but the resultant meshwork of fibers forms a wonderfully strong and compact basket. When the female has disposed of all the loose material on the rim, she tugs at any loose ends in sight, especially on the outside of the nest. She reaches far over the edge and pulls such strands over the rim, if possible, and thrusts them into the inside of the wall."

For shaping the inside of the nest, she "supports herself by her head and tail on the rim and stamps her feet alternately in rapid succession on the bottom and sides of the cup. The nest is usually so wet that the stamping can be heard several meters away. After a few seconds she rises and settles down at a slightly different angle and duplicates the performance. This procedure may be repeated a number of times in quick succession. * * *

"After the outer wall is high enough the female adds material to the inner side of the wall and in the bottom in order to make the cavity the right size and shape. This may be called the inner cup. She does not loop the strands around the supports but drops each load directly in front of her as she enters the nest. Her breast is then applied to this newly brought material while she stamps her feet in the manner already described, thus making the nest compact, and the inside smooth, round, and of the correct diameter. The general appearance of the inner wall when finished is different from the outer wall. The direction of nearly every strand is in an arc, parallel to the circumference, and the brim is generally smooth and on a horizontal plane.

"A final stage of construction that is by no means universal is the addition of some fine, dry grasses which serve as a lining. When present they are usually confined to the wall and often only immediately below the rim on the inside, thus constricting the opening."

Ira N. Gabrielson (1914), writing of a Nebraska swamp, says: "The Yellow-headed Blackbirds were by far the most abundant breeding form of the swamp. In the part examined there were probably several hundred nests; in the remaining half of the swamp the number is only a matter of conjecture. The nests which we examined were practically identical in location, being built in the wild rice growing some distance from the shore. They were woven in basket shape about three or more stems from eighteen inches to two and one-half feet above the water. The water in the region of the nests was about hip deep and they seemed to be confined to a belt of this depth around the part of the swamp studied."

Eggs.—The yellow-headed blackbird lays from three to five eggs to a full set, most commonly four, and only very rarely five. Of 504 nests examined by Ammann (MS.), only 8 contained 5 eggs, while 282 held 4, and 110 sets consisted of 3 eggs.

Major Bendire (1895) writes:

The eggs of the Yellow-headed Blackbird vary in shape from ovate to elliptical and elongate ovate; the shell is finely granulated, strong, and rather glossy. The ground color varies from grayish white to pale greenish white, and this is profusely and pretty evenly blotched and speckled over the entire surface with different shades of browns, cinnamon rufous, ecru drab, and pearl gray. The markings are usually heaviest about the larger end of the egg, and sometimes a specimen is met with which shows a few fine, hair-like tracings, like those found on the eggs of the Orioles.

The average measurement of 134 eggs in the United States National Museum collection is 25.83 by 17.92 millimetres, or about 1.02 by 0.71 inches. The largest egg in the series measures 28.96 by 19.81 millimetres, or 1.14 by 0.78 inches; the smallest, 23.11 by 17.53 millimetres, or 0.91 by 0.69 inch.

Incubation.—Incubation is performed entirely by the female with no help from the male, except that he sometimes feeds her on the nest. The period of incubation has been reported by different observers within rather wide limits.

Roberts (1909) says that in "seventeen nests the period of incubation, inclusive of the day on which the last egg was laid, to the day on which the first egg hatched, was nine days in one instance, ten days in twelve, eleven days in three, and twelve days in one. Thus ten days may be considered the usual period of incubation. The nine-day period was in the case of the only set of five eggs that hatched." The eggs hatched irregularly, though in three nests all hatched on the same day, and in three others one hatched each day.

Reed W. Fautin (1941b), who made some very extensive studies of the nesting of the yellow-headed blackbird in Utah, writes:

The females were not assisted by the males in any way in the incubation of the eggs, 56.6 percent of them beginning incubation at the time the second egg was laid, with a tendency for the beginning of incubation to be delayed longer the larger the clutch. The length of the incubation period varied from 12 to 13 days, 74.6 percent of the eggs hatching in 12 days.

The attentive periods during incubation ranged in length from 1 to 41 minutes, with an average of 9.1 minutes. These periods were longest during mid-day when the females were seemingly protecting the eggs from the sun. During 83 hours of observation the females spent an average of 63.9 percent of their time on the nest, with a range from 53.1 to 69 percent.

The inattentive periods ranged in length from 1 to 18 minutes, with an average of 5.4 minutes. These periods tended to be longest during the morning and evening hours when feeding was most intensive.

The hatching success of the larger Provo River colony amounted to 75.7 percent, while that of the smaller Lakeview colony was only 60.6 percent, giving an average of 70.9 percent for the two. Wind and predation were responsible for the destruction of 90 (20.3 percent) of the eggs before the time of hatching, and 39 (8.8 percent) failed to hatch because of being addled or infertile.

Eighty-three females nested in the Provo River colony and 40 in the Lakeview colony. There were about 35 males in the former colony and only 12 in the latter, suggesting promiscuity or polygamy. There were no yearling males in either colony, but plenty of them were seen in the surrounding regions.

Young.—In another excellent paper, Fautin deals with the development of young yellow-headed blackbirds. Both studies were conducted in the same two colonies, near Provo, Utah, during the spring and summer of 1937, from April to September, some 128 nests being kept under observation. He (1914a) found that: "The average weight of the nestlings at the time of hatching was 3.3 grams and at 10 days of age was 51 grams; the greatest percentage of increase in weight occurred during the first day after hatching, while the greatest actual increase in body weight occurred between the fifth and sixth days, amounting to 6 grams at that time." He noticed that nestlings of the same age varied as much as 15 to 20 grams in weight at the time of leaving the nest, though the smaller ones were as well feathered and as active as the larger ones; inasmuch as adult males are much larger than females, averaging about 35 grams heavier, it is likely that the larger nestlings were males.

Feather development began soon after hatching; the sheaths of the primaries appeared the second day. At eight to nine days of age the contour feathers were sufficiently developed to cover all the apteria except possibly the one on the abdomen.

The males aid very little in caring for the nestlings. Only two males were observed to make any attempt to feed the young. One of these fed the young eight times during a period of eight hours and six minutes while the female fed them 102 times during the same interval. The other male fed another brood of nestlings eight times while the female fed them 92 times during the same period. * * *

Food of the nestlings consisted principally of insects and spiders. The spiders and smaller insects constituted the greater part of the diet during the first few days after hatching, while larger insects such as dragonflies and grasshoppers together with some vegetable matter formed the bulk of the food as the young became older. * * * For the first day or two after the young are hatched they are fed either by regurgitation or else on food materials so small that they escaped notice, for during that time the females were seldom seen carrying food in their mouths although the young were visited six to seven times per hour. Probably they were fed by regurgitation during that time. * * *

The nestlings left the nests when nine to twelve days of age and remained among the dense vegetation of the nesting area until they were able to fly. * * * The young are unable to fly at the time they leave the nest but they are very adept at making their way through the vegetation. After abandoning the nest they never return to it but are to be found among the vegetation down near the surface of the water, sometimes sitting on the dead floating vegetation. * * * For the first four or five days they move about by hopping from one stem or leaf to another with remarkable agility. Following this hopping stage they make short flights of about two to four feet and thus gradually develop their ability to fly. By the time they are three weeks old they are frequently seen to make short flights of about 25 yards. From this stage on, their ability to fly develops very rapidly and they are soon seen pursuing their parents, coaxing noisily for food."

Mortality among the nestlings was very high, due largely to a heavy rainstorm accompanied by high wind which destroyed the nests, eggs, and young. Many eggs and young birds were devoured by predators, largely unknown but probably snakes, small mammals, and perhaps birds of prey or crows. "Out of 314 nestlings, hatched from 443 eggs, 215 were destroyed before they were old enough to leave the nest. This gives a percentage of success (i.e., young fledged from the total number of eggs laid) of 22.4." In the Minnesota colony studied by Roberts (1909), all the young disappeared, 100 percent loss, from some unknown cause.

Fautin (1941a) says that: "A partial post-juvenal molt occurred about the last of July when the plumage of the fledgling was changed to that typical of the first-year birds. During this time the birds left the nesting areas and remained in seclusion in the dense cattail marshes. After most birds had completed their autumn molt they wandered about the fields in large flocks during the day, and returned to the marshes at night."

Gabrielson (1914) made two interesting observations:

The method by which the young left the nest was interesting. At 5:38 a. m. one of the young clambered to the edge of the nest, seized one of the supporting reeds with each foot and climbed up them a short distance above the nest, advancing each foot alternately. After going about eighteen inches, the bending of the stalks under his weight brought them in contact with others onto which he went. After travelling in the tops for a little way, he commenced to work toward the water, and reaching a broken reed rested a while. In a few moments he proceeded along this reed to another and was soon out of sight. * * * I had one glimpse of some of the dangers to which the young Yellowheads are exposed. One of the young from a neighboring nest was sitting on a reed about two inches above the water when the jaws of a hungry pickerel rose from the water and the nestling disappeared. It was done so quickly that if I had not been looking directly at the bird it would never have attracted my attention.

Roberts (1909) says of the food of the nestlings: "Grasshoppers, various insects and a large black larva of some sort which the birds obtained from among the decayed vegetation in the shallow water along the edges of the slough formed the chief food supply. These larvae were ugly and formidable objects and were thrust down the throats of the young birds with considerable difficulty. On one occasion a female was seen carrying a large flat object, squirming and curling about her bill, which was evidently a leech."

Mrs. Wheelock (1905) writes: "The young are fed by regurgitation for two days, afterwards by both methods for two days, then entirely by fresh food. Examination of the crops of the broods reared in late June showed, on the first day, snails, waterslugs and larvae all partially digested. On the second day, insects denuded of wings, legs, and all hard parts, and thoroughly crushed as well as predigested, were found mixed with occasional water moss. The third day showed little change in the menu, but the food was less digested and, on this day, occasional meals of fresh food began to supplant the regurgitated."

The nesting success in the 504 nests studied by Ammann (MS.) was not so good as that reported by Fautin (1941b). The 504 Iowa nests contained 1,565 eggs, an average of 3.1 eggs per nest. "Of the 173 successful nests, 40 were completely and 133 partially successful, an average of 2.5 young were raised. Eggs hatched in 44.2 percent of the nests and young were fledged in 34.3 percent. Of all the eggs laid, 53.6 percent hatched and 27.5 percent became successfully fledged young. In comparing the nesting success of this species with others it is found to be much lower in every respect. The percentage of eggs hatched and young fledged of 481 nests of six other species of passerine birds is 61.4 and 43.0, respectively."

Fred G. Evenden, Jr., writes to me that he found a yellow-headed blackbird's nest in a swamp near Corvallis, Oreg., that had been thoroughly torn up by northwestern redwings that nested in abundance in the swamp, and says that "the yellow-heads were not tolerated by the redwings, being chased and attacked whenever they were in the swamp area."

Plumages.—The small nestlings are only thinly covered with buffy down on the feather tracts of the head and back, but the first plumage soon begins to appear, pushing the down out on the tips of the feathers, where it persists longest on the top of the head. Chapman (1921a) gives the best description of the juvenal plumage of the yellow-headed blackbird as follows: "The whole head and breast are warm buff, giving the effect of a brown-headed bird; the abdominal region whitish; the back blackish, both more or less fringed with buff; the tail and wings black, the wing-coverts tipped with white. At the post-juvenal molt the tail and wing-quills are retained, while the rest of the plumage is exchanged for a costume which resembles that of the female, but is usually without streaks on the breast, or if streaks are present, they are yellow." I think this description must refer to a young male, for the female has no white in the wings.

Fautin (1941a) says: "The first-winter plumage of the young is acquired by a partial post-juvenal molt as a result of which the buffy feathers of the head, neck, and breast regions of the fledglings are replaced in the males by yellowish feathers tipped with brownish on the sides of the head, throat, and breast, with a collar sometimes extending around the back of the neck. The feathers of the back nape, crown and wings are a deep brown while those of the under parts and especially those of the belly and crural regions are somewhat paler around the edges. The autumn plumage acquired by the juvenal females is much the same as that of the adult females."

Ridgway (1902) describes the immature male in first-winter plumage as "similar to the winter female, but larger; general color darker (nearly black on pileum, auriculars, and orbital region); superciliary stripe deeper ocher yellow; malar region, chin, and throat chrome yellow, and chest dull cadmium yellow or orange-ochraceous; no white streaks on breast; primary coverts narrowly tipped with white."

This plumage is worn without much change until the first post-nuptial molt the following summer. Apparently, young males do not breed in this plumage.

Young females in first winter plumage are much like the adults, but colors and more veiled; the breasts are streaked with dull whitish; they evidently breed the following spring, when less than a year old.

The prenuptial molt of adults and young, is apparently very limited, confined mainly to the region of the head and neck, the nuptial plumage being produced chiefly by the wearing away of the dusky tips of the autumn plumage. A complete molt occurs in late summer, at which the fully adult plumages are acquired. In the adult male the bright yellow, or orange, of the head and neck is obscured, sometimes nearly concealed, by dusky tips; and in the adult female the colors are duller, less distinct, and the white streaks on the breast are less clear.

The adult male in his nuptial plumage is a handsome bird; Ridgway (1902) describes a high-plumaged male as having "head, neck, and chest yellow or orange (varying from canary yellow to almost cadmium orange, rarely to saturn red); lores, orbital region, anterior portion of malar region, and chin black; rest of plumage uniform black, relieved by a white patch on the wing, involving the primary coverts (except their tips and shafts) and portions of the outermost greater coverts; anal region yellow or orange."

Food.—Beal (1900) analyzed the contents of 138 stomachs of the Yellow-headed blackbird:

As indicated by the contents of these stomachs, the food for the seven months [April to October, inclusive] consists of 33.7 percent of animal (insect) matter and 66.3 percent of vegetable matter. The animal food is composed chiefly of beetles, caterpillars, and grasshoppers, with a few of other orders, while the vegetable food is made up almost entirely of grain and seeds of useless plants. Predaceous beetles (Carabidae) constitute 2.8 percent of the season's food, . . . other beetles a little more than 5 percent . . .

Caterpillars constitute 4.6 percent, but nearly two-thirds of them are taken in July, and in that month they form 21.5 percent of the month's food. Remains of the army worm (*Leucania unipuncta*) were identified in 6 stomachs.

Grasshoppers are eaten to the extent of 11.6 percent for the season, but mainly after August. "The remainder of the animal food, 9.7

percent, is made up of other insects, chiefly Hymenoptera (ants, wasps, etc.), with a few dragon-flies and an occasional spider and snail."

Of the vegetable food, grain collectively amounts to 38.9 percent, more than half of the vegetable food and more than one-third of all the food.

Of grain, oats hold first place, as in the food of the redwing, and are probably eaten in every month when they can be obtained, although none were found in the 5 stomachs taken in September. The 3 October stomachs contained an average of 63 percent, but a greater number of stomachs would in all probability give a smaller average. August, apparently the next month of importance, shows 43.2 percent. Next to oats corn is the favorite grain, and was eaten to the extent of 9.8 percent, nearly all in the months of April, May and June, with a maximum of 48.8 percent in April, when no wheat was eaten. Wheat appears from May to August, inclusive, and is the only vegetable food that reaches its highest mark in August. The average for the season is 3.5 percent.

Beal (1900) found weed seeds to be an important item in the food: "Beginning with 18 percent in April, it increases to 34 percent in June, drops to 6.6 in July (to make room for caterpillars and grasshoppers), rises to 36.1 percent in August and finally to 64.4 percent in September. * * * The weeds found in the stomachs are almost precisely the same as those eaten by the redwings, and in practically the same proportions. Barngrass (*Choetochloa*), *Panicum*, and ragweed (*Ambrosia*) are the leading kinds, supplemented by *Polygonum*, *Rumex*, and others."

The yellow-headed blackbird is mentioned by La Rivers (1941) as one of the birds seen eating the Mormon cricket. Kalmbach (1914) records it as feeding on the alfalfa weevil. "Of 21 stomachs collected in June, only 4 failed to contain the weevil. The insect formed 43.48 percent of the yellow-head's food and was taken at an average of more than 6 adults and 47 larvae per bird. The largest number taken by any of this species was 190 larvae and 2 adults. Another record was 160 larvae and 2 adults. Three adults and 117 larvae were eaten by one bird, while five others had taken more than 170 individuals apiece."

Linsdale (1938) says of the feeding habits in the marsh: "Forage places varied, but nearly all the marshy parts of the pond were explored for food. Both males and females spent much time feeding close to the water among the plants (cattails, sedges, *Hippuris*). A favorite food hunting place was the mud or shallow water close to the shore line. As soon as the air warmed sufficiently for flying insects, the blackbirds spent much time capturing the insects in the air. Females flew into the air after insects as often as or more often than did males."

In the spring, these and other blackbirds are often seen following the farmer as he plows his fields, to pick up the grubs and insects turned up by the plow. W. J. McLaughlin of Centralia, Kans., writes (Am. Naturalist, vol. 3, p. 493): "During their stay they make themselves very valuable to the farmers by destroying the swarms of young grasshoppers. On the writer's land the grasshoppers had deposited their eggs by the million. As they began to hatch, the yellow-heads found them out, and a flock of about two hundred attended about two acres each day, roving over the entire lot as wild pigeons feed, the rear ones flying to the front as the insects were devoured."

Economic status.—The foregoing remarks on food throw considerable light on the economic status of this bird, for although the yellow-headed blackbird destroys a few useful predaceous beetles and shows a fondness for dragonflies that help destroy other annoying insects, to its credit is the fact that the bulk of its insect food consists of injurious species. It does, however, along with other blackbirds, cause considerable damage to the grain crops, pulling up the seedlings to eat the kernels, feeding on ripening grain, attacking grain in shocks, and injuring corn on the ear while it is in milk. But, as the records show that the various grains were eaten throughout most of the spring and summer, much of this must have been waste grain of no economic importance. On the whole, the bird is probably more beneficial than harmful, except in a few places where it is sufficiently numerous to cause appreciable damage to crops.

Behavior.—DuBois (MS.) noticed that at certain nests containing young, the parents chirped and hovered over their nests when approached, showing much more solicitude than the birds which had only eggs; the latter usually sat off at a little distance and looked on, without any demonstration whatever. Fautin (1941b) found the females very shy about their nests, leaving very silently as the nest was approached, but they never hesitated to drive away another bird from the immediate vicinity of the nest. "The emitting of an alarm call by one of the members of the colony would also cause them to leave their nests and fly to the assistance of the one that had sounded the alarm. Such cooperative behavior was witnessed on several occasions. On one occasion, when an American Bittern (*Botaurus lentiginosus*) visited the marsh, it was so severely attacked that it could not escape by flight and crawled down among the dead bulrush stems to avoid the onslaught until the confusion subsided and part of the Yellow-heads had retired from the scene of the conflict."

While I was watching a colony of these blackbirds breeding in a North Dakota slough, a marsh hawk which had a nest not far away happened to fly over the colony; whereupon the blackbirds, yellowheads, and redwings, arose in a cloud all over the marsh and flew about for a few minutes, cackling and squealing, until the hawk departed; this happened several times, whenever the hawk appeared. Others have noticed similar behavior. Linsdale (1938) saw a blackbird fly after a marsh hawk, "but the pursuit was spiritless." He noted that, when a prairie falcon circled overhead, they gave the alarm and "hurried to the cover offered by a bush." They also gave alarm and flew at a Swainson's hawk that flew over; they were disturbed by a nighthawk, but did not attack it; two crows were driven away. Wetmore (1920) saw them driven to shelter by a marsh hawk.

On the ground, the yellowheads walk sedately, seldom hopping, or run rapidly in pursuit of a moving insect. Of their flight, Linsdale says:

"The flight of the yellow-headed blackbirds contrasted markedly with that of the red-winged blackbirds. It was slow and deliberate and seemed to reflect the whole manner of the species. The dull whistle made by the wings could be heard distinctly for 50 yards or farther as the birds flapped heavily from one perch to another."

Wetmore (1920) comments on the perching ability of the birds as follows:

The feet of the Yellow-head are relatively very large with long, strong toes and the birds use them to advantage in walking about on floating aquatic vegetation or soft mud. In the rushes they prove themselves expert gymnasts. Often they alighted near the tips of the tall round-stemmed tules and as they swayed under their weight the birds supported themselves by their wings while they slid their feet quickly down to a new hold, trying several grips until finally they were low enough so that the rush supported them. This was done with great quickness as the birds shifted from grip to grip rapidly. At times instead of sliding down they reached out and grasped a second stem with one foot, dividing their weight between the two and standing suspended with the feet five inches or so apart.

On the subject of combativeness, Ammann (MS.) writes: "Judging from the behavior of nesting Yellow-heads toward humans, the male is more pugnacious and aggressive than the female. On several different occasions while I was banding four- or five-day-old young the male darted at my head and narrowly missed me. Once after I had picked up a fledgling the male flew at me quite forcefully, striking the side of my head with his bill. On another occasion I was in a blind and saw an adult male molesting a fledging. Much to my surprise another adult male immediately attacked the intruder and a short combat in mid-air ensued. Both feet and bills were brought into action. In a few seconds the assumed father of the fledgling got the better of the intruder and while holding him down, half submerged on some floating vegetation, pecked viciously at the back of his head.   The blows were delivered slowly, deliberately, and sharply with the aid of body and neck movements.   This lasted fully eight minutes.   The subdued male continuously uttered alarm calls, and whenever he turned his head around to offer resistance he pecked him about the eyes.   The one-sided battle ended when some other males were attracted to the scene; they, however, did not join in the combat.   The victim was slightly bloody about the nape, had lost a number of feathers, and I supposed that he was almost dead, but he got up, shook himself, and flew weakly away. * * * Females were never seen fighting among themselves, nor attacking men."

Voice.—My impressions of the striking song of the yellow-headed blackbird, as heard many years ago while my hearing was good, are mentioned in the beginning of this story; the *oka wee wee, oka wee wee, oka wee wee* notes were the dominant sounds in the slough, and I can seem to hear their rhythmic swing even now.   But I cannot find in print any rendering of the song that is quite like what I wrote in my notes at the time.   What Dawson (*in* Dawson and Bowles, 1909) calls the alarm cry "uttered with exceeding vehemence, *klookoloy, klookoloy, klook ooooo*," seems to have a similar rhythm and may be a variation of what I heard.   Then he adds: "*Ok-eh-ah-oh-oo* is a musical series of startling brilliancy, comparable in a degree to the yodelling of a street urchin, a succession of sounds of varying pitches, produced as tho by altering the oral capacity. * * * The last note is especially mellow and pleasing, recalling to some ears the liquid gurgle of the Bobolink."   Mrs. Bailey (1928) quotes from some manuscript notes from Merrill, of Mesilla Park: "While nothing can be more raucous than the note of a single individual, the united voices of a few hundred * * * produce an effect very pleasing, if not strictly harmonious."   These are all the words I can find of even faint praise of the song.

Everybody else condemns it as unmusical and unattractive.   Aretas A. Saunders tells me that, from his memory of it in Montana, "the form and length of the song is quite like that of the red-wing, there being several short notes at the beginning and a more prolonged note at the end.   The quality is most unmusical, however, and the last note sounds like a ludicrous squawk."

The severest condemnation comes from P. A. Taverner (1934):

The song of the Yellow-headed—if song it can be called, as it lacks every musical quality—is like that of no other Canadian bird.   Climbing stiff-leggedly up a reed or tule stalk the male, with wings partly raised, lowers his head as if to be violently ill, and disgorges a series of rough, angular consonants, jerkily and iregularly, with many contortions and writhings, as if their sharp corners caught in the throat and they were born with pain and travail.   They finally culminate and bring satisfied relief in a long-drawn, descending buzz, like the slipping of an escapement in a clock spring and the consequent rapid unwinding and futile running down of the machinery. The general effect of the performance may be somewhat suggested by the syllables '*Klick-kluck-klee—klo-klu-klel—kriz-kri-zzzzzzz-zeeeeee.*'

Jean M. Linsdale (1938) describes the song as follows:

The number of notes in the song of the males varied; sometimes it was only one drawn out, harsh call. However, the most usual song was composed of 5 notes. The first one was explosive and loud, the next two lower and shorter, followed by 2 long drawn out notes at slightly higher pitch. When the males were at the pond this song was given at rather regular intervals and from habitually used singing perches. These were most often at exposed points where the announcing bird could be seen from, and could see in, many directions. The song appeared to be useful as much to repel invasion by other males as for any other possible service. * * *

Other types of notes were heard, as follows. A series of high-pitched notes, with a few guttural sounds when heard at close range, was given on the circular flight made when a march hawk came near. When potential danger first appeared, a plaintive whistle much like that of the red-wings was given. In flight the females gave single chucks, much like the notes of red-wings. About the nesting sites they had a variety of harsh, screeching notes.

Wetmore (1920) noted that the song "was subject to much variation, but ordinarily resembled the syllables *Klee Klee Klee Ko-Kow-w-w*, the last low and much drawn out."

Ammann (MS.) recognized two distinct types of song, the buzzing and the accenting. He describes the former as follows: "The buzzing song is practically the same for all males. It is begun with several short, slightly descending, comparatively low-pitched, melodious introductory notes (uttered with the bill closed), followed by a loud, very harsh, drawn-out wavering buzz or wail, rather suddenly increasing in volume at the first and held to the end. The most peculiar contortions of the body accompany both parts of the song. During the introductory notes the head is always turned to the left so that the bill is pointing at right angles to the front. At the beginning of the buzz or wail the angle of the bill to the axis of the body is decreased about half and held thus throughout the rest of the song; the neck is extended, bill pointed upward, the wings slightly opened, tail widely spread, and the whole body made to vibrate slightly, the entire procedure giving the general impression that the bird is in great agony."

Of the accenting song, he says: "This song is totally different in general character from the buzzing song although it may often include a short buzzing note such as that described for the latter; even similar introductory notes are used. The head and neck are not twisted much and the entire performance is usually shorter and more precise, seeming to be delivered with less strain or agony to the bird. It is also harsh and not musical but more pleasing to me than the buzzing song. The various syllables are nearly always clearly defined since they are usually separated by short intervals. At the beginning of the main part, the throat is swelled, tail spread wider than usual, and sometimes the wings are slightly opened; at the final note, the breast is thrown forward, the neck stretched upward and the head snapped back, so that the bill is pointing almost straight up."

Field marks.—The adult male yellow-headed blackbird is too conspicuously marked to be mistaken for anything else; the head, neck, and upper breast are bright yellow, in marked contrast to the black of the rest of the plumage, and the white patch in the wing coverts shows plainly in flight and slightly when at rest; in fall the yellow of the head is partially obscured by dusky tips. Females and young males are dark brown, instead of black, with much dull yellow or yellowish buff on the throat and chest, even in the juvenal plumage. The females are always much smaller than the males, and have no white in the wings.

Yellowheads can sometimes be recognized at a considerable distance in flight. Their flight is somewhat undulating, like that of redwings and not like the straight-line flight of grackles; they differ from the redwings in their flock formations, which are long, irregular, loose flocks, like those of the grackles and not like the wide, company-front flocks of the redwings; they can also be distinguished from the grackles by their shorter tails.

Enemies.—The eggs and young of the yellow-heads are preyed upon by various forms of furred and feathered enemies. Small mammals that can swim are likely to climb to the nests and rob them. Crows, and perhaps grackles, sometimes steal the eggs or small young, which are found in abundance in the colonies. The defensive response in the colony to the appearance of a falcon, marsh hawk, or even a harmless nighthawk or bittern, shows that almost any large bird is regarded as a potential enemy, to be driven away by concerted action.

The nests are not uncommonly invaded by cowbirds; Friedmann (1929) cites several authentic cases in various parts of the bird's range, and mentions one case in which six eggs of the cowbird and four of the blackbird were found in a single nest. It would seem that a young cowbird would have small chance of survival in the nest of a species of this size; there seems to be no record of such survival.

Dawson (1923) once found a large "blow snake" coiled just below a nest full of young blackbirds.

According to information given to Ammann (MS.) by Paul L. Errington, he names the chief predators on young and adult yellow-headed blackbirds, in the probable order of their importance, as mink, great horned owl, marsh hawk, red fox, and muskrat. Said Errington, "I would judge that the heaviest pressure by mink upon the blackbirds occurs in late summer and early fall, probably to a considerable extent upon immatures at night." Errington reported that from the spring of 1933 to July 1935, 280 great horned owl pellets were collected, many of them taken when no yellowheads were present; of these, 27 contained a minimum of 36 yellowheads, of which at least 6 were young. From the gullet collections of the young from 12 marsh hawks' nests, during three seasons, 26 specimens of yellowheads were identified, 5 of which were young. Of the other two predators, Errington has this to say: "Foxes take a variable number of Icteridae other than meadow larks, but I believe that redwings are more apt to occur in their diet than yellow-heads. * * * The muskrat often has a meat tooth and may very well eat blackbirds it finds freshly dead or may even kill an occasional cripple or a very immature bird that it may find in the water or in some similarly accessible place. However, as an active predator upon blackbirds, I would not say that it rates at all."

Ammann (MS.) adds: "The three largest of the known destructive agencies were a rise in water level of the lake, a cold rainstorm, a short, violent windstorm. They accounted for the loss of 28.7 percent of all nests and 31.7 percent of all eggs and young (using the total number of eggs as a basis for the latter figure). * * * Internal parasites were found in the alimentary tract of 21.4 percent of the 117 specimens examined. Acarina were found in 17 percent and Mallophaga of four species on 59 percent of the 122 specimens examined."

Fall.—Fautin (1941a) writes:

During the molting period which began in July the Yellow-headed Blackbirds left the nesting areas and congregated in large flocks in marshes where the growth of cattails, *Typha latifolia*, and bulrushes was most dense. Here they remained very much in seclusion during the greater part of the day, coming out only in the mornings and evenings to feed. Very often the males were found in one part of the marsh and the females and juvenals in another. This association of the females and juvenals may have been due to the greater attentiveness of the females to the young during their nestling period. * * *

When the autumn molt was near completion, about August 1, the Yellow-headed Blackbirds, together with other species of blackbirds, came out of hiding and roved about in the fields during the day, returning to the cattail marshes to roost at night.

Migration began about September 1. By September 7 only three females could be located in the vicinity of the study areas. One week later a single juvenal male in a flock of about fifty Brewer's Blackbirds, *Euphagus cyanocephalus cyanocephalus*, was all that could be found and by September 17 all had left the vicinity of the study area.

Mrs. Bailey (1902) says of the fall wanderings: "From their breeding grounds in the sloughs and tule marshes the yellow-headed blackbirds scatter out and wander over the whole of the western plains country, appearing in flocks with grackles, red-wings, or cowbirds in the characteristic hordes of the fall migration, or in flocks by themselves in fields and meadows, along the roadsides, often in barnyards and corrals, and sometimes in city streets, flocks with pompous, yellow-caped males strutting about among the dull-colored females and young, talking in harsh, gutteral tones."

At this season the handsome adult males are often seen in flocks by themselves, and the females and young in larger separate flocks.

P. A. Taverner (1934) writes:

The days are spent on the bountiful stubble fields, and the nights in the marshes. A blackbird roost just before sunset is an interesting place indeed. The birds come in from every direction, talking and croaking loudly, in vast black clouds, looking, on the horizon, like wisps of smoke blowing before the wind. They pitch into a bed of reeds already occupied by earlier arrivals, until each stalk seems strung with big, black beads. At the onslaught of the incoming contingent, birds are dislodged right and left, there is a babel of protesting voices and a fluttering of many wings that whirr loudly in the still air as the surface of the green marsh boils with black forms seeking new resting places. The confusion gradually subsides until the next arriving flock starts the hubbub over again.

Thus it goes on as the sun sinks, until all are in, and then the evening wind chases waves over the soft green surface of the reed beds, without revealing a hint of the hordes of black bodies beneath that are resting through the stillness of the night.

Winter.—The yellow-headed blackbirds, having withdrawn from the northern portions of their breeding range, spend the winter in the southern United States and northern México. They are still to be found, however, in some of the extreme southern parts of their summer range in more or less reduced numbers. In their winter range, they roam about over the fields and plains in enormous mixed flocks, visiting the ranches, barnyards, and poultry farms, much as they did in the fall.

DISTRIBUTION

Range.—Western Canada to central Mexico.

Breeding range.—The yellow-headed blackbird breeds from central Washington (Yakima Valley, Bumping River), central British Columbia (Vernon, Cranbrook, Tachick Lake), central-western and northeastern Alberta (Clairmont, Fort McMurray), north-central Saskatchewan, central and southeastern Manitoba (Grand Rapids, Winnipeg), northern Minnesota, north-central Wisconsin, northeastern Illinois, and northwestern Ohio (locally); south to southern California (Potholes, San Jacinto Lake), southwestern Arizona (near Yuma, Imperial Dam), northeastern Baja California (Colorado River Delta), south-central Nevada (Pahranagat Valley), southwestern Utah (formerly Virgin River Valley), central and central-eastern Arizona (Mormon Lake, Marsh Lake), southern New Mexico (Mesilla, Carlsbad), northern Texas, Northwestern Oklahoma (Cimarron County), and northeastern Missouri (Sarcoxie, Clark County), central Illinois (Quiver Lake), and northwestern Indiana (Lake and Porter Counties). There are summer records which may indicate breeding in western Texas, central-eastern Missouri, southern Illinois, Michigan, and central Ohio.

Winter range.—Winters north to central California (Sacramento Valley), central Arizona (Clarkdale), southern New Mexico (Socorro, Carlsbad), central and southeastern Texas (Medina, Port Arthur), and southern Louisiana (Calcasieu Parish, Octave Pass); south to southern Baja California (San José del Cabo), Jalisco, Michoacán, Guerrero, Puebla and central Veracruz.

Casual records.—Casual in southwestern British Columbia and central Mackenzie, and from northern Michigan, southern Ontario, and western Pennsylvania south to southern Louisiana, and along the Atlantic seaboard from Maine to northern Florida.

Accidental in the Arctic Ocean (100 miles west of Point Hope, Alaska), northern Manitoba (Churchill), central Quebec (Rupert House, Godbout), Nova Scotia (Sable Island), southern Florida (Royal Palm Hammock, Key West), Cuba (Havana—market specimen, Guantánamo) Barbados, and Greenland (Sardlog, Nanortalik).

Migration.—Early dates of spring arrival are: Arkansas—Rogers, March 10. Kentucky—Meade County, April 19. Missouri—Chillicothe, March 8 (median of 4 years, March 13); New Haven, March 11. Illinois—Hinsdale, March 12; Paris, April 2; Chicago region, April 16 (average, May 3). Indiana—Goshen, April 17; Kokomo, April 19. Ohio—Sandusky, April 15. Michigan—Schoolcraft County, April 9; Monroe County, April 29. Ontario—Middlesex County, April 29. Iowa—Sioux City, March 10 (median of 21 years; April 23). Wisconsin—Madison, March 23; St. Croix County, April 5. Minnesota—Hutchinson, March 20 (average of 19 years for southern Minnesota, April 16); Foreston, April 7 (average of 9 years for northern Minnesota, May 6). Texas—El Paso, March 7; Taylor County, March 26. Oklahoma—Oklahoma City, February 27; Custer County, March 5. Kansas—Bendena, February 18; Harper, March 2. Nebraska—Hastings, February 24; Red Cloud, March 6 (average of 14 years, April 12). South Dakota—Vermillion, April 1; Sioux Falls, April 2 (average of 7 years, April 30). North Dakota— Jamestown, April 14; Cass County, April 21 (average, May 4). Manitoba—Margaret, April 14; Treesbank, April 18 (median of 42 years, May 3). Mackenzie—Fort Chipewyan, May 24. New Mexico—Rincon, February 16. Arizona—Tucson, February 27; Phoenix, March 8. Colorado Denver, February 15 (median of 23 years, April 23); Walden, March 30 (median of 11 years, April 14). Utah—Bear River Refuge, Brigham City, March 30. Wyoming—Wheatland, April 10; Laramie, April 11 (average of 8 years, April 25). Idaho—Deer Flat, March 5; Rupert,

March 29. Montana—Choteau, April 8; Billings, April 15; Fortine, April 23 (median of 7 years, May 3). Saskatchewan—Qu'Appelle, April 6 (median of 14 years, May 4); Wiseton, April 15 (median of 20 years, April 27). Alberta—Stony Plain, April 15. California—Siskiyou County, March 17; Fresno, March 23. Oregon—Klamath Falls, March 1; Weston, March 15. Washington—Spokane County, March 20. British Columbia—Okanagan Landing, April 22.

Late dates of spring departure are: Michoacán—Quiroga, April 29. Sonora—Guirocoba, May 10. Arkansas—Huttig, May 6. Kentucky—Guthrie, May 20. Missouri—Palmyra, May 20. Illinois—Chicago, May 21. Texas—Corsicana, June 4; Hidalgo, May 24 Oklahoma—Cleveland County, May 29; Oklahoma City, May 28. New Mexico—Glenrio, May 18. Arizona—Tucson, May 13. California—Orange County, May 26; Death Valley, May 20.

Early dates of fall arrival are: California—Daggett, July 19. Colorado—Yuma, July 14; Weldona, July 15. New Mexico—Clayton and Glenrio, July 14. Oklahoma—Oklahoma City, July 13. Texas—Somerset, July 12; Waco, July 16. Michigan—Bruce Crossing, August 24. Indiana—Indianapolis, August 22. Missouri—Concordia, July 23.

Late dates of fall departure are: Alaska—100 miles west of Point Hope, October 12 (only record). British Columbia—Peniction, October 19. Washington—White Bluffs, October 14. Oregon—Harney County, November 20. Nevada—Indian Springs, October 25. California—Stockton, November 19; San Geronimo, October 17. Alberta—Andrew, September 28. Saskatchewan—Indian Head, October 20; Wiseton, September 28 (median of 6 years, September 25). Montana—Huntley, November 15; Fortine, October 6. Idaho—Rupert, September 18. Wyoming—Laramie, October 20 (average of 5 years, October 7). Utah—Ashley Creek, Unitah County, September 30. Colorado—Fort Morgan, November 10 (median of 12 years, September 7); Boulder, October 30. Arizona—Tombstone, November 21. New Mexico—Clayton, November 22. Manitoba—Winnipeg, October 28; Treesbank, October 20 (median of 14 years, September 10). North Dakota—Rice Lake, September 17. South Dakota—Mellette, October 26; Sioux Falls, October 9 (average of 5 years, September 26). Nebraska—Neligh, November 26; Gage County, November 5. Kansas—Clearwater, October 14; Harper, October 10. Oklahoma—Oklahoma City, October 17. Texas—Houston, November 28; Commerce, November 24. Minnesota—Hutchinson, November 14 (average of 6 years for southern Minnesota, October 6); St. Vincent, October 25. Wisconsin—Appleton, October 24; LaCrosse, October 11. Iowa—Elkader, November 3; Northwood, October 23; Sioux City, October 17 (median of 11 years, September 15). Michigan—Detroit,

October 12. Ohio—Toledo, October 22. Indiana—East Chicago, October 15. Illinois—Chicago region, October 30 (average, September 15); Rantoul, October 23. Missouri—New Haven, November 6. Kentucky—Guthrie, October 18. Arkansas—Rogers, October 11. Maine—Monhegan Island, September 11. Massachusetts—Watertown and Northampton, October 15. New York—Orient, October 4. Pennsylvania—Chester County, September 15. Maryland—Baltimore, October 1.

Egg dates.—Alberta: 8 records, June 4 to June 19.

California: 98 records, April 21 to June 28; 53 records, June 2 to June 10.

Illinois: 20 records, May 20 to June 21; 10 records, May 25 to June 8.

Minnesota: 24 records, May 19 to June 12; 14 records, May 27 to May 31.

Nevada: 15 records, May 22 to June 3.

Utah: 8 records, May 16 to June 4 (Harris).

AGELAIUS PHOENICEUS PHOENICEUS (Linnaeus)

Eastern Redwing

Plates 9 and 10

HABITS

Everyone who notices birds at all knows the red-winged blackbird, or redwing as it is now called; at least they recognize it as a black bird with red on its wings. It is very conspicuous and self-revealing whenever one approaches its haunts. It could hardly be overlooked by even the most casual observer, as the male flies up to announce his presence and display his colors.

The numerous subspecies of the redwing are widely spread all over the continent of North America, except in the arid desert, the higher mountain ranges, the forested and the Arctic regions, wherever they can find suitable marshes in which to breed. The presence of water, or at least its proximity, is essential; and the birds must have certain types of dense vegetation in which to conceal their nests. Marshes or sloughs supporting extensive growths of cattails, bulrushes, sedges, reeds, or tules are their favorite breeding haunts; but where similar types of vegetation, or water-loving bushes or small trees, grow in ponds, around the shores of lakes or along the banks of sluggish streams, the redwings find congenial homes. Wherever such conditions exist throughout this continent, from Central America nearly to the Arctic Circle and from the Atlantic to the Pacific, some form of this species is likely to be found.

Spring.—The redwings are among our earliest spring migrants; the eastern redwing leaves its winter haunts in the southern States before the end of February, reaches New England in March (rarely earlier), and arrives in eastern Canada in April or earlier. In Massachusetts, we look for the first of these harbingers of spring about the second week in March. I wrote in my notes for March 22, 1900: "The first interesting sight that met our eyes, as we walked down the country road, was a detached flock of some ten robins in an old stubble field, the first I had seen that year; it was a welcome sight and their bright red breasts seemed to reflect the warmth of coming spring. A flock of about fifteen redwings, adult males, also arose from the same field and circled about, wheeling with better precision than the best of trained soldiers, their jet black uniforms and scarlet epaulets flashing in the sunlight as they turned. All their movements seemed to be governed by the same impulse, instantly obeyed, as they swooped down upon a small apple tree and alighted with every head pointing toward the wind. Our approach started them off again toward some swampy woods, where they scattered and alighted among the tops of the taller trees."

William Brewster (1906) says: "For several weeks after their first appearance in early spring Redwings are usually found in flocks composed wholly of males. At this season they are seldom seen about their breeding grounds excepting in the early morning and late afternoon. At most other hours of the day they frequent open and often elevated farming country, where they feed chiefly in grain stubbles and weed-grown fields. When disturbed at their repasts they fly to the nearest deciduous trees and immediately after alighting burst into a medley of tumultuous song, inexpressibly wild and pleasing when heard at a distance, but rather overwhelming if the flock be a large one and close at hand."

Chapman (1912) writes attractively of this early spring behavior: "A swiftly moving, compact band of silent birds, passing low through the brown orchard, suddenly wheels, and, alighting among the bare branches, with precision of a trained choir breaks into a wild, tinkling glee. It is quite possible that in the summer this rude chorus might fail to attract enthusiasm, but in the spring it is as welcome and inspiring a promise of the new year as the peeping of frogs or the blooming of the first wild flower."

No better life history [1] of the redwing has ever been published than that written by Arthur A. Allen (1914), based on an exhaustive study of the bird near Ithaca, N. Y. I regret that space will not permit quoting from it as fully as it deserves. His study throws new

---

[1] The reader should also consult the valuable paper by Robert W. Nero, "A behavior study of the red-winged blackbird," Wilson Bull., vol. 68, pp. 5-37, 129-150 (1956).

light on the migratory movements of the species, and suggests that similar studies of other species might be equally enlightening.

As a result of his studies at Ithaca in 1911 and 1910, he divides the migratory waves into seven classes as follows: "Vagrants" arrived from February 25 to March 4; migrant adult males from March 13 to April 21; resident adult males from March 25 to April 10; migrant females and immature males from March 29 to April 24; resident adult females from April 10 to May 1; resident immature males from May 6 to June 1 (1910); and resident immature females from May 10 to June 11 (1910).

The "vagrants" come during the first warm days of spring, although the marshes may still be frozen and the ground still covered with snow; they are supposed to be birds that have wintered not very far south; they do not appear every year, but when they do come they are seen in February; they "are for the most part adult males, but immature males or females may be found among them. They are never in large flocks, and often occur singly. The reproductive organs are very small. * * * They do not frequent the open marsh."

Of the arrival of the migrant males, Allen (1914) says:

The first true migrants arriving in the spring are adult males. They appear in flocks, some of which contain a hundred or more birds, and ordinarily are first noted in the marsh, although occasionally seen in tree tops or stubble fields on the uplands. * * * At this season of the year, about 4:30 in the afternoon, let us take a stand at the upper end of the marsh and gaze southward up the Inlet Valley. Presently we discern what appears like a puff of smoke in the distance, drifting in at a considerable height. After a minute or two the smoke is resolved into an aggregation of black specks, and then, as it drops lower and lower, it takes on that irregular form so characteristic of Redwinged Blackbirds. With one last swoop and flutter of wings, they alight on the more prominent of the few scraggly trees at the southern end of the marsh. The migration has begun. For a few moments they shake out their feathers and give vent to their feelings in song. It is but a short time, however, before they start again for the north.

A few birds drop out of the passing flocks and settle down into the marsh for a while, but they soon rise again and join another migrating flock. Flocks coming in late in the day fly low and settle for the night in the scanty shelter of the still dormant flags.

Every available perch, not so high as to be conspicuous, is filled with birds down to the water's surface, but were it not for the unspeakable din that arises from the hundred of throats, one would scarcely be aware of their presence, so inconspicuous are they against the dark water. If one disturbs them now, there is a rush of wings, but they do not fly far. Raillike they drop back into the marsh a short distance away, and soon resume their indescribable discord. * * *

This period of the migration, which I have termed the arrival of *migrant adult males*, continues for about two weeks before the *resident* birds begin to arrive. Each evening there is a well-defined flight into the marsh; each night the birds all roost together; and each morning they all leave for the north. The marsh to them at this period is a shelter for the night only, and the entire day is spent on the uplands.

The birds referred to at the beginning of this chapter by Brewster, Chapman, and the author probably belonged in this class, migrating adult males. Allen (1914) says of the arrival of resident adult males:

The arrival of resident males is first made clear by the actions of the birds themselves. To one unfamiliar with their habits the *exact* time of arrival is not apparent. Up to this time the birds, for the most part, have kept in more or less well-defined flocks. They have been difficult to approach, the slightest annoyance starting them off. * * * About the end of March, however, certain birds arrive, in whose actions a difference is noticed. They do not fly away at one's approach, or, if frightened, soon return to the same spot. These birds do not associate with the migrating flocks, and they roost alone. If one is enabled to identify an individual bird among them by such characteristics as abnormal feet or the loss of its tail or a primary feather, as has frequently been done in this study, one finds that it never changes its station in the marsh after its arrival. * * * From their first arrival, they assume all rights to the domain in which they have established themselves. Frequently these domains adjoin one another closely, but the birds seldom trespass on one another's rights. When they do so, they seem to recognize the owner's prerogative, so that serious quarrels never ensue.

The resident males have been at their stations only a few days before the first females and immature males appear among the migrating flocks. The last days of March and first of April usually usher them in. Says Allen (1914): "Within a few days, as their numbers increase, small flocks made up entirely of females are observed. It is about this time—the end of the first week in April—that the males begin to show a slight interest in the presence of the females. The former now spend more of their time in the marsh, and resent intrusion into their domains. By this time their reproductive organs show considerable increase in size. Among the migrating birds at this time there is an increasing preponderance of immature males and of females. The latter shun the presence of the males, and whenever they do approach one of the residents, they are immediately driven off."

During the early part of the third week in April, another group arrives, the resident adult females. According to Allen (1914):

The flocks break up and the single birds scatter over the marsh, as did the resident males upon their first arrival. Usually they select a place near some male or group of males. They are much more retiring than the latter, however, and keep mostly near the water's surface, where they are inconspicuous. Whenever they appear on the tops of the cat-tails, or more especially, when they attempt to fly, they are immediately pursued by one or more of the males. Occasionally a male drives a female in great circles over the marsh and even to a considerable height. Eventually, however, he relinquishes the pursuit and returns to his post. The earlier migrant females, when pursued in this way, immediately leave the marsh. But now, as the male ceases pursuit, the female checks her flight and is soon again at her station near the male.   Such maneuvers announced the arrival of the resident females.

About the first week in May, after most of the adult resident birds have begun to nest, the resident immature males begin to appear in numbers.   From the second week in May until the last of the month, these flocks continue to arrive   The resident immature females begin to appear with the immature males about the middle of the month. They increase in numbers until the first of June, when they far out-number the males, and by the second week, when the last migrating birds are recorded, they compose the entire flocks.   Says Allen (1914), "It is doubtless through some of these birds, at a time when unattached males are difficult to find, that many of the cases of polygamy arise."

Probably the movements of the different classes of migrants are not always as clearly defined as indicated by Allen.   Fred M. Packard (1937), while banding redwings at the Austin Ornithological Research Station, at North Eastham, Mass., found "unsuspected variation in the behavior of the migrating birds on Cape Cod.   Some were appar-ently true migrants; they were caught but once, and did not repeat. Others lingered for a few days or even a month, repeating during that period, and then left; these also were migrants.   A third group stayed in the vicinity from the time of their arrival through the nesting period, as true residents.   A large fourth group was composed of individuals that were trapped once or twice on arrival, and then disappeared, exactly like migrants; but these returned after an interval varying from 2 months to 2 weeks, some to nest nearby, others to disappear again."   Cape Cod is a long, narrow, curving peninsula pointing northward at its terminus and facing a broad expanse of water.   Perhaps the returning birds of the fourth group preferred to turn back, rather than risk the long flight over the water.

Territory.—As indicated above and as noted by all observers, the resident adult male, on his arrival on the breeding grounds or soon after that, "stakes out his claim" to the territory that he has decided to estab-lish and to defend.   This claim may be large or small, depending on the size of the marsh and the density of its population; in a large marsh with few redwings nesting in it, the territories may be extensive and well outlined; but in a dense colony, the claims are close together and the boundaries are not so well marked.   The male stands his ground and defends his territory against intruding male redwings and other trespassing birds; he even drives away female redwings until he is ready to mate.

Ernst Mayr (1941) writes as follows on territorial behavior: "Early in the season, when the weather was still cold and the males had just recently established themselves in their territories, they spend a good deal of their time sitting on the top of small bushes or old cat-tail stalks and calling softly *chuck-chuck*, particularly when migrating blackbirds flew overhead. They were rather fluffed up and only the yellow margin of their shield showed. As soon as a singing spell 'overcame' one of the birds his whole attitude changed, and he displayed his red brilliantly—only to fall back into his former lethargic condition when the singing was ended."

Courtship.—While the male redwings are defending their territories and driving away migrating redwings of both sexes, the resident females come, between April 10 and May 1 at Ithaca, according to Allen (1914). They select their own territories, where they plan to build their nests, and these are usually near the station of some established male or group of males. At first the male drives away the newcomer and chases her about over the marsh, but she returns to the spot she has selected. Eventually he is ready to select his mate and may be seen following her about. "He never allows her to escape from his sight, and as she hunts about near the water's surface, he vaunts himself on the nearest cat-tail. They now may be considered mated."

Probably most redwings are mated in pairs that are true to each other, but this is a matter that is not easily determined in a large colony. Allen (1914) says: "Certain pairs have been observed throughout the season, and found to be mated as steadfastly as are most birds, while in others the tie seems to bind only so long as the male is watchful and able to exert his lordship in driving away other males. A female has been observed to receive one male with spreading wings and quivering feathers, and in the next moment, when this bird had been driven off, to welcome the victor with the same freedom and display."

Females sometimes take a more active part in the courtship performance as observed by Thomas Proctor (1897), who says: "And very amusing indeed it was to watch these comedians in sober brown, but in extemporized ruffs, puffs and puckers, pirouette, bow and posture, and thus quite out-do in airs and graces their black-coated gallants. Their shrill whistle, the meantime continually vied with, or replied to, the hoarse challenges of their admirers, while in noisy chattering, and in teasing notes, they were excessively voluble."

It is generally believed that the redwing is often polygamous, though by no means always so. It is often evident that there are more females and more occupied nests in a marsh than there are males. In one swamp studied by Mayr (1941), he shows 12 nests in his sketch in what he supposed were 6 territories; he was unable to determine the exact number of males, but says that "there were not less than four and not more than six." In another swamp, two males had two females each and another had only one.

Mabel Osgood Wright (1907) writes: "When Redwings live in colonies it is often difficult to estimate the exact relationship between the members, though it is apparent that the sober brown, striped females outnumber the males; but in places where the birds are uncommon and only one or two male birds can be found, it is easily seen that the household of the male consists of from three to five nests each presided over by a watchful female, and when danger arises this feathered Mormon shows equal anxiety for each nest, and circles screaming about the general location."

Numerous banding records have indicated the males far outnumber the females, but this is probably due to the fact that the males enter the traps more readily than the more retiring females, and so are more often recorded. What is probably a more reliable conclusion as to the actual sex ratio was found in the careful studies of J. Fred Williams (1940). He states in his summary:

In a study of nestling Eastern Red-Wings made at Indian Lake, Ohio, from June 18 to July 22 it was found that the young could be sexed by dissection at any time after hatching.

With the age of nestlings known to the nearest day it proved possible to distinguish between the sexes by means of weights after the fifth day, and by means of tarsal lengths after the eighth day.

The following sex ratios were found:

Among 119 young, representing the full egg complements of 35 nests, 57 males: 62 females.

Among 94 young which were successfully fledged, 47 males: 47 females.

Among 21 young which died during the nesting period, 9 males: 12 females.

The apparent deviation of the first and third of these ratios from the expected 50:50 could easily be due to random variation in sampling.

To assume that the even 50:50 birth rate, or nearly that, is the rule, does not agree with the well-known fact that the females outnumber the males on the breeding grounds, unless we also assume that the females begin to breed when less than one year old and that the males, at least most of them, do not mate until they are nearly 2 years old. This is true of the yellow-headed blackbird, and probably also of the redwing. With the sexes as unbalanced, as they are in the breeding colonies, polygamy is likely to be quite prevalent and promiscuity, or even polyandry, may often occur, through the latter is probably rare.

Dr. Charles W. Townsend (1920) gives the following excellent account of the courtship display of the male:

The courtship of the Red-winged Blackbird centers as distinctly about the display of the scarlet epaulettes as does the courtship of the Peacock about the display of his train. The adult male Red-wing when absorbed in feeding is a plain blackbird with a pale yellow stripe on his shoulder or one with a narrow band of red. The color may even be entirely covered up by the prevailing blackness of his costume. When, however, his love passions are excited he spreads his tail, slightly opens his wings, puffs out all his feathers, and sings his *quonk-quer-ee*, or his still more watery and gurgling song, appropriate to an oozing bog, his *ōgle-ŏggle-yer*. Now when he puffs out his body feathers he especially puffs out, erects, and otherwise displays to their best advantage the gorgeous scarlet epaulettes. These patches become actually dazzling in their effect as he slowly flies toward the object of his affections, for these beauty spots are most effective when seen from in front.

While admiring the gorgeous display of brilliant scarlet and gold set in its framework of glossy black, one is apt to overlook the awkward posture of the bird; standing on some prominent perch, he leans forward, pointing his bill toward his tail beneath the branch, with his back hunched up, as if he were to become violently nauseated, suggesting the ludicrous performance of the cowbird.

Allen (1914) mentions another form of courtship:

In addition to the ordinary display and erection of feathers, a method of soaring is now indulged in. In comparison with that of the Lark, it is rather crude, but undoubtedly it is akin to it. Mounting in a rather irregular spiral, the male bird attains a considerable height, where he hovers, oftentimes for long periods, while his wings barely flutter. Song is not generally indulged in. Eventually, with half-closed wings, the bird drops down in a zigzag course to the marsh. A dozen or more birds may frequently be seen in the air at once, as they perform these evolutions. At this time, also, hovering at a much lower height is frequently indulged in. With a few quick strokes of his wings, the male vaults from his post into the air, and with quivering wings and flaming shoulders, gives vent to his pent-up passion in the "scolding song" described above.

Nesting.—Redwings build their nests in a variety of situations, though usually in a marsh, swamp, or wet meadow, where the nests are placed in cattails (*Typhus*) dead or living, rushes (*Scirpus*), sedges (*Carex*), tussocks of marsh grass, or such water-loving bushes as button bushes (*Cephalanthus*), alders (*Alnus*), or willows (*Salix*). Such associations in shallow ponds, or along the shores of lakes or the banks of sluggish streams, afford suitable nesting sites. Although the birds prefer the vicinity of water, their nests are often found on dry uplands, sometimes at a considerable distance from any water, in fields of tall grass, clover, and daisies, where they must be built close to or even on the ground. Nests in bushes and trees also have been reported by several observers.

A. D. Du Bois has sent me the data for 42 nests found in four Northern States; 3 of these were in trees or bushes from 8 to 9 feet above the ground; 2 were in clumps of nettles on the margin of a marsh, 2 feet above the dry ground. Of 24 nests, reported to me by T. E. McMullen, found in New Jersey, 6 were in bayberry bushes near marshes and among sand dunes near the ocean; one was 9 inches up in a clump of goldenrod in a clover field; and another was 8 inches up in a wild rose bush standing in 8 inches of water. Alexander F. Skutch tells me that he found two nests in upland alfalfa fields near

Ithaca, N. Y. "The two were built in exactly similar situations, in the midst of the stalks of an alfalfa plant, with the bottom in each instance three inches above the ground." Witmer Stone (1937) records redwings' nests in privet hedges, marsh elders (*Iva frutescens*), and one in a small cedar bush, in New Jersey. A. Sidney Hyde (1939) found a nest in a clump of vetch (*Vicia*) and another in a wild cherry shrub, in northern New York.

William Brewster (1937) says of the nesting habits of redwings at Lake Umbagog, Maine: "Most of them breed on small, floating islands moored not within areas permanently covered by the lake but in bordering marshes which have every appearance of thus belonging to it, whenever completely submerged. The islands float only at such times but they keep ever level with the surface of the water, however quickly it may rise or fall, yet seldom shift otherwise than vertically, being too firmly anchored to solid ground beneath by tough, flexible roots which proceed from living bushes—and perhaps also medium sized trees—that overspread what are essentially buoyant rafts of vegetable matter for the most part long since dead."

Althea R. Sherman (1932) refers thus to tree nesting in Iowa: "It is 25 years since Red-winged Blackbirds began nesting in the tops of our trees, which grow more than half way up the hillside from a brook frequented by others of their species. Since 1907, when four females built nests at heights of 18 to 22 feet from the ground in separate plum trees, there has been great increase in growth of wild currant, wild gooseberry and elderberry bushes in our house yard of about an acre in extent. In these bushes more frequently than in the tops of plum trees do the Red-wings nest."

C. J. Maynard (1883) adds the following: " "I have found the nests on an island in the marshes of Essex River, placed on trees twenty feet from the ground! In one case, where the nest was placed on a slender sapling fourteen feet high, that swayed with the slightest breeze, the nest was constructed after the manner of our Baltimore Orioles, prettily woven of the bleached sea-weed called eel-grass. So well constructed was this nest, and so much at variance with the usual style, that had it not been for the female sitting on it, I should have taken it for a nest of *I. Baltimore*. It was six inches deep."

Dr. George M. Sutton (1942) published a photograph of another pensile nest, found by Malcom W. Rix in Oneida County, N. Y. It was suspended "at the end of a grape-covered willow branch, about three feet above water several feet deep. * * * The inside depth of the nest was only slightly greater than that of the general average of the species, and not comparable to that of a Baltimore Oriole's nest. The color of the nest was distinctly that of a Red-wing's, although the materials apparently were somewhat finer than usual."

W. E. Clyde Todd (1940) mentions two Pennsylvania nests that were more than 30 feet from the ground in willow trees, the highest I have seen recorded. Brewster (1906) reports a nest in a vertical fork of a small apple tree in an orchard not far from a pond. Harold M. Holland (1923) found a redwing's egg and a cowbird's egg in a Bell's vireo's nest; and later an egg of the redwing and two cowbird's eggs, in another Bell's vireo's nest, were so much like the eggs in the other nest that they appeared to have been laid by same interlopers.

Allen (1914) describes the progress of the nesting at Ithaca:

The first nests built are located in the dead stubs of the cat-tails that have been burned over during the previous fall. At first they are not sheltered by any vegetation of any kind, for the new growth is barely above the water. * * * As the season advances and the vegetation grows, green stalks are included in the support. At first these are not sufficiently strong to serve alone as a support, and consequently the nests are always attached on one side to the dead stub. * * * This is true of most of the nests constructed in early May, and it generally results in disaster. So firmly are the nests fastened by the strands of milkweed fiber, that the side attached to the green blades is carried upward by their growth, while the other, attached to the dead stubs, remains fixed. As a result, the one side is lifted at the rate of almost an inch a day until the nest is inverted. The birds continue to incubate until the last egg is rolled out. * * * By the end of the third week in May, most of the vegetation in the marsh is sufficiently strong to support a nest, and as a result, nests built at this season are located rather indiscriminately in cat-tail, sedge, burreed, water horsetail, dock, or arrow arum. By the first of June the cat-tails and sedges are matured, and have become very dense and harsh. The Redwings now desert them for the softer vegetation, such as the dock and smartweed, which by this time fill most of the small ponds.

The time required for building a complete nest is usually 6 days. Of this time, 3 days are spent on the outer basket and "felting," and 3 days on the lining. Many of the later, more poorly built nests require much less time for construction, some of them being completed in as few as 3 days. * * * The construction of the nest, in all cases observed at Ithaca, has been entirely by the female. The male has never been seen with nesting material in his bill. He is very attentive, however, during the process.

* * * The adult birds commence building again, often before the first young have left the nest. The second nest is located in the immediate vicinity of the first, frequently within a distance of 10 feet. This is true also when the first nest has been robbed or destroyed. One pair, which was experimented upon, built 4 nests within a radius of 25 feet between April 25 and May 18.

Nuttall (1832) gives us the most complete description of the nest of the redwing as follows:

Outwardly it is composed of a considerable quantity of the long dry leaves of Sedge-grass (*Carex*), or other kinds collected in wet situations, and occasionally the slender leaves of the flag (*Iris*) carried round all the adjoining twigs of the bush by way of support or suspension, and sometimes blended with strips of the lint of the swamp *Asclepias*, or silk-weed (*Asclepias incarnata*). The whole of this exterior structure is also twisted in and out, and carried in loops from one side of the nest to the other, pretty much in the manner of the Orioles, but made of less flexible and handsome materials. The large interstices that remain, as well as the bottom, are then filled in with rotten wood, marsh-grass roots, fibrous peat, or mud, so as to form, when dry, a stout and substantial, though concealed shell, the whole very well lined with fine dry stalks of grass or with slender rushes (*Scirpi*). When the nest is in a tussock, it is also tied to the adjoining stalks of herbage; but when on the ground this precaution of fixity is laid aside.

Harold B. Wood sends me the following note: "A dissected nest, which had been built around 18 burreed stalks, was composed of 142 cattail leaves, up to 21 inches in length, and lined with 705 pieces of grasses. It also contained 34 strips of bark of water willow, up to 34 inches in length, which made 273 laps around the reeds, with only one making a complete loop around a stalk. The tensile strength of the matting was tested by placing in the nest increasing weights until a weight of four pounds was held before the nest began to slip down the reeds. Eleven of 42 nests were completed and never used; no nest was ever used for a second brood. Red-wings will not abandon eggs merely because they are discovered, as will robins."

Eggs.—The eastern redwing lays from three to five eggs in a set, usually four. Bendire (1895) describes them as follows:

The eggs of the Red-winged Blackbird are mostly ovate in shape; the shell is strong, finely granulated, and moderately glossy. The ground color is usually pale bluish green, and this is occasionally more or less clouded with a pale smoke-gray suffusion. They are spotted, blotched, marbled, and streaked, mostly about the larger end, with different shades of black, brown, drab, and heliotrope purple, presenting great variation in the amount, character, and style of markings. Occasionally an entirely unspotted egg is found.

The average measurement of 380 eggs in the United States National Museum collection is 24.80 by 17.55 millimetres, or about 0.98 by 0.69 inch. The largest egg in the series measures 27.94 by 19.05 millimetres, or 1.10 by 0.75 inches; the smallest, 20.57 by 15.75 millimetres, or 0.81 by 0.62 inch.

Young.—Allen (1914) has this to say about the incubation of the eggs: "During the days when the eggs are being deposited, frequently both birds continue their excursions to the uplands. With the laying of the third egg, incubation begins, and thenceforth both birds remain in the marsh. Incubation, so far as observed, is performed entirely by the female. In one instance the first egg hatched in ten days, and frequently one or more of the eggs requires twelve, but the usual period is eleven days."

Of the development of the young, he writes:

At *hatching* the young are blind and helpless. The skin is scarlet, with but a scant covering of buffy or grayish down along the principal feather tracts. They are at first exceedingly helpless, scarcely able to raise their heads for food, but they gain strength rapidly after the first feeding. During the *first* day there is considerable increase in size. On the *second* day feather sheaths of the primaries and secondaries show distinctly. By the *third* day these feather sheaths appear distinctly along all of the tracts. On the *fourth* and *fifth* days there is a great increase in the size of the body and in the length of the quills. On the *sixth* the feather sheaths of the wing break open. On the *seventh* the wing feathers have grown considerably, and those of the other tracts begin to break.   On the *eighth* all of the sheaths have broken, and the wing feathers have attained considerable length.   On the *ninth* the feathers have grown still further, but do not yet cover all of the bare spaces.   The young can fly short distances, however, and can not be kept in the nest if once frightened or removed.   If the nest has become polluted, as frequently occurs when it has become greatly compressed by the growing vegetation, they may leave of their own accord on this day.   On the *tenth* the stronger of the young leave and climb to near-by supports.   If the nest is approached, all leave, but otherwise the weaker remain until the *eleventh* day, when all scatter to the vegetation in the immediate vicinity.   They all remain in this neighborhood for at least ten days, even after the parents have ceased caring for them and have started a second brood.

He quotes from F. H. Herrick as follows: "In the space of four hours on the first day * * * fifty-four visits were made and the young were fed forty times.   The female brooded her young over an hour, fed them twenty-nine times, and cleaned the nest thirteen times.   The male made eleven visits, attending to sanitary matters but twice.   * * * On the following day, * * * in the course of nearly three and one-half hours, 55 visits were made, and the young were fed collectively or singly 43 times.   * * * The male bird served food eleven times and attended to sanitary matters once. In the course of forty-two minutes the first young bird to leave the nest was fed eight times, seven times by the mother and once by the father."

Allen continues: "The principal insects eaten are May flies, caddis flies, and lepidopterous larvae.   Generally three or four insects are brought each time, and one delivered to each young.   This is not always the case, however, for sometimes the entire mass is given to one bird.   There seems to be no order in this distribution, the young bird with the longest neck and widest mouth always getting fed first. The food is delivered well down into the throat of the young, and if not immediately swallowed is removed and given to another."

Ira N. Gabrielson (1914) listed the following items given to a brood of young redwings during 51 feedings: 12 unidentified items, 11 wireworms, 1 cricket, 3 beetles, 2 May flies, 3 other flies, 4 green worms, 20 grasshoppers, 3 moths, 1 spider, 4 tomato worms, and 1 measuring worm.

Wood says in his notes: "Of the 37 nests which were followed through the season, 16 had successful broods; 23 contained 73 eggs, of which 53 hatched (72 percent).   From these 73 eggs only 35 full-grown young birds left the nests, a productivity of 48 percent.   Two out of 94 eggs were infertile."   In his published paper (1938), he writes: "The ability of a nestling redwing to take care of himself was tested.   A nestling less than two or three days old would be apt to drown if it should tumble out of the nest.   As they grow older they become more able to save themselves.   Placed in water, the half-grown nestling will float and can swim, but in a very excited manner. They will swim to the reeds and hold on, calling for their parents. When well covered with feathers, but yet a few days before being ready to vacate the nest, they readily swim, but excitedly, and can climb up the cattails to the nest. They are not combative and can not protect themselves against enemies."

Probably two broods are normally raised in a season, and perhaps often three.

Plumages.—The early nestling plumages are described above. Dwight (1900) describes the juvenal plumage of the young male as follows: "Above, including sides of head, wings, tail, and lesser coverts (i. e., the so called 'shoulders') dull brownish black (no red at this stage), the feathers edged with buff, palest and narrowest on primaries, rectrices, head and rump, and richest on scapulars and secondaries. Below pinkish buff, ochraceous on the chin, thickly streaked (except on the chin) with brownish black. Obscure superciliary line ochraceous-buff."

A complete postjuvenal molt, beginning in August, the time varying for the earlier and later broods, produces the first winter plumage of the male, in which the "entire plumage, including wings and tail," is "greenish black much veiled with buffy and ferruginous edgings, palest below and faint or absent on primaries and rectrices. Lesser wing coverts ('shoulders') dull orpiment-orange, each feather with subterminal bars or spots of black. Median coverts rich ochraceous buff usually mottled with black subterminal areas chiefly on the inner webs, the shafts usually black."

The first nuptial plumage is "acquired by wear, which is considerable, birds becoming a dull brownish black by loss of the feather edgings and by fading. The mottled 'shoulder patches' are characteristic of young birds, the amount of orange varying greatly. The wings and tail show marked wear."

A complete postnuptial molt occurs in August, at which young and old become practically indistinguishable. Dwight describes this adult winter plumage of the male as "lustrous greenish black, feathers of head and back, greater wing coverts and tertiaries edged more or less (according to the individual) with buff and ferruginous brown. Below, the edgings are paler or absent. The bright scarlet-vermilion 'shoulders' are acquired together with the rich ochraceous buff median coverts."

The full brilliancy of the spring plumage is produced by wear, the buff and brown edgings disappearing; the wings and tails of the adults show less wear than in the young birds.

Of the plumages of the female, Dwight (1900) writes; "In natal down and juvenal plumage females differ little from males, the juvenal dress perhaps averaging browner above with less buff below and the chin narrowly streaked. The first winter plumage is acquired by a complete postjuvenal moult as in the male, from which the female now differs widely, being brown and broadly streaked. The first winter plumage is hardly distinguishable from the adult winter and passes into the first nuptial by wear, which produces a black and white streaked bird, brown above. A pinkish or salmon tinge is often found in females in any of these plumages, especially about the chin and head, and an orange or crimson tinge may show on the 'shoulders' of the older birds."

Food.—Beal (1900) prepared an extensive report on the food of the redwing, based on an examination of 1,083 stomachs collected during every month in the year from most of its range in the United States and Canada. In spite of the prevailing impression that red-wings are very injurious to the farmer's interests, his diagram shows no very decided foundness for grain, as most of the birds' food consisted of weed seeds and insects. Unfortunately, no stomachs were examined from the rice-growing region during sowing and harvesting of this crop, where considerable damage is claimed. "The food of the year was found to consist of 73.4 percent of vegetable matter and 26.6 percent of animal." His table shows the following average percentages for the 12 months: Animal food—predaceous beetles 2.5, snout-beetles 4.1, other beetles 3.5, caterpillars 5.9, grasshoppers 4.7, other insects 4.1, spiders and myriapods 1.3, other animal food 0.5, total 26.6 percent; vegetable food—fruit 0.6, corn 4.6, oats 6.3, wheat 2.2, other grain 0.8, weed seeds 54.6, other vegetable food 4.3; total—73.4 percent. The consumption of weed seeds amounts to 97 percent in November.

Another table shows the frequency with which certain vegetable foods were taken. Among the larger items, oats were found in 190 stomachs and corn in 117. Weed seeds of some kind were apparantly found in all the stomachs, panic grass in 168, bear grass in 271, ragweed in 189, and smartweed in 200. Small fruits were seldom eaten, blackberries being found in 7 stomachs, blueberries in 2, and gooseberries, strawberries, and currants were found in only one stomach each.

Of 84 specimens examined by F. H. King in Wisconsin, 37 had eaten corn and weed seeds, 31 only seeds, 7 only corn, 3 rye, 2 oats, 8 wheat, and 2 tender herbage; five had eaten 7 beetles, four 7 grasshoppers, one a moth, and one a caterpillar; eight had eaten small mollusks. Bendire (1895) includes small mollusks and newts in the food. Forbush (1907) writes: "They forage about the fields and meadows when they first come north in the spring. Later, they follow the plow, picking up grubs, worms, and caterpillars; and should there be an outbreak of cankerworms in the orchard, the Blackbirds will fly at least half a mile to get cankerworms for their young. Wilson estimated that the Red-wings of the United States would in four months destroy sixteen thousand, two hundred million larvae."

During the nesting season, much of the redwings' food is obtained in the marshes, but they resort regularly to the uplands to glean insects, grain, and seeds in the plowed fields, cultivated lands, and recently cut hay fields. They even resort to trees at times. Du Bois says in his notes: "From an upstairs window I watched a female redwing, as she searched the foliage of the nearby basswood for the small, smooth, green caterpillars which infest these trees. Her method was similar to that of the vireos, though she lacked some of their skill and grace. She hopped from twig to twig, eating the caterpillars from the leaves; and once she made a little flight to take a caterpillar from the under side of a leaf while hovering in the air. I had seen a female redwing at the same business in this tree before."

Francis H. Allen writes to me: "In October the redwings feed on the seeds of a white ash behind my house. They come there day after day, sometimes for a week at a time. I notice the manner of feeding of a small flock composed of both sexes. After reaching up and picking off a samara, the bird held it against the twig on which it perched and in this way evidently detached the wing, or perhaps shelled the seed. They seemed to be unable to cut off the wing with the bill alone without a solid twig to aid them. My neighbor, Mr. John S. Codman, has seen redwings eating seeds from white pine cones in the tops of the trees, perching on the cones as they picked them out."

Southerners have complained that redwings pull up the long-leaf pine seedlings to eat the seeds. But they are useful in destroying the cotton boll weevil in the south and the alfalfa weevil, two of our most destructive weevils. They also eat the larvae of the gypsy moth and the tent caterpillar.

Economic status.—On its northern breeding grounds the eastern redwing is almost wholly beneficial, and comparatively few complaints are made of serious damage to crops. Its food while here consists almost entirely of insects, very few of which are useful species, and weed seeds, which form by far the largest proportion of its food. The young are fed almost exclusively on insects. It does some damage to sprouting grain in the spring, and to sweet corn in the summer, while the kernels are soft and milky, by tearing off the husks and ruining the ears for the market. Other grains are also attacked to a limited extent, but much of the grain eaten is waste grain picked up from the ground.

In the Middle West, where the redwings are much more abundant and where the cereal crops are more extensively cultivated, these and other blackbirds, in late summer and fall, swoop down in vast hordes on the grain fields and do an immense amount of damage to the grain both while it is ripening and while it is being harvested. Even there, the redwing has some good points in its favor. Lawrence Bruner (1896) writes from Nebraska: "Even when it visits our corn fields it more than pays for the corn it eats by the destruction of the worms that lurk under the husks of a large percent of the ears in every field. Several years ago the beet fields in the vicinity of Grand Island were threatened great injury by a certain caterpillar that had nearly defoliated all the beets growing in many of them. At about this time large flocks of this bird appeared and after a week's sojourn the caterpillar plague had vanished, it having been converted into bird tissue."

In the Southern States, it does great damage to the rice crop by pulling up the seedling rice plants in the spring and by eating the soft grain as it ripens. In this respect the redwing is almost as bad as the bobolink. It does some good, however, by destroying the seeds of the so-called "volunteer" rice, which, if allowed to grow, would injure the value of the crop.

S. D. Judd (1901) says that on the fall migration, bobolinks and redwings converge and swarm into the limited area of the rice districts so as to destroy annually $2 million worth of the crop. And B. H. Warren (1890) quotes T. S. Wilkinson as saying: "The rice crop in Louisiana, from the time the rice is in the milk till harvest time and during harvesting, is much damaged by birds, principally the Red-shouldered Blackbird. Shooting is the only remedy thus far resorted to which is at all effective, and it is only partially so. I have known rice crops to be destroyed to the extent of over 50 percent, which is a loss of say $13 per acre. While this is an extreme case, a damage and expense of from $5 to $10 per acre is very common."

Beal (1900) says in conclusion: "In summing up the economic status of the redwing the principal point to attract attention is the small percentage of grain in the year's food, seemingly so much at variance with the complaints of the bird's destructive habits. Judged by the contents of the stomach alone, the redwing is most decidedly a useful bird. The service rendered by the destruction of noxious insects and weed seeds far outweighs the damage due to its consumption of grain. The destruction that it sometimes causes must be attributed entirely to its too great abundance in some localities."

Behavior.—On the ground the redwing walks deliberately, or runs, or hops rapidly when trying to keep up with a feeding flock. In late summer or early fall, one may occasionally see immense flocks of redwings mixed with grackles, cowbirds, and starlings feeding in the open fields. Such flocks sometimes contain hundreds or even thousands of birds. I have seen flocks that covered as much as an acre or more in a broad expanse of meadow or pasture land, densely spread over the ground like a great black mantle. The flock moves along steadily as it feeds, all moving in the same direction; at intervals those in the rear rise, fly over the main flock, and settle in front of the advancing horde, to resume their feeding; this happens again and again, giving the impression of a vast rolling cloud of black birds. When the edge of the field is reached the whole mass rises in a body, to rest in the treetops for a time, or to swoop down into another field.

In the air the flight of the redwing is characteristic; it flies with bursts of rapid wingbeats, between which are slight intermittent pauses, producing a somewhat wavy motion. The flocks are in orderly formation, wheeling and turning in unison, but the individual birds in the flock are constantly changing their positions, rising and falling more or less independently. The vast flocks that travel about through the Southern States in fall and winter are most impressive. Pearson (1925) writes: "At this time they may be seen in flocks numbering tens of thousands, and they present a marvelous spectacle as they fly with all the precision of perfectly trained soldiers. I have seen fully thirty thousand of them while in full flight suddenly turn to the right or the left or at the same instant swoop downward as if they were all driven by common impulse. They perform many wonderful feats of flight when on the wing. Sometimes a long billow of moving birds will pass across the fields, the ends of the flying regiment alternately sinking and rising, or even appearing to tumble about like a sheet of paper in a high wind."

Wilson (1832) says: "Sometimes they appeared driving about like an enormous black cloud carried before the wind, varying its shape every moment; sometimes suddenly rising from the fields around me with a noise like thunder; while the glittering of innumerable wings of the brightest vermillion amid the black cloud they formed, produced on these occasions a very striking and splendid effect."

Redwings are very aggressive in driving away any large bird that approaches their nesting places; crows, hawks, and even ospreys are vigorously attacked and pursued sometimes far beyond the boundaries of the territories; even the bittern is driven to cover in the marsh. Francis Allen tells me that he once saw a redwing "riding on a crow's back for an appreciable length of time."

If a man approaches a nesting colony, even within a hundred feet, the male redwing rises from his lookout perch and flies out to meet him with loud cries of alarm or harsh *chacks*, hovering over his head and threatening to attack him, but seldom actually striking him. Alexander F. Skutch says in his notes: "As I crossed one large meadow where several redwings apparently had nests, I had an escort of guardian males all the way; for as soon as I passed beyond the bounds of the domain of one of them and he dropped behind, another vigilant bird would take over, hover over me, and shriek down imprecations."

Du Bois writes to me of a most pugnacious redwing, saying: "He would hover directly over my head, where I could not see him, and from that advantageous position would strike the top of my head, pecking so hard through a thin summer cap that the blows were quite stinging. After he had struck repeatedly, I hoisted a bamboo staff that I was carrying, directly under him, thus forcing him upward; but he alighted on the top of the staff and sat there, temporarily, looking down at me. Three days later, when I had stooped over, near his nest, he struck me on the back and on the arm, and even alighted for an instant on my back. He attacked the camera, also, when I left it standing on its tripod covered with a focussing cloth."

The great fall and winter roosts of redwings and other blackbirds are well known, but few have noted the early summer roosts of the males alone while the females are busy with their nesting. Dr. A. K. Fisher (1896) has told us about this as observed in southern New York in June: "The red-winged blackbird is another species which appears to leave its mate and family to spend the night in company with other males. While watching in this marsh during the early summer evenings the writer has seen flocks composed wholly of males flying in, from an hour before sunset until dusk. Some of these bands contained a hundred or more noisy fellows, while others were made up of only eight or ten individuals. It is probable that all of the males of a given inland marsh band together toward sunset and come to the great rendezvous to spend the night."

Experiments were conducted by Reginald D. Manwell (1941), at Syracuse, N. Y., in April and May, to determine the strength of the homing instinct in the redwing. He released 133 males at distances varying from 2 to 210 miles from the place of capture; of these, 47 birds were recaptured after their return. "The proportion of birds recaught after any given liberation did not exceed 50 percent and was generally not over 33 percent." Some others may have returned, but were not captured. Most of them returned within a week or two, but some did not appear until the following spring.

Voice.—Aretas A. Saunders contributes the following full account of the song: "The song of the red-wing, well known to bird lovers as *conqueree*, is actually much more variable than this simple rendition. It generally consists of from 1 to 6 short notes, followed by a somewhat longer trill. The quality is pleasing, and the presence of prominent liquid and explosive consonant sounds give it a gurgling sound.

"The *conqueree* song, to my ear more like *ko-klareeee*, is by far the commonest form, the first note being lowest in pitch, the second medium, and the trill highest. Of 102 records of red-wing songs, 46 have 2 notes followed by a trill, and 19 are as described above. A good many songs of different individuals are apparently just alike, beginning on A″, the second note on C‴, and the trill on E‴. On one occasion I listened to 8 birds singing in chorus: 6 of them sang this song, another ended with the trill on D‴, and the other began on C‴ and ended on G‴, but all sang the simple 2 notes and a trill.

"Of my records, 10 have only 1 note before the trill, 29 have 3 notes, 9 have 4 notes, 1 has 5, and 1 has 6; 6 other records do not end in a trill, but follow the trill by a low-pitched terminal note *ko klareeee tup*. While it is common for the trill to be the highest pitch of the song, I have 14 records in which the note before the trill is about 1 tone higher than the trill. A peculiar variation, of which I have 9 records, has the trill made up of notes slow enough to be heard separately and counted. In such cases the number of notes in the trill varies from 5 to 7. Such songs usually have but 1 note before the trill, so that such a song sounds like *ka-lililililip*.

"Red-wing songs are short, varying from ⅝ to 1⅜ seconds. The range of pitch, however, is great, from A′ to G‴. Individual songs are very variable in range of pitch, from half a tone to 8½ tones. The commonest range, and about the average, is 3½ tones; 20 of my records have this range; 10 other records range the 6 tones of a full octave, and 10 more range over an octave. Songs with the greater ranges have 4 to 6 notes before the trill.

"When a male red-wing sings, it commonly spreads the tail, half-spreads the wings, ruffles up the feathers on its back, and lifts the red feathers on its 'shoulders,' so that they flash brilliantly with the coming of the *conqueree*. At times it sings in flight, and often, when flying from one perch to another, hovers a foot or so above the contemplated perch and sings just before alighting. In the spring migration one may find a flock of male red-wings in the tree-tops, nearly every one singing at short intervals, so that the result is a loud continuous chorus. In May, in the nesting season, in a cat-tail marsh well populated with red-wings, there is a chorus of song just as daylight is beginning. Each male sings his song two or three times a minute, and each female continuously emits a high-pitched, sharp call. I do not remember to have heard this call at any other time.

"The common call of the red-wing, usually written *chack*, often sounds to me more like *tsack*. An alarm note, used when one nears the nest, is a downward slurred *peeah*, and another, less frequently heard, is a mournful sounding downward slide, like *peeiiaoh*.

"The season of song begins with the first arrival in spring, which in Connecticut is March or sometimes late February. It terminates in late July or early August. The average of 17 years is July 25, the earliest July 16, 1917, in Connecticut, and the latest August 5, 1940, in Cattaraugus County, New York. Ordinarily red-wings do not sing at all in the fall, but once, October 31, 1937, I found a small flock of males, several of them singing."

Du Bois writes to me: "On April 28 and 29, 1930, I heard a thrush-like song suggestive of the veery coming from somewhere beyond a house; and on May 2, I definitely saw a female red-wing singing this song at the edge of the marsh by the road." I can find no other mention of a female song.

Witmer Stone (1937) gives his impression of the voices of the pair when their breeding ground is invaded as follows:

As one approaches the nesting site the male launches into the air and begins to call *sheep; sheep; sheep; sheep;* each call separated from the next by an interval. Then as the excitement increases there is a long drawn *zeeet* interpolated irregularly thus. *sheep; sheep; sheep; sheep; zeeet; sheep; sheep; sheep; zeeet; sheep; sheep; zeeet,* etc., the bird all the while poised on rapidly beating wings directly overhead, and now and then swooping down still closer. The female, arising from her perch on a cattail, has a similar note but less harsh than the *sheep* of the male, and she also utters a much more rapid and differently pitched series of notes; *chip-chip-chip-chip; chip-chip-chip-chip-chip,* etc., then both birds alight on a bayberry bush and call together, the female seeming to relieve the male entirely from the first part of his cry and to her repeated *chip-chip-chip-chip,* etc., he contributes only the long drawn *zeeet* at regular intervals so that the combination is almost like his opening effort.

In recording the vibration frequencies of passerine song, Albert R. Brand (1938) found that the highest note in the song of the eastern redwing had a frequency of 4,375 vibrations per second, the lowest note 1,450, with an approximate mean of 2,925 vibrations per second.

Enemies.—Probably more redwings have been killed by man than by any other one agency, for when they swoop down in clouds on the corn fields, grain fields, and rice plantations they have been slaughtered in multitudes to protect the crops. Wilson (1832) gives the following graphic account of how they used to be killed in great numbers, while roosting at night in the marshes. In some places—

when the reeds become dry, advantage is taken of this circumstance, to destroy these birds, by a party secretly approaching the place, under cover of a dark night, setting fire to the reeds in several places at once, which being soon enveloped in one general flame, the uproar among the Blackbirds becomes universal; and, by the light of the conflagration, they are shot down in vast numbers, while hovering and screaming over the place. Sometimes straw is used for the same purpose, being previously strewed near the reeds and alder bushes, where they are known to roost, which being instantly set on fire, the consternation and havoc is prodigious; and the party return by day to pick up the slaughtered game.

Before it was made illegal to sell game in the market, redwings were killed in large numbers in the fall and sold in markets as "reed-birds"; when fattened on grain or rice, their little bodies served as delicious morsels for the gourmand's table; few could distinguish them from bobolinks.

The high mortality rate in the nestlings has been mentioned above; probably 50 percent of the eggs laid fail to produce young large enough to leave the nest. The large nesting colonies are fruitful hunting grounds for furred and feathered predators. Crows and grackles eat the eggs, and even the small nestlings, if they are left unguarded. Dr. Allen (1914) accuses the long-billed marsh wren as being accountable for the greatest devastation, which is rather strange since they live so close together in the marshes. He says:

While I was standing near a nest containing two eggs, I noticed a peculiarly acting Marsh Wren about 30 feet away. The vivacious notes so characteristic of the species were not uttered. It made its way through the vegetation directly toward the nest until within about 10 feet of me, when it began to circle. After I had retired to a distance of about 15 feet, the Wren went without hesitation straight to the nest, hopped upon the rim, and, bending forward, delivered several sharp blows with its beak upon one of the eggs. It then began to drink the contents much as a bird drinks water. After a few sips, it grasped the eggshell in its beak and flew off into the marsh, where it continued its feast. * * * That cases are not isolated is shown by the fact that of 51 nests of the Redwing observed in a limited area, the eggs of 14 were destroyed in this or in a similar way, and it is not at all uncommon to find one or more of the eggs of a nest with neat, circular holes in one side, such as would be made by the small, sharp beak of a Wren.

J. A. Weber (1912), of Palisades Park, N. J., tells of seeing a bronzed grackle causing a great commotion in a colony of redwings. He shot the grackle and found a young redwing in its bill; the skull of the young bird, which was large enough to have been out of the nest for about a week, had been crushed. An investigation of the nests in the vicinity showed them to contain only one or two young in each, indicating that the grackles may have robbed them. Usually the grackles take only the eggs or the very small nestlings.

The reactions in a redwing colony to the presence of hawks and other large birds show that they are regarded as potential enemies; great horned owls could do considerable damage to the adults and also to the larger young, as could marsh, sharp-shinned, and Cooper's hawks; even the apparently inoffensive bittern might not object to eating a tender nestling. Minks, foxes, and weasels, and in the drier spots squirrels, could easily climb to the nests and destroy the eggs or young. Wood says in his notes that "water snakes, *Natrix sipedon*, seen in the swamp, gave evidence of having destroyed some nests." The damage done to nests, eggs, and young by predators is, however, not always a total loss to the productive capacity of the colony, for the redwings will continue to build new nests and make repeated attempts to raise their broods until well into midsummer, when their reproductive urge wanes.

Friedmann (1929) calls the redwing "a fairly common but rather local victim" of the cowbird. At Ithaca, Allen (1914) found hundreds of nests but never any cowbirds' eggs. On the other hand, Walter A. Goelitz (1916), of Ravina, Ill., writes: "Until this year I have never found the eggs of this bird in Red-wing nests, but in a little colony of some twenty-five pairs of Red-wing Blackbirds, I destroyed eleven Cowbird eggs on June 17th and six on June 27th of the present season."

Robert H. Wolcott (1899) never saw a cowbird's egg in a redwing's nest during his collecting in Michigan, but found it not unusual in Nebraska. He says: "The owners of the nest, in case eggs of their own have already been deposited, apparently peck holes in all, including that of the intruder, and desert the nest. But in one instance a nest was found where the single, still fresh Cowbird's egg which it contained had been almost entirely buried beneath a new floor, and above this were four Blackbird's eggs."

Out of hundreds of nests, found at Buckeye Lake, Ohio, by Milton B. Trautman (1940), "a Cowbird's egg was found in each of 4 nests. These nests were isolated. Apparently, it was sometimes possible for a Cowbird to lay its egg in a solitary nest without discovery, whereas if it attempted to lay an egg in a nest in a colony, it was driven away. Once eggs were in the nests the Cowbird was not tolerated about the nesting colonies."

Redwings are afflicted with a number of external and internal parasites; Allen (1914) lists four species of *Acarina* and three of *Mallophaga*; and Harold S. Peters (1936) names three species of lice, one fly, three mites, and two ticks that infest the eastern redwing.

In spite of their many enemies, some redwings seem to live for a reasonable number of years. From his study of banding records on Cape Cod, Mass., Packard (1937) has this to say about longevity: "Averages compiled from the 266 returns show that 16 percent of the total number of Red-wings banded survived one year, 7 percent two years, 4 percent three years, 2 percent four years, and 0.3 percent five years after banding." This takes no account of any survivors that did not return to the traps; and the ages of banded birds is not always known. He continues: "The oldest males in the records are two banded as adults in April 1931, and taken yearly through 1936. As it requires at least two years to attain to adult plumage, these birds were hatched in 1929, or earlier, thus being at least seven years old. Several females lived five years after banding." Banding records published by May Thacher Cooke (1937) show that 6 redwings lived for 5 years after banding, 2 for 6 years and 1 for 8 years; only 2 of the 5-year-old birds were banded as young birds, so that some of the others may have been 2 years older than the records indicate.

Field marks.—The male redwing, with his gaudy epaulets, is unmistakable; but the female, with her brown back and streaked breast, is much less conspicuous. At a distance, redwings in any plumage can often be recognized by their flight and flock formations, as described above and as suggested by the field marks of the yellow-headed blackbird.

Fall.—After the young of the second brood are strong on the wing, sometime in July, the females and young gather in flocks and feed on the uplands during the day, returning to the marshes to roost at night. The adult males form separate flocks and follow the same plan. But early in August, all the redwings seem to disappear, during the molting period, and are not much in evidence until the middle of September or later, all in fresh plumage and ready to migrate. Allen (1914) explains this disappearance as follows:

The adult males, which begin molting about two weeks earlier than the females or young, are the first to go, and shortly they are followed by the females and young. To the ordinary observer they have completely disappeared. No longer are they seen leaving the marsh in the morning or returning at evening. Along the ponds, streams, and lake shore there are none to be seen. They are apparently gone from the neighborhood. If at this time, however, one penetrates into the heart of the marsh, where the flags wave four and five feet over his head, he may hear a rush of wings ahead of him as a flock of birds breaks from cover and drops again into the flags a short distance beyond. He may hear this again and again, and yet never see a bird, so impenetrable is the thicket of flags. A few vigorous "squeaks," however, such as frequently draw birds from cover, and the secret is disclosed. A flock of tailless, short-winged birds hover above his head for a moment, and then is off again into the tangle. If specimens are collected, the disappearance of the Red-wings is no longer mysterious. Aside from the loss of the tail, which is obvious, one finds that the outer primary feathers are but just breaking their sheaths. With such handicaps, it is no wonder that the long flights to the uplands are not attempted, and that they seek protection in the effectual shelter of the marsh.

About the middle of September, the males appear again on the uplands, 2 weeks ahead of the females and young. Says Allen (1914):

Well defined migration begins about the middle of October. At that time all loitering ceases, and the evening and morning flights in and out of the marsh are very regular, scarcely a bird lingering during the day. Beginning about three-fourths of an hour before, and continuing about half an hour after the sun has disappeared behind the hills, they can be seen in flocks of from ten to a thousand continually dropping into the marsh. * * * The form of the flock is rather irregular, but always with the long axis at right angles to the direction of flight, thus differing from the characteristic form of the flocks of Grackles which sometimes extend for over a mile in length, although only a few rods wide. The maximum flight occurs at sundown. The morning flight is not so regular as that in the evening, and it extends over a shorter period. Beginning a few minutes before sunrise, flocks are continually in sight for about thirty minutes. Their formation is open and they vary in numbers from a few to over ten thousand birds, the largest flocks extending to the east and to the west as far as the eye can see, but generally not more than a hundred birds deep. * * * The method of segregation of these birds in the morning flight is interesting. A single male or a small group of males, finding themselves in a flock of females, drop out of the ranks and await the appearance of a flock of their own sex, or until their own numbers are sufficiently augmented to form a flock of some size, which they are again up and away. * * * The fall migration continues until about the middle of November. The last birds seen are generally scattered flocks of females."

The southward migration from Cape Cod, and perhaps from other localities in southern New England, apparently starts much earlier than from Ithaca, N. Y., as described above, due to different conditions in the marshes. Fred M. Packard (1936) writes: "The swamps of Cape Cod differ considerably from those about Ithaca. Cat-tails are few at the station, and the marsh plants rarely grow taller than four feet, affording but little shelter. * * * While the marsh studied by Dr. Allen is an ideal place for birds to remain undisturbed during the molting period, the swamps of Cape Cod seem poorly suited for such a purpose." From "the almost complete absence of Red-wings in September and later at the station," and from the dates and localities of recoveries of birds banded at the station, he concludes that they "begin the southward migration in July and August before the summer molt is started, and that they probably complete the molt in swamps after their migration has begun. Unlike the swamps of Cape Cod, many of the marshes on the flight route, such as those found near Newark and Salem, New Jersey, afford suitable protection for molting, comparable to that provided by the marsh at Ithaca."

His map, showing fall and winter recoveries of banded birds, indicates that the flight route from Cape Cod follows along the north shore of Long Island Sound to northern New Jersey, across that State to the Delaware River, avoiding the seacoast of New Jersey, and then along the coastal marshes to South Carolina. All but 1 of his 18 recoveries came from these marshes.

Milton B. Trautman (1940) has this to say about the migration of redwings at Buckeye Lake, Ohio: "During fall the species was more numerous than it was at any other season, and many thousands were present daily. On September 10, 1927, Edward S. Thomas took a picture of a small part of a flying flock. There were more than 400 birds in the picture, and we estimated that there were at least 10,000 in the flock. Undoubtedly, there were days during each fall when 20,000 to 50,000 were present."

Winter.—The winter range of the eastern redwing includes much of its breeding range in the southeastern and southern States. Most of the birds spend the winter south of the Ohio and Delaware Rivers, and from northern Florida to northern Louisiana and northeastern Texas. But some few are to be found occasionally in winter considerably north of these limits, even as far north and east as southeastern Massachusetts, locally and chiefly along the coast.

Their winter habits are much like those of the fall months, when they travel about in large mixed flocks with cowbirds, rusty blackbirds, grackles, and starlings. Milton P. Skinner (1928) says that, in North Carolina in winter, they show a tendency to join with meadowlarks and pipits. He says further: "During the winter from Christmas until March 1927, there was a flock of 200 Red-winged Blackbirds almost constantly with the Cowbirds about the Pinehurst stock-yards. Although they were usually on the ground, they often alighted on low oaks, sapling pines and even on tall gums, clustering close together on the very top in compact flocks. Occasionally flocks of Red-winged Blackbirds were seen elsewhere, particularly about old cowpea fields. Early in the winter, and again after the winter was over, I found these blackbirds about old cornfields, freshly planted oat fields, and swampy places, but I did not see them there during the winter."

DISTRIBUTION

Range.—Southeastern Canada to Florida.

Breeding range.—The eastern redwing breeds from eastern Nebraska, Missouri, eastern Iowa (Johnson and Clayton Counties), northern Wisconsin (Danbury), central Ontario (Sault Ste. Marie, Lake Abitibi), southern Quebec (Saint Félicien, Gaspé), New Brunswick, Prince Edward Island, and central Nova Scotia; south to northeastern Texas, northeastern Louisiana (Mer Rouge, Tallulah), northern Mississippi, south-central Alabama, southwestern Georgia (Newton), central-northern Florida (Cherry Lake, Gainesville), and southern (except the extreme southwest) Georgia (Savannah).

Winter range.—Winters rarely north to Kansas, southern Ontario (Chatham, Ottawa), southwestern Quebec, Connecticut and southeastern Massachusetts; casually to New Hampshire (Warren); regularly south to southern Texas (Brownsville, Tivoli), southern Louisiana, southern Mississippi (Gulfport, Saucier), and Florida.

Casual records.—Casual in southeastern Quebec (Piashti Bay) and northern Nova Scotia (Cape Breton, Sable Island).

Migration.—The following data refer to the species as a whole. Of the 14 races, 5 are migratory, 9 are resident. Many of the migration records are unidentifiable as to race.

Early dates of spring arrival are: Alabama—Scottsboro, February 18. Georgia—Macon, March 9. South Carolina—Walhalla, March

12. North Carolina—North Wilkesboro, February 26. Virginia—
Blacksburg, February 9. West Virginia—Bluefield, February 13.
District of Columbia—January 23 (average of 29 years, March 1).
Maryland—Laurel, February 5 (median of 7 years, February 21).
Pennsylvania—Berwyn, February 3. New Jersey—Cape May, Feb-
ruary 9. New York—Branchport and Shelter Island, February 17.
Connecticut—New Haven, February 10. Rhode Island—Block
Island, February 7. Massachusetts—Harvard, February 22 (average
of 7 years, March 11). Vermont—Putney, February 24. New
Hampshire—Hollis, March 3. Maine—Lewiston, March 7. Que-
bec—Kamouraska, March 8 (median of 14 years, April 7); Montreal,
March 9. New Brunswick—Summerville, March 23. Nova Scotia—
Halifax, April 9. Prince Edward Island—Alberton, March 30 (median
of 9 years, April 11). Tennessee—Nashville, February 10; Athens,
February 12 (average of 7 years, March 1). Kentucky—Guthrie,
February 10. Missouri—Charleston and Warrensburg, February 8.
Illinois—Chicago region, February 19 (average, March 10). Indi-
ana—Waterloo, February 14 (average of 19 years, March 1). Ohio—
Buckeye Lake, February 4 (median, February 23). Michigan—
Washtenaw County, February 12; Blaney Park, March 15. Ontario—London, February 18 (average of 12 years, March 17); Ottawa,
March 15 (average of 30 years, April 2). Iowa—Iowa City, February
17. Wisconsin—Dane County, February 21. Minnesota—Northfield, February 21 (average of 28 years for southern Minnesota, March
14); Stearns County, March 13 (average of 15 years for northern
Minnesota, March 20). Texas—Wichita Falls, February 15. Oklahoma—Caddo, January 22. Kansas—Harper, January 28. Nebraska—Lincoln, February 1; Red Cloud, February 5 (median of
23 years, March 7). South Dakota—Yankton, February 21; Sioux
Falls, March 6 (average of 7 years, March 14). North Dakota—Cass
County, March 6 (average, March 19); McKenzie County, March 31
(average of 10 years, April 14). Manitoba—Killarney, March 22;
Treesbank, March 23 (average, April 9). Saskatchewan—Sovereign,
March 12; Wiseton, March 25. Mackenzie—Fort Simpson, April 28.
New Mexico—Clayton, February 14. Arizona—Grand Canyon Na-
tional Park, February 19. Colorado—Durango, February 6. Utah—
Bear River Refuge, Brigham City, February 10. Wyoming—Carey-
hurst, February 9. Idaho—Rathdrum, February 12 (average of
9 years, March 2). Montana—Charlo, February 10; Fortine, Febru-
ary 13. Alberta—Belvedere, March 18. California—Berkeley,
February 3. Oregon—Klamath Basin, February 8. Washington—
Camas, February 1; Spokane, February 4. British Columbia—
Okanagan Landing, February 12.

Late dates of spring departure are: El Salvador—Lake Olomega, April 6. Sonora—Tesia, April 5. Baja California—San Jose del Cabo, April 5. District of Columbia—May 18. Maryland—Laurel, May 18 (median of 6 years, May 6). New York—New York City, May 15. Massachusetts—Essex County, May 20. Mississippi—Gulfport, April 4. Ohio—Buckeye Lake, median, April 15. New Mexico—Apache, April 27. Arizona—Tucson, May 19. California—Cima, May 12.

Early dates of fall arrival are: California—Lathrop, October 3. Arizona—Grand Canyon National Park, September 15. New Mexico—Colfax County, August 7. Oklahoma—Caddo, September 10. Texas—El Paso, August 7. Ohio—Buckeye Lake, median, September 15. Massachusetts— Springfield, July 5. Maryland—Baltimore County, July 11. District of Columbia—July 8. Baja California—San Jose del Cabo, August 28. Sonora—Hermosillo, October 22. Chihuahua—Chihuahua, November 6.

Late dates of fall departure are: British Columbia—Okanagan Lake, November 22. Washington—Yakima Indian Reservation, November 7. Oregon—Prospect, October 31. California—Lafayette November, 26. Alberta—Whitford Lake, October 29; Belvedere, October 27. Montana—Kirby, December 1. Wyoming—Laramie, December 9; Yellowstone Park, November 2. Utah—Ogden Valley, December 4. Colorado—Ramah, December 6. Saskatchewan—South Qu'Appelle, December 1. Manitoba—Treesbank, November 22 (average, October 24). North Dakota—Argusville, November 20. South Dakota—Mellette, December 22; Sioux Falls, November 15 (average of 6 years, November 8). Nebraska—Cortland, November 30. Kansas—Onaga, November 29. Texas—Dallas, November 28. Minnesota—Minneapolis, December 16 (average of 12 years for southern Minnesota, November 15); Elk River, November 14 (average of 17 years for northern Minnesota, October 28). Wisconsin—Beloit, December 11. Iowa—Winthrop, December 11. Ontario—Point Pelee, December 10; Ottawa, November 10 (average of 20 years, October 18). Michigan—Mount Clemens, December 14; McMillan, December 6. Ohio—Toledo, December 26; Buckeye Lake, November 24 (median, November 20). Indiana—Hobart, December 19; Bicknell and Carlisle, December 11. Illinois—Murphysboro, December 12; Chicago, December 5 (average of 16 years, October 16). Missouri—Palmyra, November 22. Kentucky—Lebanon, November 10. Tennessee—Nashville, November 30. Prince Edward Island—Alberton, October 1 (median of 5 years, September 29). New Brunswick—Memramcook, October 22. Quebec—Montreal, November 14, (average of 9 years, October 23). Maine—Lewiston, November 28; Hudson,

November 15. New Hampshire—Hanover, October 24. Vermont—Shelburne, November 22. Massachusetts—Harvard, November 26 (average of 8 years, October 20). Rhode Island—Block Island, December 4. Connecticut—New Haven, December 7. New York—New York City, December 7; Watertown, December 6. New Jersey—Elizabeth, December 10. Pennsylvania—Warren, November 28. Maryland—Laurel, December 15 (median of 4 years, November 25). District of Columbia—average of 6 years, November 19. West Virginia—Bluefield, December 26. Virginia—Blacksburg, December 14. North Carolina—Raleigh, December 3. Georgia—Americus, November 13.

Egg dates.—Alberta: 30 records, May 28 to July 5; 15 records, June 1 to June 9.

Arizona: 27 records, April 4 to June 25; 14 records, May 18 to June 4.

California: 360 records, March 26 to June 26; 180 records, May 1 to May 31.

Florida: 47 records, April 15 to June 29; 24 records, May 16 to May 20.

Illinois: 74 records, May 10 to June 14; 38 records, May 22 to May 31.

Massachusetts: 103 records, May 16 to June 21; 80 records, May 27 to June 6.

New Jersey: 24 records, May 13 to June 22; 12 records, May 19 to May 31.

North Dakota: 32 records, June 2 to July 5; 20 records, June 6 to June 9.

Oregon: 24 records, May 1 to June 6; 14 records, May 14 to May 18.

Texas: 44 records, May 1 to July 5; 25 records, May 15 to June 5.

Washington: 12 records, April 8 to June 21; 6 records, April 26 to May 18.

Wyoming: 16 records, April 20 to June 9; 11 records, April 20 to April 27 (Harris).

AGELAIUS PHOENICEUS MEARNSI Howell and van Rossem

Florida Redwing

HABITS

Some confusion has existed in the past and some differences of opinion have been expressed as to the proper nomenclature to be applied to the redwings of the eastern United States and as to the distribution of the subspecies. This need not be discussed here, as it is fully explained in a study of the redwings of the southeastern United States by Arthur H. Howell and Adriaan J. van Rossem (1928). They demonstrate that the old name for the Florida redwing, *A. p. floridanus*, should be restricted to the birds of extreme southern Florida, that the eastern redwing (*A. p. phoeniceus*) breeds as far south as Gainesville in northern Florida, and they propose the above scientific name for the redwings that breed over the greater part of the Florida peninsula.

In describing this new race, named in honor of Edgar A. Mearns, they assign to it the following subspecific characters:

Compared with *phoeniceus:* Size smaller; bill longer and more slender, both actually and relatively; coloration of upper parts in females more brownish (less blackish); under parts more buffy (less whitish), the dark streaks more brownish. * * * In the present race, the maximum brownish suffusion found in *Agelaius phoeniceus* is attained; this character at once distinguishes *mearnsi* from all the other races occupying the Caribbean area (*bryanti, floridanus, littoralis, megapotamus,* and *richmondi*).

Specimens from the Gulf Coast of Florida, particularly from the northern portion, have somewhat thicker bills than those from central and eastern Florida, thus indicating a gradual approach in this character to *littoralis* of the western Gulf Coast. Specimens from the Caloosahatchee Valley, (Alva and Ft. Myers) show approach in paler coloration to *floridanus,* of south Florida.

Breeding material is lacking from the lower St. Johns Valley, hence the area of intergradation with *phoeniceus* is not definitely known; quite probably this race will be found to range northward nearly or quite to Jacksonville.

They give the range of *mearnsi* as the "greater part of the Florida peninsula, south to the lower Kissimmee Valley and the Caloosahatchee River; north at least to Putnam County (San Mateo) and Anastasia Island; west on the Gulf coast to Apalachicola."

All through such parts of central Florida, within the limits named above, as I have visited, and these include most of the State, we have always found this redwing to be an abundant resident bird in all suitable places—around ponds, marshes, sawgrass sloughs, wet places in the flat pine woods, or open grassy savannas where there is sufficient moisture. During the two winters that I spent at Pass-a-Grille, Pinellas County, it was a common dooryard bird. A. H. Howell

(1932) says: "The Everglades, before drainage operations were begun, supported an immense population, and even now, with large areas drained and under cultivation, the birds breed there abundantly, as also on the extensive marshes bordering the upper St. Johns and Kissimmee Rivers."

Nesting.—Howell (1932) says: "A number of pairs are usually found nesting near together, their nests being placed in small bushes growing in shallow water or on marshes at a height of 1 to 8 feet above the water or ground. The nests are compactly woven of the stems of saw grass or similar materials and firmly bound to the bush in which they are placed."

Donald J. Nicholson, of Orlando, Fla., has sent me three sets of eggs of the Florida redwing, and tells me that eggs can be found in his vicinity from the last week in March until late in July. The sets all consisted of three eggs each. One of the nests was 3 feet up in a buttonwood growing in a pond; another was similarly located, 2½ feet up, in one of several buttonwoods on swampy land among bunch-grass and sawgrass; the third was attached to the stems of a water-myrtle, 4 feet from the ground, near a pond. Another set in my collection was taken in Nassau County by W. W. Worthington; the nest was suspended 3 feet up among grass in a salt marsh, and held four eggs. There is a set in the T. E. McMullen collection that was taken from a nest in a grass field on a farm. While I was hunting for gallinules' nests in a deep-water pond near Zephyrhills with Oscar Baynard, we found two redwings' nests in ty-ty bushes, not far above the water; the water was so deep that we had to use a boat; there was a broad border of pickerel weed all around the pond, with boggy, or floating islands of flags, *Sagittaria*, small willows, and ty-ty bushes scattered over it, and with bonnets and white pond lilies in the deeper parts.

Eggs.—Three eggs seems to form the usual set for the Florida redwing, but four eggs are not unusual. They are apparently similar to the eggs of the species elsewhere, except for size. The measurements of 50 eggs of the five southern races of this species average 23.5 by 17.1 millimeters; the eggs showing the four extremes measure 27.7 by 17.3, 23.4 by 18.8, 21.3 by 16.8, and 21.8 by 15.5 millimeters.

Howell's (1932) account of the food of the Florida redwing was evidently taken from Beal's (1900) bulletin on the food of this species, as quoted from under the eastern redwing; but Beal distinctly said that the different subspecies were not considered separately.

Philip A. Du Mont (1931), in his paper on the birds of Pinellas County, says of the status of this race in that region: "A few are permanent residents. The bulk of the breeding birds winter farther south and arrive in Pinellas County about the middle of April. I found this species abundant in Collier County in January. * * * The song of this bird seems to differ consistently from that of the eastern bird. An extra descending note is added at the end which makes the song of the Florida bird *conk-a-ree-a*. This was called to my attention first by the late Maunsell S. Crosby."

Holt and Sutton (1926) observe: "The Florida Red-wings are much more graceful than the northern birds. Often they were seen swinging and climbing about the willow or bay-berry bushes, like Baltimore Orioles searching for insects."

DISTRIBUTION

Range.—The Florida redwing is resident from northern Florida, except in the extreme north-central section (Apalachicola, Cedar Keys), and extreme southeastern Georgia (Okefenokee Swamp, Saint Marys); south to south-central Florida (Fort Myers, Jupiter).

Winter range.—In winter wandering to southwestern Georgia (Grady County).

AGELAIUS PHOENICEUS FLORIDANUS Maynard

Maynard's Redwing

HABITS

This form of redwing is resident in the extreme lower part of the Florida mainland and the Florida Keys, north to Lake Worth on the east coast, and the town of Everglade in Collier County on the west, including tropical Florida and much of the everglades. This bird was considered at one time to be the Bahama redwing, *A. p. bryanti*, but it has been shown to be a distinct race worthy of recognition under the name given to it by Maynard.

He (1896) gives it the following subspecific characters: "Form and general coloration similar to that of the Red-wing but smaller, with the plumage more velvety black, and the buff edging to the scarlet shoulder, deeper. The bill is a little longer and much more slender."

Of its nest, Maynard (1896) writes:

The wide-spread marshes of the everglades of Florida are covered with a luxuriant growth of tall grass which attains to the height of five or even six feet. These vast plains form the homes of hundreds of Red-winged Blackbirds and there they also breed. As the grass is submerged in at least a foot of water in the spring, the Blackbirds are obliged to suspend their nests near the top of the stout stalks, of which they bring several together weaving the leaves in the nests and around them in order to make them secure. The everglades are seldom free from wind which often blows a gale, waving the grass back and forth furiously, so that the birds are forced to build exceedingly compact structures or they would be blown to pieces. The nests are therefore made of the leaves of the coarse saw grass which abounds, neatly and firmly woven together. The swaying motion to which their domiciles are constantly subjected, has a tendency to throw the eggs out, and would, were it not that the birds who have doubtless been taught by the experience of former generations, build their nests very deep and, not content with this, they make them more secure by contracting the entrance so much that it is impossible for the eggs to fall out, even when the grass bends so that the tops touch the water. * * *

May first of that same season found me standing on one of the small outer keys, about a hundred miles south of the point last described. This islet, like many others, contained a small lagoon in the center, around which was a belt of land that supported a number of trees, mainly the kinds known as Buttonwood and Mangrove. There were a large number of Red-winged Blackbirds breeding on this Key but I was puzzled to find the nests, for I could not see them in the trees and there were no bushes or grass. After watching them attentively for a few moments, I saw a female emerge from a small hole in a Buttonwood tree not far from the ground, and climbing up to it discovered the nest which was built like that of a Blue Bird. I afterward found several in similar places all containing eggs. For a time I could not understand why the birds had chosen these novel situations for homes, but the *ha-ha* of a passing group of Fish Crows helped to enlighten me, for I knew that the predatory habits of this latter named species renders the eggs of all birds unsafe if exposed, unless the owners are sufficiently strong to protect them, and what the Red-wings lacked in strength they made up in cunning, as they placed their treasures where it was impossible for their enemies to get at them.

Howell (1932) says: "In the Everglades near Royal Palm Hammock, June 12, 1918, I found nests with eggs and young in a saw-grass marsh and in low bushes." Earle R. Greene (1946) found a nest, with one egg on July 24, 1942, on Boca Chica Key, that was 6 feet up in a mangrove bush.

DISTRIBUTION

Maynard's redwing is resident in southern Florida (Everglades, Miami, Key West).

AGELAIUS PHOENICEUS LITTORALIS Howell and van Rossem

Gulf Coast Redwing

HABITS

A. H. Howell and A. J. van Rossem (1928) have given the above names to the redwings that are resident along the Gulf coast region, from Choctawhatchee Bay, in northwestern Florida, westward along the coast at least to Galveston, Tex. Following is their description of it:

Compared with *Agelaius phoeniceus phoeniceus* of northeastern United States: Coloration of females darker, both above and below, particularly on the rump; general tone of upper parts in breeding plumage fuscous-black, with median crown stripe and buffy edgings on nape and interscapular region nearly obsolete; ground color of under parts less buffy (more whitish), the dark streaks broader and averaging more blackish; wing and tail slightly shorter; bill slightly more slender in lateral profile. Compared with *A. p. mearnsi*: Coloration of females throughout very much more blackish (less brownish), the brown and buff edgings to the feathers of the head, nape, interscapular region, and wings very much reduced; streaks on under parts decidedly more blackish, the ground color less buffy (more whitish); bill shorter, and thicker at base; wing averaging slightly longer.

This subspecies, the darkest of all the eastern races, apparently ranges little, if any, above the tidewater region. It appears to be more closely related to *phoeniceus* than to *mearnsi* or *megapotamus*, but material is lacking to show with certainty the area of intergradation with any of these races."

Dr. H. C. Oberholser (1938) calls this redwing "an abundant permanent resident in southern Louisiana. * * * Throughout all the marshes in the Gulf Coast region in the southwestern part of the State, even in the winter, it is one of the most abundant birds."

Nesting.—In May and June 1910, I spent about a month cruising with Warden Sprinkle among the islands off the coasts of Louisiana and Mississippi, mainly in the Breton Island Reservation. On all the islands that we inspected, wherever there was a little moisture and suitable vegetation, we found redwings common and in some places abundant. Their favorite nesting sites were in the black mangrove bushes; the nests were placed 3 or 4 feet above the ground, and usually held three eggs.

Francis M. Weston has sent me some notes on the Gulf coast redwing, as observed in the region of Pensacola Bay, Fla. He says that the nests "are usually built in clumps of the needle rush (*Juncus roemerianus*) and in the scattered bushes and low trees that grow in or adjacent to the areas covered by this rush."

Young.—Weston relates the following experience: "A young bird, barely able to fly, fluttered out of the march at my feet and headed out across a large salt-water pond. I saw at once that it could not possibly reach the far shore.   As a matter of fact, it fell into the water before it had gone 20 feet.   Immediately, it turned back toward the shore it had just left and, disregarding my presence (although my approach had been the occasion of its flight), fluttered to safety along the surface of the water.   It seemed to travel in a sitting posture on the water with the forward part of the body held high and the wings beating the surface without seeming to submerge.   When it regained the shore at my feet, it was not bedraggled—appeared to be perfectly dry—and seemed none the worse for its experience."

Fall.—Weston writes in his notes: "By the end of October, the redwings resort to the high lands in large flocks during the day, although they always return to the marshes in the late afternoon to pass the night.   At this season, too, they can often be found among the sand dunes along the Gulf beaches, where they feed on the seeds of the sea oats (*Uniola paniculata*)—the birds perch on the swaying stalks and pick the seeds from the ripened 'heads.'   Their fondness for the seeds of the long-leaf yellow pine (*Pinus palustris*) seems not to be generally known.   My notes for October 31, 1943, recount my discovery of this preference:

"I have often noticed flocks of redwings in the pine woods in fall, but not until today have I been able to get near enough to find what the attraction was.   Today, I succeeded in driving my car almost under a pine tree in which a flock of about a hundred birds was very active.   Certainly, the birds were eating the pine seeds, though I could not see just how they extracted them from the cones—seed 'wings' were raining down around and upon the car as long as the birds were in the tree."

DISTRIBUTION

Range.—The Gulf coast redwing is resident in southeastern Texas (Brenham, Galveston); southern Louisiana (north, at least, to Crowley, Clinton), central western and southern Mississippi (Saucier, Vicksburg), southern Alabama (Mobile), and northwestern Florida (Pensacola Bay, Whitfield).

Casual records.—Casual farther west in Texas (Tivoli, Eagle Lake).

AGELAIUS PHOENICEUS MEGAPOTAMUS Oberholser

Rio Grande Redwing

HABITS

Oberholser (1919a) describes this redwing as "similar to *Agelaius phoeniceus richmondi* from southern Vera Cruz and Tabasco, Mexico, but larger; female more grayish above and less ochraceus below." He gives as its distribution—

central southern Texas and northeastern Mexico. Breeds north to central Texas; west to eastern Coahuila; south to Nuevo Leon and northern Vera Cruz; and east to Tamaulipas and the Brazos River in Texas.

* * * This new subspecies differs from *Agelaius phoeniceus phoeniceus* in somewhat longer wing, rather shorter bill, and much lighter coloration of the female; from *Agelaius phoeniceus sonoriensis* and *Agelaius phoeniceus fortis* in very much smaller size; and from *Agelaius phoeniceus neutralis* in greatly inferior size and paler female. Birds belonging to this geographic race have hitherto been referred to *Agelaius phoeniceus richmondi*, but they are so different from typical representatives of the latter that subspecific separation seems desirable. It is a larger and less brownish edition of *Agelaius phoeniceus richmondi*, and replaces that form in Texas, Tamaulipas, and Nuevo Leon. It seems to be more or less permanently resident, as no specimens have been taken outside of its breeding range.

Southward it passes into *Agelaius phoeniceus richmondi* somewhere in the northern part of the State of Vera Cruz; westward through central western Texas into *Agelaius phoeniceus neutralis;* northward in central northern Texas into *Agelaius phoeniceus predatorius;* and along the coast of southeastern Texas beyond the Brazos River into *Agelaius phoeniceus phoeniceus* of the southeastern United States.

In considering Oberholser's names, as used above, allowance must be made for some changes that have been made since his paper was written.

Nesting.—Although he did not recognize it as a subspecies, George B. Sennett (1878) was the first to give any information on the nesting habits of this blackbird in the vicinity of Brownsville, Tex.; he writes;

"I found this species breeding in great numbers along the Lower Rio Grande. They usually build their nests low, among the rank growth of weeds and willows that spring up in the resaca beds after the annual overflows of the river. One nest, however, I found at least 20 feet high in a mezquite-tree. It was composed of bleached grasses and attached to a leaning branch; it was partly pensile, and looked like a large nest of the Orchard Oriole, *Icterus spurius*. I was deceived into climbing for it."

On May 23, 1923, near Brownsville, I found a number of nests of the Rio Grande redwing, containing from three to four eggs, placed from 3 to 10 feet above the ground in some slender willows growing along the edge of a pond.

Referring to the same general locality, Herbert Friedmann (1925) says: "Some 15 nests were examined and all were in bushes or trees in dry locations and varied from within five feet of the ground to over 20 feet above it. In all his years of field work in this region Camp has never found a Red-wing's nest built over the water."

George Finlay Simmons (1925), referring to the Austin region, which is probably near the northern limit of the breeding range of this race, says that the nests are placed "1 to 20, usually 6, feet up, firmly woven to limbs and twigs of willow or ligustrum trees or bushes, to cattails, blood-weeds, reeds, rushes, tules, cane or saw-grass; along creeks, sloughs, river margins, draws, edges of pasture ponds, and about artificial lakes."

DISTRIBUTION

Range.—The Rio Grande redwing is resident from central Texas (Del Rio, Kerrville, Giddings) south to southeastern Coahuila, México, and northern Vera Cruz.

AGELAIUS PHOENICEUS ARCTOLEGUS Oberholser

Giant Redwing

HABITS

Oberholser (1907) characterized this large northern form as "similar to *Agelaius phoniceus fortis*, but female decidedly darker below, the streaks more blackish and more extensive, about as broad as the white interspaces; above more blackish. Male with wing and tail averaging shorter; bill larger; and buff of wing-coverts somewhat paler. He reported its geographical distribution as "Montana, North Dakota, Minnesota, and northern Michigan, north to Keewatin, Athabaska, and Mackenzie; in migration south to Colorado, Texas, Illinois, and probably Ohio," and says further: "This new form is much like *Agelaius phoeniceus phoeniceus* in color, the male in this respect being practically indistinguishable, and the female barely less blackish above and below; but in size *A. p. arctolegus* is much greater, as the subjoined measurements will show. It differs from *Agelaius phoeniceus neutralis* in larger size; in more blackish upper parts, broader and darker streaks on the lower surface of the female; and paler buff on the shoulder of the male."

He did not give it the common name "giant," which does not seem to be warranted, inasmuch as his tables of measurements show that the average measurements of the thick-billed redwing are somewhat greater than those of the present form; it would seem that the name "northern" would be more appropriate. In the same tables, *arctolegus* seems to have a thicker bill than *fortis*, the so-called thick-billed redwing! However, these common names are much more likely to stand the test of time than are the so-called scientific names, which are subject to change at the whim of any "specialist in speciation." For a thorough study of the status of this subspecies the reader is referred to a very enlightening paper on the subject by P. A. Taverner (1939), which well illustrates the difficulty of recognizing some of these microscopic subspecies when taken away from their breeding grounds.

The series of redwings that we collected during the breeding season in southwestern Saskatchewan proved very puzzling; they were not quite typical of either *fortis* or *arctolegus*, but the measurements of my birds seem to agree rather closely with those given for the latter, to which form they should probably be referred.

Spring.—As the giant redwing cannot be recognized in life, it is almost impossible to trace its migration through the ranges of other forms. Wetmore (1937) reports a specimen taken in Nicholas County, W. Va., as late as May 11, 1936. It probably reaches the southern portions of its breeding range in late February or early in March, in much the same way as the eastern redwing does, the passing flocks coming first and the resident birds later. Ian McT. Cowan (1939) thus describes the arrival in the Peace River district in British Columbia: "When we reached Tupper Creek on May 6 male blackbirds were abundant, seemingly defending territories and in full song. Large flocks of migrating birds composed entirely of males were moving through daily and it was noticed that little mixing took place between the 'residents' and the migrants. * * * A very few females were seen on May 9 and subsequently, but not until the 18th did the females start to arrive in numbers. Just after dawn on this date a flock of between twenty and thirty females came down and joined the males."

Nesting.—Although not recognized as such at the time, this was undoubtedly the form of redwing we found nesting so abundantly in North Dakota in all the reedy sloughs. In the large sloughs, where there were hundreds, perhaps thousands, of nests of yellow-headed blackbirds in the tall reeds in the deeper parts, we found the redwings almost as common, with nests around the edges in the shorter vegetation over shallower water, and in the long grass on the borders.

O. A. Stevens (1925), of Fargo, N. Dak., published a short paper on the redwing population of a ditch that drained the marshes of the Red River of the North, in which cattails and marsh grasses were growing. "The early nests, which include most of them, were built in last year's cat-tail stalks, from 1 to 2 feet above the water. On

July 13, the new growth of cat-tails had reached its full height and flower stalks were present. The nests containing eggs at that date were new and placed 3 to 4 feet above the water. Of the nests found on June 11, three were lined with a handful of fluff from the last year's seed-stalks. These were the only such seen. A few nests were in small willow trees."

He has sent me some notes on the same colony, made the following year: "The first birds seen were two males on March 29. On April 26, three females were present and the number of males was increasing rapidly. No signs of nests appeared until May 10 when five had been begun. The first eggs were found May 21 and the first young on June 3. Repeated rains the first of June converted the ditch into a river. Several nests were flooded on the seventh and by the tenth all of the earlier ones were covered. On the latter date ten new nests were found, high up in the small willow trees, except one which was in a last year's sweet clover plant."

Approximately 70 nests were in use at the date of flooding. No more nests were built in the cattails until July 19, when two nests with eggs were found. Most of his new nests were built in 4 days and the first egg was laid from 1 to 4 days thereafter; but in one case the birds took 7 days to build the nest and waited another week before laying the first egg.

Geographically, the blackbirds that we found nesting abundantly in southwestern Saskatchewan in 1905 and 1906 are referable to this race, though we doubtfully recorded them at that time as *fortis*, as explained earlier in this account. They were very common around the sloughs and along the creeks, nesting in the flags and long grasses on the edges of the sloughs and over the water in the shallower portions. I collected a nest, containing four fresh eggs, at Crane Lake on June 5, 1905; it was placed 10 inches above the water in a bunch of reeds (*Scirpus*) on the edge of a slough; the nest was well made of dry reeds and was lined with dry grasses.

In the Peace River district of British Columbia, Cowan (1939) found giant redwings breeding rather late in the season. "Egg laying commenced about the end of May at Austin's Pond where the birds were all building in dead sedges before the new growth was well under way. Many pairs on the shore of Swan Lake did not complete nest building until about June 23. Here the nests were mostly in the dense stands of *Equisetum* growing in the shallow water and in consequence nest building had to await growth of these early in June."

At Austin's Pond, he says: "Repeated observation led us to the conclusion that there were but four males with the six females. On one occasion one male was observed to mate with two different females within the space of ten minutes." He found two males that were breeding in the immature plumage characteristic of the first winter.

Eggs.—The giant redwing lays larger sets of eggs than the southern races, from four to six. In a series of 30 sets in the collection of A. D. Henderson, of Belvedere, Alberta, there are 7 sets of five and 2 sets of six. The eggs are indistinguishable from those of the eastern redwing.

Winter.—The giant redwing ranges in winter to Kansas, Arkansas, Louisiana, Alabama, Texas, and Illinois, and as a straggler as far east as Connecticut; but many winter farther north.

Roberts (1932) says: "Flocks of Red-wings, often of considerable size, may remain through the winter in southern Minnesota, feeding in weed-grown corn-fields, around barns and strawstacks and open springy marshes and brooks, and spending the nights in the sheltered lowlands." Stragglers are often found even farther north, enduring temperatures below zero.

This redwing seems to be a common, and perhaps an abundant, winter resident in Ohio. Milton B. Trautman (1940) estimated that about 20 percent of the redwings migrating through or wintering at Buckeye Lake are referable to *arctolegus.* Louis W. Campbell (1936) writes: "During the past 8 years flocks of from 20 to 300 Red-winged Blackbirds have been found wintering about Toledo. * * * In an effort to determine the composition of these flocks of wintering birds, twenty-three specimens were collected during 1934, 1935, and 1936, between the dates of December 27 and February 29. Twenty-one of these proved referable to *Agelaius phoeniceus arctolegus.* * * * The earliest spring specimen of *Agelaius phoeniceus phoeniceus* was taken on March 12, 1933. The evidence thus indicates that the common wintering Red-winged Blackbird of the Toledo region is *Agelaius p. arctolegus.*"

DISTRIBUTION

Range.—Yukon and Mackenzie to Louisiana.

Breeding range.—The giant redwing breeds from southeastern Yukon, central Mackenzie (Fort Norman, Fort Simpson), northwestern Saskatchewan, north-central Manitoba (The Pas, Oxford House), and western and northeastern Ontario (Lake Attawapiskat, Moose Factory); south to central British Columbia (Williams Lake, Tachick Lake), southwestern Alberta (Waterton Lakes Park, Milk River), eastern Montana (Powder River County), southern South Dakota (Menno, Vermillion), and Iowa (east to Tama and Van Buren Counties).

Winter range.—Winters casually north to southern British Columbia (Okanagan Landing), southeastern Saskatchewan (Estevan), southern Manitoba (Brandon), northern and central Minnesota (Hennepin County), northeastern Illinois (Waukegan), southeastern Michigan (Erie), southern Ontario (Brankford), central Ohio (Licking County) and western West Virginia (Mason County); regularly south to north-central Colorado (Semper), central Texas (Boerne, Edge), and Louisiana (Belcher, Jefferson Parish).

Casual records.—Casual in southeastern Alaska (Mole Harbor, Sergief Island), central Yukon (Mayo Landing), west-central British Columbia (Kispiox Valley), northern Manitoba (Churchill), extreme northeastern Ontario (Cape Henrietta Maria), central New York (Cayuga and Tompkins Counties), Connecticut (North Haven), and Georgia (Tifton).

Accidental in northern Alaska (Cape Price of Wales, Barrow) and northern Mackenzie (headwaters of the Dease River).

AGELAIUS PHOENICEUS FORTIS Ridgway

Thick-Billed Redwing

HABITS

The thick-billed redwing seems to be very closely related to the giant redwing and so much like it in measurements that Ridgway (1902) did not separate the two forms. He called *fortis* the northern redwing and assigned to it the far northern range of the bird we now call *arctolegus*. Both of these two forms are about the same size, considerably larger than the eastern redwing, and both have thick bills, as mentioned under the preceeding race.

The thick-billed redwing, as it is now understood, breeds from Idaho, Wyoming, and South Dakota to Colorado and northern Texas. Its breeding range extends well into the foothills of the Rocky Mountains, where it has been detected at elevations of 7,500 and 9,000 feet in Colorado.

Young.—At an Iowa nest of the redwing, evidently of this race, Dr. Ira N. Gabrielson (1915) made the following observations on the female feeding the young:

Altogether during the 170 feeding visits she brought 203 morsels of food. Of these, grasshoppers were 34.97%, moths 9.37%, larvae 9.35%, unidentified 17.24%, and the remaining 29.09% was composed of various insects. The unidentified were mostly small insects captured among the arrowhead lilies but we could not identify them. A very small frog was fed on one visit. As far as numbers were concerned the distribution of food to the nestlings was very equal, A receiving 34.97% of the insects fed, B, 32.51%, and C, 31.51%. It is not so easy to estimate the percentage by bulk on account of the varying sizes of the insects fed. * * *

The position of the blind and the surrounding vegetation exposed the nest to the sun from 8:30 to 10:10 while it was shaded during the remainder of the day. On July 1, the day on which we watched during this period, she spent 50 minutes or exactly one-half the time in shading the young while not a minute was so spent at any other time of the day. In shading the young she always assumed the same position with her head toward the sun and broadside to the blind. One foot was placed on each side of the nest, the beak held wide open, the wings half spread and slightly drooping, and the feathers of the head and neck elevated. This resulted in entirely shading the young and is the most perfect development of this brooding position yet noticed in an individual bird.

Voice.—The song and call notes of the thick-billed redwing are generally considered to be similar to those the eastern redwing, but Francis H. Allen tells me of a song that he heard in Colorado that "ended with a peculiar turn something like *conqueree-ee-lyoo*."

Enemies.—Cowbirds are probably more abundant throughout the range of this redwing than they are in the east, hence this blackbird sometimes is often imposed upon. L. R. Wolfe wrote to Friedmann (1934): "Probably ninety percent of the redwing nests [in Decatur County, Kans.] contained one or more eggs of the cowbird and I remember frequent extended searches to find a nest without eggs of the parasite."

Winter.—While large numbers of thick-billed redwings remain in winter throughout the southern portions of their breeding range, especially in Colorado, there is a heavy southeastward movement in the fall toward their winter quarters in the Southern States, from New Mexico to Louisiana.

Harry Harris (1919) writes of their coming to the region of Kansas City, Mo.: "They began arriving in small numbers about the middle of November and continued coming in increasing numbers until during the intense cold periods of late December and January there were countless thousands resorting to common roosts in the timbered bottoms along the Missouri River. In the early mornings when the birds scattered to feed, great flocks flew over the city to their feeding grounds on the prairie regions many miles to the south and west. It is estimated that some of the flocks covered daily from thirty to fifty miles on these journeys."

W. E. Lewis (1925) gives the following graphic account of the immense flocks of redwings, with a few Brewer's blackbirds and cowbirds, as seen flying to and from their winter roosts in Oklahoma:

They could be seen coming for three or four miles, in a column that resembled at that distance the line of smoke given off by a distant locomotive, except that it was constantly writhing and twisting like a sinuous serpent. As the dark band approached, the individual birds could be distinguished. The band was perhaps thirty feet across and there were usually about ten to fifteen birds to the rod of cross section. Sometimes there are fewer than this, but sometimes many more. The column was not continuous. Possibly there would be a mile or two of blackbird ribbon, then a gap of half a mile, then a longer section. On February 13,

I saw a practically continuous stream about seven miles long. It is hard to accurately estimate the total number of individuals, but I think thirty thousand would be conservative.

DISTRIBUTION

Range.—Montana and western Kansas to Louisiana:

Breeding range.—The thick-billed redwing breeds east of the Rockies in western Montana (Teton and Gallatin Counties), western Nebraska, and western Kansas (Decatur County); south through southeastern Idaho (Bear Lake County), central and central-eastern Utah (Salt Lake City, Spanish Fork, Moab), and Colorado to southwestern Utah (Pinto, Saint George), southern Nevada (intermediate toward *sonoriensis*), central and central-eastern Arizona (San Francisco Mountains, McNary), central and southeastern New Mexico (Fort Wingate, Carlsbad), and (probably) northern and western Texas (Boise, Canadian, Ysleta).

Winter range.—Winters from northern Utah (Morgan County), Colorado (Barr, Colorado Springs), and eastern Nebraska (Lincoln); south to western and central Texas (El Paso, Hot Springs, Eagle Lake); casually east to Arkansas (Fayetteville, Arkansas County), Tennessee (Reelfoot Lake), Mississippi (Rosedale), and Louisiana (Belcher).

AGELAIUS PHOENICEUS NEVADENSIS Grinnell

Nevada Redwing

HABITS

This Great Basin redwing breeds from southeastern British Columbia and northern Idaho, through much of northeastern California and southward on the east side of the Sierra Nevada to San Bernardino County, and through Nevada to eastern Arizona, New Mexico, and western Texas.

A. J. van Rossem (1926) in his study of the California races gives this form the following diagnosis: "Bill stouter than in *caurinus* or *sonoriensis*, but still decidedly more slender than in *neutralis*. Males with exposed portions of middle wing coverts usually clear buff, but frequently with a small amount of black present, and occasionally with the exposed black even predominant over the buff. Females decidedly less buffy than *caurinus* and with darker and broader ventral streaking than in *sonoriensis*. Not always distinguishable from *neutralis* in coloration, but streaking below averaging narrower and sharper, and bill diagnostic if similar ages are compared."

His seems more comprehensive and clearer than the original

description by Grinnell (1914a) which follows: "In shape of bill and other general characters closely similar to *A. p. sonoriensis;* male scarcely distinguishable, but female conspicuously darker colored, on account of the great relative breadth of black streaking both above and below; in this respect similar to female of *A. p. caurinus,* but bright rusty edging on back and wings replaced by ashy and pale ochraceous; bill in male of *caurinus* more slender than in either *sonoriensis* or *nevadensis.*"

Spring.—Claude T. Barnes has sent me the following account of the spring behavior of the redwing: "During the spring of 1942 I frequently visited Farmington bay, Utah, for the purpose of recording migration dates, especially those of the Nevada redwing (*Agelaius phoeniceus nevadensis*). Despite the severe cold winter and stormy late spring, the male redwings appeared on the creek-willows on February 20. They sang perfunctorily, and, while sitting, none showed their red epaulets. Day after day there was little change in the male flock, except that it grew more vociferous, on March 19, for instance, the male chorus being very pronounced, with much clucking as well as song. Still only the yellowish crescent showed on the wing.

"For the next few days the male flock was dispersing, each male selecting his favorite locale, a brook-footed post here, a marsh fence there, or a reeded patch where slow water ran, always apparently a spot where fresh water was near and perches such as willows and wires either existing or in the making, such as ungrown rushes.

"And then, on March 23, a flock of 30 drab little females appeared on the scene, staring curiously about the fen from fence wires and manifesting no interest in the scattered males, who, indeed, reciprocated their indifference. A male atop a fence post beside them treated them as harmless strangers. When the females flew it was in a flock together.

"On April 9 the yellow-headed blackbirds appeared; but the redwings were still in status quo—female flock, isolated males.

"On April 13 the male's epaulets showed brilliant red in sitting posture; and for the first time the female flock began to disperse. Males began chasing females, and by April 25 no sign of a female assemblage remained."

Jean M. Linsdale (1938) made the following interesting observation on the nests:

A feature of the nests of red-winged blackbirds of special interest was noticed in Smoky Valley. This is that the lining in nearly every instance was pale yellow or whitish in color. This contrasted especially with the almost invariably dark color of the lining of the nest of the Brewer blackbird. In these two species as in others which had light or dark colored nest linings, the whitish lined nests were in open situations often exposed to the sunlight, the dark lined nests were in shaded places usually protected from direct sunlight. This contrast extended also to the color of the down on nestlings. Nestling red-wings had conspicuously whitish down, nestling Brewer blackbirds were decidedly blackish. These seem fairly obvious examples of adaptions to exposure to sunshine—the whitish nests and young to reflect sunrays, the dark ones to absorb them. Apparently it is desirable for both the eggs and young birds to be thus protected.

Nesting.—In general the nesting habits of the Nevada redwing do not differ materially from those of other races of the species, the nests being placed low down in tufts of grass, in marsh vegetation, in various shrubs near water, or as high as 5 or 10 feet from the ground in willows. Robert Ridgway (1889) "found a colony which had built their nests in 'sage bushes' (*Artemisia tridentata*) growing in and about a shallow alkaline pond, on Antelope Island, in the Great Salt Lake." J. S. Rowley has sent me the following account of an especially dense colony in an isolated locality: "I found an old reservoir on the desert between Mojave, Kern County, and Little Lake, Inyo County, Calif., on a deserted farm. Since there was no surface water for miles around, these redwings had taken this place over. The tule patch was only about 50 feet square and there must have been at least 200 nests occupied there on April 19, 1934. I had to use great care in going through the tules so as not to trample redwing nests."

Eggs.—The Nevada redwing ordinarily lays four or five eggs, probably more often four than five. These are indistinguishable from eggs of adjacent races.

Food.—E. R. Kalmbach (1914) gives this redwing credit for eating large numbers of alfalfa weevils in Utah. "Of 42 birds examined, only 2 had failed to eat at least a trace of the weevil, and it was taken on an average of 5.24 adults and 27.16 larvae per bird. In bulk it amounted to 40.76 percent of the stomach contents."

George F. Knowlton (1944) says that a redwing, probably of this race, "was collected in an alfalfa field southeast of St. George, Utah. Microscopic examination of its stomach contents revealed that it contained a great mass of pea aphids (*Macrosiphum pisi*) estimated to exceed 1,400 individuals. The pea aphid population in this field was high enough to cause moderate crop injury. A second male red-wing was collected approximately one-half mile away along an alfalfa-field fence line and near to sugar-beets. This stomach contained 85 pea aphids; one of four additional aphids it contained was a green peach aphid (*Myzus persicae*), a species that causes some damage to nearby sugar-beets intended for seed production."

Behavior.—Walter P. Taylor (1912) writes of the behavior of this redwing with relation to other species:

On more than one occasion was the belligerent disposition of this blackbird in evidence. Flocks of four to eight individuals were frequently seen pursuing some distressed raven; they swooped at the fleeing bird with every appearance of intent to do bodily harm, but I was not able to observe that they did actually strike the fugitive. Individuals do not seem to be particularly timid about attacking a raven, even when no other redwings are about. Magpies come in for a share of abuse. Apparently redwings do not confine their attacks to birds of their own size or larger, for one was observed driving a Savannah Sparrow from a grass stem. Upon the flight of the sparrow, the blackbird settled down on the vacated perch.

Linsdale (1938) noted considerable evidence of polygamy: "Just as in the yellow-headed blackbird a great disproportion was noted in the numbers of males and females at each nesting colony. This was not always apparent upon casual watching, but close study revealed it to be the condition practically everywhere. At 1 or 2 places where there was only a single nest there was 1 male, and 1 female, but usually there were several females and several nests for each male in the colony. Once on May 14, 1932, 1 such group composed of 1 male and 6 females flew up from a marsh in Smoky Valley and lit on a buffalo-berry bush."

DISTRIBUTION

Range.—British Columbia to Nevada and Arizona.

Breeding range.—The Nevada redwing breeds from central-southern and southeastern British Columbia (Kamloops, Newgate) south through central Washington (Conconully, North Dalles), northern Idaho (Coeur d'Alene, Lewiston), west-central Oregon (Gateway, Prospect), and central-northern and eastern California (Seiad Valley, Yosemite, Little Lake) to central-southern California (Victorville; Death Valley) and southern Nevada (Ash Meadows).

Winter range.—Winters north to south-central British Columbia and northern Idaho; south to western and southern California (Palo Alto, Oro Grande) and southern Arizona (Lochiel).

AGELAIUS PHOENICEUS CAURINUS Ridgway

Northwestern Redwing

HABITS

Ridgway (1902) describes this redwing of the humid northwest coast as "similar to *A. p. phoeniceus* but wing and bill longer, the latter more slender; adult male with buff of middle wing-coverts deeper (deep ochraceous-buff or ochraceous in winter plumage); adult female rather more heavily streaked with black below and, in winter plumage, with upper parts much more conspicuously marked with rusty."

Comparing it with other California races, A. J. van Rossem (1926) gives it the following diagnosis: "Bill longer and more slender than in *nevadensis* or *sonoriensis*, and slightly different from either race in shape. Adult males with middle wing coverts clear buff, unmarked with black except in examples from northwestern California and southwestern Oregon, where intergradation with *mailliardorum* has left its impress. Females richly marked in strongly contrasting colors, the plumage being suffused with buff and the feathers edged with rich browns and buffs at the expense of gray tones; the scattered feather edgings of the interscapular region usually light, contrasting strongly with the rest of the plumage."

The northwestern redwing, which seems to be nowhere especially abundant, is mainly migratory, though a few spend the winter as far north as western Washington. Its summer range extends from southwestern British Columbia to northwestern California, at least to Humboldt Bay. Van Rossem (1926) says that it "winters much farther south than is generally supposed. It is of common occurrence in the San Francisco Bay district * * * and in the San Joaquin Valley."

I cannot find that it differs materially in its habits from the other California races.

DISTRIBUTION

The range of the northwestern redwing lies west of the Coast Ranges from British Columbia to California. It breeds along the coast from southwestern British Columbia (Courtenay, Abbotsford) to northwestern California (Eureka, Requa), and in land along the lower Trinity River in California. It winters throughout its range and south to central-western California (Palo Alto) and the Great Valley of California (Gray Lodge State Game Refuge, Buena Vista Lake). It is accidental in northern Sonora (Sonoyta).

San Francisco Redwing

HABITS

This local race is evidently a bicolored redwinged blackbird, recently separated from the more widely spread *A. p. californicus*, the coastal representative of that subspecies, once regarded as a species.

In naming it, A. J. van Rossem (1926) describes it as "similar to *Agelaius phoeniceus californicus*, but bill smaller and less swollen at base. Females with wing averaging slightly longer, coloration darker and posterior underparts rarely streaked. Males with exposed portions of middle wing coverts usually entirely black." He gives the range as: "Central coast region of California from central Monterey County north at least to Sherwood, Mendocino County; east to include Suisun Bay and the western slopes of the inner coast ranges." And he adds: *Mailliardorum* is the darkest of the races of *Agelaius phoeniceus* found in the United States and probably represents in the least diluted form the formerly widespread stock which has so plainly left its mark throughout the west on the invading *'phoeniceus'* strain. Females of the streaked type occur rarely. In San Benito County there is, as would be expected, a tendency toward streaking which reflects the proximity of *californicus;* and in Mendocino County, where an approach to *caurinus* takes place, the same condition is observed. These streaked females are darker than the corresponding type of *californicus*, and they are of course distinguishable by smaller bill."

This race apparently does not differ at all in its habits from the closely related bicolored redwing, its nearest neighbor.

DISTRIBUTION

The San Francisco redwing is resident in central coastal California (Sherwood, Lower Lake) south to Carmel River, Soledad, and Paicines.

AGELAIUS PHOENICEUS CALIFORNICUS Nelson

Bicolored Redwing

HABITS

The history of the above name is interesting. We older naturalists can remember when there were only three kinds of red-winged black-birds, all full species, recognized in North America. These were the red-and-buff-shouldered blackbird (*Agelaius phoeniceus*) in the east, the red-and-white-shouldered blackbird (*Agelaius tricolor*) in California, and the red-and-black-shouldered blackbird (*Agelaius gubernator*) in California and Mexico. Audubon (1842) figured these three species and used the above three names, which survived for half a century, in the 1886 and 1895 A.O.U. Check-Lists.

Agelaius gubernator is a Mexican species, and it was not long after the publication of the 1895 Check-List that E. W. Nelson (1897) discovered, in comparing specimens of this species from the tablelands of Mexico with those from California, that "certain differences are found which warrant the naming of a geographical race. As *A. gubernator* was described from the tablelands of Mexico it follows that the California bird is the new one.

"The breeding females of typical *gubernator* from the plains of Puebla lack nearly all of the light streaking on the entire upper surface, including the wings, and the light streaks are less marked on the lower surface.

"Among other differences from true *gubernator* are the notably smaller size and slenderer bills of the northern birds."

He proposed calling the California bird *Agelaius gubernator californicus*, and this name was adopted in the 1910 Check-List.

The discussion that followed, as to whether *gubernator* was specifically distinct from *phoeniceus*, at least as shown in the California races, finally led to another change in the name, for which Joseph Mailliard (1910) was mainly responsible. In his long and exhaustive study of large series of specimens of the California races, he seems to have demonstrated satisfactorily that the *gubernator* and *phoeniceus* types are connected by every degree of intergradation, and are therefore not specifically distinct; he proposed to call the California bird the bicolored redwing, and this name was officially adopted in the 1931 Check-List, making the bicolored redwing a subspecies of *Agelaius phoeniceus*. For the steps which led to this conclusion, the reader is referred to Mr. Mailliard's illustrated paper.

For comparison with other California races, A. J. van Rossem (1926) gives the following diagnosis for *A. p. californicus:*

Bill similar in shape and size to *Agelaius phoeniceus neutralis*, but males with exposed portions of middle wing coverts more extensively black, rarely clear buff, sometimes entirely black, but usually with a small amount of buff visible, particularly on distal middle coverts. Females averaging much darker throughout and less streaked (more blackish) below. Differs from *Agelaius phoeniceus mailliardorum* in much heavier bill in both sexes. Males with longer tails, and with middle wing coverts less frequently entirely black. Females with slightly shorter wings, under parts usually more streaked, and coloration paler throughout. * * * Range—Tejon Pass, in extreme northwestern Los Angeles County, north through the San Joaquin-Sacramento Valley to about 4 miles south of Red Bluff, Tehama County, Calif. East in suitable localities into the Sierra Nevada foothills; west to the eastern slopes of the inner coast ranges and to, but not including Suisun Bay.

The specimen figured by Audubon (1842) was supposed to have been taken on the Columbia River, but this is far beyond its present known range; if the locality is correctly given, it must have been a straggler.

Courtship.—Grinnell and Storer (1924) describe the courtship performance as differing only slightly from that described for the eastern redwing:

As soon as the flocks begin to break up, the males commence courting and their displays are carried on with little cessation from daylight to dark throughout the nesting season. For this they seek some open situation, never far from the favorite swampy haunts. The male lowers and opens his tail in wide fan shape, spreads and droops his wings until the tips reach to or below his feet, raises his red wing patches outward and forward like a pair of flaming brands, and having swelled out as large as possible, utters his curious throaty song, *tong-leur-lee*. Usually this is done while he is perched; less often he mounts into the air and flies slowly over a circling course without departing far from the object of his attention.

Nesting.—Of the nesting sites chosen by the bicolored blackbird in the Fresno district, John G. Tyler (1913) writes:

Almost every clump of tules in the various sinks and ponds is made use of by nesting blackbirds, while in many instances a colony will take possession of a grain field, building their light, basket like structures amid the swaying wheat or barley stalks, from six inches to two feet above the ground.

Not infrequently this species departs from the usual customs that have been followed for so long, and nests in treetops. One such colony found May 25, 1906, was occupying some willows along a canal, one nest was fully thirty feet from the ground and resembled a kingbird's home, except that several long streamers of dry tule strips were left dangling and swaying in the breeze, making the nest very conspicuous. That this site was chosen from preference and not from necessity was clearly evident, as there was a growth of tules all along the edge of the canal, and a half section of wheat adjoining. Another colony chose nesting sites among the thick foliage of a long row of fig trees, the nests being situated from twelve to twenty feet above the ground. In driving along the road after the leaves had fallen from the trees I counted eighteen nests in a short section of the row. Almost under these trees was a small ditch in which water stood nearly all summer, and which was partly concealed by willows, tules, and sedges; but perhaps the close proximity of a schoolhouse had taught the birds to elevate their nests and conceal them as well.

The nests of the bicolored blackbird are usually built in tules at various heights above standing water. A typical nest of this race is thus described by Grinnell and Storer (1924):

The nest consists of three parts: (1) An outer loosely woven framework of tule leaves fastened to the standing (dead) stems and growing leaves of the tule thicket. The attachment of this outer framework to the tules is very loose, an arrangement which undoubtedly saves some nests from being tipped over when one side is attached to growing tules and the other to a dead stem. (2) Next comes the body of the nest, a firm structure comprising some tules, but chiefly of finer material. This material is worked in while wet, either while it is green or, perhaps, after it has been taken to the stream-side and moistened. Some foxtail grass of the current season and still partly green was incorporated in this layer of one of the nests examined. Some of the material, in the particular nest here described, had a coating of green algae suggesting that tules broken down into the water had been used. This middle, wet-woven layer when dried and ready for use is so strong as not to break on moderate pressure with the hands. This is the important structural element in the nest. (3) Finally there is an inner lining of fine dry grass stems of the previous year's growth. The fibers of this layer are chiefly interwoven with each other, but some extend into the middle layer and hold the two layers together. This inner layer forms the soft lining on which the eggs and later the newly hatched young rest. Later still it gives a holdfast for the sharp claws of the growing young who can thus secure themselves against being tumbled out of the nest during high winds or when the nest is beset by marauders.

On this point they say that "a single young bird, nearly fledged, was found in one of the nests examined at Lagrange. When an effort was made to lift this bird from the nest, he clung tenaciously to it and each of his sharp claws had to be released in turn from the lining material. Later, when released over dry ground, he flew in a direct line toward the nearest patch of green, a willow tree, and the instant he touched the foliage he seized the latter with clenching claws and hung there until disengaged again."

Eggs.—Four eggs seems to be the usual number for the bicolored redwing; sets of five are rare, and Grinnell and Storer (1924) report one set of six. They say: "The ground color of the eggs is pale blue, and the scattered markings of dark brown or black, chiefly at the larger end of the egg, consist of dots, spots, streaks, and lines, the latter often running around the pole of the egg." Bendire (1895) says: "The average measurement of forty-four specimens in the United States National Museum Collection is 24.07 by 17.35 millimetres, or about 0.95 by 0.68 inch. The largest egg in the series measures 26.42 by 17.78 millimetres, or 1.04 by 0.70 inches; the smallest, 21.34 by 16.76 millimetres, or 0.84 by 0.66 inch."

Food.—F. E. L. Beal (1910) made a comprehensive study of the food of the bicolored redwing based on the examination of 198 stomachs collected in every month in the year. The food was found to consist of 14 percent animal matter and 86 percent vegetable matter. The greatest amount of animal food, insects, was eaten in May, amounting to nearly 91 percent. Beetles, mostly leaf bettles and weevils, aggregated about 5 percent. Wasps and ants were eaten very sparingly in summer, as were certain bugs, less than 1 percent of each for the year. Grasshoppers constituted over 15 percent of the food in July, but only 1.5 percent for the year. Caterpillars aggregated 5.5 percent for the year, but amounted to over 45 percent of the food in May. The vegetable food consists of grain and weed seeds.

Grain amounts to 70 and weed seed to 15 percent. The grain consists of corn, wheat, oats, and braley. Oats are the favorite. They amount to over 47 percent of the yearly food, and were eaten in every month except February, when they were replaced by barley. The month of maximum consumption was December, when nearly 72 percent was eaten, but several other months were nearly as high. Wheat stands next to oats in the quantity eaten, nearly 13 percent. It was taken quite regularly in every month except March and May. Barley was found only in stomachs taken in February, October, and November, and nearly all of it was taken in February. The average for the year is 5.5 percent. Corn is eaten still less than barley, and nearly all was consumed in September, when it reached 46 percent of the month's food. A little was eaten in May, August, and October, but the aggregate for the year is only slightly more than 4 percent.

Fruit is not eaten by the bicolored redwing. Among the weed seeds, amounting to 15 percent of the total food, he lists 12 species of troublesome weeds and other useless plants. Of the food of the young, he says: "The food was made up of 99 percent of animal matter and 1 percent of vegetable, though most of the latter was mere rubbish, no doubt accidental. Caterpillars were the largest item, and amounted to an average 45 percent. Bettles, many of them in the larval state, stood next with 32 percent. Hemiptera, especially stinkbugs and leafhoppers, amounted to 19 percent. A few miscellaneous insects and spiders made up the other 3 percent."

As to the economic status of this blackbird, he writes: "In summing up the facts relating to the food of the bicolored redwing, the most prominent point is the great percentage of grain. Evidently if this bird were abundant in a grain-raising country it would be a menace to the crop. But no complaints of the bird's depredations on grain have been made, and it is significant that the grain consumed is not taken at or just before the harvest, but is a constant element of every month's food. As the favorite grain is oats, which grows wild in great abundance, it must be admitted that, with all its possibilities for mischief, the bird at present is doing very little damage."

Tyler (1913) says that in the Fresno district, "farmers regard this bird with considerable disfavor on account of its fondness for newly planted grain, and because of its attacks upon ripening Kaffir, or Egyptian, corn. In districts where large fields of alfalfa are under irrigation these birds are of much service in destroying various bugs and worms."

Harold C. Bryant (1912) made a thorough study of the relation of birds to a grasshopper outbreak that occurred in the San Joaquin Valley in 1912. Although the number of grasshoopers taken per day by each individual bicolored redwing was exceeded by the daily numbers taken by several other species (this bird ranking sixth in this respect), by reason of its much greater abundance the total number of grasshoppers destroyed by the species as a whole far exceeded that for any other species. He figured that the total population of bicolored redwings consumed 78,590 grasshoppers per day; western meadowlarks came next with a daily score of 24,720 for the total population. "The bicolored redwing was the bird most abundant. Large flocks of from one to four hundred individuals were often seen busily engaged in catching grasshoppers. At times these flocks were seen at a considerable distance from their usual habitat. They appeared to feed almost wholly in the infested districts, and more often in alfalfa fields than in pasture land."

DISTRIBUTION

Range.—The bicolored redwing is resident in the Great Valley of California from Fouts Springs, Red Bluff and Columbia Hill south to Los Banos, Cuddy Valley, and Visalia.

Casual record.—Casual in southeastern California (Calipatria).

AGELAIUS PHOENICEUS ACICULATUS Mailliard

Kern Redwing

HABITS

Joseph Mailliard (1915a) described this scarce and extremely local subspecies as "similar to *Agelaius phoeniceus neutralis*, but of larger size, feet averaging somewhat larger; but chiefly characterized by a longer, and comparatively more slender bill than any other form of this genus in the United States."

Of its range, he said: "So far this form has only been found in east-central Kern County, Calif., in the Walker Basin, just north of the town of Caliente, and on the South Fork of Kern River, between Isabella and Onyx, thus probably being restricted to a very small range." And speaking of its coloration and markings, he said that this form "seems to be between *neutralis* and *nevadensis*, both racially and geographically, and appears to have been developed by some unknown factor in the small area it must occupy among the foothills of the southern Sierra. Specimens of *Agelaius* taken at Buena Vista Lake, thirty or forty miles west of this area, and across the plains, are indistinguishable from the general run of *neutralis*, while the form on the east is *sonoriensis*, and that on the northeast is Grinnell's new form, *nevadensis*."

A. J. van Rossem (1926) gives the following diagnosis for the Kern redwing:

Size larger and bill longer than in any other California race. Males very similar to *californicus* both in individual and average amount of black present on exposed portions of middle wing coverts. Females also paralleling *californicus* in variability, but coloration richer; feather edgings, where present, stronger in tone, with rich browns and buff at a maximum; grays at a minimum. * * *

In view of its coloration *aciculatus* is obviously of "*gubernator*" origin, and because of its isolated habitat it has not been affected by the thick-billed "*phoeniceus*" stock which is now dominant in the San Diegan Faunal Area and in parts of the San Joaquin Valley. Such modification as has taken place has come from the east, from the slender-billed chain, as is at once apparent from bill proportion and shape. * * *

Aciculatus departs entirely from its breeding grounds directly after the nesting season. The bulk of the individuals probably winter in the San Joaquin Valley, but because of their comparatively limited numbers the collecting of one is a matter of chance. There is at hand a female taken at Buena Vista Lake on December 30, a young male from the same locality April 14 (not breeding) and an adult male from Corona, Riverside County, December 8. The Corona male is not typical but is best referable to this race.

Mailliard (1915b) in a later paper makes these further remarks on the Kern redwing:

That the habitat of the Kern Red-wing is extremely limited seems, from our present knowledge, to be a reasonable conclusion, even though it is known to inhabit two districts rather widely separated topographically. The first place where it was found was the "Walker Basin," which is a meadowlike valley of only a few thousand acres in extent, separated from the San Joaquin Valley by a range of mountains over four thousand feet high, its only outlet being by way of a narrow gorge through which the Walker Creek flows into the Kern River, whose bed is at the bottom of a narrow canyon for miles below the point of intersection. The marshy portion of the Walker Basin is so limited that but few individuals exist there. In fact we saw none at all while passing along the edge of this district, but van Rossem took some there in 1914.

As far as we know, the next, and only other, spot where these birds are to be found is on the South Fork of the Kern River, some four or five miles above its junction with the North Fork, twenty-five or thirty miles farther inland than the Walker Basin and separated from it by two fairly high ranges of mountains, the river itself being probably at an elevation at this point of some 3,000 feet. Here the narrow valley opens out a bit, to half a mile or more in width, with "fans" covered with desert vegetation running up into the steep canyons that cut into the masses of shattered rock which constitute the mountains on either side. In the comparatively level bottom are small marshy spots and lagunas where bunches of tules or cat-tails grow, while in places water has been brought in from the river and alfalfa or barley is grown.

We found the red-wings mostly in the lagunas, or near them, though some were seen among the hundreds of Brewer Blackbirds (*Euphagus cyanocephalus*) which were following the water as it spread over the fields and feasting on the insects among the alfalfa. The red-wings were usually in small groups or colonies, and far from numerous. In fact we came across but few spots they seemed to favor by their presence. This irrigated strip extends some eight or ten miles up the river to where the valley contracts again and it seems to be the only likely locality in which to expect these birds in all that neighborhood.

DISTRIBUTION

The Kern redwing is resident in the mountain valleys of east-central Kern County, south-central California (Bodfish, Isabella, Weldon, Onyx). In winter probably near breeding range; recorded at Buena Vista Lake.

AGELAIUS PHOENICEUS NEUTRALIS Ridgway

San Diego Redwing

HABITS

Ridgway (1902) describes this race as "similar to *A. p. sonoriensis*, but smaller, the adult female darker, with streaks less strongly contrasted above, those on lower parts rather broader and grayer, the upper parts with little if any rusty, even in winter plumage."

Comparing it with other California forms, A. J. van Rossem (1926) calls it "similar to *Agelaius phoeniceus californicus* in size and shape of bill. Males with exposed portions of middle wing coverts more extensively buffy, often unmarked with black. Females more streaked (less blackish below) and with coloration paler throughout. Differs from *Agelaius phoeniceus nevadensis* in heavier bill in both sexes, and in broader streaking on underparts of the females." He finds its range to be the Pacific drainage from Sierra Juárez, in Baja California, to west-central San Luis Obispo County, in California; and adds:

Neutralis is a common resident in all suitable localities in the San Diegan Faunal Area. Along the southeastern border of its range there is, because of environmental conditions, no intergradation with *sonoriensis*. Intergradation may occur in the San Gorgonio Pass region of Riverside County, but there is no direct proof of this possibility. The easternmost station for neutralis in this region is Redlands, while a tongue of *sonoriensis* extends up into the Coachella Valley on the desert side. * * *

Neutralis is resident in the sense that the breeding area is coextensive with the winter range. A single exception to this statement is an adult male, No. 8205, Museum of Vertebrate Zoology, taken six miles west of Imperial, Imperial County, May 6, 1909, which is unquestionably referable to *neutralis*.

This race does not differ materially in its habits from neighboring races.

DISTRIBUTION

The San Diego redwing is resident in southwestern California (Santa Margarita, Redlands, Jacumba) and northwestern Baja California (Sierra Juárez, El Valle de la Trinidad, El Rosario). It is casual in winter in southeastern California (Imperial).

AGELAIUS PHOENICEUS SONORIENSIS Ridgway

Sonora Redwing

HABITS

A. J. van Rossem (1926) describes the Sonora redwing as being—

Of the slender-billed *sonoriensis-nevadensis-caurinus* chain. Bill longer and more slender than in *nevadensis* and of different shape than in *caurinus*. Males with middle wing coverts more often and more extensively marked with black than in *nevadensis*, and therefore not to be confused in this respect with *caurinus* which is, except in the extreme southwest corner of its range, essentially an immaculate buff-winged form. Pale tipping of feathers in fall plumage more extensive and paler than in the other California races, and very frequently persisting (even in fully adult males) on the interscapular region until the bird is in worn (late May) plumage. Females by far the palest of the California races. Paler and with narrower ventral streaking than in *nevadensis*; paler and less buffy than in *caurinus*, with markings more diffused (less contrasted) than in that form. * * *

After examining the type, a young female in first winter plumage taken at Camp Grant, 60 miles east of Tucson, Ariz., February 10, 1867, van Rossem concludes:

This locality is east of the established breeding range of *sonoriensis* as now understood and in a region occupied by both *fortis* and *nevadensis* in winter. Mr. Ridgway [1902] himself gives the type locality as "Mazatlan, w. Mexico." It is unfortunate that the type was not selected from the latter locality, for Mazatlan birds are essentially the same as Colorado River valley specimens. In color, the type is not quite like the average from the metropolis of the race and its bill is shorter than any other female *sonoriensis* so far examined. It recalls certain young females of *fortis* in some particulars and its identity may yet be shown to lie in that direction. However, the case demands further material for final solution and I continue to apply the name, for the present, to the birds inhabiting the lower Colorado River and its tributaries and the coastal districts of Sonora and Sinaloa.

E. W. Nelson (1900) was the first to call attention to the unfortunate selection of a type for what we now call *sonoriensis;* his paper throws some light on what has caused considerable confusion as to the propriety of the name, as well as to the distribution of the subspecies.

Harry S. Swarth (1929) makes the following comments on Ridgway's type and the status of this form in Arizona:

It [the type] differs from the mode of the *Agelaius* of the lower Colorado Valley, to which the name *sonoriensis* has been generally applied, in having a distinctly heavier, stubbier bill, in which particular it can not be matched in a large series of Colorado River birds. In coloration, however, it is closely similar to some females from the Colorado River, and correspondingly different from the mode of *nevadensis* and *fortis*. Altogether, I am disposed to let the name *sonoriensis* continue to stand for the Colorado River form, and to regard the type specimen as a stray or migrant, a winter-taken bird from beyond the normal breeding range of the subspecies. There has already been such a confusion of the names applied to this race, as well as to the proper type locality, that I am unwilling to suggest a change that might cause further trouble.

The point I wish to emphasize here is the fact that there are two subspecies of *Agelaius phoeniceus* breeding in southern Arizona, one occupying the valley of the lower Colorado River and its tributaries as far east as Tucson, the other, the region east from the Santa Catalina and Santa Rita mountains. Breeding birds from Phoenix and Tempe are mostly indistinguishable from Colorado Valley specimens. Breeding birds from near Tucson are intermediate, some of them having distinctly heavy and stubby bills, as compared with the slender-billed western race, but on the whole they are best associated with the Colorado Valley subspecies.

If we accept the conclusions of van Rossem and Swarth, which seem reasonable in view of our present knowledge, we must revise our ideas of the breeding range of the Sonora redwing. For a number of years in the past this form was supposed to breed in suitable localities entirely across the arid portions of southern Arizona and even in extreme southwestern New Mexico, but now its breeding range seems to be limited to the area cited under "distribution," below.

DISTRIBUTION

The range of the Sonora redwing lies in southeastern California and Nevada to western Mexico. It is resident from southeastern California (Indio), southern Nevada (opposite Fort Mohave, Arizona), central-western, central, and southeastern Arizona (Mohave, Wikieup, Safford); south to northeastern Baja California (Colorado Delta) and northern Sonora. It winters south to southern Baja California (Santiago, San José del Cabo), southern Sinaloa (Mazatlán, Escuinapa), and central Durango (Papasquiero).

AGELAIUS TRICOLOR (Audubon)

Tricolored Redwing

HABITS

This handsome blackbird was discovered by Nuttall near Santa Barbara, Calif., in 1836. He sent a male specimen to Audubon, who described it in his Ornithological Biography (1839) and figured it in his other great illustrated works as one of the only three forms of redwings recognized at that time. His specific name has stood on the A. O. U. Check-List ever since as a binomial; it has not been split into subspecies, nor has it been shown to integrate with other forms of *Agelaius*. Nuttall wrote to Audubon at that time: "Flocks of this vagrant bird, which, in all probability, extends its migrations into Oregon, are very common around Santa Barbara in Upper California, in the month of April." Its range is now known to extend from southern Oregon, west of the Cascade Range, southward through California, west of the Sierra Nevada, to northwestern Lower California. Its center of abundance seems to be in the San Joaquin Valley in California.

Coues (1874) questioned the status of this bird as a distinct species on the grounds that its bill is similar in shape to that of some of the races of *phoeniceus*, and "the difference in the shade of red is no greater than that observable in specimens of *phoeniceus* proper, while the bordering of the red in the latter is sometimes nearly pure white."

Baird, Brewer, and Ridgway (1874), however, point out certain differences which seem to substantiate the tricolored redwing's claim to specific status:

Immature males sometimes have the white on the wing tinged with brownish-yellow, as in *A. phoeniceus*. The red, however, has the usual brownish-orange shade so much darker and duller than the brilliantly scarlet shoulders of the other species, and the black has that soft bluish lustre peculiar to the species. The relationships generally between the two species are very close, but the bill, as stated, is slenderer and more sulcate in *tricolor*, the tail much more nearly even; the first primary longer, usually nearly equal to or longer than the fourth, instead of the fifth.

Two strong features of coloration distinguish the female and immature stages of this species from *gubernator* and *phoeniceus*. They are, first, the soft bluish gloss of the males, both adult and immature; and secondly, the clear white and broad, not brown and narrow, borders to the middle wing-coverts.

The lesser wing coverts ("shoulders") of the adult male are colored a much darker red than in any of the subspecies of *A. phoeniceus*, a dull crimson, or the color of venous blood, very different from the bright vermilion or scarlet of the other species.

Ralph Hoffmann (1927) says of the haunts of this redwing: "In the San Joaquin and Sacramento Valleys there are many small irrigation reservoirs fringed with a dense growth of tules. From these in spring and early summer issues a medley of droning and braying sounds, and lines of blackbirds fly out in all directions to the neighboring fields or fly back with food for the young."

Courtship.—Lack and Emlen (1939) made extensive studies of the breeding behaviour of tricolored redwings: "Intensive watching in an *uncrowded* portion of the Willow Slough colony showed that each male held a territory some six feet square, to which it usually confined its movements when in the colony, in which it sang and courted, and from which other males were driven out. An immature male was once tolerated in an adult's territory for two minutes, but it was promptly driven out when a female arrived. Fighting never seemed serious and boundary demonstrations, so common in typical territorial birds, were not seen."

They were unable to distinguish separate territories in the central part of the colony, where the population was denser.

Both male and female often showed similar raising and lowering of expanded wings and tail as a preliminary to copulation. When inviting the male, the female usually arched the body and pointed the beak vertically upward, sometimes quivering the wings or raising and lowering the beak. * * * Once, from a tree in the colony, a male flew down in song and with expanded wings to copulate with a female below; this is the only case we observed of an aerial song-flight. * * *

As in *phoeniceus*, polygamy seems the rule. Of three males in contiguous territories with known boundaries, two had three and the other had two building females; laying occurred in seven of the eight nests. Occasionally a male displayed to two females in quick succession. All the females laid at about the same time. The females usually ignored each other, but occasionally chased each other short distances.

At times, two males were seen displaying to the same female, but usually one, the trespasser, was chased away. As already noted, in an owner's absence, his neighbor trespassed and courted one of his females. One female, which was individually distinguishable, returned with building material when her own male was absent; the next-door male postured sexually, whereupon she flew over to his territory and both displayed. Her own male then reappeared, and she returned and displayed with him. In neither case did copulation follow. These incidents suggest that promiscuity may occur at times, but polygamy, not promiscuity, would seem the rule where we watched; we do not know that this is true for the denser parts of the colony.

Nesting.—The tricolored redwing is one of our most highly gregarious species. It nests in enormous, most densely populated colonies, the nests being placed more closely together than in any other colonies of marsh-nesting blackbirds. Estimates of population density have been made by many observers; these estimates are subject to wide variation and some of them are evidently inaccurate or were carelessly made. Johnson A. Neff (1937) devoted six seasons to a careful and thorough study of the nesting colonies of this species over most of its breeding range in California, and his counts and estimates seem to have been made more accurately than some others. He makes the following general statement: "The writer has noted almost every possible variation in density of population. Twelve nests were observed in one small willow, and thirty-six were counted in one clump of about four tall willows growing from the same root. In cattails, nests have been noted at least as numerous as one to each three square feet; from one stand in thick cattails, without moving the feet except to rotate, we counted from sixteen to thirty-six nests; the average of many counts ran well over twenty. A count made in a marginal colony averaged one nest to each nine square feet. In another colony sample counts, in a number of ten-foot squares, ranged from sixteen to thirty-four nests."

His tables, showing the variation in sizes of different colonies, are quite enlightening, the numbers running from less than a hundred to over two hundred thousand nests in a colony. Some of his descriptions of various colonies follow:

About twenty miles east of Sacramento a reservoir, on what is known as the Nimbus Ranch, owned by the Natomas Company, was dammed or dug, about 1912, as a source of water supply for gold dredgers. Cattail and tule developed about 1916, and since 1920 or 1921 blackbirds have inhabited the area in great numbers. Marsh growth in 1932 covered 30 to 40 acres. On March 4, 1932, the roosting population of this area estimated at "nearly a half-million birds," fed over an area fully forty miles in diameter. By April 25, 1932, nesting was under way, and by May 1 many of the nests held full sets of eggs. In May 1932, many trips were made to this marsh, and the estimate of several cooperators was placed at 100,000 nests. * * * By 1935, dredgers had so changed the terrain that only 2,000 to 3,000 returned to this place; the feeding area was too far away. In 1936 this locality was deserted; three smaller marshes a few miles away were densely occupied by a population totaling about 100,000.

On April 30, 1932, at a point five miles west of Watsonville, Piper found a colony of about 1,000 Tri-colors nesting in a rather dry marshy area; there was no standing water, but there was a thick tangle of blackberry vines, nettles, and rather sparse cattails. Nests were uniformly in early stages of construction, with no eggs.

On May 14 and 15, 1932, Gabrielson and Jacobsen found a nesting colony in a patch of thistles on a small slough about fifteen miles northwest of Merced on the Crane Ranch road. The thistle patch was from 75 to 125 feet wide, forming an almost impenetrable jungle. Nests held eggs or young. These observers estimated that the birds numbered between 60,000 and 75,000 pairs.

On May 19, 1933, the writer discovered a huge flight of Tri-colors on the holdings of the Dodge Land Company and the Perriott Grant ranch which overlap the Glenn-Colusa county line northeast of Butte City. Here there are a number of sloughs which are not continuously filled with water; their width varies greatly and it is virtually impossible to estimate the total area. On May 30, 1933, tens of thousands of birds were flying back and forth into the cattails and tules in these sloughs, carrying nesting materials. The birds were active over an area roughly four miles east and west by six miles north and south. The number of birds, apparently all nesting in the slough area, was so far beyond comprehension that after spending parts of three days here the writer gave up in despair with the thought that an estimate of 250,000 adults was ridiculously low.   On July 18, 1933, another visit to the section disclosed a general area of about forty square miles centering around these sloughs which literally teemed with squalling young Tri-colors and adults hustling for food for the immense aggregation.

On May 10, 1934, a nesting colony was noted in marshes which extend from the Culver Ranch into the Cross Ranch, four miles east of Norman, Glenn County. About two weeks later, after nesting was under way in the entire marsh, an irrigation company official, practiced in judging land areas, estimated that nesting covered virtually sixty acres.   During the nesting period many nest counts were made on sample areas; all averaged close to one nest for every five square feet. Even at one to ten square feet, the nests in this marsh would number about 260,000.   As the estimated number of nests listed in this report is 200,000, this permits sufficient allowance for any parts of the marsh not so heavily populated.

Rollo H. Beck wrote to me on May 22, 1944: "In last month have destroyed 850 nests in one farmer's grainfield, and in a small ditch filled with tules; found several spots where two separate nests were placed on same tules.   Of interest to me was the grainfield nesting, nests in barley or attached to mustard stalks and barley.   Last year 10 colonies were all in tules in water."

Dawson (1923) says of the nests: "The nests, I say, are *everywhere*, now at middle levels, 2 or 3 feet above the water, where one may peep into them, now overhead where we must thrust in exploratory fingers, now hung perilously close to the water where a change in level may overwhelm them.   Now and again they crowd each other, when two or three birds select the same stems.   Here are two nests side by side, and here one above another.   Here a bird has lashed her foundation too high, and the top will not go on because of a neighbor's foundation."

Elsewhere (1927), he describes the nest as follows: "Each nest is lashed firmly within a group of upright cattail stems; and an art which anchors an edifice midway of such unencouraging rods is a high art.   The sides of the nest are both woven and coiled, but the bottom is coiled only, and that most ingeniously.   I have seen a dead cattail leaf five feet long reduced to a single close-set spiral.   The body, or matrix, of the nest is made of macerated leaves, or vegetable waste, laid on wet.   Occasionally a little mud finds its way into the composition, but this is not essential.   And, finally, after the matrix is well dried, a smart lining of coiled grasses is added, and egg-laying begins."

In the same paper he writes of the zoning system in the colony:

Nesting commences on schedule and proceeds with the regularity of clockwork. We do not know where the High Council convenes which assigns quarters to the incoming citizens, but we do know that first comers, to the number, it may be, of a thousand, gather in the center of the swamp.   Days pass and nothing is done. Then as at a given signal, all fall to work and begin nest-building.   This central group, of those who have received building licenses, works thenceforth unremittingly, and with such uniformity of success that a visitor can determine the very day when first eggs, second eggs, and so on, laid by practically every female member of the commune. Newcomers—and there is from now on a constant stream of influx—in like manner, group themselves in a section immediately adjoining the central colony. These first tarry for recruits, and then set to at a given signal. Thus, in contiguous but distant sections of a large swamp, one may find the nest-under-construction group, the one-egg-laid group, the just-hatching group, and so on, all on the same day. In some smaller swamps, there are concentric rings of activity.

John G. Tyler (1907) tells of a colony in a dense growth of nettles in a low, damp sink at the end of an abandoned slough:

In the lowest land the nettles were very dense and some of them were six feet or more in height; but toward the border where the ground was higher and dryer they gradually became smaller until at the outer edge they were scarcely six inches high and were finally replaced by a rather thin growth of foxtail grass. On two sides of the nettle patch was a more or less dense fringe of willows. * * *

Before reaching the nettles I was somewhat surprised when a female blackbird fluttered up from the grass and revealed a nest built on the bare ground. A rather hasty search resulted in the finding of several other nests in like situations. These were all built out in the short, thin grass and not concealed at all or protected from the rays of the sun and would certainly have made a rich harvest for some prowling egg-eater. There was nothing, however, to indicate that they had been disturbed in any way.

After entering the nettles, he found that there "were nests everywhere: in some instances three or four built one on top of another, tho in such cases only the upper one appeared to be occupied. The average height from the ground was between one and three feet, but many were seen that were ten and twelve feet up in the willows. They were all built almost entirely of grass stems that had been freshly pulled, giving the nests a bright, green appearance. Some of them had a few coarse brown weed stems woven into the framework but in the majority no other material but the grass was used and none contained any lining. As the heads of the grass had not been detached, the nests presented a ragged, fuzzy appearance."

A fact that impressed him more, perhaps, than anything else was that "in the center of the colony where the nettles were thickest, nearly all of the nests contained small young birds and doubtless it was the parents of these that I first saw. A little farther out, however, there were full sets of badly incubated eggs while near the outside were incomplete sets of fresh eggs."

Eggs.—Dawson (1923) says that—

"four eggs being the stern rule of *A. tricolor*, sets of five or six were pretty sure to contain an egg structurally weak. The lime had played out. Of the only set of seven found, one egg collapsed in the nest, and another in being transferred to the collecting box. * * * The eggs of the Tricolored Redwing are normally of a pale niagara green tint, sharply and sparingly marked—small-blotched or short-scrawled—with an intense brownish black pigment. The variation, not in the quality but in the application of this single pigment, determines the highly varied results secured. Often the pigment is shadowed, or "washed," along its edges, revealing thus its brown character. Not infrequently a tinge of the pigment is suffused throughout the shell, and we get such basic tints as glaucous, yellowish glaucous, "tilleul buff," and even deep olive-buff. Again, and more rarely, the pigment is spread about superficially, in whole or in part, paling thus to vinaceous buff, or fawn-color. In two instances in the M. C. O. collections the color appears as a uniform vinaceous clouding on a warm buff ground; and in one of these the freckling is so minute and so uniform as to render the egg almost indistinguishable from that of a Yellow headed Blackbird."

According to Bendire (1895) the measurements of 201 eggs in the United States National Museum average 27.75 by 20.35 millimeters; the largest egg measures 30.78 by 22.61, and the smallest egg 21.59 by 18.29 millimeters.

Young.—Lack and Emlen (1939) state that "the incubation period, determined by comparing the stage of development on various dates through May and early June in each of 4 colonies, is about 11 days, the fledgling period 13 days." They said that "both sexes fed the young," but Grinnell and Storer (1924) observed, in another colony, that "the females did all the work of feeding the young."

Joseph Mailliard (1914) writes:

After hunger fear seemed to be one of the first sensations developed in the young nestlings. So much was this the case that the youngsters, say a week old, would flop out of the nests on the approach of a human being and fall into the water. * * * As the young left the nest and took to the tules their feeling of fear did not diminish, and they would flutter or scramble away so fast in the thick high tules that it was a difficult matter to procure a few for specimens. * * *

By June 15 the colony was greatly scattered, many of the young accompanying their parents abroad in search of food. * * * Those old enough for flight seemed to return to the tules every night, and often for the purpose of finding rest and shade in the daytime as well. By July 1 the colony was beginning to disintegrate, and even before that date small flocks of old and young together could be seen working toward the north, while but few were noticed returning from that direction.

Plumages.—In a general way the plumages and molts of the tricolored redwing are similar to those of the other redwings, with a few specific differences, some of which are shown in descriptions by Ridgway (1902), who said: "Young (sexes alike) much like summer female, but general color browner and under parts of body narrowly streaked with dull grayish white; middle and greater wing-coverts margined terminally with dull buffy whitish, producing two narrow bands; tertials narrowly margined with dull buffy whitish.

"Immature female (in first winter [plumage]) similar to the adult female in winter, but much browner, the pileum, hindneck, and back strongly tinged or washed with brown, and the superciliary and malar stripes, lighter streaks of anterior under parts, and margins of wing-coverts brownish buffy."

He does not describe the young male in first winter plumage, but Dawson (1923) describes the young male in his first spring as "like adult, but lesser wing-coverts tawny or brownish red, variously admixed with black; the middle coverts wholly black, or variously mixed black and white."

After the first complete postnuptial molt, young birds become indistinguishable from adults. Ridgway says that the adult male in winter is "similar to the summer male, but plumage still softer and more glossy and middle wing-coverts more or less tinged with brownish buff."

The adult male in summer he describes as "uniform glossy blue-black, the plumage with a silky luster; lesser wing-coverts brownish carmine or dull crimson; middle coverts white, in abrupt and conspicuous contrast." The adult female in winter is "similar to the summer female, but plumage softer, more glossy, and of a more grayish cast, with pale (light buffy grayish) margins to feathers of lower parts much broader."

Dawson (1923) describes the adult female in spring as—

Similar to that of *Agelaius phoeniceus*, but more uniform in coloration and much darker; above sooty black, nearly uniform, from back posteriorly, but with some obscure skirtings of brownish gray on head and nape; below sooty black, nearly uniform from breast posteriorly, although with faint skirtings of lighter, or whitish—these skirtings sharply defined on lower tail-coverts; breast mingled black and whitish in about equal proportions, clearing anteriorly to white, sparingly flecked with black on throat; an obscure whitish line over eye; lateral coloration throughout blending the characters of upper and lower plumage; a dull ruddy element often present in the whites, and (in older examples ?) the lesser wing-coverts more or less skirted with dark red.

Food.—Beal (1910) examined the stomachs of 16 tricolored red-wings, of which he says: "From the examination of so small a number, final data on the food can not be obtained, but so far as the testimony goes, it indicates that both species (*phoeniceus* and *tricolor*) consume more insects and less grain than the bicolored. The stomachs of the tricolored contain 79 percent of animal matter to 21 of vegetable. The animal matter consists mostly of beetles and caterpillars, with a decided preponderance of caterpillars. The vegetable food is nearly all weed seed. One stomach alone contained barley."

Mailliard (1914) tells a slightly different story:

By the time incubation was completed in the majority of nests and vast numbers of young beaks were opening wide for needed nourishment, the barley in the neighborhood was just reaching the pulpy stage, being "in the milk," as it is called, when the kernels of grain are much relished by the redwings on their own account and much prized as food for the young. Hence a large amount of damage is done by these birds when the grain is in this state, and this keeps up even when the grain becomes quite hard. But, while thousands of the redwings were visiting the barley fields, as many more were bringing in grasshoppers, cutworms, caterpillars and various sorts of insects in various stages of growth, and probably the harm done to the grain is more than offset by the good work of destroying injurious pests of the insect world.

Nuttall wrote to Audubon (1842) that tricolored redwings were seen in the suburbs of Santa Barbara, feeding "almost exclusively on the maggots or larvae of the blow-flies, which are generated in the offal of the cattle constantly killed around the town for the sake of the hides."

Economic status.—As can be seen from the above statements, the economic status of the tricolored redwing is not as bad as it has been painted. Examination of stomachs has shown that a surprisingly small amount of grain is consumed, which is largely offset by the immense numbers of injurious insects eaten. Where the birds are especially abundant, however, hordes of them may be seen flying to and from the fields of barley, rice, and other grains, leaving many stripped heads of ripening grain on the stalks, resulting in considerable losses, especially in fields near the large breeding colonies.

Behavior.—The outstanding characteristic of the tricolored redwing is its highly gregarious behavior at all times, the density of its nesting colonies, the immensity of its flocks, and its social habits. Dawson (1923) puts it very aptly as follows: "*Agelaius tricolor* is intensely gregarious, more so perhaps than any other American bird. Every major act of its life is performed in close association with its fellows. Not only does it roost, or ravage grain fields, or foregather for nesting, in hundreds and thousands, but the very day of its nesting is agreed upon in concert. In continuous procession the individuals of a colony repair to a field agreed upon in quest of building material; and when the babies are clamoring the loudest for food, the deploying foragers join their nearest fellows and return to the swamps by platoons and volleys, rather than as individuals."

And Grinnell and Storer (1924) remark: "Zealous guarding of the nesting precincts, which is so marked a trait in the behavior of the male Red-wing, is not practiced by the Tri-color. There is not the need for each and every male to remain at the nest while the female is absent; the nests are located so very close together that there are always enough adult birds about the colony to sound an alarm should an enemy appear. It would seem as though the Tri-colored Blackbirds had attained to a more successfully communal stage of development in their domestic affairs than have the Bi-colored Red-winged Blackbirds."

Tyler (1913) says that—

It is not unusual to find a few of this species associating with the large flocks of mixed blackbirds that are so often seen in winter, but for the most part the Tricolors seek no company aside from that of their own kind.

During the month of March great hordes of Tricolored Blackbirds fly northward in what is evidently a local migration. Every morning, from daylight until after sunrise, they pass over at frequent intervals; sometimes half a dozen birds together and again in large compact flocks. If the weather is clear they fly at a height of over a hundred feet from the ground, but on foggy mornings they whiz along skimming just over the surface of the earth, in a flight that is very rapid for blackbirds. At such times they are entirely silent, in surprising contrast to the loose, straggling bands of Bicolors that go creaking along before dark on many a fall evening.

Dawson (1923) says: "The normal flock movement is in itself distinctive. The birds fly silently, with not so much as a rustle of wings; and they pass close to the ground, or at most at an elevation of fifteen or twenty feet. Each member of the flock rises and falls with each recurrent effort of the wings, quite independently of his fellows; but there is no vacillation or disposition to break away. Each bird is solely and ominously intent upon 'getting there'."

Voice.—Tyler (1913) evidently did not admire the song of the tricolored redwing when he wrote: "I have yet to hear the bird that can produce a more unmusical, strident series of notes than the Tricolored Blackbird, and when two or three hundred unite to vociferate in concert, the result absolutely defies all description—yet I would willingly listen to them for hours. The very harshness seems to appeal to a bird lover, when more musical bird songs would pass as commonplace."

Dawson (1927) describes the song as follows: "Instead of the hearty *konqueree*, or the lively *keyring* of the swamp redwing, *Agelaius phoeniceus*, we have, *Look awaay choke, awaay awaay choke*, or *awaak* or *chuaack choke*, as though sound were being squeezed out of nearly empty bellows. An anxious *jup* note reminds us rather of the crow blackbird than of cousin *phoeniceus;* while, if we were to retire to the oak-clad foothills, where belated courtships are still in progress, we should hear the curious 'stomach-ache song' of the yellow-headed blackbird only stopped down and subdued."

According to Ralph Hoffmann (1927): "The song of the Tricolored Redwing lacks the liquid quality of the preceding species. The song may be written *oh-kee-quáy-a*, with a braying quality. The common call note is a nasal *kape*."

Field marks.—The adult male may easily be recognized by the broad and conspicuous band of white, the median wing coverts, in sharp contrast with the glossy black plumage and the dark red "shoulders"; this shows plainly even when the bird is perched. The female is darker than the females of the other neighboring redwings, the lower parts from the breast posteriorly being solid dark, sooty brown, almost black. Other details are described under plumages.

The behavior of the tricolored redwing is quite distinctive, as mentioned above, and its voice is quite different.

Enemies.—The densely populated breeding colonies of tricolored redwings, with thousands of nests filled with eggs or small young, offer tempting chances for predators, furred or feathered, to enjoy a "field day"; many colonies have suffered heavy predation, and some have been almost, or quite annihilated. Lack and Emlen (1939) report two cases of mass desertion of nests, with destruction of eggs; of one of these, they say: "One colony near Marysville was reported to contain about 60,000 birds up to May 12. At the time of our first visit on May 16 only a few hundred were left. An examination of about one hundred nests revealed that more than the three-fourths contained freshly broken eggs or minute shell chips; only a few were undisturbed and these latter contained freshly laid eggs. On June 2 no adult birds were seen in the vicinity; of 114 nests examined, 62 contained shell chips, 46 others were empty and 6 contained newly hatched, but dead, young. Some of these nests showed small holes in the lining, as if made by birds' beaks."

Mailliard (1900) found a number of deserted nests in a colony in Madera County: "Those in the more exposed situations seemed to have been robbed, probably by the *Buteo swainsoni*, which were numerous in the neighborhood and one pair of which had a nest in a tall poplar tree but a few yards away, and possibly by some of the many *Nycticorax n. naevius* which simply swarmed in the most attractive spots. * * *

"The crop of one *Buteo swainsoni* contained two young just hatched and also the remains of two others with portions of the shell still sticking to them and which must have been just on the point of hatching. These were apparently the young of *A. tricolor.*"

Neff (1937) writes: "Heermann wrote in 1853 of the large numbers of Tri-colored Red-wings shot for the market. This practice still continues, and during the past 5 years it is probable that fully 300,000 blackbirds of the combined red-winged group have been marketed from the Sacramento Valley, with no apparent change in the status of any of the kinds involved. During the winter of 1935–36, 88,000 blackbirds were shipped from Biggs alone. * * *

"Destruction of the birds by man, of nesting sites through drainage or reclamation, of nests by predators or by the elements, and other factors, have played their part. All combined, however, they have made only fractional inroads on this species during the period covered by this report."

This last statement is quite reassuring in view of the fears, expressed only 5 years previously, that the wholesale poisoning of rodents with thallium and the still more destructive campaigns against this blackbird might result in a serious reduction in its numbers, or even practical extermination of the species. Describing this campaign and its results, Thomas T. McCabe (1932) wrote at that time:

I do not care to record in detail the minutiae of the technique employed, further than to say that grain poisoned with strychnine was placed on a small area of clean plow-ground close to the swamp, following several baitings with clean grain, which had attracted the birds and accustomed them to feeding on the spot. When the poison was finally placed, the effect was appalling. Great numbers died at once on the poison-ground, where within a very small radius 1,700 dead birds were tossed into a central pile. Later the surface of the shallow water beneath the willows became an almost solid floor of floating bodies where the observers hesitated to enter because of the stench which hung in the quiet air. Weeks later the bases of the cattails were awash with countless dead. At the time of our visit, May 21, the remainder of the grain was still doing its work, for fresh as well as decayed birds were still in evidence, often hanging, caught by the branches or clinging with the death grip of one foot, from the trees and from the nests in the rushes.

The destruction of adult birds, however, was much the smaller fraction of the total effect. As is often the case in large Tri-color rookeries, the nests were roughly divisible into groups. Only in two extremely small areas in the rushes had the eggs not hatched. Elsewhere the vast majority contained either new-hatched young or fledglings nearly ready to leave the nest. The enormous number of nests in the willows (a single tree contained 34) were not closely investigated. In the rushes, one might have spent a day forcing his way through the tall dense greenery, with from two to five or six nests continually within reach, yet leave untouched larger areas where no locomotion but swimming was possible. Yet judging from the small fractions I had time to cover he could hardly have found a dozen nests in which the young were alive and vigorous. Of the hundreds of broods I saw, all, practically speaking, were either dead (the vast majority) or feebly alive in some stage of starvation or grilling and parching by sunburn. A few evidently healthy adults were still passing in and out of the swamp, but the usual noisy cloud of enraged parents no longer hung over the invader's head.

After making several very careful surveys and counts, a total of 30,000 birds destroyed seemed to him "very conservative."

Winter.—The tricolored redwing is practically resident the year round in most of its breeding range. It has no regular north and south migration. After the breeding season and the molting period is over, the birds leave their breeding grounds and wander about the open country in search of good feeding places in the grain and stubble fields and about the ranches; they travel mostly in large flocks of their own species, but a few may mingle in the mixed flocks of other blackbirds.

DISTRIBUTION

Breeding range.—The tricolored redwing breeds east of the coast ranges from southwestern Oregon (Agency Lake, Klamath Falls) south through California, west of the Sierra Nevada (Modoc Plateau, Great Valley, Walker Basin, San Bernardino, and along the coast from Bodega Head to Chula Vista), to northwestern Baja California (San Rafael Valley, El Rosario).

Winter range.—Winters throughout its range in California (at least north to Glen County) with winter specimens reported from Baja California (El Rosario).

Egg dates.—California: 304 records, April 1 to June 17; 200 records, April 22 to May 20.

AGELAIUS HUMERALIS (Vigors)

Tawny-Shouldered Blackbird

HABITS

This well-known Cuban species is entitled to a place in the A. O. U. Check-List because of the capture of two specimens at Key West, Fla., by William W. Demeritt (1936), who describes the interesting event as follows:

In the course of my bird banding operations there were trapped at my station at Key West, Florida, two black birds, at the time unfamiliar to me. They proved to be Tawny-shouldered Blackbirds (*Agelaius humeralis* (Vigors)) which species is native to the island of Cuba, and has also been found on Haiti. These individuals were taken on February 27, 1936, on the Key West Lighthouse Reservation. They had been about for several days associated with Red-winged Blackbirds, of which there was a considerable number present at that time. They were kept in captivity until April 7, when they were shipped alive to the Biological Survey at Washington, D. C. There the previous tentative identification as *Agelaius humeralis* was confirmed by Dr. Harry C. Oberholser of that Bureau. They have been deposited as specimens in the Biological Survey collection in the United States National Museum, as proof of the record.

This is evidently a very common bird in Cuba, where it is known as the Cuban redwing or mayito. Thomas Barbour (1923) says of it: "The Mayitos abound in winter in great tame swarms, and haunt dooryards and gardens, whispering and wheezing metallically, and the volume of sound is very great. In the spring the males seek mates and the pairs split off and nest in April and May. They build, on palm fronds or on clumps of air plants, a nest of grass and Spanish moss lined with hair and vegetable wool. Formerly they did great damage in the rice fields, but today, beyond raising an unconscionable racket, they are very pleasing and ornamental neighbors.

"This is the black bird with tawny shoulder-marking and with the female black also, but still having a shoulder patch, though less extended and often much invaded with black feathers."

Wetmore and Swales (1931) say of its status in Haiti:

The tawny-shouldered blackbird was unknown in Hispaniola until its discovery near the mouth of the Artibonite River, a short distance from St. Marc, Haiti, in

the summer of 1927 by Stuart T. Danforth and John T. Emlen, jr. Five specimens, an adult and an immature male, and three females, were taken near some
sloughs. * * * Danforth and Emlen report that they observed about twenty of
these blackbirds on the date mentioned near sloughs along the Artibonite River,
about 8 miles from St. Marc, where they were in flocks of five to ten, resting in
trees standing in water. Some were feeding young birds on the wing. * * *
The limited area from which this blackbird is known in Haiti, and the fact that it
has not been recorded earlier suggest that it may have been established recently
on the island by individuals come from Cuba. Abbott did not secure it during
extensive travels on the island nor did Wetmore observe it during his work in
the field so that it can hardly be widespread in distribution since it is a bird that
is conspicuous and easily seen when its haunts are visited.

The tawny-shouldered blackbird measures 200 mm. or a little more in length
and is glossy black in color, with the bend of the wing, or "shoulder," deep brownish buff. Male and female are alike in color.

DISTRIBUTION

The tawny-shouldered blackbird is resident throughout Cuba and
locally in west-central Haiti (Port de Paix, lower Artibonite River).
It is accidental in Florida (Key West).

ICTERUS SPURIUS (Linnaeus)

Orchard Oriole

Plates 11, 12, and 13

HABITS

The origin of the name *spurius*, which is decidedly undeserved and
inappropriate, is discussed at considerable length by Wilson (1832),
who tells how a female Baltimore oriole was thought to be the male
of this species; this error resulted in the name spurious, or bastard,
Baltimore oriole, which at one time was applied to our orchard oriole;
and the name *spurius* still clings to it.

The orchard oriole enjoys a wide distribution in the central and
eastern United States, breeding from the northern tier of the Central
States, extreme southern Ontario, and extreme southern New England,
southward to northern Florida and the Gulf States. Its center of
abundance during the breeding season seems to be in the States
bordering on the Mississippi Valley, especially to the southward,
where it is really abundant in some places. It is comparatively rare
in the northern portions of its range, so very rare in southeastern
Massachusetts that I have seen only one nest in over 60 years.

As its name implies, the orchard oriole shows a decided preference
for orchards in rural districts near human dwellings, where apples,
pears, or peaches are cultivated; and when these colorful trees are in
bloom in spring, we are likely to find these orioles gleaning among the

opening foilage or preparing to build their basket nests. But it is by no means confined to such habitats even in the breeding season, for it is equally at home in the shade trees about houses or along village streets or country roadsides. In the prairie regions it lives in the timber belts along the streams or in the tree claims about farms and ranches; and in the south it is especially common about the plantations and in the shade trees about the planter's home. Everywhere it shuns the forests and the heavily wooded regions, preferring the open and cultivated lands, especially near human dwellings. In the north, where orchards are not as common as they were, the orchard oriole seems to find a satisfactory substitute in the nurseries, where trees and shrubs of many kinds are cultivated. And H. C. Oberholser (1938) writes: "One of the interesting and rather surprising ornithological experiences in southeastern Louisiana, particularly in the region of the Mississippi River Delta and the coastal areas west of that point, is to find the Orchard Oriole so common an inhabitant of the marshes, occurring even in the grasses and reeds as well as in the bushes and trees that fringe the bayous and ditches."

Spring.—Alexander F. Skutch writes to me: "The orchard oriole disappears from Central America during April. My latest record for Costa Rica is April 6, when a solitary male was seen at El General. In the Motagua Valley of Guatemala, where the species is so abundant during the winter, the last individual for the season, a female, was seen on April 21. Their sojourn here covers 9 of the 12 months."

Alexander Wetmore (1943), while collecting birds in southern Vera Cruz, Mexico, observed a heavy migratory flight of these birds, of which he says: "During the end of March and early April I saw more orchard orioles near Tres Zapotes than I had observed in all my previous years of observation of this species in its northern home. Some days they fairly swarmed, so that it was necessary to scrutinize carefully every bird collected to avoid shooting them."

These were probably birds that would migrate northward through eastern Texas and the western part of the Mississippi Valley. According to George F. Simmons (1925), the orchard oriole arrives in the Austin region around the middle of April, where it is also a common summer resident.

Some individuals, probably many, migrate straight northward across the Gulf of Mexico, from Yucatán to Louisiana and other Gulf States at least as far east as northwestern Florida. George H. Lowery, Jr. (1946), recorded an immature male that came aboard his ship on April 30, 1945, 94 miles south of the Louisiana coast and approximately halfway across the Gulf from the coasts of Texas and Florida; and he mentions two males and a female seen by Joseph C. Howell near the middle of the Gulf, May 3–6, 1945.

Francis M. Weston writes to me from Pensacola, Fla.: "The orchard oriole is an early migrant, usually arriving during the last week of March in northward flight across the Gulf of Mexico. Normally, it is common; but when incoming flights meet adverse weather conditions—rain, heavy fog or strong northerly winds—and several successive days' arrivals are halted and weather-bound in this coastal area, they become unbelievably abundant. At such times, I have seen hundreds of orioles in city parks and gardens or in a single small patch of woods. Under these conditions, and when their sojourn happens to coincide with the bloom period of the black locust (*Robinia pseudo-acacia*), the orioles show marked preference for this species of tree. Whether they actually feed on the flower parts or are attracted by the insects that swarm in the scented blossoms, I do not know; but on April 15, 1934, I counted 40 orioles busily feeding in two locust trees that stood side by side in a city garden.

"On the first day that the weather becomes propitious for a continuation of the interrupted northward flight, the nonresident orioles prepare to leave. They become restless late in the afternoon and resort to the tops of the tallest trees, where their bright colors glow in the last, level rays of the setting sun. Frequent tentative starts are made by small groups, which circle a time or two and then return to their perches. The birds are still there when the light fails and the observer on the ground can no longer distinguish them against the darkening sky. The actual 'take-off' may come shortly after dark. Certainly, by morning not an oriole is left in a patch of woods that harbored hundreds the evening before."

As a migrant in Cuba, according to Barbour (1923), "The Orchard Oriole appears occasionally in spring in company with Baltimore Orioles or alone. It seems possible that they are regular migrants, and have been overlooked among the native Orioles in immature dress." Earle R. Greene (1946) reports it as a "fairly common spring migrant" along the lower Florida Keys from April 9 to 22. A. H. Howell (1932) says: "On the Tortugas, migrants were reported April 11, 1890, April 14, 1909 (abundant), and April 26 to 28, 1914. There is but one record from Key West—April 29, 1887." He goes on to give a number of dates for the west coast of Florida, but none for the east coast, where this oriole seems to be an extremely rare migrant south of St. Johns County. From the above it appears that the orchard oriole advances from its tropical winter home on a broad front from eastern Mexico and Texas to the Gulf coast of Florida, diminishing in numbers in the latter region.

When the migrating birds leave the Gulf States, they advance northward and rather rapidly on a similar broad front, though more abundantly in the Mississippi Valley than on the Atlantic coast, reaching the northern limits of their breeding range early in May. In Missouri, according to Widmann (1907), the "first to arrive are the old males followed after a few days by the first females and the first males of the second year. It is from 1 to 2 weeks after the first males have come before their full strength is reached and their song heard everywhere."

Nesting.—Although there are a few scattered breeding records for Massachusetts, I have seen only one nest here. During the month of June 1915, a pair of orchard orioles built a nest and reared a brood of three young in Berkley, about 8 miles from my home, in a farmyard and close to a house. The nest was suspended from the end of a long, drooping branch of an apple tree and fully 15 feet from the ground. It was well concealed among the leaves and was made almost wholly of freshly dried yellowish grasses, with a few leaves of the tree woven into it; it was deeply hollowed, thin-walled on the sides but with a thickly wadded bottom, and was lined with very, very fine white, silky, woolly substances. I collected the nest after the young had left it, but neither the old nor the young birds were ever seen again.

T. E. McMullen has sent me the data for four New Jersey nests, ranging from 6 feet up in an elder bush to 10 and 15 feet up in apple and pear trees, and for a North Carolina nest that was 20 feet from the ground in a maple. A. D. Du Bois' notes record a nest found in Lake County, Ill., that "was about 8 feet from the ground, hanging at the end of a branch of a small, lop-sided apple tree in an old abandoned orchard on a hill. It was constructed of fresh grasses, gray-green in color, fragrant like new hay. The grasses appeared to have been green when first woven into the nest—a wonderfully woven cup, contracted at the top. This little deserted orchard of barely a dozen trees also hid the nests of kingbird, Baltimore oriole, catbird, robin, yellow warblers, chipping sparrows, a phoebe and a vireo."

A. C. Reneau, Jr., has sent me his records of 23 nests of the orchard oriole, found near Independence, Kans., of which 8 were in elms, 8 in button-bushes, 5 in willows, and 1 each in a cottonwood and a maple. The lowest nest was only 4 feet up in a buttonbush and the highest 30 feet from the ground in a willow. The dates ran from May 14 to July 3. Ten of the nests, six of which were in the same tree, were near kingbirds' nests, one being within 5 feet of such a nest; another was within 25 feet of a marsh hawk's nest.

In some notes he sent to me on Georgia nests, Frederick V. Hebard says: "The orchard oriole individuals show some tendency to nest at the same time and then to gather in flocks up to 18. Out of five nests, two were built in live oaks, two in pecans and one in a long-leaf pine sapling. The last was lodged in pine needles near the top

of the trunk, about 11 feet up. The young had left the nest between
June 24 and 26, 1942. The nest was removed and examined June
29. It was a pendulous affair of wire grass with its bottom still
green. Outside it was 3⅝ inches deep and 3 by 3¾ inches across.
Inside it was 2½ inches deep and 2¼ by 3 inches across. It was
difficult to understand why it had not fallen as had another, blown
out of a pecan in a storm about May 23, 1946. This nest, when
examined June 8, was composed of golden wire grass, and measured
outside of 2½ inches deep and 3½ by 4 inches across. Inside it was
2¼ inches deep and 2½ by 2½ across. The proportions of the latter
were the same as the three other nests observed. One could see
through the bottom of all five nests."

M. G. Vaiden writes to me from Rosedale, Miss.: "In this immediate
area the orchard oriole prefers the country district to the small town;
it is just the opposite with the Baltimore oriole, a bird found almost
exclusively breeding within town limits. The orchard oriole can be
found nesting over the water in the small cottonwood and switch-
willow growths usually found in shallow to deep barrow-pits. It is a
most numerous nesting bird in such areas, constructing its semipensile
nest near the top of the swinging treetop, or out at a short distance
from the top on a limb. Most of the nests have been found in switch-
willow or finger-cottonwood growth, but not all are found over water.
Occasionally a very large pecan, cottonwood, sycamore, or elm will
be found to contain a nest of the orchard oriole. Two nests have
been found within the town limits."

The following account of the nesting habits of the orchard oriole
in northwestern Florida comes to me from Francis M. Weston: "The
typical nest of the orchard oriole is suspended from a forked terminal
twig, usually of a large tree, after the manner of the much pictured
nest of the Baltimore oriole, but it is never deep enough to be described
as 'pensile'—its depth is usually less than its outside diameter. Invariably, the nest is woven of long blades of green grass that later
turn yellow and give the nest of this species its characteristic color.
I was thoroughly familiar with the style and normal situation of this
nest before I came to Pensacola, so it was a source of surprise and
disappointment to me that I succeeded in finding only a few nests
here where the birds are so common. It was years before I could
account for my failure. Then, on May 20, 1923, I saw a female
oriole disappear into a dense festoon of Spanish 'moss' (*Dendropogon
usneoides*) that hung from a low branch of a deciduous oak (*Quercus*
sp.). Within the festoon, I found a nest, typical in structure, size,
shape and color, but unique in that it was not attached to a twig of
the tree but was wholly supported by the strands of the 'moss'—after
the manner of the nests of the parula and the yellow-throated war-

blers, which nest exclusively in such situations. The nest contained four eggs, well along in incubation. Guided by this discovery, I soon found other nests similarly concealed in Spanish 'moss.' I now conclude that at least 50 percent of the oriole nests in this central Gulf Coast region are so situated, while the remainder are in the normally exposed locations on terminal twigs.

"The earliest nest I have ever known contained a full set of eggs on April 29, 1929, but the average for complete sets is the latter half of May. Late nests, probably second or even third attempts by birds that failed the first time, have been seen as late as the latter half of June and even in July. The latest nest I have ever known still contained well-grown young birds on July 14, 1937."

In Duval County, northern Florida, S. A. Grimes (1931) finds this oriole nesting in the Spanish "moss" very commonly, but also in pecans, other orchard trees, longleaf pines, black gums, oaks, buttonwood saplings, live oaks, sweet gums, hickories, and chinaberry trees, at heights ranging from 4 to 50 feet above the ground. Arthur T. Wayne (1910) has found the nest as high as 70 feet, in South Carolina.

H. H. Kopman (1915) regards the orchard oriole as "the most conspicuous summer visitor in the fertile alluvial section of southeastern Louisiana. * * * Its abundance as a breeder in the southeastern portion of the State, however, can scarcely be comprehended by those whose acquaintance with it is confined to its appearance in more northern localities. In one live oak in a plantation yard where there were many more trees of this kind I once counted nearly twenty nests of this species."

Near Brownsville, Tex., George B. Sennett (1878) says "it likes to build in mezquite, wesatche, and willow trees." Farther north, near Austin, Tex., George F. Simmons (1925) lists the following nesting trees: Hackberry, mesquite, cedar elm, winged elm, peach, pear, huisache, retama horse-bean, honey locust, eastern live oak, black willow and pecan trees. Probably many other trees are selected in other parts of its range, for the orchard oriole does not seem to be at all particular in its choice of a nesting tree; weeping willows seem to offer favorite sites.

The nest is beautifully and compactly woven in the shape of a semiglobular cup with a contracted rim. Wilson (1832) says: "I had the curiosity to detach one of the fibres, or stalks of dried grass, from the nest, and found it to measure thirteen inches in length, and in that distance was 34 times hooked through and returned, winding round and round the nest!" The materials used in the construction of the nest vary but little in character; R. C. Tate (1926) says that, in Oklahoma, these consist of "fresh blades of Mesquite grass and gramma grass, yucca fibres, fibers from tree cactus and prickly pear."

Bendire (1895) describes a large well-built nest, taken on Shelter Island, N. Y.: "The outer diameter at the widest part, a little below the middle of the nest, is 4½ inches; the outside depth is 4 inches. The upper rim of the nest is somewhat contracted; the inner cup is 3 inches deep by 2½ inches in diameter. The sides are thick and securely fastened to several branches, but the bottom does not come within 2 inches of the fork of the crotch in which it is placed." It was placed "in an upright fork of a small branch in a thorn pear tree." Ora W. Knight (1908) watched the building of a nest in Texas: "A nest which was discovered in its very first stages of construction was completed in 6 days and an egg was laid daily until a set of five was completed, when incubation commenced. Both birds help to build the nest and aid in the incubation."

The orchard oriole is a friendly, sociable bird and is often found nesting in orchards with kingbirds, robins, chipping sparrows and other species, with all of which it seems to be on good terms. The eastern kingbird seems to be a favorite companion, from which it may gain some protection. This companionship is referred to above and several observers have mentioned it in print. H. C. Campbell (1891), for example, mentions seven such cases; in one case the oriole's nest was within 7 feet of the kingbird's and in another instance the two nests were only 3 feet apart. He says further: "In 1887 I found a nest of the Orchard Oriole in an apple tree. When the nest contained five eggs I collected it. While at the nest a pair of Kingbirds came and made even more demonstration than the Orioles. I found the Kingbird's nest in a rotten apple tree about 200 feet distant from the tree containing the Oriole's nest."

John V. Dennis has sent me some full notes on his interesting experience with what might be considered as communal nesting of the orchard oriole in the Delta National Wildlife Refuge in Louisiana, located between the Mississippi River and the Gulf of Mexico, some 70 or 80 miles below New Orleans. "Lying on the east bank of the Mississippi, two miles above Pilotstown, is the refuge headquarters area, comprising approximately seven acres. Seventy-eight shade trees, as well as numerous shrubs and ornamental plantings, are in the area. Forty-five of the trees are hybrids of live oak and water oak. The other trees are mainly camphor, willow, and magnolia.

"Nest building began about May 1. The peak of nesting activity occurred during the first half of June. The last nest to be observed under construction was one begun on July 4. A total of 114 nests were counted during this single nesting season on the seven-acre tract under study."

Dennis counted 45 oaks, containing a total of 80 nests; 15 camphor trees, 8 nests; 8 magnolias, 5 nests; 5 willows, 4 nests; 2 elms, 6 nests;

2 cottonwoods, 2 nests; the only pecan held 4 nests; there was 1 nest in 1 of the 2 loquat shrubs; in over 50 ornamental shrubs there were only 2 nests; and in over 100 black elderberries there were only 2 nests.

"The most surprising discovery occurred about a month after nesting had begun at the headquarters area. I had frequently seen orchard orioles in the vast marshes which extend eastward from the Mississippi for a distance of some 10 miles to the waters of the Gulf. I hadn't suspected nesting in such an unusual habitat for the orchard oriole until I found some very agitated adult birds in a cane break near the mouth of Dead Women Pass. A search revealed their nest. It was built in roseau canes, *Phragmites communis*. The nest was woven about three stalks, which acted as its support. This nest, and others which were discovered later, was located on the outer edge of a cane break overlooking a body of water.

"On all subsequent visits to the marsh I made every effort to find new nests. Eventually some 10 were found in widely separated areas of the marsh; one was less than a hundred yards from the mud flats of the Gulf of Mexico. All were built in roseau cane, usually at a height of about 7 feet. Some nests were completely built of various grasses, while others were almost entirely constructed of salt meadow cordgrass, *Spartina patens*. Those furthest from willows, sometimes as far as 5 miles from a tree of any kind, were lined with cattail down. Otherwise the chief item used was the down from willow catkins. The only exception to this was in areas where thistles (found only on filled-in land) grew nearby. Then thistledown was used copiously in lining nests."

Referring to the nests in trees, he noted that the lowest nest was 2½ feet from the ground, and the highest nearly 40 feet. "As often observed, the orchard oriole showed preference to trees occupied by the eastern kingbird. Two kingbird nests were in the study area. One of the nests was in a small hybrid oak. Nesting concurrently in the same tree were four pair of orchard orioles."

Of seven nests under daily observation, five were built in 3 days, one in 4 days and one in 5.

Eggs.—The orchard oriole lays from three to seven eggs to a set, four and five being the commonest numbers; Bendire (1895) says from four to six, mostly five. He describes them as follows:

The eggs are mostly ovate in shape, but occasionally a set is found which is decidedly elongate ovate. The shell is moderately strong, close grained, and without gloss. The ground color is usually pale bluish white, and this is sometimes faintly overlaid with pale pearl gray or grayish white. The markings, which are nearly always heaviest about the larger end of the egg, consist of blotches, spots, scrawls, and tracings of several shades of brown, purple, lavender, and pearl gray, varying in amount and intensity in different specimens. In the majority of the eggs before me the darker markings predominate, but the lighter-colored and more neutral tints are nearly always present to a greater or less extent.

The average measurement of one hundred and thirty-three specimens in the United States National Museum collection is 20.47 by 14.54 millimetres, or about 0.81 by 0.57 inch. The largest egg in the series measures 22.35 by 15.24 millimetres, or 0.88 by 0.60 inch; the smallest, 18.03 by 14.22 millimetres, or 0.71 by 0.56 inch.

Young.—Bendire (1895) says: "Incubation lasts about 12 days, and I am of the opinion that this duty is exclusively performed by the female. I have never seen the male on the nest, but have seen him feed his mate while incubating. I believe as a rule only one brood is raised in a season. Both parents show equal solicitude and devotion in the care and defense of their young from prowling enemies, and will boldly and furiously attack any intruder."

Grimes (1931) doubts if the period of incubation exceeds 12 days, says that only one brood is raised each year, and that "male and female share the task of incubating the eggs, and both feed and brood the young, which leave the nest when ten or twelve days old. Food for the young, which I have noticed consists largely of katydids, is usually secured at a distance of one hundred yards or more from the nest. * * * The female is by far the more solicitous of the pair when the eggs or young are in danger."

In the seven nests under daily observation, Dennis found the period of incubation, from the first egg laid to the first hatched, to be 12 days in one nest, 15 days in one, and 14 days in the other five. The young were fledged in from 11 to 14 days.

In her observations on a family of orchard orioles in Wisconsin, Winnifred Smith (1947) noted that "the female did all the incubating. * * * During three hours the male fed the young 23 times, the female 14 times. Feces were carried off or eaten by the male eight times and only once by the female." After the young had left the nest, the "male undertook the care of two fledglings while the female took care of one. The male chased the female when she attempted to feed the two in his charge. * * * The family remained in the vicinity until July 30 after which they were not seen again in 1946."

Plumages.—Dwight (1900) describes the juvenal plumage of the orchard oriole as "above, including sides of head and neck, pale grayish olive-green, buffy on rump. Below, pale sulphur-yellow. Wings pale clove-brown, the primaries and secondaries narrowly edged with dull white, the median and greater wing coverts with pale buff forming two indistinct wing bands. Tail yellowish olive green."

The first winter plumage is acquired by a partial postjuvenal molt, beginning late in July, at which time only the wing-quills and the tail feathers are retained. In this plumage both sexes are much like the adult female in winter. A limited prenuptial molt, mainly about

the head and throat, occurs in late winter or early spring, before the
birds come north, at which time the young male acquires the black
throat, or a number of black feathers in that region, and a few chestnut feathers more or less scattered over the body. Young males
are known to have bred in this plumage. Young females have an
even more limited molt at this season. Adults have a limited
prenuptial molt about the head and throat.

Young birds and adults have a complete postnuptial molt in early
fall after they have gone south, at which time the adult winter plumage is assumed. This is like the spring plumage, except that the
brown and black colors of the male are heavily veiled and nearly
concealed by buff or yellowish tips which wear away before spring.
Old females sometimes have a few black feathers in the throat.

Dickey and van Rossem (1938) throw some light on the molts of
the orchard oriole, based on specimens collected in El Salvador:

> On arrival in mid-August these orioles are in fearfully abraded plumage, for
> they have, contrary to the usual custom, completed the migration before the
> annual molt has taken place. This is true of adults and young alike, and when
> the latter arrive they are still in soft, juvenal feather. The process of annual
> renewal is a relatively slow one, and not until the latter part of October (in one
> case November 4) is the new plumage completely acquired. Males in their
> second year, that is, those which have molted from the black-throated, greenish
> plumage of the first year to the first, brown, subadult plumage, are characterized
> by broad buffy tipping to the feathers of the body plumage. This tipping makes
> such males superficially more or less like *Icterus fuertesi*, but most of the lighter
> color wears off by midwinter. During early April some 1-year-old spring males
> show a limited spring molt involving both the chin and throat, and some new
> lack feathers appear on these parts.

Todd and Carriker (1922) collected an adult male in Colombia on
October 15, 1915, that was completing the postnuptial molt. "The
rectrices are about two-thirds grown, and the wings retain only the
two outermost primaries of the old dress."

Food.—Judd (1902) studied the contents of 11 stomachs of the
orchard oriole, collected on a Maryland farm in May and June; the
food "was composed of 91 percent animal matter and 9 percent vegetable matter. The latter part was nearly all mulberries; the former
was distributed as follows: Fly larvae, 1 percent; parasitic wasps,
2 percent; ants, 4 percent; bugs, 5 percent; caterpillars, 12 percent;
grasshoppers, including a few crickets, 13 percent; beetles, 14 percent;
May-flies, 27 percent; spiders, 13 percent. Thus beneficial insects—
parasitic wasps—formed only 2 percent of the food, and injurious
species—caterpillars, grasshoppers, and harmful beetles—amounted
to 38 percent."

Bendire (1895) writes: "Few birds do more good and less harm
than our Orchard Oriole, especially to the fruit grower. The bulk of

its food consists of small beetles, plant lice, flies, hairless caterpillars, cabbage worms, grasshoppers, rose bugs, and larvae of all kinds, while the few berries it may help itself to during the short time they last are many times paid for by the great number of noxious insects destroyed, and it certainly deserves the fullest protection."

Barrows (1912) says that "two specimens were killed in an orchard overrun with canker worms in Tazewell County, Ill., in 1881, and the contents of their stomachs studied by Professor S. A. Forbes. He found that nearly four-fifths of their food was cankerworms, while other caterpillars formed all but three percent of the remainder, this being ants. Butler states that in Indiana when the young leave the nest the whole family go into the cornfields and feed upon the insect enemies of the corn."

According to A. H. Howell (1924), this oriole "is a persistent hunter of boll weevils, and is one of the few birds that has learned to seek out and destroy this pest which hides in the cotton squares. Nearly one-third of the stomachs of this species taken in the Texas cotton fields contained boll weevils; the average number of weevils found in a stomach was 2 and individual birds had eaten as many as 13 weevils at a meal."

Economic status.—The only arguments that can be advanced against the orchard oriole as an economically valuable bird are the claims that it eats the stamens in the blossoms of the fruit trees; that it occasionally helps itself to various small fruits such as cherries, strawberries, and raspberries; and that it does some damage to grapes and ripening figs. But the slight damage done is insignificant when compared with the great good that it does in destroying harmful insects, which make up 90 percent of its food.

Behavior.—The orchard oriole is a gentle, friendly, and sociable bird that lives in perfect harmony with many other birds in more or less close association and seems to enjoy human environments. It is a restless, lively bird, and not particularly shy, but since it spends most of its time flitting about in the trees in search of insects, or keeping out of sight among the foliage, it is not as easily observed as some others. When it is singing freely in the spring, we are often attracted to it by its voice and can catch a glimpse of its pretty colors and its graceful, slender form as it hops from twig to twig, or makes short flights among the branches, or hangs head downward to pry under a leaf in search of its prey. During the courting season, an ardent male may sometimes be seen to rise high above the treetops and to pour out an ecstasy of song as it descends to its leafy shelter.

Voice.—Aretas A. Saunders contributes the following study: "The song of the orchard oriole consists of a series of rather rapid, musical notes, exceedingly variable in time and pitch. As I study my 32 records of this song, I can find no fixed pattern for the song and no general rule that does not have exceptions. One marvels that under such conditions the song, when well known, is always recognizable in the field. The quality, mainly musical but with occasional harsh notes, is more like that of the robin than that of the Baltimore oriole. While most of the notes are distinctly separated, there are two-note phrases, connected by liquid consonant sound, and slurs. The notes vary up and down in pitch, but there are occasional series of notes on the same pitch.

"The number of notes in the songs I have recorded varies from 7 to 19, averaging about 12. The songs vary from 1⅚ to 3⅓ seconds in length. The pitch varies from E″ to B‴, 3½ tones more than an octave. Individual songs range from 2½ tones to an octave, or 6 tones. The average range is about 3.85 tones.

"A common characteristic of many of the songs is that they end in a downward slurred note, distinctly harsher than the other notes of the song, that suggests the quality of the scarlet tanager, rather than that of the robin; 22 of my records have such an ending.

"Another characteristic is a series of very short notes, all on one pitch, usually near the end of the song. I have heard this described as a trill, but I use the term 'trill' only for series of notes so rapid that they cannot be separated by ear and counted. Under that definition, I have never heard an orchard oriole sing a trill. The number of notes that are thus rapidly repeated varies from 3 to 6 in my records, and 18 of them contain such a series of notes, while only 3 are without this series or the downward slur; 11 records contain both. Downward slurs are common, though not always terminal; 27 of my records contain them and 14 contain 2 or more. Upward slurs are rarer, only 6 records containing them.

"The orchard oriole sings from the time of its arrival to the earlier part of July. In eight seasons when I was able to observe this species, the last song averaged July 10, with July 5, 1944, and July 17, 1941, as earliest and latest dates. This bird has a long, rattle-like call, and a shorter one very similar to the *chack* of the redwing."

The vivacious, attractive song has been compared to the rollicking outburst of the bobolink, the rich spring song of the fox sparrow, and the warbling songs of the purple finch or the warbling vireo. It is not as loud, nor as rich as that of the Baltimore oriole and is quite unlike it, but it is equally pleasing. Chapman (1912) says of it: "His voice is indeed unusually rich and flexible, and he uses it with rare skill and expression. Words can not describe his song, but no lover of bird music will be long in the vicinity of a singing Orchard Oriole without learning the distinguished songster's name." C. W. Townsend (1920) writes: "The full song of the Orchard Oriole is given with great abandon from a perch and especially on the wing. I have heard one sing six times in a minute and have tried to express his song by the words *Look here, what cheer, what cheer, whip yo, what cheer, wee yo.*" Witmer Stone (1937) represents it with the syllables "*teetle-to—wheeter-tit-tillo-wheetee, chip, chip, cheer.*" Another bird called: "*Choop, choop, choolik* as if trying to start a song and failing in the effort." Francis H. Allen tells me that the food calls of the young resemble those of the Baltimore oriole, but are higher pitched and more rapid. Young males in first year plumage sing enthusiastically; and sometimes females sing a little.

Field marks.—The adult male orchard oriole is unmistakable in his black and chestnut plumage. The young male is like the female, but has a black, or partially black, throat and usually more or less chestnut scattered through his plumage. The female might easily be mistaken for the female of some other orioles, but she differs from the Baltimore oriole by being olive-green above, instead of brownish olive, and having less of an orange tinge on the under parts, which are dull yellow.

Enemies.—The orchard oriole is a not uncommon host of both the eastern and the dwarf cowbirds. There is a set in my collection containing a cowbird's egg. Dr. J. C. Merrill (1877) mentions a nest that contained three eggs of the red-eyed cowbird, "while just beneath it was a whole egg of this parasite, also a broken one of this and of the Dwarf Cowbird." The nest was, of course, deserted.

Harold S. Peters (1936) mentions only one external parasite as found on this oriole, a louse, *Myrsidea incerta* (Kell.).

Grackles, which sometimes nest in the same trees with the orioles, probably rob some of the nests of eggs or small young. And young birds that leave the nests prematurely fall easy prey to various predators.

Fall.—The orchard oriole spends only about 10 weeks in the northern part of its breeding range, arriving early in May and leaving soon after the middle of July. It lingers through August and occasionally into September in some of the Southern States; Howell (1932) gives one very late date, October 13, 1917, for Royal Palm Hammock in southern Florida. The fall migration is started and, apparently, often finished before the annual molt is accomplished; some young birds arrive in Central America while still in juvenal plumage, and many adults are still molting when they arrive.

As soon as the young are able to fly, old birds disappear with their families, forming into flocks, and are seen no more in their breeding haunts. In Missouri, according to Widmann (1907), "after the young are grown the species roams in July and August in troops through the country living mostly on wild cherries, wild grapes and other wild fruit, sometimes visiting orchards. After August 20 the species is seen only occasionally, though we may come upon a few later in the month, or in early September, exceptionally later (September 17, 1903, New Haven; September 21, 1903, Kansas City)."

Kopman (1915) says: "This species becomes inconspicuous at Gulf coast latitudes after the middle of August, though little companies of them may be in evidence for a few days at a time at intervals until Sept. 10 or 15. Such transients usually form part of slight waves including other species. The latest date of departure is Sept. 26, 1914, near Poydras, St. Bernard parish Louisiana. The average date of departure is about Sept. 15."

F. M. Weston writes to me: "The orchard oriole is the first summer resident to disappear from the Pensacola region. It becomes rare early in July and, in some years, it is not seen after July 15. Ordinarily an occasional bird appears in August and, twice in a period of thirty years, I have recorded occurrence in September—September 1, 1940, and September 3, 1944. In both cases these were apparently family groups of young birds. This suggests that the birds that are successful in their family affairs leave early in July, and it is only the few that are delayed by having to 'try, try again' that make up the sparse August population

"The fall migration route of the orchard oriole is certainly not a reversal of the spring route. Then, as noted above, they come across the Gulf in tremendous numbers and pass through this region on their way to more northerly breeding grounds. In fall, when a successful breeding season must have at least doubled their numbers and even a poor season would not have diminished them, few orioles are seen. During 30 years of continuous field observation, I have never under any circumstances of favorable or adverse weather conditions seen any concentrations of orioles, though it is a common experience to find thousands of migrants of other species weatherbound on this coast on several occasions every fall. Our local breeding population of orioles merely withdraws from this region, and no birds from more northerly areas come in to replace them."

There seem to be few fall records for southern Florida, none for the Florida Keys and none for Cuba. By what route these orioles migrate to their winter homes in Central America and northern South America does not seem to be known. Some may migrate across the Gulf farther west, or they may follow the coasts of Texas and México, but conclusive data are lacking.

Grimes (1931) gives the following account of their disappearance from northeastern Florida:

Late in June the oriole's singing begins to diminish in force, and discordant notes creep in as the song becomes broken and unmusical. After the first week in July it is unusual to hear the song at all, and I have noticed that the gathering flocks are composed entirely of plain yellow birds—females and young or perhaps only young birds—and that the males have suddenly disappeared. Sometimes these flocks consist of as many as twenty-five or thirty individuals, but more commonly of ten or twelve. It seems that several families of orioles from a certain breeding area congregate in a selected stretch of woods and fields after the nesting season and spend a month or six weeks there getting acquainted and organized before starting on their southward migration, but this is a conjecture that would be difficult to substantiate.

Winter.—Alexander F. Skutch contributes this account: The orchard oriole is one of the very first of the visitors from the north to reach Central America as fall approaches. On July 20, 1932, I found a male and female together in a bushy pasture beside the Motagua River in Guatemala—only 3 months earlier, on April 21, I had seen the last of the spring migrants lower in the same Valley. Cherrie recorded the species at San José, Costa Rica, on July 31; while still farther south it has been met at El Pozo de Térraba, Costa Rica, on August 10, and in the Canal Zone on the same date. Many adults arrive in northern Central America in worn breeding plumage, having left their nesting area before completing the postnuptial molt. Considering these facts, Osbert Salvin long ago surmised that the orchard oriole might breed in the Guatemalan highlands; but subsequent intensive exploration has failed to reveal its presence there during the summer months.

"Like the Baltimore oriole, the orchard oriole is widely distributed over Central America during the period of the northern winter. It is found in midwinter from Guatemala to Panamá, and along both the Caribbean and Pacific coasts. I have met it at this season in regions of such contrasting climate and vegetation as the wet Caribbean lowlands of Guatemala and Honduras and the arid coast of El Salvador, where in early February these birds were abundant amid cacti and low, thorny trees at Cutuco. But although as widely, the orchard oriole is by no means so uniformly distributed over Central America as the Baltimore oriole. It seems to winter in far greater numbers in Guatemala and Honduras than in Costa Rica and Panamá. In the former countries it equals or exceeds the Baltimore oriole in abundance, at least at lower altitudes, while in Costa Rica the Baltimore oriole is certainly the more common bird—on this last point my own experience is quite in accord with that of Carriker (1910), whose ornithological work in the country antedated my own by a third of a century. In altitudinal range, the orchard oriole is far more restricted than the Baltimore oriole. Even in northern Central America, where it is so abundant in the lowlands, it is rarely met above 4,000 or 5,000 feet. In the Térraba Valley of Costa Rica, from 1,500 feet upward, the orchard oriole is a very rare winter visitant, while the Baltimore oriole is fairly abundant.

"Although my few Costa Rican records of the orchard oriole are all of single individuals, in northern Central America, where the species is far more abundant, it is more sociable during the winter months, wandering in straggling flocks through the riverside trees, the plantations and shady pastures, but rarely entering heavy forest. In the banana plantations, these oriole hang head downward beside the huge, red flower buds and push their sharp bills into the long, white, tubular blossoms to sip the abundant nectar. In the pasture lands they straggle along the fence lines, where living trees of the madre de cacao form the posts, and investigate the pink, pealike blossoms which in February or March cover the long, leafless branches. The single orchard oriole that in four years I have seen on my farm in southern Costa Rica was visiting the madre de cacao blossoms in a hedgerow. When they find groves of introduced eucalyptus trees, the orioles probe the clusters of long white stamens, either for nectar or for the small insects attracted to the flowers.

"I have twice found the roosts of wintering orchard orioles. They seem to prefer stands of tall grass of one sort or another. In the Lancetilla Valley, on the northern coast of Honduras, many roosted nightly in a patch of introduced elephant grass, *Pennisetum purpureum*, which formed an impenetrable thicket 6 or 8 feet high. They went early to roost, sometimes retiring an hour before nightfall. By the time the crowds of small resident finches joined them there, they were completely hidden from view amid the tall grass, whence would issue a few snatches of their breezy song, audible above the chatter of the garrulous seedeaters. Here the orchard orioles slept with seven other species of birds, both resident and migratory, including a few Baltimore orioles, as told in the section devoted to that species. I found the orchard orioles roosting in this patch of grass in September and October, and again in February of the following year. In March 1932, a small flock roosted in a dense stand of young giant canes, *Gynerium sagittatum*, that were colonizing a sandy flat left by a shift in the channel of the Río Morjá, a small tributary of the Motagua in Guatemala. The canes, still only 10 feet or less in height, had attained only a fraction of their full stature, and resembled some tall, coarse grass, like the elephant grass in which I had found the orioles roosting in an earlier year, rather than a mature stand of *Gynerium*.

"I have heard no other winter visitant sing so much during its sojourn in Central America as the orchard oriole. Upon arriving in Honduras and Guatemala in August, the males often delivered fragments of hurried, whistled song. In September their songs came more rarely; but toward the end of October, more than two months after the arrival of the first-comers, I still occasionally overheard them deliver brief, subdued refrains. From October to March they were practically songless, but before the vernal equinox they began to sing sweetly again. In the Motagua Valley in Guatemala, where orchard orioles winter in great numbers, at the beginning of April I heard no bird's voice so much as theirs; for the Gray's thrushes, so abundant in the cleared plantation lands, had not yet come into full song. The orioles that roosted in the dense stand of young canes beside the Río Morjá raised a delightful chorus when they awoke at dawn. Their music increased in both quality and abundance up to the time of their departure; and the young, black-throated, yellowish males, eager to use their newly acquired singing voices, performed as much if not more than the mature males in chestnut and black. They captured my heart as no other birds, they whistled so often and so cheerily on the eve of their long migration, when most other birds of passage sing little or none."

In El Salvador, according to Dickey and van Rossem (1938), the orchard oriole is a "common winter visitant and abundant fall and spring migrant in the lowlands throughout the country." They refer to some of its habits as follows:

About the middle of February, in 1926, the ceiba trees on the coastal plain at Rio San Miguel were a solid mass of pink bloom, to which came unbelievable numbers of orchard orioles in search of the swarming insects. Until this sudden concentration we had noticed no sex segregation, but now it was suddenly apparent that these great flocks, composed of hundreds of individuals, were made up almost exclusively of old males. On February 20, a great ceiba standing alone in a grass pasture was watched for over an hour. No accurate estimate could be made of the number of birds present, but it certainly ran into many hundreds. The wide-spreading mat of blossoms was at least one hundred feet from the ground, and the darting restless swarm of old males packed it literally to a point where there was no room for more. * * * Orioles of the smaller species (particularly of the genus *Icterus*) are not, as a group, noted for their flocking tendencies, but *spurius* while in winter quarters is very much of an exception to this general rule. Not only does it spend the day in small groups, but it frequently concentrates still further at sundown and roosts in good-sized flocks. Such a night roost, composed of about fifty birds, was seen on many occasions in a tangle of mimosa and vines in a barranca at Divisadero. Others were observed at Barra de Santiago in the low scrub of a sand spit between the ocean and lagoon.

DISTRIBUTION

Range.—Manitoba and southern Ontario to northern South America.

Breeding range.—The orchard oriole breeds from southern Manitoba (Cypress River), central and southeastern Minnesota (Nisswa, Stillwater), central Wisconsin (northern Wood County), southern Michigan (Greenville, Port Huron), southern Ontario (Lambton, Gananoque), north-central Pennsylvania (Punxatawney, Lock Haven), central and central-eastern New York (casually to Ithaca

Wilmington), and central and northeastern Massachusetts (Amherst, Fitchburg); south through eastern and central-southern North Dakota (Devils Lake, Bismarck), central South Dakota (Stamford, Grass Creek), central Nebraska (Fort Niobrara Refuge, North Platte), northeastern Colorado (Wray), central-northern and western Texas (Amarillo, Marfa) to central Durango, central Nuevó Leon, northern Tamaulipas, southern Texas (Hidalgo, Brownsville), the Gulf coast, and northern Florida (Aucilla, Saint Augustine).

Winter range.—Winters from Colima, Guerrero, Puebla (Huexotitla), central Veracruz (Jalapa), Yucatán, and Quintana Roo (Cozumel Island); south to southern and central-eastern Colombia and northwestern Venezuela; in migration west to southern Sinaloa (Labrados, Rosario) and Nayarit (San Blas); and east through peninsular Florida, the Florida Keys, and western Cuba.

Casual records.—Casual in New Mexico (Hagerman), central Colorado (Denver, North Creek), Wyoming (New Castle), western South Dakota (Buffalo Gap, Grand River Agency), south-central Manitoba (Lake Saint Martin), northern Michigan (McMillan, Onaway), southern Quebec (La Colle), northern Vermont (Middleburry, Orleans), central New Hampshire (Grafton County), southern Maine (Auburn, Calais), New Brunswick (Grand Manan) and Nova Scotia (Cape Sable Island).

Accidental in California (Eureka) and Nevada (Halleck).

Migration.—Early dates of spring arrival are: Costa Rica—El General, February 24. Cuba—Havana, April 10. Florida—De Funiak Springs, March 15; Pensacola, March 22 (median of 40 years, March 30); Fort Myers, March 28 (median of 7 years, April 12). Alabama—Mobile, March 30; Decatur, April 6. Georgia—Savannah, March 18; Tifton, March 29. South Carolina—March 27; Aiken, April 3. North Carolina—Raleigh, April 16 (average of 30 years, April 25). Virginia—Norfolk, April 21. West Virginia—French Creek, April 21 (median of 9 years, May 1). District of Columbia—April 25 (average of 40 years, May 3). Maryland—Laurel, April 7 (median of 9 years, April 30); Denton, April 19. Pennsylvania—Philadelphia, April 19; Beaver, April 22 (average of 13 years, May 1). New Jersey—Moorestown, April 15; Demarest, April 23. New York—Bronx County, April 28; Buffalo, April 30. Connecticut—Jewett City, April 29 (average of 21 years, May 10). Rhode Island—Providence, May 5. Massachusetts—Newton Highlands, April 17; Chatham, April 20. Maine—Springvale, May 9. Louisiana—Thibodaux, March 20; Bains, March 21. Mississippi—coastal Mississippi, March 21. Arkansas—Helena, March 30 (average of 24 years, April 13). Tennessee—Memphis, April 2; Athens, April 6 (median of 7 years, April 13). Kentucky—Guthrie, April 14. Missouri—St. Louis, April 15 (average of 11 years, April 22); Columbia, April 18 (median of 20 years, May 2). Illinois—Murphysboro, March 20; Chicago region, April 1 (average, May 15). Indiana—Silverwood, Fountain County, April 4. Ohio—central Ohio, April 3 (average, May 3); Columbiana, April 8. Michigan—Ann Arbor, April 23 (average of 22 years, May 6). Ontario—Hyde Park, May 2; Port Dover, May 4. Iowa—Wall Lake, April 23. Wisconsin—Waukesha, May 2. Minnesota—La Crescent, May 3 (average of 16 years in southern Minnesota, May 12); Polk County, May 11 (average of 10 years in northern Minnesota, May 20). Texas—Olmito, March 24; Dallas, April 9. Oklahoma—Oklahoma City, March 23; Ardmore, April 13. Kansas—Bendena, April 10; Harper, April 21 (median of 13 years, April 29). Nebraska—Bladen, March 21; Red Cloud, April 30 (average of 24 years, May 8). South Dakota—Yankton, April 28. North Dakota—Fairmount, May 23. Manitoba—Winnipeg, May 18. Colorado—Beulah, May 15; Yuma, May 17.

Late dates of spring departure are: Colombia—Aracataca, March 2. Panamá—Gatún, March 31. Costa Rica—El General, April 10. Nicaragua—Eden, March 28. Guatemala—Quiriguá, April 21. El Salvador—Barra de Santiago, April 12. Veracruz—Omealca, May 15. Nuevo León—Montemorelos, May 21. Cuba—Havana, April 19. Florida—Bradenton, May 19; Tortugas, April 30. New Jersey—New Brunswick, May 30. New York—New York City region, June 8. Illinois—Chicago, May 18 (average of 6 years, May 15).

Early dates of fall arrival are: Texas—Tivoli, August 15. South Carolina—Greenwood, July 5. Tamaulipas—Pano Ayuctle, August 6. Durango—Papasquiaro, August 8. Veracruz—July 16. Michoacán—Apatzingán, August 12. El Salvador—Lake Olomega, August 16. Honduras—Cantarranas, August 3. Nicaragua—Escondido River, August 20. Costa Rica—San José, July 31. Panamá—Permé, August 3. Colombia—Fundación, October 15.

Late dates of fall departure are: California—Eureka, October 6 (only record). North Dakota—Fargo, September 10. South Dakota—Faulkton, October 15; Yankton, September 9. Nebraska—Omaha, September 12. Kansas—Hays, October 12; Onaga, September 9 (median of 9 years, September 1). Oklahoma—Oklahoma City, September 18. Texas—Austin, October 10. Minnesota—St. Paul, September 3 (average of 5 years in southern Minnesota, July 31). Wisconsin—Milwaukee, October 14. Iowa—Woodbury County, September 24. Ontario—London, September 16. Michigan—Grand Rapids, September 28. Ohio—Toledo, September 24; central Ohio, August 28 (average August 19). Indiana—Richmond, September 20. Illinois—Deerfield, October 10; Freeport, September 24. Missouri—Concordia, September 26 (average of 8 years, August 23); Kansas

City, September 21. Kentucky—Versailles, October 2. Tennessee—Athens, September 22 (average of 6 years, August 24). Arkansas—Rogers, October 1. Mississippi—Edwards, October 21. Louisiana—New Orleans, October 10. Nova Scotia—Sable Island, September 28. Massachusetts—Lexington, September 30. Rhode Island—Kingston, November 8. Connecticut—Hartford, September 18. New York—Massapequa, September 27. New Jersey—Long Beach, September 8. Pennsylvania—McKeesport, October 19; Berwyn, September 21 (average of 14 years, August 31). Maryland—Gibson Island, October 13; Charles County, September 21. District of Columbia—September 14 (average of 5 years, August 27). West Virginia—Bluefield, September 10. Virginia—Richmond, September 22; Lexington, September 19. North Carolina—North Wilkesboro, October 27; Raleigh, August 22 (average of 10 years, August 6). South Carolina—Spartanburg, September 26. Georgia—Augusta, September 7. Alabama—Smelley, October 13. Florida—Royal Palm Hammock, October 13; Fort Myers, October 11 (median of 7 years, September 28); Pensacola, September 14 (median of 18 years, August 25). Sinaloa—Escuinapa, October 25.

Egg dates.—Illinois: 12 records, May 13 to June 20; 7 records, May 22 to June 10.

Kansas: 8 records, June 5 to June 21.

New York: 4 records, May 29 to June 13.

South Carolina: 11 records, May 16 to June 6; 6 records, May 24 to May 31.

Texas, 30 records, April 29 to July 2; 18 records, May 8 to May 30.

ICTERUS GRADUACAUDA AUDUBONII Giraud

Audubon's Black-Headed Oriole

HABITS

Bendire (1895) writes:

This is one of the sixteen new species of birds described by Mr. J. P. Giraud in the Annals of the New York Lyceum of Natural History, in 1841, from specimens collected in Texas in 1838. Some time afterwards Mr. John H. Clark, the naturalist attached to the Mexican Boundary Survey, obtained several specimens near Fort Ringgold, Texas. He reported it as not abundant, and its quiet manners and secluded habits prevented it from being very conspicuous. It was most frequently observed by him feeding on the fruit of the hackberry, but whenever approached while thus feeding it always showed signs of uneasiness, and soon after sought refuge in some place of greater concealment. Usually pairs were to be seen keeping close together, apparently preferring the thick foliage found on the margins of ponds or on the old bed of the river. They did not communicate with each other by any note, and Mr. Clark was struck by their remarkable silence. Their habits seemed to him very different from those of any other Oriole with which he was acquainted.

George B. Sennett (1878) says: "This large Oriole cannot be said to be very abundant on the Rio Grande, although it is by no means rare. I think it is by far more retiring in its habits than any other of the family. If I were to go in search of it I should seek dense woods, near an opening, with plenty of undergrowth, where the Rio Grande Jay loves to dwell."

As all other naturalists who have visited this fascinating region along the lower Rio Grande have found this handsome, black-headed oriole far from abundant and not conspicuous among the many interesting Mexican species that there extend their ranges into the United States, it is not strange that we saw very few when I visited Brownsville in 1923. What few we saw were in the dense forests along the resacas, or stagnant water courses, the former beds of streams, or in other wooded regions; these usually contained large specimens of mesquite, hackberry, ebony, huisache, and a few palms, with a heavy undergrowth of shrubs, small trees, persimmons, granjenas, coffeebeans, and bush morning-glories.

At early morning and again after sunset, these woods resounded with the weird chorus of loud screams from the chachalacas; all day long the white-winged and white-fronted doves filled the air with their tiresome cooing; and the noisy Derby flycatchers often proclaimed their presence with loud, clamorous notes from the treetops; but we did not hear, or failed to recognize, the song of Audubon's oriole. Frequent glimpses were had of the brilliant green jay and the lovely little Texas kingfisher, but the oriole kept mostly hidden in the foliage.

Our black-headed oriole, as it was formerly called, is a northern race of a Mexican species, which ranges from the lower Rio Grande Valley southward into Tamaulipas and Nuevo León in México. The type race (*A. g. graduacauda*) ranges over the southern portions of the Mexican plateau; it is a smaller bird, with the bill much stouter, shorter, and the culmen more curved; the black of the head and neck is more extensive; there is no white in the wings, and the middle coverts are black instead of yellow; and the tail is entirely black. In our bird the outer edges of the wing quills are edged with white, broadly so on the innermost secondaries, and the greater wing coverts are usually edged with white near the tips; the outer tail feathers, also, are more or less edged and tipped with white.

Nesting.—Sennett (1879) was the first to discover the nests and collect authentic eggs of Audubon's oriole, about which he writes: "This year I was fortunate in obtaining, within our limits, nests and eggs of this large Oriole. Two incomplete sets were found early in May, which enable me to identify a complete set of four obtained last year. The latter set was taken at Hidalgo, Texas; the two former, at Lomita. The three nests were found in heavy timber, some ten or twelve feet from the ground, are half-pensile, something like those of the Orchard and Bullock's Orioles, and attached to upright terminal branches. They are composed of dried grasses woven among the growing twigs and leaves, so as to form a matting light and firm. They measure on the inside some three inches in depth and rather more in width."

Since then, the National Museum has received from William L. Ralph a fine series of the eggs, taken near Brownsville, Texas. Based on this material, Bendire (1895) describes the nest as follows:

The nest of this Oriole is usually placed in mesquite trees, in thickets and open woods, from 6 to 14 feet from the ground. It is a semipensile structure, woven of fine, wire-like grass used while still green, and resembles those of the Hooded and Orchard Orioles, which are much better known. The nest is firmly attached both on the top and sides, to small branches and growing twigs, and, for the size of the bird, it appears rather small. One, now before me, measures 3 inches in depth inside by about the same in inner diameter. The rim of the nest is somewhat contracted to prevent the eggs from being thrown out during high winds. The inner lining consists of somewhat finer grass tops, which still retain considerable strength, and are even now, when perfectly dry, difficult to break. Only a single nest of those found was placed in a bunch of Spanish moss, and this was suspended within reach of the ground; the others were all attached to small twigs. * * *

Nidification begins sometimes early in April, but usually about the last week in this month. Fresh eggs have been taken on April 23 and as late as June 8. Attempts are probably frequently made to rear two broods in a season, but many of them are unquestionably destroyed each year by the Red-eyed Cowbird, as well as through other causes.

Eggs.—Bendire (1895) gives the following good description of the eggs:

The number of eggs to a set varies from three to five. Sets of one or two eggs of this Oriole, with two or three Cowbirds' eggs, seem to be most frequently found, some of the first-named eggs being thrown out to make room.

The eggs differ somewhat in the character of their markings from those of the remainder of our Orioles; they are ovate and elongate ovate in shape, and the shell is rather frail and lusterless. The ground color is either pale bluish or grayish white, and occasionally the egg is only slightly flecked with fine markings and a few hair lines of different shades of brown and dark purple, these being nearly evenly distributed over the surface. In others the ground color is partly obscured with a pale purple suffusion, and more profusely blotched and streaked with different shades of claret brown, purple, ferruginous, and lavender, resembling somewhat certain types of Brewer's Blackbirds' eggs, while an occasional set is profusely blotched with coarse, heavy markings of cinnamon rufous and numerous finer spots of the same tint, these almost completely hiding the ground color. The markings are generally heaviest about the larger end of the egg.

The largest egg of the series measures 26.42 by 18.80 millimetres, or 1.04 by 0.74 inches; the smallest, 23.62 by 17.78 millimetres, or 0.93 by 0.70 inch.

Plumages.—Chapman (1923b) describes the plumages briefly as follows: "In nestling plumage, Audubon's Oriole is olive-green above, greenish yellow below, the wings and tail being externally brownish. The black head of the adult is acquired at the postjuvenal (first fall) molt, but the wings and tail are still those of the young bird. This plumage is worn throughout the first nesting season, at the end of which the black wings and tail are acquired and the bird resembles our figure. The female closely resembles the male and often cannot be distinguished from it in color, but usually the back is more olive-green, less pure yellow than in the fully adult male."

The above description is correct as far as it goes. In a large series that I have examined, the postjuvenal molt seems to begin early; I have seen a young bird in juvenal plumage that was acquiring a black throat on July 19, and another that had a nearly complete black head on July 14. As stated above, this first winter plumage is worn without much change all through the spring; I have seen birds in this plumage in March, April, May, and June, and as late as July 7. The postnuptial molt of adults is apparently not completed until some time in September; one taken September 3 was still molting wings and tail.

Behavior.—The following quotations are taken from some papers sent to John Cassin (1862) by Lieutenant Couch:

The Black-headed Oriole was seen for the first time on the third of March 1858, at Santa Rosalio rancho, eight leagues west of Matamoras. It had paired, and both male and female were very shy and secluded, seeking insects on the *nopal* (a species of prickly pear), or among the mimosa trees, never seeming to be at rest, but constantly on the look-out for their favorite food.

At Charco Escondido, farther in the interior of Tamaulipas, this bird was well known to the *rancheros*, who were disposed to give it a bad reputation, stating that it often came to the rancho to steal the freshly-slaughtered beef, hung up to dry in the sun. Whether this was true or not, I had no opportunity of ascertaining; but my acquaintance with the Black-headed Oriole, at this place, I have a particular reason for remembering. * * * It was the day after a severe *norther*, and the whole feathered kingdom was in motion. My guide soon called my attention to two *calandrias*, as these birds are called by the Mexicans, which were quietly but actively seeking their breakfast. The male having been brought down by my gun, the female flew to a neighboring tree, apparently not having observed his fall; soon, however, she became aware of her loss, and endeavored to recall him to her side with a simple *pout pou-it*, uttered in a strain of such exquisite sadness, that I could scarcely believe such notes to be produced by a bird, and so greatly did they excite my sympathy, that I felt almost resolved to desist from making further collections in natural history, which was one of the principal objects of my journey into the country. * * *

My stay in Mexico was not sufficiently protracted to enable me to study the habits of this interesting bird as fully as I could have wished. Generally, its flight is low and rapid, and it seemed to prefer the shade of trees. It was observed almost invariably in pairs, and the male and female showed for each other much tenderness and solicitude. If one strayed from the other, a soft *pou-it*, soon brought them again together.

Voice.—Couch (Cassin, 1862) observed further: "I have never heard the lay of any songster of the feathered tribe expressed more sweetly than that of the present Oriole. At Monterey, it is a favorite cage-bird. The notes of the male are more powerful than those of the female."

Sennett (1878) says: "It is a sweet singer, never very generous with its music, and only singing when undisturbed.

"I remember once sitting in the edge of a woods, watching the movements of some Wrens just outside, the only sounds to be heard in the woods being the discordant notes of the Rio Grande Jay, when suddenly, from over my head, there burst upon my ear a melody so sweet and enchanting that I sat entranced, and, listening, forgot all else. I soon discovered the whereabouts of the singer, and watched him as he flitted about from branch to branch, singing his wonderful song. I have no power to describe a bird's song, least of all this Oriole's."

Field marks.—Audubon's oriole should be easily recognized as a large oriole, with a wholly black head and neck, and with black wings and tail, the rest of the body being yellow, rather more greenish yellow on the back and clearer yellow below, but without any orange tinge. The sexes are practically alike.

Enemies.—Audubon's oriole probably has as many enemies as other birds, but the cowbirds seem to be as troublesome to it as any of which we have record. Bendire (1895) says that it "seems to be greatly imposed upon by the Red-eyed Cowbird; half of the sets in the collection contain from one to three of these parasitic eggs; but none of the equally common Dwarf Cowbird have, as far as I am aware, yet been found in them."

Herbert Friedmann (1929) writes: "Near Brownsville, Texas, I found two nests of the Audubon's Oriole; both of them containing eggs of the Red-eyed Cowbird. One had two eggs of the Oriole and one of the Red-eye. The other contained one Red-eyed Cowbird's egg and one of the Dwarf Cowbird and one of the owner. Both northern races of the Red-eyed Cowbird are parasitic on the Audubon's Oriole."

Winter.—Audubon's oriole is mainly resident throughout the year within its breeding range, but it is said to occur in San Luis Potosí in winter, and to wander casually as far north as San Antonio, Tex. H. P. Attwater (1892) called it a "rare winter wanderer" in the latter locality; he secured a fine male there on March 27, 1890, and, on

February 13 of the next year, he obtained three specimens out of a flock of 8 or 10; the next day they were all gone, and he did not see them again.

DISTRIBUTION

Range.—Southern Texas to Central México.

Breeding range.—Audubon's oriole breeds from southern Texas (Rio Grande City, Hidalgo, Brownsville) and possibly casually north to Pleasanton and Austwell south at least to central Tamaulipas (Realito, Río Cruz).

Winter range.—Winters throughout breeding range south to Nuevo León (Mesa del Chipinque, south of Monterrey), San Luis Potosí (Hacienda Angostura), and southern Tamaulipas (Victoria, Tampico).

Casual records.—Rare in south central Texas (San Antonio area).

Migration.—Largely a permanent resident. Early spring date north of normal range: Texas—Lytle, Atascosa County, March 4.

Egg dates.—Texas: 15 records, April 23 to June 15; 8 records, May 7 to May 28.

ICTERUS CUCULLATUS SENNETTI Ridgway

Sennett's Hooded Oriole

HABITS

This is a northeastern race of a Mexican species, ranging from Tamaulipas, in northeastern México, into the lower Rio Grande Valley of Texas. According to Ridgway (1902), it is "similar to *I. c. cucullatus,* but lighter in color; adult males less decidedly orange, the color of pileum, chest, etc., deep cadmium yellow, never cadmium orange; adult females much lighter in color, the yellow of the under parts dull or pale gamboge instead of saffron or ochreous, the back and scapulars lighter grayish, and light olive-greenish of pileum, rump, etc., clearer; wing and tail averaging decidedly shorter."

In the lower Rio Grande Valley, Sennett's hooded oriole is an abundant and familiar summer resident. George B. Sennett (1878) reported it as "very common in the vicinity, and among timber of any respectable growth." He found it "more plentiful than all the rest of the genus combined." James C. Merrill (1878) says: "This is perhaps the most common Oriole in this vicinity during the summer, arriving about the last week in March. It is less familiar than Bullock's Oriole, and, like the preceding species, is usually found in woods." When I visited Brownsville in 1923 we found this oriole very common in the woods and about the ranches and towns, where we found several nests. On his visit, a year later, Herbert Friedmann (1925) found it very common, "close to houses at times; in fact they seem not to mind human presence at all." He found 16 nests.

Nesting.—The early accounts of the nesting habits of Sennett's hooded oriole differ considerably from what recent observers have noted.   Sennett (1878) wrote:

Their usual nesting places are the hanging trusses of Spanish moss, everywhere provokingly abundant on the larger growth of trees.   I have also found their nests on the lower limbs of trees and the drooping outer branches of undergrowth; but wherever found, the inevitable Spanish moss enters largely or wholly into their composition.   So durable is this moss that it lasts for years, and as a consequence there are everywhere ten old nests to one new one.   The heart of the moss when separated from its white covering becomes the "curled hair" of commerce.   The Hooded Oriole takes this dry vegetable hair, and ingeniously weaves it into the heart of a living truss of moss, making a secure and handsome home. I took one no higher than my head, and others thirty feet or more from the ground.

Later, he wrote (1879): "One nest was discovered, in a corn-field, made of Spanish moss, which was interwoven with a couple of leaves of two corn-stalks, which it thus bound together; another was found in a truss of Spanish moss, having dried grasses for lining, instead of the usual dead and black hair-like moss.   In several nests were horse-hair and tufts of goats' wool."

Merrill's (1878) account is somewhat similar:

The nests of this bird found here are perfectly characteristic, and cannot be confounded with those of any allied species; they are usually found in one of the two following situations: the first and most frequent is in a bunch of hanging moss, usually at no great height from the ground; when so placed, the nests are formed almost entirely by hollowing out and matting the moss, with a few filaments of a dark hair-like moss as lining; the second situation is in a bush (the name of which I do not know) growing to a height of about six feet, a nearly bare stem throwing out two or three irregular masses of leaves at the top; these bunches of dark green leaves conceal the nest admirably; it is constructed of filaments of the hair-like moss just referred to, with a little Spanish moss, wool, or a few feathers for lining; they are rather wide and shallow for Orioles' nests, and, though strong, they appear thin and delicate.   A few pairs build in Spanish bayonets (*Yucca*) growing on sand ridges in the salt prairies; here the nests are built chiefly of the dry, tough fibres of the plant, with a little wool or thistle-down as lining; they are placed among the dead and depressed leaves, two or three of which are used as supports.

Bendire (1895) says of a nest built in a yucca: "One now before me, in an excellent state of preservation, measures exteriorly 3½ inches in depth by 3 inches in width; the inner cup is 2½ inches wide by 2 inches deep.   It is built throughout of yucca fiber and contains no lining.

"Nidification begins in April, and the earliest record of a full clutch of eggs having been taken is April 17, a set of five; the latest was July 5; probably two broods are raised in a season."

Perhaps the Spanish "moss" had largely disappeared from the vicinity of Brownsville at the time of our visit, for all of the nests we saw were in palms or palmettos. We found several of their nests in palms 25 or 30 feet from the ground and generally inaccessible. On May 25, 1923, we found three nests in a grove of palmettos near a house; a man and his boys helped us to climb to these by means of a ladder; and from one of them I collected a nice, fresh set of four eggs. The nests were all neatly woven cups, made entirely of the fibers of the palmetto leaves; they were securely fastened to the under side of the leaf, which generally was green, the supporting fibers being sewn through some strong portion of the leaf; as a result they were well shielded from either rain or sun.

Friedmann's (1925) experience was similar; all his 16 nests were "sewn on to the under side of the palm or banana leaves"; they were much shallower than nests of the Baltimore oriole, but deeper than those of the orchard oriole. Neither of us saw any nests in Spanish "moss" (*Tillandsia*); in fact, I cannot remember seeing any lichen in that vicinity; but some of the trees on the edges of the resacas supported more or less *Usnea*.

Eggs.—Bendire (1895) writes:

The number of eggs laid to a set varies from three to five, sets of four being most common, and an egg is deposited daily. They are mostly ovate in shape; the shell is delicate, rather frail, and without luster. The ground color is dull white, occasionally this has a pale buffy and again a faint bluish tint. The eggs are blotched and spotted, principally about the larger end, with irregularly shaped markings ranging from dark seal brown to claret brown, purple, mixed with ochraceous, mouse, and pearl gray, and these rarely run into lines and tracings, so prevalent in the eggs of most of our Orioles. Some eggs are fairly well marked, others only faintly; the lighter shades mentioned largely predominate over the darker ones, and in some the latter are entirely wanting.

The average measurement of ninety-three specimens in the United States National Museum collection is 21.59 by 15.24 millimetres, or 0.85 by 0.60 inch. The largest egg in the series measures 22.86 by 16 millimetres, or 0.90 by 0.63 inch; the smallest, 18.80 by 15.24 millimetres, or 0.74 by 0.60 inch.

Plumages.—I have seen no very young orioles of this species, but Chapman (1923a) describes the sequence of plumages briefly as follows: "Nestlings of both sexes resemble the adult female [based on the subspecies *Nelsoni*], and the female wears essentially similar colors for the remainder of her life. After the post-juvenal molt the male apparently continues to resemble the female during the first part of the winter or even early spring when it acquires a black throat and lores. This constitutes its first breeding plumage and it is worn until the post-nuptial (second fall) molt at which the bird passes into adult winter dress."

Young males and females, during their first winter, are both somewhat duller in color than are the adult females at that season.

Ridgway (1902) describes the adult winter plumage of the male as similar "to the summer plumage, but the orange or orange-yellow duller, especially on upper parts, where more or less obscured by a tinge or wash of olivaceous; scapulars and inter-scapulars margined terminally with light olive or olive-grayish; tertials more broadly margined with white."

Behavior.—Sennett (1879) noted that these orioles "were continually peering about the thatched roof of our house and arbors adjoining for insects. They were more familiar than any of the other Orioles about the ranch." And Baird, Brewer, and Ridgway (1874), state on information received from Captain McCown:

When met with in the woods and far away from the abodes of men, it seemed shy and disposed to conceal itself. Yet a pair of these birds were his constant visitors, morning and evening. They came to the vicinity of his quarters—an unfinished building—at Ringgold Barracks, and at last became so tame and familiar that they would pass from some ebony-trees, that stood near by, to the porch, clinging to the shingles and rafters, frequently in an inverted position, prying into the holes and crevices, apparently in search of spiders and such insects as could be found there. From this occupation they would occasionally desist, to watch his movements. He never could induce them to partake of the food he offered them.

Enemies.—Sennett's hooded oriole is often imposed upon by the dwarf cowbird and the red-eyed cowbird, principally the latter, and the eggs of both species are sometimes found in the same nest. Out of the 16 nests observed by Friedmann (1925) near Brownsville, one held an egg of the dwarf cowbird and two of the oriole, and three contained eggs of the red-eyed cowbird.

DISTRIBUTION

Breeding range.—Sennett's hooded oriole breeds from Southern Texas (Rio Grande City, Port Isabel) south along the Gulf coastal plain to southern Tamaulipas (probably Paso del Haba).

Winter range.—Winters throughout its range south to northern Guerrero (Taxco, Iguala) and Morelos (Cuernavaca).

Migration.—The data deal with the species as a whole.

Early dates of spring arrival are: San Luis Potosí—Valles, March 24. Tamaulipas—Rancho Rinconada, March 5. Nuevo León—Linares, March 5. Sonora—Tesia, March 21; Baja California—Santo Domingo, February 28. Texas—Hidalgo and Brownsville, March 12 (median of 9 years in Cameron County, March 15). New Mexico—Carlsbad Cave region, March 24. Arizona—Tucson, March 14 (median of 13 years, March 25). California—Los Angeles County, March 5 (median of 32 years, March 20); Santa Barbara, March 14.

Late date of spring departure: Sonora—Guirocoba, May 20.

Early dates of fall arrival are: Sinaloa—Escuinapa, October 24. Guerrero—Taxco, November 1.

Late dates of fall departure are: California—Los Angeles County, November 1 (median of 5 years, September 13). Arizona—Tucson, October 8. New Mexico—Guadalupe Canyon, October 4. Texas—Brownsville, November 12.

Egg dates.—Arizona: 22 records, May 22 to Aug. 25; 11 records, June 6 to July 18.

California: 72 records, April 7 to Aug. 12; 36 records, May 9 to June 12.

Baja California: 15 records, May 1 to Aug. 10; 8 records, July 14 to July 25.

Texas: 44 records, April 8 to July 5; 22 records, May 1 to May 29.

ICTERUS CUCULLATUS CUCULLATUS Swainson

Swainson's Hooded Oriole

Similar in its habits to Sennett's hooded oriole, this race is somewhat darker, more orange, less yellowish in the males, and notably darker in the females.

Swainson's hooded oriole ranges from western Texas south to Guerrero and Veracruz, and its breeding range extends from western Texas (Boquillas, Del Rio), Chihuahua (Sabinas), Nuevo León (Monterrey, Linares), and Tamaulipas (Gómez Farías); south to San Luis Potosí (probably San Luis Potosí), northern Guerrero (Iguala), and southern Veracruz (Orizaba, Catemaco). Its winter range is uncertain; the northernmost winter specimens are from Morelos (Cuernavaca) and central Veracruz (Mirador); in spring and fall specimens have been taken in western Texas (Marathon, Langtry), Nayarit (Santiago), Michoacán (Lake Pátzcuaro, Tacámbaro), central Guerrero (Chilpancingo), and Veracruz (Puerto México).

ICTERUS CUCULLATUS NELSONI Ridgway

Arizona Hooded Oriole

Plates 14 and 15

Contributed by ROBERT S. WOODS

HABITS

Despite its shy, quiet ways, probably few birds of the Southwest have impressed themselves upon the average human consciousness more definitely than the Arizona hooded oriole. This is due not only to the eye-arresting coloration of the adult male, but to the fact that it finds its most congenial surroundings among plantings of palms and flowering shrubs, the former furnishing nesting sites and material, and the latter a favorite food. In spring and summer it is a common inhabitant of city parks and gardens, though it manifests none of the boldness and assurance that characterize some of our dooryard birds.

In the United States, the Arizona hooded oriole is a summer visitant to the southern portions of New Mexico and Arizona, and southeastern California. Typically a species of the Lower Sonoran Zone, the hooded oriole is seldom seen among the yuccas and junipers frequented by Scott's oriole. Previous to the large-scale development of irrigation, it appears to have been confined mainly to woodlands bordering the watercourses of the lower country. Much of its territory is shared by Bullock's oriole, which sounds its ringing notes from the tops of eucalyptus or cottonwood trees while the hooded oriole makes its more silent way through the shrubbery and branches below.

Concerning the haunts and habits of the hooded oriole in southern Arizona, H. W. Henshaw (1875) said: "It shuns the arid districts, and is found only in the fringes of deciduous trees along the streams. Here it seeks its food among the foliage of the cottonwoods, and flies from thence to the low bushes on the cañon sides, spending much of its time among them, gleaning insects from the branches, or even descending occasionally to the ground. I did not hear the song; the birds, at the time of my acquaintance with them, being busy in providing for their young, and seeming to find their time too fully occupied to devote any to music. Their common notes are a rolling chatter, which somewhat resembles that of our common Baltimore Oriole, but is much weaker and fainter."

Also referring to conditions of an earlier day, Bendire (1895) wrote: "Within our borders it is more common in southern Arizona than anywhere else, and I found about twenty of its nests here during the spring and summer of 1872. * * * I rarely saw one far away from water at any season of the year. The dense, shady groves of cottonwood and mesquite trees in the creek bottoms appeared to be its favorite haunts. It is a shy, restless creature, nearly always on the move, looking for insects of various kinds and their larvae, including hairless caterpillars, and small grasshoppers." It may be doubted whether the first statement of the foregoing quotation is still true, as suitable habitats have increased greatly.

Spring.—While Dawson (1923) found the hooded oriole beginning to arrive in California late in March, corresponding dates for southern Arizona may be somewhat later, W. E. D. Scott (1885) stating that they arrive about the middle of April. For a time the males are more frequently seen than the females.

Courtship.—"During the mating season, beginning about the latter part of April," says Bendire (1895), "several males may sometimes be seen chasing a female and scolding and fighting each other for the coveted prize." Little if anything has been published regarding any characteristic courtship practices of this species, but I have seen an adult male execute a series of exaggerated bows as he advanced slowly along a horizontal limb of a tree in which a female was perched. Again, in midsummer, a male in second-year plumage was observed hopping round and round his mate in a tree, singing softly and repeatedly posturing with open bill directed toward the zenith, while the female faced him, also with open bill.

Nesting.—For a bird which spends most of its time comparatively close to the ground, the Arizona hooded oriole chooses surprisingly high nesting sites. With one exception hereinafter mentioned, the numerous nests observed by Bendire (1895) and Scott (1885) in Arizona ranged from 12 to 45 feet from the ground.

As in the case of the cactus wren, a species having a somewhat similar geographical range in the United States, the nesting materials used by the hooded orioles seem to differ as between Arizona and California, a difference hardly to be accounted for by the relative availability of the materials. However, whereas the former seems to construct grass nests only on the Pacific slope, the present species uses such materials mainly at points farther east, and seldom in California. The various nest descriptions quoted below plainly bring out this difference.

Bendire (1895) says: "In southern Arizona nidification begins rather late, rarely before May 20, and sometimes later. In southern California, however, it commences fully a month earlier, and a full set of eggs was taken by Mr. Theodore D. Hurd, near Riverside, California, on April 23." Referring to a nest containing three fresh eggs, found in Arizona on June 5, he continues:

It was suspended from a bunch of mistletoe growing on a limb of a cottonwood tree, about 40 feet from the ground, and was hard to get at. This, like nearly all of the nests found by me, was woven of a species of slender wiry grass growing in moist places, which was used in a green state. It contained a little cottonwood down for lining. Its green color, closely resembling the surrounding foliage, made it very difficult to see. It was securely fastened to several mistletoe twigs among which it was placed. Fully three-fifths of the nests found by me were placed in similar situations; the others were suspended in mesquite (excepting one found in an ash tree), at various heights from 12 to 45 feet from the ground. The majority of these nests were woven of this green wire grass, which seems admirably adapted for this purpose, and a few only were made of dry yucca fibers; the latter were much more easily seen. In some instances this material was also used for the inner lining, mixed with willow down or a little wool, rarely with a few feathers, or a small quantity of horsehair.

While some of the nests were semipensile and slung somewhat like a hammock, so that they rocked like a cradle with every breeze, in the majority some of the surrounding slender twigs among which the nest was placed were incorporated into its walls and sides, securing it almost immovably in position. None of the nests seen by me in any way resembled those of Bullock's Oriole, which was also common here. They were always much brighter colored, not nearly so deep, and were constructed of entirely different materials. Neither do the grass-woven nests of the Arizona Hooded Oriole resemble the common type of its near relative found in Texas. I refer to the nests built of tree moss, which are usually located in bunches of the same material. But those of either form of the Hooded Oriole, when built of yucca fibers, might be readily mistaken for each other. Besides the trees already mentioned, Mr. Scott found it breeding in sycamores, and in California it nests in walnut, cypress, gum, and fan palms, the fibers of which, according to Mr. Theo. D. Hurd, are almost exclusively used as nesting material in that locality.

Hurd (1890) published the following interesting notes on the nesting habits of this oriole, as observed by him in that vicinity: "For the rearing of the first brood the nests are usually suspended in overhanging branches of the blue gum (*Eucalyptus globulus*), but it is a noticeable fact that the second nests are more commonly attached to the leaves of the palm tree. Why this is I do not know, unless they want to begin laying as soon as possible, and therefore build where material is most easily obtained. When in palms the nests are fastened directly to the under side of a large leaf, leaving a small opening on one or more often on either side, for the bird to enter."

Says Bendire (1895): "Two and possibly even three broods are sometimes raised in a season. I found slightly incubated eggs in Arizona on August 25. From three to five eggs are laid to a set; in Arizona usually only three or four; but Mr. Hurd reports taking a set of seven on May 6, 1890. An egg is deposited daily until the set is completed."

He reports having seen the male carrying nesting materials, and adds: "The nest is well built, it is basket or cup shaped, with a very thick bottom and strong sides. It averages about 4 inches in height externally. The inner cup is oval, about 2½ inches deep and 3 by 2 inches wide, and it takes about 4 or 5 days to complete it."

Henshaw (1875) thus describes the nests observed in Arizona: "I saw quite a number of what I took to be the nests of this species, suspended low down from the branches of the cottonwoods and various deciduous trees; one or two being not more than ten feet from the ground. These were made of grasses, and woven and interwoven in such a manner as to make a very firm durable nest, and shows that this species is not inferior to its allies in the art of construction." Scott (1885) says: "Two broods are raised, and not infrequently three, during their stay here, and a new home is built for each brood. The old birds are great workers when building their nests, and the rapidity with which so elaborate a structure is completed is astonishing. Three or four days at most generally suffice to complete the structure." He then describes in considerable detail 10 nests in a canyon, presumably near Tucson—

All taken from three kinds of trees, cottonwood, sycamore, and a kind of ash; and, considering that the location of all were not a mile apart, it would seem that taste or fancy had much to do with producing in the same locality, where the materials used by all of the builders are abundant and easily obtained, structures varying so widely in general appearance, in the materials of which they are built, and in their method of building, as well as in mode of attachment to the tree.

Some of the nests, it will be seen, are as truly pensile as those of *Icterus galbula;* others are more like those of *Icterus spurius;* while one at least rests on a stout twig and is hardly to be regarded as a hanging nest at all.

Of the 10 nests, 8 are described as composed mainly of grasses, either coarse or fine, 1 of yucca fibers, and 1 of a combination of these 2 materials. In addition to these, he mentions nests built in clumps of mistletoe in mesquite trees, and also an unusual nesting site at a height of only 8 feet on the trunk of a yucca in the open desert.

Eggs.—The eggs are indistinguishable in size and appearance from those of other races of the species, as described under Sennett's hooded oriole.

Young.—Bendire (1895) gives the incubation period as 12 to 14 days. Irene G. Wheelock (1904) says: "Incubation lasts thirteen days, and in this the male takes no part." Since the nesting sites usually chosen do not readily lend themselves to observation of the interiors of the nests, statistics relative to the development of the young are not plentiful. According to Mrs. Wheelock (1904), "The young Orioles are born naked except for flecks of down on the crown and along the back. They are fed by regurgitation for four or five days. The eyes open on the fourth day, and pinfeathers soon begin to darken the skin. In two weeks the nestlings are fully fledged, looking much like the mother, and are ready for their début. Nevertheless they are very helpless, and are fed and cared for by both parents for some time after leaving the nest."

Plumages.—The plumages and molts are similar to those of the species elsewhere, described under Sennett's oriole, to which the reader is referred.

Food.—In general, the food of this oriole consists of a combination of insects and the nectar of flowers, but also some fruits, such as berries and cherries. In addition to the fruits mentioned above, hooded orioles are fond of loquats, but in my experience they pay little attention to peaches, grapes, or other later ripening soft fruits. Nectar undoubtedly fills a larger place in their diet than is recognized by some writers. Where suitable flowering plants are present in abundance, the birds will spend much time in diligently probing the blossoms of agaves, aloes, hibiscus, lilies, and other tubular forms. In procuring nectar from large flowers, the favored method is to perch on the stem of the blossom and puncture the base of the tube with the sharp bill. While a certain amount of insect food would naturally be obtained from the flowers, the fact that nectar is the primary object is indicated by their custom of occasionally slitting unopened lily buds, a habit by no means popular with gardeners.

As might be expected from their fondness for nectar, orioles enthusiastically respond to offerings of sugar sirup, of which they will consume relatively large quantities, drinking deeply and often. They appear rather more tolerant of dilution of the sirup than do hummingbirds. An originally saturated solution seems to be as readily taken when diluted to half strength.

Behavior.—Except with respect to its nests, this species seems to have received little detailed study. Of its general habits, Mrs. Wheelock (1904) says: "Like the orchard oriole, he haunts the heavy foliage, flitting through the open only *en route* to a fresh pasture. Restless, shy, ever on the move, searching for caterpillars on the under sides of the leaves chickadee fashion, picking in the crevices for larvae like a nuthatch, and snapping up grasshoppers with a little jump as do young meadowlarks, he is usually to be found within 12 feet of the ground." While these statements are true as generalizations, the hooded oriole does not hesitate to risk a more exposed situation when necessary in order to explore the flowering stalk of an agave or aloe, and the males sometimes sing from the tops of tall trees.

Their agility on the wing is apparently not such as to encourage them to attempt the capture of flying insects, though the flight is fairly strong and swift. It seems, however, to be used solely as a means of getting from one place to another, and never as a method of expressing exuberance of spirit or expending surplus energy. In going about through the trees and shrubbery, the orioles are likely to climb along the branches with minimum use of the wings. When approached, they lean forward and lower their heads in a characteristic attitude while peering nervously at the intruder.

While preferring well-watered situations, the orioles do not seem greatly interested in the water itself, though they occasionally bathe. The nectar which forms a part of their diet doubtless makes the drinking of water unnecessary. Though not rated as a gregarious species, there seems to be a certain desire for companionship, and in spring before the nesting activities are under way, two of the brilliantly hued males may often be seen feeding at the same flower stalk. The young also remain in rather close association for some time after attaining self-support. They are not quarrelsome, either among themselves or with other species. In spite of the fact that they so commonly frequent low shrubbery, these birds rarely alight on the ground, though on occasions they may be seen hopping over a lawn, presumably in search of insects.

Voice.—The presence of the Arizona hooded oriole is usually first betrayed by a liquid chirp repeated at intervals, or by a chatter like that of Bullock's oriole, but lighter and softer in tone. Contrasting strongly with the buglelike notes of the latter species is its pleasant but unpretentious warbling song, which is neither loud nor frequent, and is interspersed with the typical chatter or rattle.

Field marks.—In flight, the male hooded oriole can most easily be distinguished from the Bullock's and Scott's orioles by the apparently solid black of the tail. The body color of this subspecies is deeper yellow than in Scott's oriole, but less orange than in Bullock's. In the western tanager, of somewhat similar size and coloration, the tail is nearly even instead of graduated, the wings have yellow patches, and the throat is without black. The entirely yellow crown of the hooded oriole is distinctive, and the bill is more slender than in the others mentioned. Descriptions of coloration are hardly adequate guides to the field identification of the female orioles, but in the present species the bill is more distinctly decurved and the tail more definitely graduated.

Bendire (1895) says: "The Arizona Hooded Oriole is imposed on to a considerable extent by the Dwarf Cowbird, and I found several nests containing one and two eggs of this parasite with one or two only of the rightful owner."

Enemies.—Herbert Friedmann (1929) lists this species among the victims of both the dwarf cowbird (*Molothrus ater obscurus*) and the bronzed cowbird (*Tangavius aeneus aeneus*) in Arizona. Its wariness, nonterrestrial habits, and the nature of its nesting sites should render it comparatively safe from most natural enemies.

Fall.—Most of the Arizona hooded orioles have disappeared

from their summer haunts by the end of August, but a few individuals, especially the immature, remain into September.

DISTRIBUTION

Breeding range.—The Arizona hooded oriole breeds from southeastern California (Colorado River Valley), central and southeastern Arizona (Topock, San Carlos, Safford), and southwestern New Mexico (Silver City); south to northeastern Baja California (eastern base of Sierra San Pedro Mártir, lat. 31° N.) and southern Sonora (Guaymas, Agiabampo). Casual in southwestern Utah (St. George, Beaver Dam Wash), where it may breed.

Winter range.—Winters from central Sonora (Hermosillo) casually to southern Arizona (Tucson); south to southern Sinaloa (Escuinapa, Río Mazatlán).

ICTERUS CUCULLATUS CALIFORNICUS (Lesson)

California Hooded Oriole

Contributed by ROBERT S. WOODS

HABITS

This race of the hooded oriole has extended its range northward on the Pacific coast in recent years casually to the San Francisco Bay region, but W. L. Dawson (1923) has placed Santa Barbara as the northern limit of its common occurrence. Undoubtedly the extensive ornamental plantings which have been made in southwestern California have greatly increased the amount of country suitable for this bird and have correspondingly increased its potential population.

Like other species, this one occasionally departs markedly from its usual routes and schedules. F. C. Lincoln (1940) reports one individual that was banded at Los Angeles on January 22, 1939, and was found dead near Garden City, Kans., about August 5 of that year.

Spring.—In Los Angeles County the hooded orioles are usually first seen during the latter half of March, but in some years their arrival in any given breeding locality may be delayed until after the beginning of April. Dawson (1923) said "[This oriole] begins to arrive in California late in March. I say 'begins to arrive' because I think it altogether probable that there are two streams or stocks of migrants, one arriving *early* and nesting in April *and* July, the other nesting only once, in late May or early June."

Courtship.—Nothing appears to have been published to indicate that the courtship of this race differs in any way from that of the Arizona bird.

Nesting.—Like the Arizona race, this hooded oriole chooses nesting sites high up in trees. Florence Merriam Bailey (1910) says: "In choosing between individual [palm] trees, the taller seem to be given the preference." In California this oriole has been found nesting, not only in cottonwoods, as in Arizona, but also in walnut, cypress, gum, and fan palms, the fibers of which are used as nesting material.

The California oriole does not, like the more eastern races, construct its nest out of grasses. Bendire (1895) speaking of the Arizona hooded oriole, noted that a full set of eggs, undoubtedly of the California race, was found near Riverdale, Calif., by Theodore D. Hurd on April 23.

A good account of the nesting of the hooded oriole in southwestern California is found in the notes of J. F. Illingworth (1901), who states "it is difficult to find two nests of the Bullock's Oriole alike in shape or material, as they use almost anything they can find in the way of fiber." He continues:

The nests * * * on the other hand are very much alike, and I have never found one made of other material than the palm-fiber. The locations, too, are similar, a tree with large leaves being usually selected and a favorite position is under the broad, corrugated leaves of the palm. These form an excellent shelter from both rain and sun. They drill holes through the thick leaves with their sharp, slender beaks and tie the nest to them with palm-fiber. Often the nest is hung between several leaves such as those of the fig tree, when holes are cut and the palm-fibers laced in and out through them, thus drawing the leaves together to form the outside of the nest. The leaves not only aid in the nest structure but also form the best possible concealment.

An average nest * * * is 3.50 inches deep and 2.50 inches wide inside measurements, while the outside is about four inches deep and four across. Nests of both the Bullock's and Arizona Hooded Orioles are frequently taken possession of by House Finches, sometimes even before the orioles have finished them, but more often after they are deserted.

From an article on "the palm-leaf oriole" by Florence Merriam Bailey (1910) the following excerpts relating to nesting are obtained:

In eight towns and three country places in the general region between Redlands and San Diego in the summer of 1907, I counted forty nests made of palm fibers and hung in fan palms, and twelve others made of palm fiber and hung in other trees. * * * The great variety of palms used for decorative purposes in southern California gives the oriole a wide range of choice in nesting sites, but with one exception, that of a yucca-like palm in Santa Ana, the nests found were in the common native Washington fan palm, or in one too nearly like it to be distinguished by the unbotanical. The wisdom of the choice is easily appreciated for the narrow leaves of the date palm offer no protection from the hot California sun while the wide leaves of the fan palms are natural umbrellas, and among fan palms the short-stemmed varieties with close-set leaves would give little of the breeziness given by this long-stemmed one whose leaves fan reasonably free from each other. * * * By the time I had listed the fifty-two nests made of palm fiber, forty of which were hung in the palm, it seemed that, in southern California at least, *nelsoni* had won its right to the name of Palm-leaf Oriole.

Dawson (1923) gives the following formal description of the hooded oriole's nest in California: "Nest: a closely woven basket, or hanging pouch, of fine vegetable fiber, usually composed externally of a single, uniform, selected material, and in California almost invariably the shredded fibers of the Washington Palm, * * * with some inner felting of vegetable down or feathers; lashed to the under side of a palm leaf or of other large protecting leaves." The nests that I found at Azusa, Calif., were in avocado, eucalyptus, and dracaena (perhaps the "yucca-like palm" mentioned by Mrs. Bailey). In the first two locations, the nests were placed in terminal clusters of leaves, so that they were not at all conspicuous. All these nests were made of what appeared to be palm fiber, although the nearest fiber-bearing palms were perhaps half a mile distant. Other suitable fibers were scarce, and one summer a specimen of "old man" cactus (*Cephalocereus senilis*) was almost denuded of its white hairs by the orioles. In some instances, at least, material is gathered by the female while the male waits near by and flies with her to the nesting site.

One noteworthy nesting site was beneath the second-story eaves of my home, where the birds had in some manner wedged one or more fibers into a crack, despite the lack of any perching place except the lower surface of shingles and sheathing. The nest when completed dangled from a single strand, swinging and twisting in the wind, but miraculously remained in position until the young were successfully fledged, shortly after which the empty nest dropped to the ground, the removal of the tension due to the weight of the young having evidently permitted the disengagement of the fiber.

It is unfortunate that we have no information concerning the habits and abundance of this oriole in southwestern California previous to the widespread introduction of the fan palm, *Washingtonia filifera*, which is native only to a few restricted localities on the borders of the Colorado Desert. It would be interesting to know whether the species has altered its nesting habits on the Pacific slope, or whether this whole area has been populated by descendants of the birds which shared the original habitat of the palm and which followed its widening distribution.

Eggs.—The eggs are indistinguishable in size and appearance from those of other races of the species, as described under Sennett's hooded oriole.

Young.—The young are similar in every way to those of the other races of this species. That they possess some aquatic ability was noted by Frank F. Gander (1927), who reports: "On July 21, 1924, I saw two fledgling Arizona Hooded Orioles leap from their nest in a eucalyptus tree and fall 20 feet into a pond. They at once swam ashore, paddling with their feet and with their wings spread out on the water."

Plumages.—The plumages and molts are similar to those of the species elsewhere, described under Sennett's hooded oriole, to which the reader is referred.

Food.—Near Los Angeles, Illingworth (1901) found that the—

Orioles are very beneficial to the horticulturist, although they eat some early fruit such as berries, cherries, etc., but no fruit man will begrudge them these if he thoroughly understands their habits. The chief food of the orioles consists of insects and injurious caterpillars, and I have often watched them while they were searching among the branches for this latter food. They are particularly fond of a small green caterpillar that destroyed the foliage of the prune trees a few years ago. The orioles are often seen in the berry patches but they are usually in search of insects as is proven by the examination of a great number of stomachs.

Voice.—Dawson (1923) describes the vocal efforts of this oriole as "exceedingly variable both as to length and quality, now a weak rasping phrase, now a succession of sputtering squeaks, half musical and half wooden, and now a wild medley wherein are imbedded notes of a liquid purity."

Fall.—Most of the birds leave their summer haunts by the end of August, but stragglers have been reported in southern California throughout the winter.

DISTRIBUTION

Breeding range.—The California hooded oriole breeds from central California (Solano County, Fresno, Clark Mountain) south to northwestern Baja California (Santo Domingo, San José); it is casual in southern Nevada (Pahrump, Ash Meadows), where it may breed.

Winter range.—Winters casually, north to southwestern California (Pasadena, Los Angeles); the southern limits of its winter range are unknown.

Casual records.—Accidental in Kansas (Garden City).

ICTERUS CUCULLATUS TROCHILOIDES Grinnell

San Lucas Hooded Oriole

HABITS

Joseph Grinnell (1927) describes this oriole as "similar in general size to *Icterus cucullatus nelsoni* Ridgway, of Arizona and southern California, but bill in both sexes longer, more attenuated in both dorsal and lateral views, and more decurved toward tip; color tone of males in summer on bright parts of plumage averaging duller, more yellow, less orange. * * * Range.—The Cape San Lucas district of Lower California. Specimens examined from many localities from San José del Cabo north to La Paz." He believed this race to be "altogether resident in the Cape district" and could find no specimens referable to it from the mainland of México.

William Brewster (1902) states that Mr. Frazar "saw only one individual on the Sierra de la Laguna, but observed many in the cañons at its base. The species was most numerously represented about Triunfo where it frequented trees near water, and began nest building late in June. The first eggs, a set of four, were found at San José del Rancho on July 14; during the following 10 days, six nests and sets of eggs were obtained."

Of the nests of the San Lucas hooded oriole Brewster (1902) says: "[They] are essentially uniform in size and shape, and in these respects similar to the nest of the Baltimore Oriole, although smaller and decidedly shallower. All are largely composed of a fine, straw-colored, jute-like fiber firmly interwoven, and four contain only this material, but the fifth is lined with horsehair, and the sixth with cotton and a few feathers. One was attached to the under side of a palm-leaf, two to the branches of orange trees, three were in bushes, and one was suspended at the end of a drooping branch of some deciduous tree. They were placed at heights above the ground varying from four to eight feet."

Baird, Brewer, and Ridgway (1874) quote the following brief notes from Mr. Xantus: "Nest and two eggs, found May 20, about ten feet from the ground, woven to a small aloe, in a bunch of the *Acacia prosopis*. Nest and two eggs, found May 22, on a dry tree overhung with hops. Nest and one egg, found May 30, on an acacia, about fifteen feet from the ground. Nest with young, found on an aloe four feet high. * * * Nest and eggs, found on a *Yucca angustifolia*, on its stem, six feet from the ground. Nest and two eggs, found in a convolvulus, on a perpendicular rock fifty feet high. Nest and three eggs, found on a acacia, twenty-five feet high."

J. S. Rowley tells me that the nests he saw "were sewed to the under side of banana palm fronds."

The measurements of 40 eggs average 23.0 by 16.0 millimeters; the eggs showing the four extremes measure 26.6 by 15.9, 22.5 by 16.9, 22.0 by 16.0, and 22.5 by 15.1 millimeters. The eggs are practically indistinguishable from those of the species elsewhere.

DISTRIBUTION

The San Lucas hooded oriole is resident in southern Baja California from San Ignacio, Comondú, and Carmen Island south to Cape San Lucas.

ICTERUS GULARIS TAMAULIPENSIS Ridgway

Alta Mira Lichtenstein's Oriole

HABITS

This brilliantly colored and well-marked oriole was added to our fauna by Thomas D. Burleigh (1939), who collected a female near Brownsville, Tex., where it was probably only a winter wanderer, as its known range is from Veracruz, Pueblo, and San Luis Potosí to Tamaulipas in eastern Mexico. Burleigh says of its capture: "The day it was collected, January 7, 1938, it was found feeding with a flock of Green Jays (*Xanthoura luxuosa glaucescens*) in rather thick woods a few miles north of Brownsville; it was restless and wary, and was approached only with difficulty."

Two other races of the species are found in southwestern México and in Yucatán, respectively. In naming it Ridgway (1902) describes it as "similar to *I. g. gularis*, but decidedly smaller and the coloration more intense, the orange-yellow more decidedly orange (usually rich cadmium orange); black at anterior extremity of malar region, broader; bill shorter and deeper through base."

Sutton and Pettingill (1943), to whom we are indebted for most of our knowledge of its habits, describe it as—

A conspicuous, orange, black and white bird of eastern Mexico's coastal plain. * * * It is larger than the other common nesting orioles of this region, the Hooded (*Icterus cucullatus*) and the Black-headed (*Icterus graduacauda*), being fully 9 inches long. Its song is loud and repetitious. It is especially notable for two reasons: (1) The male and female are so much alike in size as well as color as to be virtually indistinguishable in the field, a resemblance that certainly is not characteristic of the Icteridae in general. (2) The nest is customarily placed in such an exposed situation as to suggest that the instinct for hiding it has been lost, or perhaps has been supplanted by an instinct for advertising it. This is hardly true of most orioles of the genus *Icterus*.

Evidently, the Alta Mira oriole is not at all secretive in its habitat nor in the selection of a nesting site, in spite of its conspicuous coloring. In southern Veracruz, according to Wetmore (1943) "these birds were found through the treetops in heavy forest, in the lines of trees bordering fields and streams, and in scattered groves through the pastures. They were the most common of the orioles and were often kept as cage birds."

Bendire (1895) wrote at some length on the frequent occurrence of orioles of this species in Louisiana, based on information received from E. A. McIlhenny; but as these may have been escaped cagebirds, the species has never been accepted as occurring naturally in that State, which is so far from its known range. It does not seem to have been reported there in recent years.

Nesting.—The Alta Mira oriole is a wonderful nest builder. Sutton and Pettingill (1943) found five occupied nests near Gómez Farías, Tamaulipas, all of which were "placed in much exposed situations. Nests of *Icterus gularis* reported from San Luis Potosí and El Salvador were placed in similarly exposed situations." The first nest was within 75 yards of the house in which they lived, and was watched daily from the beginning of the construction to the laying of the first egg. This nest was in "a living, though leafless, 50-foot-high ear tree (*Enterolobium cyclocarpum*)," about 35 feet from the ground, not far from the end of a slender branch and attached to a two-tined fork.

Building the nest required at least 18 days (April 7–24) and possibly as many as 26 days (April 7–May 2). From April 7 to 14 the work progressed irregularly; from April 14 to 17 much material was added; from April 17 to 22 the structure took on its final shape; but from that date on, work was desultory. We believe the first egg was laid on May 2. * * *

The nest's greatest outside length, from the fork to the bottom, was 25 inches. The greatest outside diameter (not far from the bottom) was 6½ inches. It was symmetrical and quite smooth, the material being well tucked in. It was made almost entirely of air-plant rootlets, most of them several inches long, and fiber stripped from palmetto leaves. The lining, which covered the bottom only, was of palmetto fiber and horsehair. Nowhere about the nest was there a feather, bit of wool or cotton or kapok fluff, or other soft material.

About 250 strands of rootlet or palmetto fiber passed over each eight-inch length of supporting twig. The remaining third of the nest-rim consisted of four or five tough rootlet "cables" hung from one tine to the other. About these, slenderer rootlets were twisted tightly, giving the edge a somewhat rope-like appearance. This third of the rim was notably thin and strong. * * *

The rootlets of the nest wall ran downward and more or less parallel to each other, as if they had purposely been allowed to dangle while the bird wove other strands about them. Some of these meridional rootlets extended the entire length of the nest, but most of the material was obviously woven in and out crosswise into a sort of rough fabric. No rootlet or fiber encircled the outside of the nest.

The wall was thickest at the bottom.   Here the material was tightly interwoven and matted.   The lining was not attached either to the bottom or to the sides. It could be lifted *en masse* without difficulty, evidently having been laid with some care and pressed into final position by the bird's body.

At this, and at all the other nests observed, only one brightly colored bird was ever seen at the nest or even bringing material; as both sexes are brightly colored and practically indistinguishable in the field, this was probably the female.   Brief notes on their other nests follow:

"Sutton discovered a partly built Alta Mira Oriole nest on April 6.   It was almost directly above one of the paths leading from the Río Sabinas to the main trail to Gómez Farías and was about 30 feet from the ground on a dead branch in a living tree at the edge of a good-sized clearing.   Here one brightly colored bird was noted repeatedly, never two."

Another nest "overhung the Río Sabinas not far from the Rancho. We found it on April 3, but we do not know how many birds worked on it.   It was in a cypress and must have been fully 50 feet above the water.   It was in plain sight for many rods both up and down stream and was not far (possibly 25 feet) from an occupied nest of the Rose-throated Becard (*Platypsaris aglaiae*) and one of the Giraud, or Social Flycatcher (*Myiozetetes similis*)."

The fourth nest "was far out on one of the uppermost branches of a large (50 feet high), completely dead tree that stood quite by itself in a well cleared field just north of the headquarters house"; and the fifth "hung from a leafless, perhaps dead branch, almost over the main highway, about 30 feet from the ground."

Sutton and Burleigh (1940) found a nest in San Luis Potosí that swung from a single telephone wire that ran above a wooded gully and was 80 or more feet from the ground.   The poles were many rods off, on ridges at either side of the gully.   Dickey and van Rossem (1938) say that, in El Salvador, "the usual sites are the tips of branches at varying heights from the ground, but sometimes the nests are hung from telephone wires, particularly if there happen to be a few tufts of epiphitic growth to provide a starting point."

It appears from the above accounts that the Alta Mira oriole purposely selects the most conspicuous nesting site that it can find, on which Sutton and Pettingill (1943) comment:

Certain it is that a conspicuous nest site is advantageous to the owner insofar as it forces enemy species to use exposed avenues of approach.   How easy it is, when we focus attention on any one bird to forget that this bird's enemies all have enemies themselves!   Any predatory creature that makes its way to an Alta Mira Oriole's nest, either by day or by night, is certain to expose itself to its own enemy species whether these happen to be enemy species of the oriole or not.

The oriole's nest must meet certain specifications if it is to be boldly advertised, of course. It must provide proper conditions of temperature and air for the eggs and young birds in spite of hanging, hour after hour, exposed to the sun. It must be tough enough, long enough, far enough out on the branch, and far enough above the ground to make a coati-mundi (*Nasua*) "think twice" before attempting a raid. It must be too deep for Brown Jays to rob easily, too tough to tear apart, too much like a trap to appeal to the female Red-eyed Cowbird (*Tangavius aeneus*). The fact that *Icterus gularis* is common proves it to be a successful species. We may believe, therefore, that its own peculiar method of nest-advertising is advantageous rather than otherwise.

Eggs.—Bendire (1895) describes an egg from Guatemala "as a pale gray, blotched and streaked with very dark brown; it measures 1 by 0.70 inch." Dickey and van Rossem (1938) say: "The eggs are similar to those of *sclateri* and *pectoralis*, that is, they are elongate ovate, with the bluish white ground color lined, scrawled, and irregularly spotted with black. A set of three eggs collected at Lake Guija May 23, 1927, measure respectively: 29.8 x 19.2; 29.6 x 18.5; and 29.1 x 19.3. Three and four eggs are the usual numbers laid."

Alexander F. Skutch mentions in his notes an egg of this species, taken in Guatemala, that "was of an extremely elongate form, in color white irregularly scrawled with lines of black and pale lilac, and measured 27.4 by 17.5 millimeters." It was on the point of hatching and was probably somewhat faded.

Young.—In his notes on another race of this species, *Icterus gularis xerophilus*, Skutch writes: "On July 19, I watched a nest which contained two young about ready to leave it. The repeated passage of the adults while feeding their nestlings had torn and enlarged the entrance until the entire side was open to within a few inches of the bottom, a not infrequent occurrence with nests of this type. When one of the parents clung to the outside to deliver an insect, two heads stretched forth, open-billed, to receive it. A third nestling had already departed, and awaited his share of the good things in the next tree. The whole time that I was within hearing, both parents, who united in feeding their offspring, uttered a continuous succession of single notes of three different kinds, and each as full of sunshine as their golden plumage. Whether they searched among the foliage for larvae and insects, or returned with food in their bills to the nest, or clung to its side in the interval between feeding their offspring and carrying away the droppings, their joy in their occupation constantly expressed itself in these happy monosyllables. Even when they interrupted their parental ministrations to scold at my intrusion, their churring protests were punctuated with these notes of gladness, which they never seemed able, or willing, to suppress. Never have I heard other birds, save vireos, sing so continuously."

He says that orioles of this race in Guatemala raise two broods in a season, and that the young of the first brood are apparently fed by the male while the female is building the second nest.

Plumages.—Ridgway (1902) describes the juvenal, or first, plumage of typical *I. gularis* as: "Head, neck, and under parts (including throat, etc.) yellow, the color duller on pileum and hindneck; back and scapulars olive; rump and upper tail-coverts dull yellow (gallstone or dull saffron), like pileum and hindneck; wings and tail as in the immature plumage, described above [see below], but greater coverts broadly tipped (on outer webs) with dull yellowish white, secondaries, broadly edged with white, primaries more broadly edged with pale gray (passing into white terminally) and with a white patch at the base."

Apparently, a postjuvenal molt renews all the contour plumage and the wing coverts, but not the rest of the wings nor the tail. Ridgway describes this first winter plumage as follows: "Head, neck and under parts as in adults, but the latter rather paler, or less orange, yellow; back and scapulars yellowish olive; lesser wing-coverts dusky, broadly tipped or margined with saffron yellowish; middle coverts dusky at base, broadly tipped with white or yellow; rest of wings dark grayish brown with paler edgings, these white, or nearly so, on greater coverts; tail yellowish olive."

Dickey and van Rossem (1938) say of this plumage and subsequent changes:

At this age the throat patches of the females are very restricted and mixed with yellow. Those of the males are larger and very much blacker, but otherwise the sexes are very similar. Some young birds have a first prenuptial molt which affects chiefly the foreparts and back, but in four out of six cases the postjuvenal plumage is worn with no discernible change for a full year or until the first post-nuptial (second fall) molt. This first postnuptial has produced in the cases of three females a livery like that of the adults except that the back is more or less mixed with yellowish green. Certainly in these cases, at least, maturity was not reached at that time, and as there is apparently no spring molt (except in a few first-year birds) the fully adult plumage could not have been acquired by these individuals until the third fall (second postnuptial) molt.

There is a large series of the various races of this species in the Museum of Comparative Zoology, in Cambridge, showing all the plumages substantially as described above. The postjuvenal molt apparently begins in August and may not always be completed before the middle of September or later. I have seen one in full juvenal plumage as late as September 12. The series contains numerous specimens in first winter plumage from December to May.

The first postnuptial molt is shown in specimens taken in August and September; in one taken September 24, this molt is nearly completed, with yellowish green edgings on the feathers of the back. The postnuptial molt of adults occurs at about the same time, the body plumage being molted first and the wings and tail last.

Food.—Of the stomachs of this species examined by Dickey and van Rossem (1938) in El Salvador, one contained small ants, one insects, one insects and berry seeds, and one berry seeds and pulp.

Referring to one of the birds shot by McIlhenny in Louisiana, Bendire (1895) says: "On dissecting the specimen he found a number of small green caterpillars and several spiders, but their principal food seemed to consist of the small purple figs, which were just ripe. While in search of food they move about exactly as the Baltimore Oriole does, swinging from slender twigs head downward, looking under limbs for insects."

Voice.—Wetmore (1943) says that the "song is a quick repetition of two or three notes without the clear tone of that of the Baltimore oriole or the troupial, though the alarm calls are like those of the northern orioles." McIlhenny (Bendire, 1895) called it "a soft, flute-like note."

Alexander F. Skutch says in his notes on the Guatemala race: "The song of this oriole consists of round, mellow whistles, uttered deliberately in a clear, far-carrying voice. In the evening, especially, it delivers single tinkling whistles spaced at rather wide intervals— crystal beads of melody strung along a thread of silence. It has a churring call, somewhat like that of its neighbor, the hooded cactus wren (*Heleodytes capistratus*), and a rather nasal note which may serve either as a call or a signal of alarm."

Field marks.—The Alta Mira oriole is a brilliantly colored and conspicuously marked bird. The sexes are almost indistinguishable in the field. The throat, interscapular region, and tail are clear black; the lesser and median wing coverts are yellow; the greater coverts are tipped with white, forming a wing bar; and the rest of the wings are black with varying amounts of white edgings. The head, neck, rump, and entire under parts (except the black throat) are rich yellow or orange-yellow. In the female, the black of the throat is somewhat more restricted and the blacks and yellows are duller than in the male. For other details, see the description of the plumages above.

DISTRIBUTION

Range.—The Alta Mira Lichtenstein's oriole is resident from central Tamaulipas (Victoria) south through eastern México to Veracruz, Tabasco, México, and Campeche.

Casual records.—Casual in southern Texas (Brownsville); nested near Santa Maria, Tex., 1951.

ICTERUS PUSTULATUS MICROSTICTUS Griscom

Western Scarlet-Headed Oriole

HABITS

The scarlet-headed oriole is the most brilliantly colored of any North American oriole that the writer has seen; except for the black throat, the head and neck of the adult male glow with intense orange, sometimes even "flame scarlet," the back is yellow streaked with black, the rump and under parts are rich orange, the tail is black, and the black wings are marked with a broad white bar and edgings; altogether it is a gorgeous bird. The species ranges widely in western and southern México. The subspecies under discussion ranges from Jalisco to Chihuahua and Sonora, in México, and has been taken as a straggler in San Diego County, Calif. Ludlow Griscom (1934) describes this race as "differing from typical *pustulatus* (Wagler) in having the spotting on the back greatly decreased in adult males small narrow lance-ovate ones instead of large round spots; this decrease in spotting equally evident in females, which are so small as to be very obscure."

Laurence M. Huey (1931) describes the capture of the record specimen as follows: "On May 1, 1931, a male Scarlet-headed Oriole (*Icterus pustulatus*), in first year plumage, was collected at Murray Dam, near La Mesa, San Diego County, California, by Frank F. Gander, a member of the staff of the San Diego Natural History Museum. * * *

"Questioning the collector regarding the capture of this unusual migrant, the writer was informed that the bird was uttering notes not unlike those of *Icterus bullocki [i] bullocki [i]*, which it was believed to be, and that its position in the sycamore tree and manner of perching were typical of that Oriole."

George N. Lawrence (1874) quotes Col. A. J. Grayson on the habits of the scarlet-headed oriole as follows:

Of the numerous species of orioles inhabiting the Tropics, this one is the most familiar about the locality of Mazatlan, and indeed all of Western Mexico. I found it as far south as Tehuantepec, Guadalajara, Tepic, and other places, where I always met with it as a well-known and common species. Its long pensile nest, its sprighly little song, and more especially the gay plumage of a fully adult male, renders it a conspicuous bird among the feathered songsters of its native woods.

The nests are generally suspended from a tough, slender branch or recumbent twig of the acacia tree, protected from the intense rays of the sun by the beautiful canopy of its fringed foliaged branches. Such a tree as the tamarind acacia is often selected, and one or two nests are sometimes seen swaying in the breeze, beneath the generous shade of this perennial beauty of the forest. The nest is composed of the thread-like or elastic fibres of the maguey plant. I have seen some in which cotton thread and twine were component parts of its elastic and firm structure. The nests are of various lengths, conformable to the material at hand for the intricate formation of the warp necessary for the weaving of this unique and airy abode, in which to rear their little family. The inside bottom is lined with the downy substance of the tree cotton, intermixed with a few feathers. In one nest I found an entire skein of yellow silk, which it had doubtless picked up where some village brunette had dropped it.

The eggs are generally five in number, rather long, of a pale blue ground, with numerous hieroglyphic scratches confluent around the larger end.

Plumages.—I have seen no small young of the scarlet-headed oriole, but Ridgway (1902) describes the young as "similar to the winter female, but without any black on throat, etc.; streaks on back obsolete, and colors duller. * * * Immature males resemble adult females in coloration." I have seen first winter birds, with the gray backs, taken in October, January, and March, and one as late as June 2.

He describes the adult male in winter plumage as "similar to the summer plumage, but white edgings to wing feathers much broader, often strongly tinged with gray; orange or yellow of back, rump, etc., more or less tinged with olive, the back often tinged or suffused with gray."

He says of the winter plumage of the adult female: "Similar to the summer plumage, but upper parts much tinged with gray, especially on back, and grayish white or light gray wing-edgings broader."

Enemies.—Friedmann (1933) mentions several instances of the bronzed cowbird (*T. a. aeneus*) laying its eggs in nests of this oriole.

DISTRIBUTION

Range.—The western scarlet-headed oriole is resident from central Sonora (Hermosillo, Ures), southwestern Chihuahua, western Durango, and Jalisco (Bolaños, Guadalajara); south to Sinaloa (Mazatlán, Escuinapa), Nayarit (San Blas, Topic), and Jalisco (Barranca Ibarra, Zacoalco).

Casual records.—Accidental in California (La Mesa) and Arizona (Tucson).

ICTERUS PARISORUM Bonaparte

Scott's Oriole

Plates 16 and 17

HABITS

The adult male Scott's oriole is a handsome bird in its striking color pattern of clear black and deep lemon yellow, but we miss the rich orange colors of some of the other beautiful orioles. In the dull-colored semidesert areas in which it largely spends the summer, however, it is one of the most attractive birds that we meet on the dry yucca plains; and not the least of its attractions is its rich melodius song, which greets us almost constantly during the nesting season.

It breeds over a wide range, from the interior of southern California, central-western Nevada, southwestern Utah, central-eastern New Mexico, and central-western Texas southward to the tip of Baja California and to Michoacán, Hidalgo, and Veracruz, in México.

Scott's oriole has been called the mountain oriole, and again it has been referred to as a desert bird; as a matter of fact it is not strictly either, for it occupies a more or less intermediate zone, or zones, such as the pinyon-juniper belt in the foothills, the desert slopes of the mountains, or the more elevated, semiarid plains between the mountain ranges, where the yuccas are widely scattered; but it seems to avoid the real desert, where the chollas and other cacti grow profusely. Ralph Hoffman (1927) says: "The bird often ranges among the junipers and pinyon pines that mingle with the tree yucca in the stony canyons along the edge of the desert, and in the Washington palms along the western edge of the Colorado Desert."

W. E. D. Scott (1885) describes an interesting canyon resort of Scott's oriole as follows:

There is a cañon that begins high up in the Santa Catalinas, and, dividing the hills and table lands on either side of it by its deep furrow, it extends for two miles or more, where it joins the valley of the San Pedro River. It is the upper and more elevated part of this cañon with which we have to do, at an altitude varying from four thousand to five thousand feet. The hills on either side are high, the cañon generally quite narrow. Live oaks are the trees of the hills and hillsides, and reach in places to the bed of the cañon. Here in parts are groves of cotton-woods and sycamores, and some cedars, and, with the exception of the very bed of the cañon, where for a part of the year is a brook, the grass covers the surface of the ground. The brook begins to dry up in its exposed parts early in May, but all summer long there is running water for at least a mile in the cottonwood grove, and in a number of places, even during the driest part of the year, the water rises to the surface, making "tanks," as they are called. Along this running water and about the "tanks," bird life is very abundant, and here, surely no desert, is the summer home of many Scott's Orioles. There is very little cactus, and none of the "chollas" that are so very characteristic of the deserts of the neighboring region.

We found Scott's orioles breeding most commonly on the semiarid valley plains between Bisbee and Tombstone, Cochise County, Ariz. These flat or rolling plains of hard, gravelly soil were bare of vegetation except for the low, scraggly, omnipresent creosote bushes so characteristic of much of the region between the mountains and the deserts. The chief attractions in this desolate region for the orioles were the widely scattered specimens of what we called the soapweed yuccas, the picturesque plants in which they were nesting.

Spring.—Scott's oriole is only a summer resident north of the Mexican border, where it arrives during the first half of April and sometimes before the end of March; the brilliant plumage of the males and their rich song make its arrival most conspicuous.

Laurence M. Huey (1926) makes the following observation on the migration in northwestern Baja California:

Many were observed on migration five miles northeast of San Quintin, February 25, 1925, although the birds were extremely shy as usual. The presence of this Oriole in numbers so near the Pacific coast offers a problem in migration routing; for the species is of extremely accidental occurrence along the coast further north, in the vicinity of San Diego, whereas inland, on the desert slope of the mountains east of San Diego, it passes regularly. Further observation of these birds will probably determine that they range up the peninsula, equally distributed from coast to coast, as far as the southern extremity of the Sierra San Pedro Martir, and that here they swing toward the Pacific, then northeastward again to the eastern slope of the mountains in southern California. A semi-arid highway, such as the Scott's Orioles prefers, is thus provided.

Harry S. Swarth (1904) says of its arrival in the Huachuca Mountains:

The earliest date at which I have seen any was March 31, 1903, when a male was secured; no more were seen until April 5, after which date they were abundant. Until nearly the end of April small flocks of from six to a dozen birds could be found along the canyons, usually below 5,000 feet, feeding in the tops of the trees, where, in spite of the brilliant plumage and loud, ringing whistle of the male birds, they were anything but conspicuous." * * * The first to arrive were the old, bright plumaged males, then a week or so later some females began to come in, and finally toward the end of April, what few flocks were seen were composed of females, and males presumably of the previous year, in every stage of plumage, most of them indistinguishable from the more highly colored females.

Nesting.—Throughout its wide breeding range Scott's oriole builds its nest in a variety of situations, depending on the environment. William Brewster (1902) records a nest, found by Frazar in the Cape region of Baja California, that was placed "among the densest foliage of a fig-tree at a height of about 8 feet, and rested on a few small twigs, but seemed to be fastened only to some twigs above, from which it was suspended." Farther north, it is said by Walter E.

Bryant (1890), on the authority of A. W. Anthony, "to prefer the low hills near the coast south of San Quintin, where it nests in the thorny branches of the candlewood (*Fouquiera columnaris*)."

Scott (1885) found five nests of this oriole in the locality described above, in Pinal County, Ariz. All the nests were within 10 minutes' walk of the house in which he lived, and all but one of them were in yuccas (*Yucca baccata*), within 10 feet of a road and about 4 feet from the ground. He gives detailed descriptions of each of them, but, as they were all much alike, the following description will suffice:

Nest of May 24. Built in a yucca, 4 feet from the ground. Sewed to the edges of five dead leaves which, hanging down parallel to trunk of the plant, entirely concealed the nest. Semi-pensile. Composed externally of fibers of the yucca and fine grasses. Lined with soft grasses and threads of cotton-waste throughout. The walls are very thin, at bottom not more than half an inch, and on the sides from one-eighth to a quarter of an inch thick. The whole nest was rather closely woven and very strong. Inside depth, 3½ inches. Inside diameter 4 inches. The whole cup-shaped. * * * I have called this nest semi-pensile, as the edges of the yucca leaves are not simply attached to the rim or top edge of the nest, but are "sewed" to the *sides* of the structure—one blade for 3 inches, three for 4 inches, and the other two for more than 2½ The nest is sewed to the blades or leaves about 7 inches from where they join the trunk of the plant, and the blades are about 22 inches long.

He describes another nest that was not so pensile, as it rested on some slanting leaves. "Inside it is lined to within half an inch of the rim with small pieces of cotton batting, some cotton twine, and a little very soft grass. * * * The walls on the sides are an inch, and at the bottom an inch and a half thick."

His fifth nest was built "in a sycamore tree, about 18 feet from the ground. Pensile, being attached to the ends of the twigs. It is composed externally entirely of the fibers of the dead yucca leaves, and there are hanging to and built into the walls four rather small dead leaves of this plant, that are partly frayed, so that the fiber is used in weaving them into the structure. The interior is lined with soft fine grasses, and only two or three shreds of cotton-waste appear here and there in the lining."

Frank Stephens wrote to Bendire (1895): "In Arizona I have seen its nest in the yucca, sycamore, oak, and pine trees; one nest found in an oak was not even semipensile, being supported at the sides and below by the upright branches between which it was placed."

Bendire saw some nests in the tall tree yuccas, or Joshua trees. One "was placed fully 10 feet from the ground, and the only way I could reach it was to stand on my horse, which I did, and secured the eggs, three in number, in which incubation had commenced. The nest was so securely fastened to the surrounding bayonet-shaped leaves that I could not pull it away, and only succeeded in cutting my hand severely in trying to do so. The nest was composed of yucca fibers, sacaton, and gramma grass, and lined with a little horsehair."

Another nest, taken by A. K. Fisher, in "Coso Valley, California, on May 11, 1891, was situated on the under side of a horizontal limb of a giant yucca (*Yucca arborescens*), about 6 feet from the ground. * * * Externally the nest measures 3½ inches in depth by 5 inches in its longest diameter and 4 inches at the narrowest point. The inner cup is oval in shape, 2½ inches deep and 3¾ by 3 inches wide." He mentions junipers as being used to a considerable extent, and says that, in Baja California, Xantus reports it breeding "in bunches of moss and in hop and other vines suspended from cacti. He mentions finding one nest in a bunch of weeds growing out of a crevice in a perpendicular rock."

On June 1, 1922, we found four nests of Scott's oriole in the Valley between Bisbee and Tombstone, as described above, each nest containing four fresh eggs. The nests were all in soapweed yuccas (*Yucca baccata?*) at heights ranging from 5 to 7 feet above ground. The yuccas were widely scattered over the open plain, which was sparsely covered with small creosote bushes. These picturesque plants (plate 16) support a dense growth of long, stiff, sharp-pointed leaves at the top of the sturdy trunk, but little higher than a man's head, and a tall flowering stalk that rises to a height of 12 or 15 feet, above the trunk. The dead, and some of the green, leaves hang down below the main cluster of living daggers, close to and parallel with the trunk or at an angle of about 45°. It is in these pendent leaves that the orioles conceal their nests, where they are protected against predators and shielded from sun or rain. The locations and the compositions of the nests were so much like those described above by Mr. Scott that it does not seem necessary to describe them further here, except that our nests were lined with fine grasses and plant down, with no cotton nor cotton-waste.

In one nest we found an egg of the bronzed cowbird.

Eggs.—Bendire (1895) describes the eggs as follows:

From two to four eggs are laid (usually three), and probably two broods are raised in the more southern parts of their range in a season. They are ovate and elongate ovate in shape. The shell is thin, rather close grained and without luster.

The ground color is pale blue, which fades considerably in the course of time, and this is blotched, streaked, and spotted, principally about the larger end of the egg, with different shades of black, mouse, and pearl gray in some specimens, and and with fine claret brown, russet, ferruginous, and lavender dots and specks in others.

The average measurement of 25 specimens in the United States National Museum collection is 23.86 by 16.98 millimetres, or about 0.94 by 0.67 inch. The largest egg in this series measures 26.67 by 17.27 millimetres, or 1.05 by 0.68 inches; the smallest, 23.11 by 15.49 millimetres, or 0.91 by 0.61 inch.

Young.—Incubation is performed by the female alone, and is said to last about 14 days. Probably the young remain in the nest for about 2 weeks, where they are fed by both parents. Mrs. Wheelock (1904) writes: "Oriole nestlings in general are proverbial cry-babies, and Scott Orioles are no exception. Insects of all sorts in all stages of development, fruit, and berries are served to them in such quick succession as to leave small time for the parent to hunt any for himself. At first the feeding is by regurgitation, but on the fourth or fifth day this method gives place to the more commonly observed one." She says that a second brood is reared in a new nest in another tree.

Plumages.—In juvenal plumage, according to Brewster (1902), "both sexes resemble the plain olive phase of the adult female, from which they differ only in having the upper parts browner, the light edging on the wing coverts and secondaries much broader and more or less tinged with yellowish." Chapman (1923b) says: "Nestling birds of both sexes are olive-green above, yellower below, with no trace of black. At the postjuvenal (first fall molt) the male usually acquires a black throat and the back is more or less streaked. These markings, particularly above, are more or less fringed with grayish and olive, and are not fully revealed until, with the advancing new year, the feathers become worn and we have the first breeding plumage." In his plate illustrating this plumage, the wide black throat patch extends upward to include the sides of the head and the forehead; the crown, hindneck, back, and lesser wing coverts are olive, spotted or streaked (on the back) with black. This plumage was evidently acquired by a partial postjuvenal molt involving all the contour plumage and the wing coverts, but not the rest of the wings nor the tail. The molt occurs in late July and August. There is apparently no prenuptial molt of consequence.

At the first postnuptial molt, the next summer, which is complete there is a decided advance toward maturity, but at least another year is required to assume the fully adult livery. During the spring migration in Arizona, Swarth (1904) secured a number of specimens of males in every stage of plumage, from those indistinguishable from the more highly colored females to those in fully adult plumage, all of which he describes in more or less detail.

The molts and plumages in the female are similar in sequence to those of the male, usually without much visible change, but some young birds acquire a little black on the throat at the postjuvenal molt, and some adult females have as much black on the throat as young males.

Food.—Like other orioles, Scott's must feed largely on insects and their larvae, but there is considerable evidence that it eats some fruit and consumes the nectar from flowers, as some other orioles are known to do. Mrs. Kate Stephens (1906) says: "In front of our sitting-room window and six feet distant are several aloes of a small species, bearing panicles of tubular orange flowers on stems about three feet high. In the latter part of April a male Scott oriole (*Icterus parisorum*) alighted many times on these stems, most frequently mornings. He would thrust his bill deeply into the blossoms and appeared to suck the nectar. * * * I got the impression that he did not gather any insects."

Bendire (1895) writes: "Their food consists mainly of grasshoppers, small beetles, caterpillars, butterflies, larvae, etc., as well as of berries and fruits. * * * I have seen them eating the ripe fig-like fruit of the giant cactus."

Grinnell (1910) says that an "apricot orchard near Fairmont was freely patronized by the Scott Orioles from the neighboring yuccas. Two shot there had their gullets distended and faces smeared with apricot pulp." And Frank Stephens (1903) found them "feeding on figs and peaches in the orchard" at Beale Spring.

A. W. Anthony (1894) writes: "In January, 1894, I found this Oriole wintering in the foothills just east of San Quintin, Lower California, and feeding extensively, if not altogether on the ripe fruit of the 'pitahaya' cactus (*Cereus gummosus*). This fruit is about the size and shape of a small orange, bright scarlet when ripe. The flesh is similar to that of a ripe watermelon but much darker with an abundance of very small dark seeds. In flavor it is not unlike raspberries, but rather acid. Unless the fruit is abundant it is almost impossible to find any that has not been torn open and the inside eaten by the birds."

Swarth (1904) says that "in feeding they sit quietly on the limbs prying and peering into such buds as are within reach, any necessary change of position being accomplished by clamboring along the branches with hardly any fluttering of the wings; and as their plumage, though bright, harmonizes exceedingly well with the surrounding foliage, they could be easily overlooked were it not for the loud notes to which the males give utterance at frequent interval."

Song.—When writing his account of this oriole Scott (1885) was impressed with the persistence of its vocal efforts. "Few birds sing more incessantly, and in fact I do not recall a species in the Eastern or Middle States that is to be heard as frequently. The males are, of course, the chief performers, but now and again, near a nest, * * * I would detect a female singing the same glad song, only more softly. At the earliest daybreak and all day long, even when the sun is at its highest, and during the great heat of the afternoon, its very musical whistle is one of the few bird songs that are ever present." In Pinal County, Ariz., he observed that this bird arrived about the middle of April, and from then until July 29 he heard the song daily, even hourly, and during the height of the breeding season often many were singing within hearing at the same time."

Dawson (1923) refers to it as "a golden song which poured down from a sycamore tree hard by. *Ly ti ti tee to, ti ly ti ti te to,* came the compelling outburst. I took it for a freak Meadowlark song at first, but once thoroughly aroused, knew it for an Icterine carol—*ly ti ti tee to, ti ly ti ti tee to*—molten notes with a fond thrill to them, more restrained than the clarion of the Meadowlark, smoother and sweeter than the tumult of a Bullock Oriole, and, of course, with the double repetition, a much longer song than either."

Grinnell (1910) says of it: "The song was loud and full, better than that of the Bullock Oriole. It reminded me of the best efforts of the latter bird, and yet bore a strong resemblance in its quality to the song of the Western Meadowlark." Others have noted this resemblance, which is a high compliment.

Field Marks.—The brilliant male in full plumage is strongly marked; the entire head, throat, neck, back, and the terminal part of the tail are black; the wings are mainly black, with yellow lesser coverts and broad white bands on the median and greater coverts; the breast, rump and much of the lateral tail feathers are bright lemon yellow, not orange as in most other orioles.

The female is yellowish olive above, mottled with dusky, and paler yellow below, but she has a black throat and two white wing bars. Other details are mentioned under plumages.

Fall.—As is the case with most other orioles, Scott's oriole is not much in evidence after the young are on the wing, and with the waning of the summer it seems to disappear from its breeding haunts. Mr. Scott (1885) writes: "After August 7 I missed the song, although the birds were abundant until the 10th of that month, and I saw a single bird or so for the following three days. Then I supposed they were all gone, but on the 14th of September, about dusk, I started one, an adult male, from a yucca where he had evidently gone to roost. He scolded angrily at me from the dead limb of a cedar near by for a few moments, when I left him to go to bed. Again, on the 18th of September, I heard a male in full song, and going closer found a party of four together, three old males and a young one of the year. This is my last note of their occurrence at this point."

Winter.—Scott's oriole spends the winter south of our border, in central and southern Mexico, as far south as Veracruz, Guerrero, Puebla, and the Cape region of Baja California.

DISTRIBUTION

Range.—Nevada, Arizona, and western Texas to central México.

Breeding range.—Scott's oriole breeds from southern Nevada

(White Mountains, Charleston Mountains), southwestern Utah (Beaverdam Mountains), north-central Arizona (Wupatki National Monument), north-central New Mexico (San Miguel County, Montoya), and western Texas (Guadalupe Mountains, Chisos Mountains); south through southeastern California (Inyo Mountains, Campo) to southern Baja California (Cape San Lucas, Victoria Mountains), central-northern and southeastern Sonora (Nogales, Rancho Santa Barbara), and southeastern Coahuila (Las Delicias). Has nested recently in central-western Nevada (Stillwater) and northeastern Utah (Powder Springs).

Winter range.—Winters regularly north to northern Baja California (San Quintín, San Fernando) and southern Sonora (San José de Guaymas, Camoa); casually to southwestern California (Garnsey, San Diego); south to southern Baja California (Miraflores), central Michoacán (Pátzcuaro), Guerrero (Chilpancingo) and Puebla (San Bartolo); east to western Nuevo Leon (Santa Catarina) and Hidalgo (Cuesta Texcueda, Pachuca).

Casual records. Casual in coastal California (Santa Barbara, San Diego), and in east-central Utah (25 miles east of Hanksville).

Migration.—Early dates of spring arrival are: Chihuahua—Carrizalillo Mountains, April 18. Texas—El Paso, April 10; Glass Mountains, Brewster County, April 13. New Mexico—Carlsbad Cave, April 6; Silver City, April 19. Arizona—Patagonia, March 9; Tucson area, March 15 (median of 11 years, March 25). Utah—Washington County, May 6; Powder Springs, May 8. California—San Felipe Canyon, March 22; Reche Canyon near San Bernardino, April 1. Nevada—Searchlight, May 5; 10 miles east of Stillwater, May 11.

Late dates of spring departure are: Michoacán—Chupícauro, April 29. Hidalgo—Cuesta Texquedo, April 7. Sonora—Tiburón Island, April 11.

Early date of fall arrival: Baja California—San Andres, September 21.

Late dates of fall departure are: California—San Diego, September 2. Arizona—Huachuca Mountains, October 9 (median of 4 years, Cochise County, October 5).[1] Texas—Hueco Mountains, El Paso County, October 17; Guadalupe Mountains, October 11.

Egg dates.—Arizona: 11 records, May 15 to June 28; 6 records, May 22 to June 4.

California: 75 records, April 24 to June 25; 40 records, May 9 to May 22.

[1] October 26, 1909 (Bird-Lore, vol. 25, p. 389), is a typographical error.

ICTERUS GALBULA (Linnaeus)

Baltimore Oriole

Plates 18 and 19

Contributed by WINSOR MARRETT TYLER [1]

HABITS

Here in New England, during the long-drawn-out spring migration, there are several red-letter days. The first of these is the day, sometimes late in February, when the wintering song sparrows, which have been long silent in the shrubbery, begin to sing their spring song, a tinkling melody which foretells the ending of winter. As the year advances, there is another welcome day, the real beginning of the migration, when the bluebird, flying over the brown fields of March, comes back to his summer home, and we hear the softest, sweetest voice of all our birds—"the herald of spring," Alexander Wilson calls him. But the greatest day of the whole year is in early May when the season is well established, when the apple blossoms are opening. Many of the birds are already here and have been singing for days. On this day, not far from the 8th of the month, the Baltimore oriole makes his dramatic entrance into New England. On every hand, in our orchards, among the high branches of our roadside elms, the little trumpeter is heard blowing his tiny bugle; all out-of-doors is animated by his buoyant personality.

Spring.—Bendire (1895), speaking of the return of the bird to its breeding ground, says: "The Baltimore Oriole usually arrives in the southern New England States, in central New York, and Minnesota, with almost invariable regularity, about May 10, rarely varying a week from this date; it arrives correspondingly earlier or later farther south or north. About this time the trees have commenced to leaf, and many of the orchards are in bloom, so that their arrival coincides with the loveliest time of the year. The males usually precede the females by two or three days to their breeding grounds, and the same site is frequently occupied for several seasons. * * * It is very much attached to a locality when once chosen for a home, and is loath to leave it."

The orioles come home with a stirring fanfare, but as each bugle is playing a different tune and the tunes are so distinctive, so characteristic of the individual bird, we can almost verify Major Bendire's statement that the birds come back to their old homes. If we take note of the peculiarities in the song of the oriole which we hear from

[1] Dr. Tyler died Jan. 9, 1954, at the age of 77. He assisted Mr. Bent with these life histories in various ways beyond the contribution of 37 complete histories.

our windows, we can recognize him, provisionally at least, as the same bird year after year until, after a time, he is replaced by another bugle, playing a different tune.

There is a banding record, however, which proves absolutely that an oriole returned three times to its breeding ground. A. Milliken (1932) banded in 1929 a female Baltimore oriole captured in an open-top Chardonneret trap baited with string and yarn. She returned in 1930 and 1931. "The bird nested very near the same place for three successive years, though the exact spot is not known."

There is a businesslike air in the returning orioles. The males go directly to our orchards, visiting the open apple and cherry blossoms, where they find food, and to the elm trees, where their nests will soon hang. The females, too, when they arrive a few days after their mates, seem eager to undertake at once the duties of the new season and begin to build so promptly that the breeding cycle is well underway, here in New England, by the end of May.

Very seldom, in a long series of years, have the orioles arrived before the Massachusetts apple trees blossom. The birds then seek their food nearer the ground, among sweet fern or other small plants.

Courtship.—We note the courtship of the Baltimore oriole chiefly in the brief period between its arrival and the laying of the eggs in early June. Forbush (1927) speaks of its behavior thus: "When their modest consorts arrive, the ardent birds soon begin their wooing. In displaying his charms before the object of his affections the male sits upon a limb near her, and raising to full height bows low with spread tail and partly-raised wings, thus displaying to her admiring eyes first his orange breast, then his black front and finally in bright sunlight the full glory of his black, white and orange upper plumage, uttering, the while, his most supplicating and seductive notes."

Francis H. Allen (MS.) says: "A male courting a female uttered a succession of low sweet whistling notes, rather monotonous and in anything but the typical oriole tone," and of another bird he says: "A male courting a female flitted about in an affected manner and sang on the wing in a longer flight."

Some years ago, after watching the courting action of a male oriole, I (Tyler, 1923) made an attempt to construct in my mind's eye how his display would appear to the eye of the female from her vantage-ground on the perch in front of him:

A male Baltimore Oriole (*Icterus galbula*) of strikingly brilliant plumage was singing loudly in a maple tree when a female Oriole took a long flight and alighted in the same tree. The male flew to her, placed himself directly before her, facing her at a distance of a few inches and here struck successively two attitudes; in one his body was nearly upright, straight and tall, in the other it was bowed downward and forward with the head at the level of the feet. The wings were held closely at the sides. In passing quickly from one attitude to the other, over and over again, he moved up and down with a sharp jerk, rather than in an easy sweeping motion, and he made a very short pause each time before changing direction.

This is a very simple motion, one may say—just an exaggerated bowing—not very different from the bowing, nodding, or swaying of many birds in the excitement of their courting displays. True enough, and it is not until we look at the action from the point of vantage of the female bird and see in our mind's eye, as nearly as we can, just what she sees, that we understand its significance.

In the first position noted above, the orange of the breast glows before her, and so near her that it fills a wide arc with blazing color. Then, as the male bird bends swiftly forward, and the head comes down, the orange is blotted out by black, as by a camera shutter, and immediately, as the bird continues to bend forward, out flashes the orange color again, now on the rump. Witnessed at close quarters, the appearance of this maneuver must be as the bursting out of a great sheet of flame, its instantaneous extinction into darkness, a flaring up again—then darkness once more.

Nesting.—In constructing its nest, a woven, hanging pouch, the oriole is perhaps the most skilful artisan of any North American bird. In southern New England we think of the little cradle as hanging most often high in the air near the end of a long drooping branch of an elm tree, where it swings and tosses in the wind, but the bird often builds here in poplars, maples, and even in the apple and pear trees of our orchards, where it is anchored to a more stable branch.

Speaking of nests in Hatley, Quebec, Henry Mousley (1916) states: "The usual nesting site selected here is near the top of some fair sized tree, generally a maple." Knight (1908) reports that in Maine, although the elm is the oriole's favorite tree, "occasionally nests are placed in maples, locust, cottonwood, poplar or other hard wood trees." Eaton (1914) writing of New York State, says: "I have found this oriole's nest hanging from Norway spruce, hemlock, and horsechestnut which one would naturally expect he never would select. In different villages of western New York the preference seems to be in this order: white elm, silver maple, sugar maple, and apple." Farther west, in Minnesota, Edmonde S. Currier (1904) remarks: "Common about the lake [Leech Lake]. * * * All the nests seen were in birch trees." A. D. DuBois (MS.) speaks of a nest in Illinois "in an oak tree, hung in a cluster of leaves at the topmost end of a branch, hidden so effectively that I should not have discovered it if I had not seen the male fly to it and chase away sparrows and other birds." M. G. Vaiden (MS.), in a letter from Mississippi, mentions pecans, sycamores, and elms as nesting sites, and includes this interesting record: "In my yard the pecan trees grow to a height of 50 to 75 feet, some of them even higher. Virginia creeper vines run up the trunk and out on most of the limbs. On May 22, a Baltimore oriole selected as a nesting site a limb of a tree which had fallen off, pulling the creeper with it and was hanging suspended in the air, the nest being attached to the creeper as well. After three eggs were laid the limb fell to the ground, but the bird, not to be outdone, built another nest in the dangling remains of the creeper, from which she fledged her young."

Usually the Baltimore oriole hangs its nest high over our heads; Eaton (1914) estimates the average height as 25 to 30 feet and he has seen a nest 60 feet above the ground. On the other hand, A. D. Du-Bois (MS.) reports "the lowest nest that has ever come to my attention was in a burr oak 7 feet 8 inches from the gound." Thomas D. Burleigh (1931) cites a still lower nest in Pennsylvania, "but six feet from the ground at the extreme end of a limb of an apple tree in an orchard."

The nest is a deep pocket hanging generally from the rim; the open- is usually at the top, rarely at the side. Bendire (1895) gives the dimensions of a nest from Ontario as follows: "It is externally 5 inches deep, and the entrance, which is oval in shape, measures 3¼ by 2 inches in diameter. The cup is 4½ inches deep by 2½ inches wide." Henry Mousley (1916) says of nests in Quebec: "The nests vary somewhat in depth, which in some cases may be as much as six inches, whilst one built in a maple opposite my house only measures three and one-half inches." M. G. Vaiden (MS.) records a nest over 8 inches deep.

The framework of the nest is made of long, pliable strips of dry plant fibers, grapevine bark, Indian hemp, silk of milkweed, and such materials as are capable of being closely woven into a fabric. Near dwellings and on farms where string, horsehair, and bits of cloth are available, these are used commonly. The nests over the streets of our country towns contain many white strings, which soon bleach to a soft gray color when exposed to the weather. At the bottom, the nest may have a lining of hair, wool, or fine grasses. Forbush (1927) speaks of an aberrant nest "chiefly composed outwardly of jet black hair from the manes and tails of horses. This nest, placed low down in a pear tree, was very conspicuous among the green leaves."

John B. Semple (1932) explains that the oriole has adapted itself to the scarcity, almost the complete absence in recent years, of horsehair. He says:

Thirty years ago the nests of the Baltimore Oriole, and those of the Chipping Sparrow as well, contained in their makeup a large percentage of long hairs from the manes and tails of horses. This material was then easily obtained along the roads and in the pastures. Even ten years ago an oriole's nest found on a farm in Monroe County, Pennsylvania, where horses were used, contained a good proportion of horsehair. But now, since automobiles and tractors have brought about a disappearance of horses which is almost complete, it has become a matter of curiosity to find out what the orioles would do. A nest taken this autumn from the same tree on a farm in Monroe County, in which the nest of 10 years ago had been built, was found still to contain a few horsehairs. These must have been quite difficult for the bird to find, for the farm is now worked only by tractors. And in a nest taken this autumn in Sewickley no horsehairs whatever were to be found. The nest was composed chiefly of fibres of the bark of Indian hemp (*Apocynum*). Felted in toward the bottom of the nest were the hair-like pappi of dandelion seed; over this was laid the fluffy, cottony covering of willow seeds (the nest was in a weeping willow tree); and the lining of the bottom of the nest was of rather stiff fibres of grape-vine bark.

Audubon (1842) says that in Louisiana the bird uses Spanish moss chiefly as a building material. In this warm climate, it weaves the walls so loosely that they permit the passage of air, and little lining is added.

It has long been a matter for wonder among naturalists how the oriole can accomplish such a finished piece of workmanship in constructing its nest, work which seems to demand a conscious planning far beyond the resources of a bird's mind. The older writers speak rather vaguely of the weaving process and have little to say about how the bird employs what Audubon calls his "astonishing sagacity."

Francis Hobart Herrick (1935) has recently made a careful, intensive study of nests of the Baltimore oriole, watching their construction from the tying of the first string to a branch until the nests were completed. The following summary embodies the main steps in the workmanship of the bird as described in Dr. Herrick's account, a comprehensive paper of absorbing interest. He says:

The first strands of bast, which are apt to be long, are wound about the chosen twig rather loosely with one or more turns, or perhaps they are passed only once across or around the branch; but subsequent modes of treatment tend to draw these threads tighter, and as their free ends are brought together, other fibers are added. From such simple beginnings a loose pendant mass or snarl of fibrous material, which I have called the primary nest mass, is slowly formed, but it is a long time before it takes on the semblance of a nest or nest-frame. * * *

Behavior at each visit, after a certain number of strands had been laid and joined, was essentially the same, the oriole usually bringing in but a single fiber and carrying it around the support and working it into the nest mass by what I have called shuttle movements of the bill. Clinging to the principal twig, hanging often with head down, and holding the thread, the bird makes a number of rapid thrust-and-draw movements with her mandibles. With the first thrust a fiber is pushed through the tangle which soon arises and forms the growing mass, and with the next either that or some other fiber is drawn loosely back. * * *

While these shuttle movements are, first and last, very similar, and almost equally rapid at all times, the number made at each visit tends to increase with the growing complexity of the product. At least one hundred shuttle movements were sometimes made at a single visit, but these were often so rapid that it was impossible to count them, and many of them must have been abortive.

In all this admirable work there was certainly no deliberate tying of knots, yet, as the sequel will show, knots were in reality being made in plenty at every visit. There certainly was no deliberate directing of the thread, as when a coat is mended or a hammock is woven in a certain way by human hands. The work was all fairly loose at first, yet naturally some of the threads became drawn more tightly than others. I do not wish to imply that the same thread that is first thrust through the nest mass or the nest wall is immediately drawn back, but only that some thread or other is blindly seized by the bill and withdrawn. * * * The irregularity of the weave of the finished work shows conclusively that the stitching is a purely random affair, though, for all that, none the less effective.

Thus in the course of 2 or 3 days' work, a loose, tangled mass of fibers, which will ultimately become one side of the nest, hangs from the supporting twigs. Many long strands dangle from this mass, their ends hanging free. When this stage of the construction is reached, the bird, working from what is to become the inside of the nest, and as Herrick describes her actions, working with "decision and feverish rapidity, with strokes of her bill pushing the threads through the nest body and then catching up the free ends of other strands and drawing them in the opposite direction; with one foot grasping a twig and the other the nest mass, thrusting and pulling, she is now astride the mass and balancing herself with spread wings, now working from above, from below, or at either side; and at each visit she is not only incorporating any new strands that are brought, but gathering up many others which, though fixed at one end, are still hanging free."

Finally, the bird takes in another twig, or other twigs, for support, outlines the framework of the other side, and then fills it in by weaving with the shuttle movements as before. The long streamers are ultimately worked into the wall of the nest. Herrick speaks of the finished nest as "an indescribable chaos of looped and knotted fibers" that is, nevertheless, "strong, durable, and adaptive." In constructing these nests "the female was the chief builder, but the male would occasionally take a share."

In the late stages of construction "the bird settles down in the nest and shakes all over in an effort to bring the pressure of the breast to bear upon its inner surface; he [in this case the bird was assumed to be the male] rises, turns, settles, and shakes again. These are the typical molding movements, and they are applied all over the lower parts of the pouch, their violence at times being such that the surrounding leaves, and even the slender tree itself, are all a-tremble."

One of Herrick's nests was completed in 4½ days. Harry C. Oberholser (1938) says the nest "is usually completed in not more than 6 days"; Bendire (1895) says: "From 5 to 8 days are usually required for its completion"; while Knight (1908) states that "nest building requires about 15 days."

The oriole begins to build so soon after its arrival on its breeding ground that the elms, a favorite tree, are barely leafing out when the nest is building. But soon, as the season advances, the leaves afford both protection and some concealment to the nest. In January 1946

A. C. Bent showed me a nest hanging, as plain as a rag on a clothes-line, on one of the elm trees in his garden. "Last summer," he said, "that nest was completely hidden in the dense foliage at the end of that long, pendant limb; I could not see it from any angle, although I often tried and knew just where it was. All the other nests I have located have been visible, from some angle at least."

It is the bird's custom to build a new nest each year. This habit is evidenced by the remains of former nests, in varying degrees of dilapidation, which sometimes hang in the same tree. An apparent exception to this rule is reported by George F. Tatum (1915) who tells of a female Baltimore oriole repairing a last year's nest. He concludes "that the old nest had been reconstructed, the only evidence of the former one being the black (old) fiber now interwoven with a little of the light (new) fiber, which bound the edge of the nest to the branch."

When we consider the sequence of steps taken in the construction of the oriole's nest, as outlined by Herrick, and recall the propensity of birds to follow a cycle in their behavior, one act following another in orderly succession throughout the year, we can conceive that it may be more in accordance with the bird's nature to progress straight through the making of a new nest from beginning to end, rather than to patch up an old nest in ill repair.

Eggs.—Bendire (1895) writes:

From four to six eggs are laid to a set, most frequently four, though sets of five are not uncommon, while sets of six are rather rare. One is deposited daily, and only one brood is raised in a season. * * *

The ground color is ordinarily pale grayish white, one of those subtle tints which is difficult to describe; in a few cases it is pale bluish white, and less often the ground color is clouded over in places with a faint, pale ferruginous suffusion. The egg is streaked, blotched, and covered with irregularly shaped lines and tracings, generally heaviest about the larger end of the egg, with different shades of black and brown, and more sparingly with lighter tints of smoke, lavender, and pearl gray. In a few instances the markings form an irregular wreath, and occasionally a set is found entirely unmarked.

The average measurement of 56 eggs in the United States National Museum collection is 23.03 by 15.45 millimetres, or about 0.91 by 0.61 inch. The largest egg of the series measures 25.91 by 16.76 millimetres, or 1.02 by 0.66 inches; the smallest, 20.83 by 14.99 millimetres, or 0.82 by 0.59 inch.

Young.—The nestling orioles are comparatively safe for the first 2 weeks, or thereabouts, of their lives, during the time they remain concealed in their little, woven pocket well above the ground at the end of a slender branchlet. Young orioles seem very quiet as nestlings; at all events we do not hear their voices until just before they leave the nest. At this time, near the summer solstice in southern New England, they begin their characteristic cry. On a certain day, over a whole township we hear it, over and over all day long, and for a week or more it continues hour after hour, a monotonous series of five or six notes, falling in pitch a little, with a ringing or resonant quality. It is a pathetic little childish cry or complaint, beseeching, yet insistent, half way between entreaty and demand, *dee-dee-dee-dee-dee*. The pitch is about F sharp, on the top line of the musical staff. This note has given the fledgling oriole the epithet "cry baby."

William Brewster (1937) gives his impression of the young oriole's note. He says: "As she [the female parent] came flying back, I was struck by the tone of mingled anxiety and interrogation of her low call. '*Where? Where?*'-she seemed to say. '*Here-we-are, here we are*' (falling inflection), both young would promptly drawl in answer and then, as she alighted near them, would repeat and extend this to: '*Here-we-are, mam-ma, here we are, mam-ma*'. It really required almost no imagination to fit these words to the calls in question and now that they have occurred to me the calling of young Orioles will no longer be to my ears, as it always has been, a disagreeable sound."

He adds in a footnote: "A week later when this call had become louder and mellower, it often bore a strong resemblance to the whistle of the Greater Yellow-leg, the form being almost exactly the same."

Audubon (1842) remarks: "A day or two before the young are quite able to leave the nest, they often cling to the outside, and creep in and out of it like young Woodpeckers. After leaving the nest, they follow the parents for nearly a fortnight, and are fed by them." William Brewster (1906) says: "After the breeding season is over both old and young resort more or less freely to bush-grown pastures and the edges of woods. On July 19, 1889, I saw upwards of forty collected within the space of half an acre in Norton's Woods, and I have met with smaller flocks at Rock Meadow and in the Maple Swamp."

Small companies of orioles in immature or female plumage are frequent here in New England up to the time when the species departs in late August. As a rule, however, there are no adult male birds in these gatherings.

Forbush (1927) and Bendire (1895) give the incubation period as 14 days; Eaton (1914) gives it as about 12 days, and DuBois (MS.) as 12 days. Bendire (1895) says that the young birds remain in the nest about 2 weeks; DuBois (MS.) reports a case in which they left in 11 or 12 days.

Plumages.—[AUTHOR'S NOTE: Jonathan Dwight, Jr. (1900), describes the juvenal plumage of the Baltimore oriole, in which the sexes are alike, as follows:

["Above, olive-brown, slightly orange tinged, brightest on head and upper tail coverts. Wings clove-brown, the primaries narrowly, the tertials broadly edged with dull white, two wing bands at the tips of greater and median coverts pale buff. * * * Tail chiefly gall-stone-yellow, centrally much darker and brownish. Below, including 'edge of wing' ochre-yellow, sometimes orange with ochraceous tinge, palest on chin and middle of abdomen, brightest on breast and crissum."

[A partial postjuvenal molt, involving the contour plumage and the wing coverts but not the rest of the wings nor the tail, begins early in July and produces the first winter plumage. Dwight says of this plumage, in the young male: "Similar to previous plumage but dull orange brown above and much brighter orange below, although lacking the black areas of the adult. The greater and median wing coverts become dull black, white tipped, the latter and the lesser coverts orange tinged."

[The first nuptial plumage is acquired by an extensive prenuptial molt involving most of the plumage except the primaries, their coverts, and the secondaries. The full orange and black body plumage is assumed at this molt, the tertials and wing coverts being broadly edged with white, and the black and yellow tail is acquired. Worn brown primaries remain to distinguish young birds from adults.

[A complete postnuptial molt occurs in July, producing the adult winter plumage. This is practically the same as the adult nuptial plumage, but the "feathers of the back are narrowly edged with dull orange (absent in older birds), which also suffuses the median and lesser coverts. The greater coverts, secondaries and tertiaries are broadly edged with white."

[The adult nuptial plumage is acquired by wear, during which the white edgings in the wings largely disappear and the orange edgings on the back are lost, producing the bright, clear orange and black spring plumage of the male.

[In the female, the molts are similar to those of the male, but there is no striking change in the color pattern from one season to another. The upper parts are yellowish olive, the wings dusky with white tipped coverts and white edged flight feathers, and the under parts are dull orange or yellowish. There is often more or less black on the chin or throat, but it is usually very restricted in extent. However, in Manitoba we collected a breeding female, now in my collection, in which the throat is wholly black, as in a male, and the whole head and neck are mainly black.]

Food.—Waldo L. McAtee (1926) gives the following comprehensive summary of the Baltimore oriole's food:

Caterpillars are the most important single element of the Oriole's food, forming over a third of the total. The Baltimore is one of the birds that decidedly are not afraid of spiny or hairy caterpillars and it has a good record against such well-known pests as the fall webworm (*Hyphantria textor*), spiny elm caterpillar (*Euvanessa antiopa*), tussock caterpillar (*Hemerocampa leucostigma*), forest tent caterpillar (*Malacosoma disstria*), and larvae of the gipsy moth (*Porthetria dispar*), and browntail moth (*Euproctis chrysorrhea*). Orioles of this species have been known to destroy entirely local infestations of orchard tent caterpillars (*Malacosoma americana*).

Beetles, ants, parasitic wasps, bugs, grasshoppers, spiders, and snails are the principal additional components of the Hangnest's animal food. Among forms injurious to woodlands that are known to be preyed upon by the bird are tree hoppers, lace bugs, scale insects, plant lice, leaf chafers, junebugs, nut weevils, adults of flat-headed and round-headed wood borers, leaf beetles including the locust leaf miner, click beetles, oak weevil (*Eupsalis minuta*), and sawfly larvae.

The wild fruits eaten by the Baltimore Oriole are mostly june berries, mulberries, and blackberries. A few vegetable galls also are consumed.

The Oriole does some damage to cultivated peas and small fruits, but has such praiseworthy food habits in general that it certainly is the best policy to take special measures to prevent access to the peas and fruits, rather than to get legal permission to destroy the birds.

Bendire (1895) eulogizes the bird's feeding habits even more enthusiastically: "Aside from its showy plumage, its sprightly and pleasing ways, its familiarity with man, and the immense amount of good it does by the destruction of many noxious insects and their larvae, including hairless caterpillars, spiders, cocoons, etc., it naturally and deservedly endears itself to every true lover of the beautiful in nature, and only a short-sighted churl or an ignorant fool would begrudge one the few green peas and berries it may help itself to while in season. It fully earns all it takes, and more too, and especially deserves the fullest protection of every agriculturist."

Additional items in the bird's diet are mentioned by Ellison A. Smyth, Jr. (1912), who says: "They frequent the potato patches with the fledged young and feed freely on potato beetles"; and A. D. Du Bois (MS.) reports seeing the bird "in a sycamore tree, picking to pieces one of the seed balls—holding it against a branchlet with his foot and apparently eating the seeds." William Youngworth (1931) notes an "unusual food"; he says: "While looking from a window on July 23, 1930, the writer saw three immature Baltimore Orioles (*Icterus galbula*) clinging to the tall hollyhock stocks that were growing along the side of the house. Close watching showed that these birds were pecking into the newly formed pericarps of the hollyhocks and were greedily eating the soft, tender seeds." Samuel Lockwood (1872) gives good evidence that Baltimore orioles decapitated countless numbers of the stingless male carpenter-bees (*Xylocopa carolina*) which were collecting honey in horsechestnut trees, and sucked out the honey. Elisha Slade (1881b) reports that he "detected a Baltimore Oriole eating the leaves [of the American aspen] with evident relish. The bird stood on a branch and picked at and tore off the leaves, eating them with as much apparent enjoyment as our domestic fowls eat the leaves of the plantain." He noted the same performance the following year. In both years the observations were made late in May.

Irene G. Wheelock (1905) gives evidence concerning the food given to the nestling orioles. She says:

On the first day, feeding by regurgitation took place at intervals averaging twenty minutes for each nestling. As the nest was not more than three feet from the window, it was possible to watch just what was being done and to make examination of the young as often as seemed expedient. * * * The food given was the soft part of grasshoppers and dragon flies, and the larvae of different species of insects mixed with green leaves—all thoroughly macerated and partially digested. No traces of fruit were found. On the third day, the male was seen to give the soft part of a dragon fly, having removed the wings in full view of the observer, without first swallowing it himself. After the fourth day all food recorded was given in a fresh condition. In the case of this brood no fruit was fed the nestlings, possibly because of the difficulty of procuring it.

Gordon Boit Wellman (1928) adds another item of food to the oriole's diet, and describes the skillful way in which the bird obtained it. He writes:

On May 13, 1928, I found a pair of Baltimore Orioles (*Icterus galbula*) feeding on the larvae of a needle miner, probably *Paralechia pinifoliella*, in a pitch pine (*Pinus rigida*). The tree could be observed closely from my study window and the Orioles were seen feeding there each day until the twenty-second of the month. Both birds worked alike; resting on one foot, the bird would pull down a needle with the other foot, tuck it under the supporting foot with the bill, remove the larva and continue to feed in this manner until the five or six needles within reach were opened and held under the foot, then a new position would be taken. The larvae were to be found about halfway down the needle, invisible from the outside. The operation of removing the larva from a needle was done with such skill that in no case did I find a needle broken or permanently bent.

In regard to the birds eating fruit and vegetables, Walter B. Barrows (1912) says: "It is true that it has a special fondness for green peas, sometimes stripping the pods so freely as to cause considerable complaint. It also punctures ripening grapes whenever it has opportunity, but particularly where vines have run up into trees or over arbors or shrubbery in such a way as to hide the bird while at work. It is rare to hear complaints from grape growers, for where the vines are numerous and properly pruned the Oriole seldom injures them. Occasionally it attacks early apples and pears, digging holes into the soft pulp and of course ruining each apple attacked."

"By watching an oriole which has a nest," says F. E. L. Beal (1897), "one may see it searching among the smaller branches of some neighboring tree, carefully examining each leaf for caterpillars, and occasionally trilling a few notes to its mate." Francis H. Allen (MS.) has seen this oriole catch flies in the air; he has also watched the birds taking nectar from trumpet creeper flowers. "They would peck into the base of the corolla and into the mouth of the calyx after the corolla had fallen. I could see the nectar glistening between their mandibles."

William Brewster (1937) speaks of an adult female oriole eating cherries: "She operated on them in a deliberate, somewhat fastidious manner, piercing the skin with her sharp bill and then slowly tasting and swallowing the juice and perhaps some of the pulp also.   In no instance was the cherry removed from the stem.   This was in marked contrast to the behavior of the greedy Robins about her, the Robins first plucking the cherry and then swallowing it whole, not without some difficulty."

Late in July, when cherries are ripe, the now fully grown young orioles come to the cherry trees alone or with their parent.   Here they lean downwards, draw up a cherry and, steadying it in some way, appear to pick out mouthful after mouthful without detaching the fruit from its stem or the branch.   They eat calmly and daintily, as a rose-breasted grosbeak eats a cherry, not like a robin who snatches it off and bolts it down.   But the orioles swallow the little cornel berries whole.

Alexander F. Skutch (MS.) sends to A. C. Bent the following account of the bird's food during the winter: "While in Central America, the Baltimore orioles subsist upon a considerable variety of both animal and vegetable foods.   In humid regions, where the boughs of the trees are thickly overgrown with moss and lichens, they find many small creatures in the mossy covering.   In the dry months of February and March, when the madre de cacao trees (*Gliricidia sepium*), which are planted for living fence posts, have shed their foliage and covered their long, slender branches with delicately pink, pealike blossoms, the orioles spend much time probing the flowers.   The bright orange-and-black birds are a lovely sight amid the cluster of pink blossoms. Whether they seek chiefly the nectar or the small insects of various sorts that swarm about the flowers, I do not know.   When the winged brood of the termites fills the air at the end of an afternoon shower, the Baltimore orioles, along with a host of other birds of the most varied kinds, take advantage of this manna and snatch the slow-flying creatures from the air.   But with their slender bills they are not particularly adept at flycatching, and often miss their intended victim. During the early months of the year, when succulent fruits are not abundant among the forests of southern Costa Rica, the Baltimore orioles eat the dry green fruits of the Cecropia tree, clinging to the slender branches of the dangling inflorescence and tearing off small billfuls; in this feasting they are joined by many kinds of toucans, honeycreepers, tanagers, thrushes, and even flycatchers.

"For several years I have maintained a feeding shelf in a guava tree besides my house in southern Costa Rica, daily placing there ripe bananas or plantains.   The Baltimore orioles were not so quick to find this new source of food as some of the resident birds; but once they made the discovery they became regular patrons, and for the past two winters have continued to visit the table in increasing numbers. They made particularly good use of it during the fortnight of almost continuous rain at the end of October 1944, when many of the local birds seemed to experience difficulty in finding enough to eat. Then birds of a dozen species came in colorful crowds and consumed the bananas and plantains faster than they ripened. In 1945, the last Baltimore oriole of the season was seen at the feeding shelf on April 20."

Behavior.—To many of us who live in the Northern States the Baltimore oriole represents the spirit of spring. He arrives at the high tide of the season's beauty when he is at the peak of his magnificent spirits. But how soon his spirits fade! A month, and he begins to step back from the footlights, leaving the stage to less dominant personalities, as the red-eyed vireo and the robin.

The oriole fits easily into the community of the breeding birds about him, often building in the same tree with one of his neighbors. M. G. Vaiden (MS.) tells of a large pecan tree in which a wood pewee, a red-eyed vireo, a wood thrush, an orchard oriole, and two Baltimore orioles had nests at the same time and lived "a fairly agreeable life together." A. D. Du Bois (MS.) reports that "a pair of kingbirds had a nest in a burr oak only 5 or 6 yards from an oriole's nest, and the two species seemed to live amicably as close neighbors."

E. H. Forbush (1907) presents a dark side of the oriole's character:

The bird, a valiant fighter, does not hesitate to attack its enemies with its sharp beak,—a weapon not to be despised. It does the fiercest battle with the Kingbird, and may be seen sometimes struggling in mid air with this doughty adversary, until both birds fall to the ground breathless and exhausted. It sometimes succumbs, however, to the swarming numbers and extreme pugnacity of the "English" Sparrow, and where the Sparrows become most numerous they often drive out the Orioles. The Oriole itself, however, is not always guiltless in respect to other birds. Occasionally it destroys other nests, either to get material for building its own, or out of pure mischief. Mr. Mosher observed a male Oriole attempting to drive another away from its nest. The stranger would make a rush at the nest, and then the owner would grapple with him. This running fight was kept up for fully 3 hours. In the meantime the rogue Oriole went to a Redstart's nest, threw out the eggs, and threw down the nest. The next day an Oriole, probably the same bird, was seen to throw out an egg from a Red-eyed Vireo's nest, when he was set upon and driven away by the owners. Three other instances have been reported to me by trustworthy observers who have seen Orioles in the act of destroying the nests or eggs of other birds; but so far as I know, few writers have recorded such habits, and they are probably exceptional.

M. G. Vaiden (MS.) remarks: "The Baltimore is a very good watchman; he defends his territory with great vigor and daring. Most of his fighting is with our red squirrels, jays, and mockingbirds, but occasionally he attacks other birds when near its nest."

Audubon (1842) speaks of the migration thus: "During migration, the flight of the Baltimore Oriole is performed high above all the trees, and mostly during the day, as I have usually observed them alighting, always singly, about the setting of the sun, uttering a note or two, and darting into the lower branches to feed, and afterwards to rest. To assure myself of this mode of travelling by day, I marked the place where a beautiful male had perched one evening, and on going to the spot next morning, long before dawn, I had the pleasure of hearing his first notes as light appeared, and saw him search awhile for food, and afterwards mount in the air, making his way to warmer climes."

Last year I was reminded of this observation of Audubon. One morning, about 10 o'clock, late in August, a male Baltimore was singing in a maple tree across the way. He had separated from his family, which had been fledged weeks before from a nest a little way down the street, and he had been singing alone each morning for a week. As I watched him, he left the tree and, rising well above the surrounding buildings, held an undeviating course slightly to the west of south until he disappeared in the distance. I did not see or hear him again.

Frank L. Farley (MS.), of Alberta, Canada, sends to A. C. Bent this interesting note: "The Baltimore oriole is one of the many species of birds that have greatly extended their range in western Canada as a result of the settlement of the country. On my arrival in central Alberta in 1892, this bird was found in fair numbers in the woodlands, and along the rivers and smaller streams where trees were present. However, on the treeless plains of the eastern half of the province it was entirely absent, except in isolated spots where trees had been spared from the ravages of prairie fires which each spring or fall swept over the country.

"At the beginning of the present century great numbers of settlers moved into this open country lying eastward of the parklands, and took up land. Shortly thereafter, large areas of the prairies were brought under cultivation, and many of the road-allowances were plowed up. Such acts spelled doom to the fires, and it was not long before small clumps of willow-poplar and various kinds of shrubs appeared. As a result, the country took on an entirely different aspect. The trees and bushes were in most cases jealously guarded by the farmers, and they grew rapidly. In a few years these oases became the home of many summer birds that until now were entire strangers to the region. Shelter, food, and nesting sites were now afforded to thrushes, flycatchers, vireos, warblers, and many other kinds of forest-loving birds. The Baltimore oriole was not long in accepting this opportunity of extending its range, and in a few years it was well represented in many of the settled districts. Even before the saplings were large enough to offer suitable branches for the suspension of their nests, the orioles learned to build them attached to small limbs close up to the main stem of the tree. On several occasions I have found them within 8 feet of the ground."

Voice.—The song of the Baltimore oriole possesses little pure beauty but it stands out prominently in the spring chorus. We are not attracted to the song as we are to the rose-breasted grosbeak's by syrupy sweetness, nor by the robin's cheerfulness, the wood pewee's artistry, or by the red-eyed vireo's almost endless singing, but by its vigor—a sort of robust manliness. Another feature of the song which attracts our interest is its infinite variety: no two orioles, we say, sing the same tune, but each bird, in the main, sticks to his own theme. It is one of the songs which, if you note it down, you must punctuate at the end with a period; the bird has said his say and stops; he has finished, for the moment anyway. The song clearly corresponds to a short sentence of half a dozen syllables or so. A point of difference between it and the songs which resemble it somewhat is that many of its single notes, often most of them, are inflected sharply downwards, as the pitch of our voice falls in pronouncing the word "yolk." We notice the same peculiarity in the loud, vibrant call of the evening grosbeak.

In the simplest form the song consists of a short series of notes on the same pitch, like blasts from a tiny trumpet, or there may be but a single blast, scarcely a song. The longer songs, with their changes in pitch and short pauses between the notes, often form rather pretty phrases, although somewhat jerky because the notes are not run together smoothly. These songs give the impression of exclamations.

Francis H. Allen (MS.) speaks of "a beautiful and unusual song, a low, sotto voce warbling, interspersed with snatches of the characteristic chattering note," and of another, "an uncommonly pretty song, containing a trill near the end, a full-voiced song, not the low warble we sometimes hear, but longer than the song usually is."

Aretas A. Saunders (MS.) sends to A. C. Bent the following analysis: "The song of the Baltimore oriole is loud, clear and of flutelike quality. It consists of a series of short notes and 2-note phrases, with short pauses between them, and commonly a somewhat longer pause somewhere in the middle of the song. It is so exceedingly variable in form that one cannot pick out any one song, or even several songs, that could be said to be more typical of the species than others. The number of notes in a song varies, according to my 102 records, from 4 to 19, but as only 1 record has 19 and no others more than 16; the 19-note one is quite unusual. The average song is about 8 notes long. It is a common habit of the bird, however, to sing single notes or short

2-note phrases between songs, and if one considered these to be separate songs, there would be many 1- or 2-note songs.

"The pitch of the songs varies from F″ to A‴. The range in pitch of single songs varies from 1 tone to 6 tones, or an octave. The average range is 3½ tones. It is interesting to note that the only record I have that is definitely the song of a female bird has a range of only 1 tone.

"In time, songs vary from ⅖ to 2⅖ seconds. Though all of the notes are rather short, there is often considerable variation in the lengths of the notes of a song, so that, though the song has a rhythm, it is not often an even rhythm."

The bird not infrequently sings while on the wing in early spring, and occasionally in August.

Ralph Hoffmann (1904) says: "The female during the mating season whistles two or three notes similar to the male's," and Tilford Moore (MS.) has heard a female sing "short, finished songs." Both sexes give a long grating chatter which often seems to indicate anxiety, and this is sometimes incorporated into the song.

The period of singing is short. The bird arrives on its breeding ground in full song and continues to sing all day long during the mating and nest-building stage of the cycle, but by the end of June there is a noticeable falling off in the singing, and during the molt the males are almost silent. Then, about mid-August, a fortnight before they leave, the males sing freely again, chiefly in the early morning.

Brand (1938) gives the approximate mean vibration frequency of the Baltimore's voice as 2,500 cycles per second, slightly below that of the robin and not far from that of the bluebird.

Enemies.—Elon Howard Eaton (1914) speaks thus of the Baltimore oriole's enemies: "In spite of the skilful placing of the oriole's nest, it is frequently visited by plunderers. I have seen crows on several occasions succeed in getting young birds from the nest and the home of the Screech owl very often shows that the young orioles have been taken and fed to the owlets. Red squirrels also descend to the nest to get the eggs and young birds, and I have seen the gray squirrels do this on one or two occasions. Generally, however, the young are reared successfully and I am inclined to think that dangers in migration and severe weather are the principal checks to the increase of this species." Referring to the incessant calling of the young birds, he says that from their notes the young "are unquestionably located by many predaceous animals and thereby destroyed."

Forbush (1927) tells of a case in which nine bronzed grackles attacked a pair of orioles, "but after two minutes of swift action the Grackles retired from the combat, leaving the 'orioles' the victors." John T. S. Hunn (1926) reports "An Oriole Tragedy." The male oriole "was caught by a small cord firmly woven into the nest structure, and so tightly twisted about his neck that he strangled to death."

The following is quoted from Herbert Friedmann's 'The Cowbirds' (1929):

An uncommon victim. Bendire, (1895, p. 486), says "this species is rarely imposed upon by the Cowbird."

Gregg, in Chemung County, New York (Proc. Elmira Acad. of Science, vol. I, no. 1, June, 1891, p. 26), records finding a nest of this bird with two young Orioles hardly fledged and one young Cowbird big enough to fly.

S. E. Parshall, (Orn. and Ool. IX, no. 11, Nov. 1884, p. 139), found a deserted nest containing three eggs of the Orioles, and three of the Cowbird and three more of the Cowbird covered up.

B. H. Warren (Birds of Pennsylvania, 1890, pp. 209–210) writes that, * * * on three occasions I have discovered the shattered remains of these eggs (Cow-birds), directly beneath the pendant nests of Baltimore orioles * * * It may be that this species sometimes * * * tosses out alien eggs."

There are other records from Indiana, Iowa, and Michigan.

Fall.—Walter Bradford Barrows (1912) summarizes the behavior of the Baltimore oriole in late summer:

Before the middle of July both old and young have disappeared from garden, orchard and park, and except for an occasional almost silent individual at rare intervals, none are seen again until about the middle of August, from which time until their departure for the south in September they are fairly common and the male frequently sings almost as sweetly as in May. This disappearance for a month or more is rather apparent than real, for a careful search of the woods and swamps will reveal a fair number of orioles, spending most of their time, however, in the leafy crowns of the higher trees, where they are hardly visible, and being almost silent are pretty sure to be overlooked. They may also be found at this season about wild cherry and service berry trees, feeding on the ripening fruit.

Francis Beach White (1937) remarks: "In July, a few Orioles are about till the second week, and occasionally a pair is seen, or an adult with young; but after that they are very scarce. These late birds are likely to be seen low in thickets—where, indeed, they spend a good deal of time feeding in the breeding season—but once in a while one will silently flash across from one tree to bury itself in the foliage of another."

Apparently the old males do not remain with their families very long after the breeding season. I find in my notes these references to this subject: "July 30, 1916—Mr. Faxon and I saw nearly a dozen birds all in female or juvenal plumage more or less associated and not a single adult male among them. I recall noting the same thing in past seasons," and "August 13, 1917—Two male orioles were feeding silently this morning in the locust trees, where they found a small, green larva. These birds, although near each other, did not follow one another about and they acted as if perfectly independent of each other. Yesterday I saw an adult male in another part of the town, feeding alone. This appears to be a habit of the male at this season, to separate himself from his family and remain alone."

Winter.—There are several references in the literature to orioles found in winter, stranded far to the north of their normal winter range. Two of these birds, one in Virginia, the other in Ohio, were feeding on apples. Doubtless most of these lost birds perish, but Robie W. Tufts (MS.) reports on an immature female bird found in Nova Scotia in such a weakened condition that he captured her on December 13, and kept her indoors over the winter, "during which time she ate chiefly grapes"; he liberated her on the following May.

Alexander F. Skutch (MS.) supplies this comprehensive report of the bird on its winter quarters: "The Baltimore oriole arrives in Central America during the second week of September, but does not become abundant before the end of the month. During the northern winter, it resides throughout the region from Guatemala to the Isthmus of Panamá, on both the Caribbean and Pacific coasts and high up into the mountains. Scarcely any other winter visitant is so widely and uniformly distributed throughout the area.

"To appreciate the wide tolerance of environmental conditions implied in the winter distribution of the Baltimore oriole, one must be familiar with the local variations in climate and the corresponding differences in the nonmigratory section of the avifauna. Thus the arid coast of El Salvador, where I found these orioles abundant among cacti and low thorny trees early in the parched month of February, has exceedingly few resident birds in common with the humid coastal districts on the opposite side of the continent, in Honduras and Guatemala, where also the Baltimore orioles pass the winter in large numbers, amid lofty rain-forests, lush thickets, and extensive banana plantations. And very few of the birds which breed in the lowlands on either coast are found in the highlands above 5,000 o. 6,000 feet. Yet in December 1933 I found a few Baltimore orioles which had apparently settled down for the winter among the oaks, pines, and alders on the Sierra de Tecpán in Guatemala, at an altitude of 8,500 feet above sea level, where at this season nights were penetratingly cold, and every clear dawn revealed all the open spaces white with frost. Thus, in Central America, this adaptable bird makes itself at home, for a period covering half the year, almost everywhere that trees supply fruits, and insects lurk amid the foliage. It is especially fond of plantations, orchards, and pastures with abundant shade trees; but it also hunts through the treetops of the tall rain-forests, although it never, in my experience, descends into the deeply shaded lower regions of the forest, and so is not often seen by the bird-watcher who wanders through the heavy woodlands.

"During the winter months, the Baltimore orioles roam about singly, or in small groups of two, three, or four, more often than in larger groups. Although a dozen or so may at times be seen feasting together in some especially attractive flowering or fruiting tree, or may share the same roost, the birds are only slightly gregarious during this season and form no big, closely knit flocks like those of wintering dickcissels and cedar waxwings. Adult males pass the winter in brightest orange-and-black plumage, and are excelled in beauty by few even of the most brilliant of the native birds; but females and young males predominate.

"Although less songful while in Central America than the orchard orioles, Baltimore orioles often voice their clear, full whistles, especially during the first weeks following their arrival in the fall, and with greater frequency for a month or so before their northward departure in the spring. Rarely, as when the sun breaks through the clouds at the end of an afternoon shower in April, they charm the hearer with a somewhat sustained performance; but for the most part they utter only single notes and brief fragments of song—tantalizing suggestions of the full, mellow verses they will soon be broadcasting from northern elm trees. A sharp *churr* is the oriole's most frequent utterance while in its winter home.

"I have often discovered the sleeping places of the Baltimore orioles, and found them as catholic in their choice of a roosting site as of habitat and food. In the cleared portion of the Lancetilla Valley in northern Honduras, a number of them roosted, during the winter of 1930–31, in an extensive stand of tall 'elephant grass' (*Pennisetum purpureum*), a tangled and impenetrable gramineous jungle higher than a tall man's head. Here they slept along with wintering orchard orioles, resident Lesson's orioles (*Icterus prosthemelas*), four species of small resident finches, and migrating kingbirds which rested here for a while before continuing southward. In the hamlet of Buenos Aires in southern Costa Rica, Baltimore orioles roosted among the broad, close-set foliage of the tall *Dracaena fragrans*—a tree of the lily family—that bordered both sides of the path leading up to the little church. The orioles began to gather shortly before sunset, and as the day waned continued to fly in, one or a few at a time. They darted into the trees very suddenly and from all sides, and then sometimes shifted about from tree to tree before they became comfortable for the night. This manner of going to roost made it impossible to make an accurate count of their numbers, but certainly a score of the orioles took shelter in these trees. When finally they had settled down, they were completely screened from my view among the broad bases of the leaves; even the glowing orange of the adult males failed to shine forth from the dark green foliage. Taken together, observations show that, in one place or another, Baltimore orioles share roosts with a large variety of other birds. Their relations with their neighbors seem always to be amicable, and I have never seen quarrels among them.

"About my home in southern Costa Rica Baltimore orioles roost amid the dark, abundant foliage of the orange trees. Sometimes they slumber upon so low a perch that I might touch them while standing on the ground. They are not always careful to conceal themselves amid the leafage. Viewed by the light of an electric torch, with their heads turned back and buried in their outfluffed plumage, they look like brighter oranges scattered among the dark, glossy leaves. In addition to the orioles that roosted in the big trees, during the early part of the year 1943, three slept in a small tree south of the house. One of these was a male in exceptionally deep orange plumage; his companions were an adult male in more yellow plumage and a female or young male. At dawn on April 15, the three birds were in the tree as usual; but during the day two apparently began to migrate; for that evening the more brilliant male came alone to the orange tree. For a week after the departure of the others, he roosted alone here, where he was seen for the last time that spring at daybreak on April 22. After his disappearance, I saw no others of his kind until the following September 10, an unusually early date, when a female or young male appeared in the trees in front of the house. At daybreak on October 10, there was an unusually handsome male in the orange tree south of the house, where apparently he had roosted. That evening he went to rest in this tree in company with an immature male. Although the bird was not banded, I like to think that he was the same brilliant male who had roosted in the same place during the preceding spring. During those wet October days he whistled enchanting fragments of song, just as he had done before his departure in April. May not the bournes of the long semiannual journeys of the Baltimore oriole be two trees—perhaps an elm tree in New England, where he nests, and an orange tree in Costa Rica, where he sleeps during the winter months?"

DISTRIBUTION

Range.—Western and southern Canada to Colombia and Venezuela.

Breeding range.—The Baltimore oriole breeds from central Alberta (Lesser Slave Lake, Lac la Biche), central Saskatchewan (Emma Lake, Yorkton), southern Manitoba (Lake Saint Martin, Indian Bay), western Ontario (Malachi, Port Arthur), northern, Michigan (Houghton, Newberry), southern Ontario (Manitoulin

Island, Lake Nipissing), southern Quebec (Montreal, Blue Sea Lake), central Maine (Avon, Dover-Foxcroft), central New Brunswick (Woodstock, Saint John), and central Nova Scotia (Berwick); south through southern Alberta (Midnapore, Brooks), southern Saskatchewan (Sovereign, Lake Johnstone), northwestern and central North Dakota (Charlson, Tokio, Bismark), central South Dakota (Faulkton, White River), Nebraska, central Kansas (Stockton, Pratt), and west-central Oklahoma (Woodward, Minco) to northeastern Texas (Marshall), northwestern, central, and southeastern Louisiana (Shreveport, New Orleans), central Mississippi (Jackson, Waverly), northern Alabama (formerly), north-central Georgia (Atlanta, Washington), western South Carolina (Greenville), western North Carolina (Asheville, Boone), central Virginia (Bedford, Charlottesville), northern Maryland (Baltimore), and Delaware. Has bred in northeastern Colorado (Dry Willow Creek). Hybridizes extensively with *I. bullockii* in western Oklahoma and western Nebraska.

Winter range.—Winters from southern Veracruz (Tres Zapotes, Cerro de Tuxtla) and Tabasco (San Juan Bautista), throughout Central America to northern and central Colombia (Rió Juradó, Chafurray, Cúcuta) and northwestern Venezuela (San Rafael, Santa Barbara). Rare in Cuba during migration. Recorded occasionally in winter, in southeastern Canada and eastern United States, from Toronto, Ontario, south to Louisiana, especially since about 1951.

Casual records.—Casual in central Ontario (Chapleau), northern Maine (Mount Katahdin, Presque Isle), Prince Edward Island, eastern Quebec (Seven Islands), Newfoundland, and Bermuda.

Accidental in Northern Manitoba (York Factory) and Greenland (Sukkertoppen).

Migration.—Early dates of spring arrival are: Costa Rica—El General, March 15. Cuba—Havana, April 12. Florida—Volusia County, March 23; Fort Myers, March 30. Alabama—Decatur, April 15. Georgia—Milledgeville, April 5; Kirkwood—April 15 (median of 13 years, April 25). South Carolina—Spartanburg, April 17. North Carolina—Weaverville, April 16. Virginia—New Market, April 19 (median of 25 April dates, April 25). West Virginia—Bluefield, April 16. District of Columbia—April 24 (average of 43 years, May 2). Maryland—Dorchester County, April 9; Baltimore County, April 18. Pennsylvania—Beaver, April 19 (average of 25 years, April 28). New Jersey—Collingswood, April 13. New York—Long Island, April 14; Altamont, April 19. Connecticut—Norwalk, April 23; Portland, April 30 (median of 46 years, May 5). Rhode Island—Providence, May 1. Massachusetts—Falmouth, April 18; Harvard, April 22. Vermont—St. Johnsbury, April 21; Bennington,

May 1. New Hampshire—Manchester, May 1. Maine—Brunswick, April 17; New Vineyard, April 22. Quebec—Westmount, May 1. New Brunswick—Scotch Lake, May 12. Louisiana—Bains and Thibodaux, April 3; Ouachita Parish, April 7. Mississippi—Edwards, March 30. Arkansas—Monticello, April 4. Tennessee—Nashville, April 7 (median of 11 years, April 15). Kentucky—Casky, April 3. Missouri—Independence, March 28; Columbia, April 2 (median of 17 years, April 22). Illinois—Hinsdale, March 27; Olney, March 28; Chicago region, April 19 (average, May 3). Indiana—Knox County, March 24; Logansport, March 28. Ohio—Canton, March 28; Buckeye Lake, April 24 (median, April 29). Michigan—Ann Arbor, April 11; Blaney Park, May 13. Ontario—Toronto, April 12; Ottawa, May 3 (average of 31 years, May 10). Iowa—McGregor, April 7. Wisconsin—LaCrosse, April 5; Superior, April 16. Minnesota—St. Cloud, April 20 (average of 34 years in southern Minnesota, May 5); Stearns County, April 28 (average of 22 years in northern Minnesota, May 10). Texas—Troup, March 30; San Patricio County, April 2. Oklahoma—Oklahoma City, March 28 (median of 10 years, April 26). Kansas—Hays, April 18. Nebraska—Red Cloud, April 23 (median of 24 years, May 2). South Dakota—Yankton, April 26. North Dakota—Devils Lake, May 4; Cass County, May 11 (average, May 16). Manitoba—Treesbank, April 30 (average, May 18); Margaret, May 7. Saskatchewan—Sovereign, April 26; South Qu'Appelle, April 28. Colorado—Lamar, May 6. Wyoming—Guernsey, May 6. Alberta—Morrin, May 11; Camrose, May 14.

Late dates of spring departure are: Colombia—Río Frío, March 10. Panamá—Jesusito, April 20. Costa Rica—El General, April 28 (median of 11 years, April 20). El Salvador—Hacienda Chelata, April 27. Honduras—Tela, April 23. Guatemala—Quirigua, April 17. Veracruz—Cuitláhuac (20 miles southeast of), May 1. Bermuda—Nonsuch Island, May 5. Florida—Pensacola, May 27; Amelia Island, May 13. Alabama—Woodbine, May 10. Georgia—Athens, May 22. South Carolina—Spartanburg, May 14. North Carolina—Highlands, June 6. District of Columbia, June 10. Maryland—Laurel, June 12 (median of 6 years, May 19). Illinois—Chicago, May 31 (average of 16 years, May 25). Texas—Commerce, May 30; Dallas, May 18.

Early dates of fall arrival are: Texas—Atascosa County, August 3. Minnesota—Minneapolis, July 2. Illinois—Chicago, August 4 (average of 12 years, August 15). Connecticut—Middletown, June 30. New York—Brooklyn, July 2. Maryland—Laurel, July 28 (median of 7 years, August 8). Virginia—Naruna, August 24. North Carolina—Chapel Hill, August 23. Georgia—Young Harris, August 14.

Alabama—Long Island, August 25. Florida—Pensacola, July 31.
Bermuda—September 16. Guatemala—San Juan Atitlán, September 9. Honduras—Truxillo, September 1. Nicaragua—Escondido
River, September 20. Costa Rica—El General, September 10.
Panamá—Changuinola, September 30. Colombia—Río Frío, October 13.

Late dates of fall departure are: Alberta—Camrose, September 7.
Wyoming—Yellowstone Park, September 4. Saskatchewan—Indian
Head, September 20; Yorkton, September 6. Manitoba—Killarney,
September 27; Margaret, September 20; Treesbank, September 5
(average, August 30). North Dakota—Fargo, September 19 (average for Cass County, September 3). South Dakota—Faulkton,
September 24. Nebraska—Hastings, October 10; Scribner, September 21. Kansas—Hays, October 24. Oklahoma—Oklahoma City,
October 16 (median of 11 years, September 16). Texas—Victoria,
October 31; Cove, October 9. Minnesota—Hutchinson, October 8;
(average of 19 years in southern Minnesota, September 5); Elk River,
September 11 (average of 8 years in northern Minnesota, August 31).
Wisconsin—Elkhorn, October 25; Superior, October 18. Iowa—
Marble Rock, October 15. Ontario—Point Pelee, September 20;
Ottawa, September 16 (average of 15 years, August 25). Michigan—
McMillan, September 26; Blaney Park, September 23. Ohio—Cleveland, October 19; Buckeye Lake, September 17 (median, September 5).
Indiana—Muncie, October 15. Illinois—Rantoul, October 4; Chicago region, October 2 (average, September 1). Missouri—Concordia,
September 26. Kentucky—Bowling Green, September 17. Tennessee—Carter County, September 30. Arkansas—Dardanelle, October 20. Mississippi—Edwards, October 6. Louisiana—Bienville,
October 10. Greenland—Sukkertoppen, September 27. Newfoundland—Newfoundland banks, September 21. Nova Scotia—Sable
Island, October 4. New Brunswick—Fredericton, August 24.
Quebec—Hudson Heights, October 1. Maine—Topsham, November 6; Bath, October 28. New Hampshire—Hanover, October 22.
Vermont—St. Johnsbury, October 17; Rutland, September 19.
Massachusetts—Edgartown, November 1; Groton, October 29.
Rhode Island—Wakefield, November 19; Kingston, October 21.
Connecticut—Hamden, November 17; Bloomfield, October 21. New
York—Ithaca, October 28; Middleport, October 20. New Jersey—
Passaic, October 7. Pennsylvania—McKeesport, October 19. Maryland—Laurel, October 20 (median of 7 years, September 21). District
of Columbia—October 15. West Virginia—Bluefield, October 29.
Virginia—Charlottesville, October 2. North Carolina—Hendersonville, October 18; Raleigh, October 4. South Carolina—Frogmore,
September 20. Georgia—Atlanta, September 16. Alabama—Greens-

boro, September 23. Florida—Miami, November 8; Pensacola, November 2. Bermuda—October 12.

Egg dates.—Illinois: 14 records, May 25 to June 14; 7 records, May 29 to June 10.

Massachusetts: 31 records, May 23 to June 9; 17 records, May 30 to June 5.

New York: 19 records, May 25 to June 10; 10 records, May 31 to June 4.

ICTERUS BULLOCKII BULLOCKII (Swainson)

Bullock's Oriole

Plate 20

HABITS

This highly colored oriole replaces the Baltimore oriole in the western half of North America, except for a narrow strip along the Pacific coast from the San Francisco Bay region to northern Baja California. Its breeding range extends from the southern parts of British Columbia, Alberta, and Saskatchewan to southern Texas and northern México, and from the western edge of the Great Plains and prairie regions to the Pacific slope. At the eastern border of its range, where it meets that of the Baltimore oriole, these two closely related species appear to interbreed, producing an interesting series of apparent hybrids, to be referred to later.

The favorite haunts of Bullock's oriole are in the growths of deciduous trees, cottonwoods, willows, sycamores, etc., that line the streams or irrigation ditches in open country, in the prairie regions, and in cultivated lands. The presence of water is not essential, for they are equally at home in some of the partially dry washes that extend down into the grasslands from the mountain canyons, where there is some underground moisture, or far from any water in the tree-claims about the ranches; they are also found living and nesting in the semiarid mesquite groves in Arizona. It is, perhaps, less intimately associated with human habitations than is the more sociable Baltimore oriole, though it does nest to some extent in villages and near houses, especially about farms and ranches.

Nesting.—Bendire (1895) describes this very well as follows:

The nest resembles that of the Baltimore Oriole, but as a rule it is not quite as pensile, and many are more or less securely fastened by the sides as well as by the rim to some of the adjoining twigs. The general make-up is similar. As many sections where Bullock's Oriole breeds are still rather sparsely settled, less twine and such other material as may be picked up about human habitations enter into its composition. Shreds of wild flax and other fiber-bearing plants and the inner bark of the juniper and willow are more extensively utilized; these with horsehair and the down of plants, wool, and fine moss, furnish the inner lining of the nests. According to my observations, the birch, alder, cottonwood, eucalyptus, willow, sycamore, oak, pine, and juniper furnish the favorite nesting sites; and in southern Arizona and western Texas it builds frequently in bunches of mistletoe growing on cottonwood and mesquite trees.

The nests are usually placed in low situations, from 6 to 15 feet from the ground, but occasionally one is found fully 50 feet up. A very handsome nest, now before me, * * * is placed among six twigs of mistletoe, several of these being incorporated in the sides of the nest, which is woven entirely of horsehair and white cotton thread, making a very pretty combination. The bottom of the nest is lined with wool. Outwardly it is 6 inches deep; inside 4½ inches. The entrance, at the top, is oval in shape, somewhat contracted, and 4 by 2½ inches wide. Another peculiar specimen before me, taken near Yreka, California, May 29, 1860, is woven among and fastened to a bunch of needles of the long-leafed pine; this nest resembles an inverted cone, and is quite unique in structure. I have also seen double nests, one placed beside and fastened to one previously built that had for some unknown reason been abandoned.

In the vicinity of Fort Lapwai, Idaho, it was especially abundant, and, although suitable nesting sites were by no means scarce, I have seen three occupied nests of this Oriole in a small birch tree close to a nest of the Arkansas Flycatcher, showing them to be very sociable birds. Near Camp Harney, Oregon, a Swainson's Hawk, an Arkansas Flycatcher, and a pair of this species nested in the same tree, a good-sized pine. A. K. Fisher tells me that he saw hundreds of these nests in a large row of cottonwoods, east of Phoenix, Arizona, in June, 1892.

In Arizona, Herbert Brandt (MS.) found this oriole often nesting close to an occupied nest of the western kingbird, in the mesquite chaparral. "At one place in a sycamore I saw them nesting within three feet of each other, and they could have used, if they wished, opposite sides of the tree, 30 feet apart. Twice they resided in the same small mesquite." In Texas, he noted a similar association between Bullock's oriole and the scissor-tailed flycatcher.

Near the Huachuca Mountains, Ariz., we took a set of eggs of this oriole, about 10 feet up in a sycamore, and a set of eggs of the western flycatcher, about 20 feet from the ground in the same tree.

On May 24, 1923, near Brownsville, Tex., I collected a set of five fresh eggs from a typical nest of Bullock's oriole about 15 feet up in a large mesquite; R. D. Camp told me that this species does not breed there, but, after watching for a long time, I plainly saw both birds go to the nest It is evidently not a common breeding bird there.

C. S. Sharp (1903) has published an interesting paper, illustrated with photographs, describing three distinct types of Bullocks' orioles' nests, one of which is wholly pensile, one semipensile, and one not at all pensile; his description of an especially beautiful pensile nest follows:

The twigs to which it was attached formed a fork, and a few inches above, another small twig extended downward in the same direction. The nest was wholly suspended from these, the twigs, with some of the leaves attached being worked into it for a little distance down the sides and back. With these exceptions and two or three long horse hairs it was composed wholly of wild oats and rather loosely woven. A few of the oat heads show on the inside where they were worked into the nest itself, but almost all are on the outside, the long stems being worked in to their heads which stood out in a beautiful and graceful fringe all around and below for from one to three inches or more. The effect was striking and unusual. * * * The dimensions in inches are as follows: Depth outside (extreme) 14; depth outside (front) to opening, 8; depth inside to opening 5½; diameter outside, 7; diameter inside, 4; circumference 21.

J. F. Illingworth (1901) writes:

Until the season of '97 I have never known the Bullock's Oriole to use palm-fiber in the construction of its home, but I found a nest May 11, 1897, in a peach tree, composed entirely of this fiber. It was well lined with chicken feathers and placed between several small branches. A pair of Bullock's Orioles built a nest this year in an almond tree near the porch, and I had an excellent opportunity to watch them while they were at work. The place chosen was in a wide fork between four small branches. Both birds worked on the nest and as soon as they had loosely formed the walls or framework, one of them worked inside and the other outside. The latter would bring a horse-hair or a piece of twine in its beak and pass the end through the wall of the nest to his mate inside who took the end and passed it out again through another place. In this way the nest was soon woven quite smooth and looked as if it had been made with a darning needle by hand.

Eggs.—Bullock's oriole lays from three to six eggs to a set; four and five are the commonest numbers, but sets of six are not rare. Bendire (1895) describes them as follows:

The eggs are mostly elongate ovate in shape, a few are ovate, and an occasional set is almost wedge-shaped or cuneiform. The shell is close grained and only slightly glossy. The ground color is generally of the same subtle grayish-white tint as that seen in the eggs of the Baltimore Oriole, but the proportion of the pale bluish white eggs is greater than with the latter. Occasionally the ground color is pale vinaceous buff. The markings are similar in color to those found on the eggs of the preceding species [Baltimore oriole], but as a rule they are not so coarse, and the fine hair lines running in irregular tracings around the larger axis of the egg are more prevalent; they are also a trifle larger.

The average measurement of 144 specimens in the United States National Museum collection is 23.80 by 15.93 millimetres, or about 0.94 by 0.63 inch. The largest egg in the series measures 25.40 by 16.76 millimetres, or 1 by 0.66 inch; the smallest, 21.34 by 15.24 millimetres, or 0.84 by 0.60 inch.

Young.—Bendire (1895) states: "Only one brood is raised in a season, and the duties of incubation, which are performed almost exclusively by the female, last about 14 days. I have often watched the sitting bird, and have never seen the male on the nest."

Mrs. Wheelock's (1904) observations confirm this statement; and she adds:

Her mate is always within calling distance, keeping a vigilant watch for squirrels, crows, and jays; and should any of these enemies appear, not only he but the mother bird, joined by all the orioles and blackbirds within hearing, will fly at the intruder and effectually banish him from the vicinity. When newly hatched, the young orioles are naked, pink babies with little tufts of thin white down on head and back. For nearly a week after they are feathered the down waves rakishly on either side of the crown and about the shoulders, gradually wearing off as they brush about through the bushes.

Like all oriole babies, these demand the constant attention of both parents, crying loudly for more the moment their mouths are emptied of the last mouthful, not in the least trying to help themselves, but following the adults about for a week or two after leaving the nest. * * * I believe the families usually keep together until late in August, when the males join flocks of their own sex for the September migration southward.

Plumages.—The natal down of Bullock's oriole is white, long, and rather scanty. The sexes are practically alike in the juvenal plumage, though the female is usually rather paler; this plumage closely resembles that of the adult female, grayish olive above with yellowish olive tail and wing coverts and dull buffy whitish below; but there is no black on the throat or wings, and no orange on the head and neck. A postjuvenal molt occurs in late summer, involving the contour plumage and the wing coverts but not the rest of the wings nor the tail; this produces the first winter plumage, in which the young male acquires black lores and a narrowly black throat, but which is otherwise like the plumage of the adult female; young birds may breed in this plumage. At the first complete postnuptial molt, the following summer, the adult winter plumage is acquired; this is like the adult spring plumage, but in the male the feathers of the back and under parts are margined with gray. Adults have a complete postnuptial molt in summer, but apparently no spring molt, the spring plumage being acquired by wear.

At my request, James L. Peters examined the large series of Bullock's orioles in the Museum of Comparative Zoology and has sent me his report, from which I gather the following additional information: The postjuvenal molt begins about the middle of August and is completed by about the middle of September. The amount of black acquired in the throat and lores at this molt seems to vary considerably and the time at which it is acquired also varies. Two males, taken September 16 and October 10, both lack it; two males, taken October 7 and 11, have whitish lores and throat patches of scattered black feathers; but a male, taken October 6, has fully developed black lores and throat patch. He says that the first prenuptial molt of the male varies greatly; "in some individuals new black feathers with olive edges appear on the crown; sometimes only half a dozen such are to be found; in others the crown is entirely covered; the first traces may appear by the end of January. Some individuals acquire new scapulars and interscapulars between the end of January and the end of March, but not over half of those examined did so."

The sequence of molts in the female is similar to that of the male. There is considerable individual variation in the amount of black acquired on the throat at the first prenuptial molt; out of 27 young females examined by Peters, 14 did not acquire it at all, 10 acquired only traces, and only 3 had the black stripe complete.

Food.—F. E. L. Beal (1910) examined the contents of 162 stomachs of Bullock's oriole, taken in every month from April to August, inclusive. The contents consisted of 79 percent animal matter and 21 percent vegetable:

The animal food consisted mainly of insects, with a few spiders, a lizard, a mollusk shell, and eggshells. Beetles amounted to 35 percent, and all except a few ladybugs (Coccinellidae) were harmful species. The coccinellids were found in 9 stomachs, but the percentage was insignificant. Many of the beetles were weevils, and quite a number belonged to the genus *Balaninus*, which lives upon acorns and other nuts. Ants were found in 19 stomachs, and 1 contained nothing else. Hymenoptera other than ants were found in 56 stomachs, and entirely filled 2 of them. Including the ants, they amount to nearly 15 percent of the food of the season. * * *

One of the most interesting articles of food in the oriole's dietary is the black olive scale (*Saissetia oleae*). This was found in 45 stomachs, and amounted to 5 percent of the food. In one stomach these scales formed 87 percent of the contents; in another, 82; and in each of two others, 81 percent. * * * Hemiptera other than scales are eaten quite regularly. They amount to a little more than 5 percent of the food. * * * They were mostly stinkbugs, leafhoppers, and tree hoppers. Plant lice (Aphididae) were found in one stomach.

Lepidoptera, moths, pupae, and caterpillars, are the largest item in the food, amounting to 63 percent in April, only 8 percent in July, and averaging a little more than 41 percent for the season. He continues:

Perhaps the most interesting point in connection with the Lepidoptera is the eating of the pupae and larvae of the codling moth (*Carpocapsa pomonella*). These were found in 23 stomachs, which shows that they are not an unusual article of diet. No less than 14 of the pupa cases were found in one stomach, and as they are very fragile, many others may have been present, but broken up beyond recognition. It is curious that the oriole should find these insects. During the greater part of their larval life they are concealed within the apple. When ready to pupate they crawl out and at once seek some place of concealment, such as a crevice in bark or among clods or rubbish, where they can undergo their changes. To find them, therefore, birds must hunt for them.

Grasshoppers amounted to a little more than 3 percent for the season, but "2 stomachs, both taken in June, contained nothing else, and another had 97 percent of them. * * * Practically all of the vegetable food consists of fruit, which amounts to a little more than 9 percent. * * * It was found in 67 stomachs, of which 16 contained cherries; 11, figs; 5, blackberries or raspberries; 1, elder-berries; and 34, fruit pulp not further identified. One stomach was entirely filled with the pulp and seeds of figs." Fruit amounted to nearly 40 percent in July.

Herbert Brandt (MS.) watched a Bullock's oriole feeding for several mornings at the blossoms of bird of paradise shrubs in Arizona. "The oriole came just at dawn. He would fly from flower to flower, often within a few inches of my face, and I could see distinctly by his throat actions that he was drinking the nectar therefrom. Never did he show indications of picking out insect life from the deep tubes. All the time he was feeding, which was usually long enough for a visit to each flower, he kept talking to himself, uttering a musical peep note. Later in the day either this oriole or another of the same kind would search the crimson flowers of the ocotillos on the other side of the house in the same manner, but I never saw the oriole among the Bird of Paradise plants at any other time than its regular dawn visits. Evidently this exotic plant does not produce nectar in its cups during daytime."

W. Otto Emerson (1904) observed these orioles feeding on honey in the blossoms of the eucalyptus trees; one that he shot had its crop so full of the honey that it oozed out of its mouth when he picked it up. Ridgway (1877) noticed that, in Nevada, in May, "they were then subsisting chiefly on the tender buds of the greese-wood," as were some other birds. Bullock's orioles also feed to some extent on apricots, persimmons, hawthorn berries, and probably many other wild fruits and berries.

Claude T. Barnes writes to me from Utah: "For half an hour I stood beneath an elm tree observing a pair of Bullock orioles. Some of the leaves were so infected with lice that they were curled edge to edge, and each oriole was busy working its bill into leaf after leaf. When standing on a branch without eating, each bird uttered a *chup* every second or so."

These orioles are helpful in destroying the cotton boll weevil. A. H. Howell (1907) writes: "These orioles are rather abundant in the regions they inhabit, and in August and September visit the cotton fields in flocks of 10 to 20 individuals. About 27 percent of those examined contained boll weevils, the largest number of weevils found in one stomach being 41. The total number of weevils eaten by 40 birds was 133, an average of over 3 weevils to each bird."

They do good work on the alfalfa weevil, also. E. R. Kalmbach (1914) says: "Although the alfalfa weevil in all its stages is found most frequently on or near the ground, it was present in each of seven stomachs collected. Two birds taken in June had fed on it to the extent of 8⅓ percent of their food, while in the following month it formed nearly twice that amount. One bird collected in July had eaten no less than 21 adults, equaling 30 percent of its food. No larvae were taken by these birds even though this form of the insect was in great abundance, so that the adults may have been captured either on the wing or upon branches of trees which had intercepted their flight."

That these orioles can catch insects on the wing is shown by the following observation by J. G. Tyler (1913): "The small yellow butterfly that is found in such numbers in alfalfa fields at certain seasons seems to be especially attractive to the orioles, and countless dozens of them are devoured. I have seen this bird in the role of flycatcher at such times, flying from a fence wire and seizing a butterfly on the wing, a rather clumsy effort but serving the purpose."

Economic status.—From the above account of its feeding habits we must conclude that, although Bullock's oriole unquestionably does some damage to cultivated fruits and berries, it more than pays for this by the large number of harmful insects that it destroys, an action that is of real benefit to the agriculturalist. For this reason, and because the beauty of its gorgeous plumage and its charming song bring so much joy to so many people, it should be rigidly protected and encouraged to live and breed about our farms, ranches, and gardens.

Behavior.—As mentioned above, Bullock's orioles are most devoted parents and staunch defenders of their nests, eggs, and young. Clarence Cottam (MS.) sends me his observations of the bird in southeastern Utah: "In this locality the orioles were quite numerous and were in the midst of their nesting season. Magpies (*Pica pica hudsonia*) also were very common. One of these omnivorous feeders, a juvenile about one-half to two thirds grown, was observed circling about an oriole's nest as though searching for a breakfast of eggs. The magpie soon alighted in the tree in which the nest was hanging and began to come closer and closer to the beautiful swinging structure. Almost at the instant the magpie settled upon the edge of the nest, the male oriole, which apparently was but a few rods away, was heard to give an abrupt and angry call of warning. A moment later the enraged male came with all his force at the intruder, striking it on the crown of the head. The magpie dropped to the ground, stunned to such an extent that I was able to pick it up, and only after 10 minutes could it regain sufficient strength to fly away."

W. L. Finley (1907) made some observations on a pair of these orioles that nested near his house, saying:

I never saw birds more in love than the orioles were. We watched them from the time they were first mated. They were always together in the trees about the orchard. * * * Just at the side of the house were three large cherry trees with wide-spreading branches almost to the windows. When the dark shades were drawn the windows made a very good mirror. One day when the pair of orioles were playing about the cherry trees I saw the female light on a low branch in front of the window. Then in a few moments she flew down and lit on the sash. The next day I saw both the orioles at the window. The male sat near on the branches and the female on the sill. As I watched she fluttered up against the window, trying her best to hang on, till she slipped down to the bottom. Then she turned her head and watched in the glass. The more she looked the more excited she seemed to get, and she fluttered against the glass till out of breath. Then the mate flew down beside her. Time after time the birds were seen at the window.

Once a strange male oriole alighted in the nesting tree, but "the new arrival had hardly lit when there was a flash of color, and the father of the nestlings darted at the intruder like a little fury. Through the branches, under trees, over the barn, and across the orchard the righteous pursuer and the invidious pursued darted."

On another occasion, "a newly mated pair of orioles were living about a grove of trees, and the male bird was in such fine plumage that a collector shot him for his cabinet. The next day the female appeared with a new husband, who was as bright and fine looking as the bird she lost the day before. At the first chance this male was also shot, partly, it was said, because he was such a fine bird and partly to see if the female would find another as readily. Two days later she appeared with a third husband, who went the way of the two former ones. The female then disappeared for a few days, but returned again with a fourth suitor. These two began building in a eucalyptus tree and soon had a family of young birds."

This incident clearly illustrates the well-known fact that the urge for reproduction is very strong in birds, also that they do not grieve long over the loss of a mate, and that there are always enough unmated birds to fill in a gap caused by accident. Though this may have been an extreme case, such happenings are very common.

A. W. Anthony (1921) tells of the strange behavior of a captive Bullock's oriole that had never shown any fear of human beings, but showed "absolute terror" whenever its mistress appeared in a new dress adorned with a string of dark beads; after the beads were removed, the behavior of the bird became normal.

Voice.—Dawson (1923) writes: "The Bullock Oriole is either musical or noisy, but oftener both together. Both sexes indulge a stirring rattle which seems to express nearly every variety of emotion. Upon this the male grafts a musical outcry, so that the whole approaches song. A purer song phrase more rarely indulged in may be syllabized as follows: *Cut cut cudut whee up chooup.* The last note comes sharp and clear, or, as often, trails off into an indistinguishable jumble. The questing note, or single call, of the male is one of the sweetest sounds of springtime, but an even more domestic sound, *chirp trap*, uttered while he is trailing about after his swinking spouse, appears ridiculously prosaic."

Grinnell and Storer (1924) describe the song of the male as a "slightly varying series of syllables, rhythmically accented, like *hip-kip-y-ty-hoý-hoy*, but with a peculiar quality impossible to describe (fide senior author); also a mildly harsh *cha-cha-cha-cha*, etc., in rapid sequence, and a single clear note, *kleek*. Female and young give simple harsh blackbird-like notes."

Ralph Hoffmann (1927) says: "In late March or early April a flash of orange or black in the delicate green of cottonwoods, a characteristic chatter, or the *kip, kit-tick, kit-tick, whew, wheet* of the song announce the arrival in southern California of the Bullock Oriole."

Mrs. Wheelock (1904) writes: "Its call-notes and song resemble those of the Baltimore, but have less sweetness and variety. Where the latter whistles half a dozen variations on his original theme of five notes, the Bullock is content to repeat the same phrase with few modifications. Nor have I ever heard him give the love song that is poured out by the Baltimore with such tenderness just at dawn when his mate is on the nest."

Field Marks.—A brilliant pattern of orange, black and white marks the adult male Bullock's oriole. The top of the head, hind neck, upper back, central tail feathers, and tip of the tail are black; the lores, a narrow stripe behind the eye, and the throat are also black. The wing coverts and edges of the secondaries are white. The rest of the plumage, including the forehead, a broad band above the eye, sides of the head and neck, and entire under parts are rich, brilliant orange; the rump, upper tail coverts, and the lateral tail feathers are also orange. The female is much more soberly colored, olive above and buffy white below, but the sides of the head and neck are more tinged with orange than in other orioles, there is a dusky streak through the eye, some black on the throat, and two white wing bars.

Enemies.—The usual enemies of small birds—crows, magpies, jays and squirrels—often attempt to rob the nests of eggs or young; the nests are often more accessible than are those of the Baltimore oriole, but the parents are good guardians and often succeed in driving the robbers away.

Friedmann (1929) records the Bullock's oriole as a rather rare victim of the dwarf cowbird, but says that this "species is frequently parasitized by the Red-eyed Cowbird." Major Bendire (1895) says; "Bullock's Oriole may occasionally rid herself of the parasitic egg; at any rate I noticed the remains of one lying under a nest of this species, with portions of one of her own. This nest contained only three eggs of the rightful owner, and the bird was sitting on these."

Fall.—Referring to the Fresno district of California, J. G. Tyler (1913) writes: "The great majority of our orioles depart about the twentieth of July, or at the close of the nesting season. No doubt a scarcity of food during the hot, dry months of August and September is responsible for the short stay of these birds. Probably they scatter out and range up into the higher hills, as many summer residents do in the southern part of the state. This species has been noted in small numbers along the San Joaquin River during August."

The fall migration in Arizona is referred to by Swarth (1904) as follows: "The only time at which I have seen Bullock Orioles at all abundant in the Huachuca Mountains was in August 1902. About the middle of the month flocks of from ten to twenty, nearly all young birds, could be seen along the canyons up to an altitude of about 5,500 feet. Most of these must have come in from other parts of the country, for I have never found them breeding at all abundantly in the mountains, being in fact, the rarest of the three species of orioles occurring there."

DISTRIBUTION

Range.—Southwestern Canada to Costa Rica.

Breeding range.—Bullock's oriole breeds from southern British Columbia east of the coastal ranges (Milner, Alkali Lake, Okanagan Landing), northwestern Montana (Flathead Lake) southern Alberta (Warner, Medicine Hat), southwestern Saskatchewan (Maple Creek, Eastend), northeastern Montana (Fairview), southwestern North Dakota (Medora), western South Dakota (Harding County, Black Hills), western Nebraska (Chadron, McCook), western Kansas (Garden City), western Oklahoma (Gate), and central Texas (Vernon, Austin); south to central and southern interior California (Mount Saint Helena, Twenty-nine Palms), southern Nevada (Charleston Mountains, Pioche), southwestern Utah (Saint George), central and central-western Arizona (Prescott, Tucson), northeastern Sonora (Saric, Pilares), probably northern Chihuahua (Casas Grandes), central Coahuila (Monclava), and southern Texas (Rio Grande City, Brownsville). Summer records to east of this range: North Dakota (Towner), South Dakota (Pierre), Kansas (Fort Riley, Manhattan, Lawrence). Hybridizes extensively with *I. galbula* in western Oklahoma and western Nebraska.

Winter range.—Winters from southern Sinaloa (Mazatlán), México. (Tlalpam), and Puebla south, west of the continental divide, to northwestern Costa Rica (Liberia); casually north to central California (Durham, Drytown) and southern Texas (Nueces), and southern Louisiana (Cameron, Baton Rouge).

Casual records.—Casual in western Washington (Tacoma, Vancouver).

Accidental in New York (Onondaga County), Massachusetts (Falmouth), Maine (Sorrento), and Georgia (Grady County).

Migration.—The data deal with the species as a whole.

Early dates of spring arrival are: Sonora—Tesia, March 19. Texas—Rockport, March 20; Cameron County, April 4. Oklahoma—

Cimarron County, May 2. Nebraska—Albion, May 7. South Dakota—Rapid City, May 21. Manitoba—Brandon, May 24. Saskatchewan—Eastend, May 26. New Mexico—Carlisle, April 14. Arizona—Santa Catalina Mountains, March 18. Colorado—Grand Junction, April 25. Utah—Saint George, April 30. Wyoming—Cheyenne, April 30 (average of 15 years for Wyoming, May 12). Idaho—Meridian, April 30. Montana—Kirby, May 8; Custer County, average, May 20. Alberta—Warner, May 15. Calfiornia—San Diego, March 1; Escondido, March 5. Nevada—Pioche, April 23. Oregon—Josephine County, April 13. Washington—Yakima, April 22. British Columbia—Okanagan Landing, May 6.

Late date of spring departure: Sonora—Guirocoba, May 12.

Early dates of fall arrival are: Guerrero—Chilpancingo, October 7. Guatemala—Finca Carolina, October 20. Costa Rica—Liberia, November 1.

Late dates of fall departure are: British Columbia—Okanagan Landing, August 29. Washington—southeastern Washington, September 8. Oregon—Klamath County, September 10. Nevada—Lee Cañon, August 25. California—Hayward, November 16; Buena Park, September 21. Alberta—Red Deer River, August 29. Montana—Huntley, September 11. Idaho—Pocatello, September 5. Wyoming—Laramie, September 16. Utah—Saint George, November 27. Colorado—Pueblo, October 24; Boulder County, September 30. Arizona—Tombstone, September 28. New Mexico—Mesilla, October 2. Saskatchewan—Eastend, August 11. Manitoba—Brandon, August 15. South Dakota—White River, August 29. Oklahoma—Cimarron, September 14. Texas—Brownsville, November 6; Victoria, October 20.

Egg dates.—California: 160 records, April 22 to June 11; 82 records, May 11 to May 25.

Texas: 22 records, April 30 to June 25; 12 records, May 15 to May 29.

Utah: 10 records, May 27 to June 17; 5 records, June 4 to June 7.

ICTERUS BULLOCKII PARVUS van Rossem

Lesser Bullock's Oriole

HABITS

Based on the study of a series of 42 adult males and 15 adult females referable to this race, van Rossem (1945) named this subspecies and described it as "similar in color to *Icterus bullockii bullockii* (Swainson) of western North America in general. Size distinctly smaller. Measurements of the type are: wing, 97; tail, 76; culmen, 18.4; tarsus 23.2; middle toe, minus claw, 16.7. The corresponding measurements of Swainson's type of *Xanthornus bullockii* (examined at Cambridge, England, in 1933, and again on July 4, 1938) are 105, 83, 20.0, 24.5, and 17.8 mm. Range.—Coastal slope of California from the San Francisco Bay region south to northern Baja California, and eastward in the extreme southern part of the range to the lower Colorado River valley. Winter range undetermined but occurs in southeastern Arizona and southern Sonora in migration."

It is of interest to note that Ridgway (1902), in a footnote, called attention long ago to the fact that orioles of this species from California, west of the Sierra Nevada, are smaller than those from the interior to the eastward of that range.

The observations made by Alden H. Miller (1931) on the song and territorial behavior of two pairs of Bullock's orioles in Contra Costa County, Calif., probably apply to birds of this subspecies. He says that—

The male Bullock's Oriole arrives on the breeding ground before the female and establishes a singing post, perhaps the entire territory. The females arrive one or two weeks later and come to occupy a territory jointly with a male. The female shares in the defence of territory. * * * The male and female of a pair do not coöperate completely in the defence of territory at least at a time before the nest is built. That is, a female during this period possesses an urge to defend a territory to the exclusion of other females, the male to the exclusion of other males. Other males during or preceding nest building are not repulsed from the territory by the female but instead may be acceptable to the female and may be courted. The converse doubtless is true of the male at other periods in the breeding season. Certainly the male before nest construction is tolerant of two females within his territory. At the beginning of nest construction the females pursue and beg from the males, posturing, fluttering the wings, and singing. At this time the males appear to be passive and consistently move away from the advances of the females. Nevertheless, in flight the males may follow after the females.

He demonstrated clearly that the females sing more or less regularly during the early part of the nesting period, and gives a chart showing the difference in the songs of the two sexes.

The utterances of female Bullock's Orioles while in defence of territory and in association with males in every way are comparable to the songs of males and may be considered as true territorial songs. The song of the female is similar to that of the male in rhythm, pitch, and quality except as regards the concluding notes of the song which in the female are slightly harsher in quality, range over lesser intervals of pitch and show important modifications of the rhythm as compared with those of the male. Before or during nest building the songs of females on occasion may be even more abundant than the songs of the males * * *.

The songs of the two females were not identical. * * * The songs of the two males always were extremely similar one to the other. The females sang repeatedly from the ground whereas the males with one or two exceptions sang only while in the trees. The females sang in the trees near their respective males.

DISTRIBUTION

Range.—California and Nevada to northwestern México.

Breeding range.—The lesser Bullock's oriole breeds from central-western and southern California (Santa Rosa, San Jacinto Mountains), southern Nevada (opposite Mohave, Arizona) and central-western Arizona (Colorado River Valley); south to northern Baja California (San Rafael, Colorado Delta) and northwestern Sonora (Colonia, San Luis).

Winter range.—Winter range largely unknown; possibly sparingly in southern California (Los Angeles) and Arizona (Parker), probably in central-western México, south to Guerrero (Chilpancingo); migrants taken in Sonora (San Javier, Tesia, Guirocoba), and Arizona (north to Camp Verde, rarely to Wupatki National Monument).

EUPHAGUS CAROLINUS CAROLINUS (Müller)

Continental Rusty Blackbird

Plate 21

HABITS

To most of the residents of the United States the rusty blackbird is known only as an abundant spring and fall migrant, for its breeding grounds are north of our border, though a few breed in northern New England and the species winters abundantly in the Southern States. Its breeding range extends northward to the limits of trees in northern Alaska and Canada and southward to the central portions of British Columbia, Alberta, Manitoba, and Ontario; it extends across our northern border into northern Maine, New Hampshire, Vermont and, New York. Frederick C. Lincoln (1935) says that "in the Stikine River Valley of northern British Columbia and southwestern Alaska" the rusty blackbird is one of several eastern species that have extended their breeding ranges to within "20 to 100 miles of the Pacific Ocean."

On its breeding grounds, the rusty blackbird seems to show a decided preference for the vicinity of water, the shores of lakes, ponds or streams, or the more or less inaccessible bogs or swamps. Bendire (1895) says:

The Rusty Grackle is much more of a forest-loving species than the other Blackbirds, and during the breeding season it appears to be far less gregarious. Its favorite haunts in the Adirondacks are the swampy and heavily wooded shores of the many little mountain lakes and ponds found everywhere in this region, and here it spends the season of reproduction in comparative solitude. I can state from personal experience that the oölogist who desires to study this species on its breeding grounds must make up his mind to endure all sorts of discomforts; millions of black flies, gnats, and mosquitoes make life a burden during his stay, while the bogs and swamps through which one is compelled to flounder in search of the nest render walking anything but pleasant.

Spring.—Large flocks of rusty blackbirds begin moving northward from their winter range in the Southern States in March, passing through the Northern States mainly in April, and reaching the northern limits of their breeding range in May. Their passage is rather rapid and the route is broadly northward along the Atlantic and Mississippi flyways, though they are sometimes seen in the Great Plains region and there is a northwestward trend in Canada toward Alaska. Some variations from the above very general statement should be noted. Milton B. Trautman (1940) says of the migration at Buckeye Lake, Ohio: "The spring vanguard of the Rusty Blackbird made its first appearance between February 18 and March 2. Its numbers were small until almost mid-March. Then a few days later a sharp increase in numbers took place, and until approximately April 12, from 50 to 3,000 individuals could be recorded daily. There was generally a decrease in numbers shortly after mid-April, and from then until May 5 only 5 to 50 individuals were observed in a day, and never more than 100 were seen. The last transients were recorded between May 8 and 22."

Referring to Manitoba, Seton (1891) writes: "April 15, 1882: Snow still deep everywhere, but melting fast. In the poplars along the slough side to-day was a large flock of Rusty Grackles. * * *

"April 21: The thousands of Grackles have been increased to tens of thousands. They blacken the fields and cloud the air. The bare trees on which they alight are foliated by them. Their incessant jingling songs drown the music of the Meadow Larks and produce a dreamy, far-away effect, as of myriads of distant sleigh bells. Mixed with the flocks of Rusty Grackles now are a few Red-winged Blackbirds."

The spring migration of the rusty blackbirds is spectacular, noisy, and ubiquitous; the birds may be seen in enormous numbers almost anywhere, following the plowman as he cultivates his land, blackening the stubble or grain fields, filling the air in passing clouds, or gathering to sing in the leafless treetops along the roadsides or in the swampy woods and roosting at night in the swamps or sloughs. As Beal (1900) puts it: "One of the most familiar sights to the New England schoolboy, and one that assures him that spring is really at hand, is a tree full of blackbirds, all facing the same way and each one singing at the top of its voice. These are rusty blackbirds, or rusty grackles, which, in their spring journey to the north, have a way of beguiling the tedium of their long flight by stopping and giving free concerts. Every farmhouse by the wayside will have its visitors, and every boy who hears them is eager to tell his mates that he has seen the first flock of blackbirds."

In eastern Massachusetts, according to William Brewster (1906): "The Rusty Blackbird comes to us from the south in early spring about the time when Pickering's hyla begins peeping. The tinkling notes of the Blackbird are, indeed, ever associated in my mind with the bell-like call of the hyla, for at this season the two sounds are usually heard together. Being pitched on nearly the same key, it is not always easy to discriminate them, especially when a score of Blackbirds and several hundred hylas are exercising their vocal organs at once."

Rev. J. H. Langille (1884) gives the following impression of the spring flight:

On the first day of May, 1880, as I stood on an iron bridge crossing a sluggish stream of Tonawanda Swamp, I saw the Rusty Grakles (*Scolecophagus ferrugineus*) constantly trooping by in immense numbers. They were moving in a very leisurely manner, immense detachments constantly alighting. The large tract of low land, covered with the alder, the willow and the osier, seemed alive with them. The sombre wave, thus constantly rolling on, must have carried hundreds of thousands over this highway in a day. Occasionally they would alight to feed in the low, wet fields in the vicinity, making the earth black with their numbers. * * * On being alarmed, either in the fields or in the bushes, these Grakles would rise in a dense, black cloud, and with a rumbling sound like that of distant thunder.

Courtship.—The following brief note by Dr. Charles W. Townsend (1920) is all that I can find on this subject: "The courtship of this bird, if such it may be called, is produced with apparently great effort, wide open bill and spread tail, resulting in a series of squeaking notes suggestive of an unoiled windmill—*wat-cheé e.* At times a sweet lower note, often double, is heard."

J. A. Munro (1947) observed that two males in the top of a tree "performed a simple display that consisted of stretching one wing downward to its full extent, then whistling a single note."

Nesting.—Frederic H. Kennard (1920) spent portions of five seasons in northern Vermont, New Hampshire, and Maine, hunting for nests of the rusty blackbird, and his excellent account of his experiences throws more light on the home life of this species than can be found elsewhere. The first two trips were made too early in the season, the last ten days in May; most of the nests held young birds, though one contained a set of five eggs that was too heavily incubated to be saved; on subsequent seasons he was more successful. He makes the following general statement about the nesting sites: "For sites they seem more apt to choose evergreens, preferably thick clumps of second growth spruce and balsam, though I have found them in dead trees or in clumps of deciduous bushes, button-bush and sweet gale, along the shores of some stream."

He says that they did not breed in colonies, the nearest nests he ever found being "a quarter of a mile apart." His lowest nest "was built about 2 feet up in a little, low black spruce, one of a clump on a floating island, in a swamp caused by raising the waters of a large lake on which it was situated." His highest nest was "about 20 feet up, in a tall, unhealthy looking spruce. It was placed in one of those thick bunches of evergreen twigs that sometimes grow close to the trunk of a spruce, and could not be seen from the ground." All the other nests were much less than 10 feet above the ground or water. One nest was in a dead spruce top that had floated down the stream in the spring floods and become stranded near its mouth. It was only a foot above the surface of the water, in a tangle of *usnea* moss, and so well hidden that "we had paddled by it in our canoe time after time without ever suspecting its presence." Another nest was beside a "brook, in a tangled growth of sweet gale overhanging a ditch, and about two feet above the water." Still another was "about 10 feet back from the edge of the stream, in a thick growth of button-bushes. The nest was placed in a crotch, a couple of feet above the water, just as a Red-wing's would have been." He shows a photograph of a nest, "built in the top of an old stump, standing in the water, out from the shore of a lake."

To illustrate the persistency of these birds in attempting to raise a brood, he took a set of eggs from a nest on May 24; 12 days later, on June 5, he took the eggs from their second nest; the birds built their third nest and laid a set of four eggs within 11 days; he took these eggs, also, but the persevering birds built a fourth nest and were allowed to raise a brood of three young.

Near Red Deer, Alberta, W. E. Saunders (1920) found several nests "in the typical location, *over water.* * * * Exceptions doubtless occur, but I have never found nests of the Rusty other than over water, and Brewer's never very near water."

Kennard (1920) gives the following excellent description of the nest:

In construction, those that I have seen, have all been particularly well built, rather bulky structures, and practically alike. A foundation is usually laid of usnea moss, sometimes in thick masses, and upon this they build their outside frame-work of twigs, usnea, lichens and occasionally a few dried grasses. In one of the nests in my collection the twigs used were mostly dead hackmetack, in another spruce, while in the remainder, twigs from deciduous trees predominated. This frame-work usually becomes thicker and more substantial as it progresses upward.

Within this outside frame they construct a well modeled hollow bowl, between five and one-half and six centimeters in depth, and between eight and one-half and nine and one-half centimeters inside diameter. This bowl, which seems to the casual observer to be made of mud, is in reality made of 'duff,' the rotting vegetable matter with which the ground of this region is covered, and which when dried becomes nearly as hard and stiff as papier mache; and shows their interesting adaptability to conditions, as real mud must at this season be hard to find. A cross-section of the nest shows the bowl to be of varying thickness, but averaging between five and ten millimetres, and so pressed into its surrounding frame as to become, when it hardens, a part of it.

After the bowl has been carefully modeled and smoothed off on the inside, it is lined with fine, long green leaves of grasses that grow in the swamps thereabouts, and is finally topped off with dried grasses and fibres of various sorts, and a few thin, bendable twigs. In recently constructed nests I have found the green lining to be absolutely constant, although as incubation progresses, these grasses, of course, gradually turn brown. The diameter of the nest when finished, just across the outside of the bowl, averages about twelve centimetres, while the diameter of the entire structure, except for a few outreaching twigs, varies from fourteen to twenty centimetres. The usual measurements from foundation to top of bowl are from eight and one-half to nine centimetres.

Bendire (1895) says that a nest taken in Herkimer County, N. Y., "measures 7 inches in outer diameter by 5½ inches in depth; the inner cup is 3½ inches wide by 2½ inches deep. One of these nests will last for several seasons, but a fresh one is usually built every year. These birds are very much attached to their summer homes, returning to them from year to year, and rarely more than two or three pairs nest in one locality; in fact, they are as often found singly."

Eggs.—The set consists of four or five eggs, and one is deposited each day. Bendire (1895) describes them very well, as follows: "The eggs of the Rusty Blackbird are mostly ovate in shape. The shell is strong, finely granulated, and slightly glossy. The ground color is a light bluish green, which fades somewhat with age; this is blotched and spotted more or less profusely, and generally heaviest about the larger end of the egg, with different shades of chocolate and chestnut brown and the lighter shades of ecru, drab, and pale gray. The peculiar scrawls so often met with among the eggs of the Blackbirds are rarely seen on these eggs, which are readily distinguishable from those of the other species."

In a series of 50 sets, reported to me by A. D. Henderson, of Belvedere, Alberta, there are 25 sets of five eggs and 3 sets of six. The measurements of 50 eggs average 25.8 by 18.6 millimeters; the eggs showing the four extremes measure 29.8 by 20.0, 26.7 by 20.1, 23.1 by 17.8, and 25.9 by 16.3 millimeters.

Incubation.—All the information that we have points to an incubation period of about 14 days, performed by the female alone. Kennard (1920) says:

The female usually starts incubation with the laying of the first egg, particularly in early spring, when the weather is cold, and sits pretty close, flying off only upon one's near approach. * * * During incubation the male is very assiduous in his attentions to the female, feeding her frequently, and seldom flies far from the nesting locality. The female at this season is usually seldom in evidence, but by watching the male, one can soon determine by his actions the approximate locality of the nest. He has the very conspicuous habit of sitting on the top of some tall dead stub or tree, often with a nice fat grub in his bill and calling to the female. This call is a two-syllabled "*conk-ee*," very similar to the three-syllabled "*conk-a-ree*" of the Redwing, but clearer and more musical, and usually distinguishable from the notes of the other blackbirds.

If disturbed by the proximity of watchers, he may delay for a while, uttering an occasional "*chip*" of alarm, but sooner or later he will fly close to the nest or to the top of some nearby stub, when the female will fly out to him, and with low "*chucks*" and much fluttering of wings, partake of the delicious morsel he has brought her.

Young.—Kennard (1920) watched a brood of young from the time they hatched until they left the nest; of this brood he writes:

"The young, when hatched, are covered with a long, thin, fuscous natal down; and fed by both parents, at frequent intervals, develop rapidly, as such young birds do. The nest is kept clean, and I saw the female frequently drop a white fecal sac in the nearby brook, as she flew away from feeding her charges. By the fifth day, the primary quills and other wing feathers are well under way, while the growths along the remaining feather tracts are starting; and slight slits begin to show between their eyelids. By the tenth day the young are well covered with feathers, through which some of their natal down still protrudes, and their eyes are nearly but not quite wide open.

A tragedy occurred to the only brood I was able to watch, for on the tenth day after hatching, one of the young was found in the water, about ten feet from the nest, dead and partially eaten. Whether he deliberately climbed from the nest, and later fell into the water, or was taken by some animal, will never be known, but the next day the three remaining young all climbed out into the adjoining bushes, it seemed to me, ahead of schedule time, for their eyes were hardly open, and they were still unable to fly.

They remained in the immediate vicinity of the nest for the next two days, climbing and hopping from bush to bush, with both parents in close attendance, till on the thirteenth day, they had learned the use of their wings; and in the evening the last one was seen to fly across the stream, followed by its mother, and to disappear in the swamp beyond.

Plumages.—As mentioned above, the young when first hatched are covered with long, thin, fuscous down. The sexes are alike in the juvenal plumage, which Dwight (1900) describes as follows: "Whole plumage slate-color washed on back and throat with sepia-brown.

Tail darker with greenish reflections. Tertials and wing coverts edged with Mar's-brown."

A complete postjuvenal molt occurs during the latter half of summer; this produces a first winter plumage, in which the sexes become distinguishable, and which is not very different from that of the adults in the fall. Dwight describes the first winter plumage of the young male as "everywhere lustrous greenish black more or less veiled above with Mar's-brown, below with wood-brown." The illustrations of these plumages in Bird-Lore, vol. 23, No. 6, opposite p. 281, seem to me to be much too highly colored.

Ridgway (1902) adds, in a footnote, the following comment: "The extent of this rusty and buffy coloring varies exceedingly in different individuals, probably according to age. In some (doubtless younger birds) the rusty is nearly uniform on the pileum and hindneck, and forms very broad tips to the scapulars and interscapulars, while the cinnamon-buffy forms a continuous broad superciliary stripe and is nearly uniform over the malar region, chin and throat. Other winter males (probably very old individuals) have scarcely a trace of this rusty and buffy coloring, being quite like summer specimens, except that the plumage is more highly glossed."

There is apparently no prenuptial molt in either young or adult birds, the spring plumage being acquired by the complete, or nearly complete, wearing away of the rusty edgings. Adults have a complete postnuptial molt in summer, beginning the middle of July.

Dwight (1900) says that the first winter plumage of the female "is very like the juvenal but with much Mar's-brown above chiefly on the head and strongly washed below with wood-brown, these colors edging slaty feathers; the lores and auriculars are dull black in contrast. The first nuptial plumage is acquired by wear and later plumages vary little from the first winter."

Food.—Beal (1900) analyzed the contents of 132 stomachs of the rusty blackbird, taken every month in the year except June and July, and reports:

The stomachs contained a larger proportion of animal matter (53 percent) than those of any other species of American blackbirds except the bobolink. This is the more remarkable in view of the fact that none were taken in the two breeding months of June and July, when in all probability the food consists almost exclusively of animal matter. While the birds are decidedly terrestrial in their feeding habits, they do not eat many predaceous ground beetles (Carabidae), the total consumption of these insects amounting to only 1.7 percent of the whole food. Scarabaeids, the May-beetle family, form 2 percent, and in April 11.7 percent. Various other families of beetles aggregate 10.1 percent, largely aquatic beetles and their larvae, which, so far as known, do not have any great economic importance. A few of the destructive snout-beetles (Rhynchophora) are also included, as well as some chrysomelids and others.

Caterpillars constitute 2.5 percent and do not form any very striking percentage at any time, except, perhaps, in May, when they amount to 11.7 percent. Grasshoppers nearly equal beetles in the extent to which they are eaten, and exceed every other order of insects, although none appeared in the stomachs taken in January, March, May, and December, and in February but a trace. In August, as usual, they reach the maximum, 44.3 percent, only a trifle higher, however, than the October record. The average for the year is 12 percent. Various orders of insects, such as ants, a few bugs, and also a few flies, with such aquatic species as dragon-flies, caddice-flies, and ephemerids were eaten in all the months except January, in which only one stomach was taken. They aggregate 13.7 percent of the whole food, but owing to the number of forms no one amounts to a noteworthy percentage, and many of them are of little economic importance. Spiders and myriapods (thousand-legs) are eaten to the extent of 4 percent and amount to 23 percent in August. Other small animals, such as crustaceans, snails, salamanders, and small fish, were found in the stomachs for nearly every month, and amount to 7 percent of the food of the year, but none of them are important from an economic point of view.

The vegetable food consists of grain, 24.4 percent, weed seed, 6 percent, and miscellaneous substances such as a small amount of fruit and a little mast, 16.6 percent of the food of the year. Of grain, corn seems to be the favorite, amounting to 17.6 percent of the year's food and averaging as much as 26.5 percent in 15 stomachs collected in November. "Wheat and oats collectively amount to only 6.8 percent of the year's food. Oats are apparently preferred and in March constitute 15.4 percent of the month's food. These March stomachs came from the Southern States, so it is probable that the grain was picked up on newly sown fields." Weed seed is not an important item, amounting to only 6 percent for the year; its "erratic distribution evidently indicates that weed seed is not sought after, but is simply taken when nothing better is at hand. Miscellaneous items of vegetable food amount to 16.6 percent of the food of the year. Fruit was found in a few stomachs, but does not appear to any important extent. Only three kinds were determined, but several stomachs contained pulp or skin that could not be identified. Several buffalo berries (*Shepherdia argentea*) were found in one stomach, hackberries (*Celtis occidentalis*) in another, and seeds of blackberries or raspberries (*Rubus*) in two or three others. Mast was found in a few stomachs, but the greater part of the miscellaneous food was indeterminable."

Francis H. Allen tells me that the rusty blackbird feeds on the seeds of the white ash. Milton P. Skinner (1928) says that, in North Carolina, "in addition to the seeds, waste grain and insects usually eaten by all blackbirds, the Rusty Blackbirds add fruits from the sour gum in December, January and February, and dogwood berries in January. In February, Rusty Blackbirds feed in cowpea fields on insects, but do not disturb any waste peas that may be present."

Economic status.—From the above study of its food habits it appears that the rusty blackbird is of no great economic importance, either one way or the other.   It does no great damage to agriculture, for the small amounts of cultivated fruits and berries eaten are insignificant; and, although it consumes considerable grain, this is mostly taken as waste grain during the late fall and winter, and does not interfere with harvesting; some newly sown grain may be picked up in the early spring.   On the other hand, as it does not spend the summer in agricultural regions, it cannot be as helpful to the farmer in destroying harmful insects as some other species.   But it does enough good to be worthy of protection.

Behavior.—Mr. Skinner (1928) writes:

Rusty Blackbirds on the ground walk, and run nimbly, with a nodding of their heads forward and backward in time to their own steps.   As compared with other blackbirds, this species is perhaps tamer and certainly more quiet, composed and dignified.   When hunting across the ground, members of the flock are continually walking and running, and frequently individual birds fly a few feet to a position at the front.   While Rusty Blackbirds fly in dense compact flocks all winter, and appear to enjoy the society of other members of their own kind, they are less apt to join other species.   When in flocks composed of several species, the Rusty Blackbirds usually split off into separate flocks composed of their own kind.   But at times they vary this and join flocks of Meadowlarks and Starlings; but on the other hand Starlings, Cowbirds and Red-winged Blackbirds more often join the Rusty Blackbirds.   During the winter these Blackbirds are also seen temporarily with Bluebirds, Juncos, Doves and Horned Larks.

While the flocks of Rusty Blackbirds are more dense and compact than most other species, they are not so much so as those of Red-winged Blackbirds.   A flock in flight moves steadily onward, but the individual birds undulate up and down, or swing from side to side, so that the relative positions constantly change and give the flock a rippling appearance.   They fly either against the wind or with it.   In the latter case, just before alighting on ground or trees they wheel and come up to their perches against the wind.   In its minor points, the flight of these birds is thrush-like.   Rusty Blackbirds are quiet during the winter, but the song also suggests a thrush rather than a blackbird.

Behavior in Ohio during migrations is thus described by Trautman (1940): "During migrations the birds were found most frequently on wet ground or near water.   Many spent the days in the cattail marshes and on the shores of the lake, where they fed while wading in the shallows.   In the inland brushy swamps they also fed in shallow water or on wet ground.   There were flocks about the 'sky ponds' and overflow puddles in fields, especially in early spring, and small groups were along the banks of the streams.   At night all except a few roosted in cattail swamps about the lake, on Cranberry Island, or in the denser and more brushy inland swamps.   Throughout the bird's entire sojourn it was a close associate of the Eastern Redwing, and to a lesser extent of the Bronzed Grackle, Cowbird, and Starling."

Mr. Brewster (1936) tells of a blackbird roost in eastern Massachusetts:

October 4, 1901. * * * A little before sunset I paddled up river to Beaver Dam Lagoon to investigate the Blackbird roost. A good many Rusty Blackbirds had already arrived and others, as well as Cowbirds, were coming almost continuously from every direction (but chiefly from the west) in small flocks or singly. Both species are roosting together in the button bushes and low, dense willows near the head of the lagoon. Into these they pitched headlong, disappearing at once among the dense foliage. They seemed to have no fear or suspicion but sought their roosts without hesitation or loss of time. A few restless birds, however, flitted from thicket to thicket before they finally settled for the night. I counted upward of 175 of which about one half were Rusties and all the others apparently Cowbirds. They made a deafening clamor, keeping it up until nearly dark.

John B. Lewis (1931) relates the following interesting experience:

About noon, November 6, 1930, in company with my friend Mr. J. Frank Duncan I was walking through a tract of partly wooded pasture land belonging to his estate. A flock of 50 or more Rusty Blackbirds (*Euphagus carolinus*) were feeding on the ground farther up the hill in the direction in which we were walking. Suddenly there was a great commotion among the Blackbirds and instantly one of them darted directly toward us, closely pursued by a Sharp-shinned Hawk (*Accipiter velox*). Mr. Duncan and I were side by side and with a space of about two feet between us. In an incredibly short time the Blackbird darted between us screaming at the top of his voice, while the Hawk, who evidently did not see us until within ten feet, frantically checked himself, noticeably fanning our faces, and when within two feet of us swerved to one side and made haste into the woods. When the Hawk began to check his speed he was within a foot of the Blackbird, and with both feet stretched forward to grasp it.

Ruthven Deane (1895) received a letter from his friend, Jesse N. Cummings, of Anahuac, Tex., telling to what extremes these blackbirds will go for food when hard pressed to find it. There had been a heavy snowfall, covering the ground to a depth of 20 inches for a period of 3 or 4 days. An artesian well had kept the ground bare on a small portion of the bay shore, where large numbers of snipe, some robins and other birds had congregated to hunt for food. The letter states: "At this small open piece of ground, the Rusty and Crow Blackbirds had collected, but I did not see them kill many Snipe the first day or two, but the third and fourth days they just went for them. I should say that I saw them actually kill ten or twelve Snipe on the ground where the snow had melted, but there were thirty or forty dead ones that I saw in other places. The Rusty Blackbirds were the principle aggressors, and it was astonishing to see how quickly they could attack and lay out a Snipe or a Robin. Both species were killed while on the ground and the Blackbirds would only eat the head, or as near as I could see, the brain, while the body was left untouched."

Voice.—Aretas A. Saunders contributes the following description of the song of the rusty blackbird, as heard on migration, based on eight records: "There are two types of song, the first a rhythmical alternation of a phrase of two or three notes, with a single higher-pitched note. This song goes on for some time, with indefinite length, and therefore may be considered a long-continued song, like songs of the mockingbirds and vireos. The first phrase is quite musical, the notes rising a little in pitch. The single note of higher pitch is rather squeaky in quality. The whole song sounds like *tolalee*——*eek*—— *tolalee*——*eek*, etc. This song is exceedingly even and rhythmic, the pauses between phrases being just twice as long as the phrases, and in my timed records a phrase and the pause following occupy from four-fifths to one full second.

"The second type of song consists of a rather rapid repetition, two or three times, of a 3-note phrase, rising in pitch. For one such song I wrote, in the field, the sound of the phrases as *kawicklee kawicklee*. This is often repeated at intervals, but less rhythmically than the first song. In one case the bird called a short *kick kick kick* between the songs.

"The pitch of rusty blackbird songs varies from A ' ' to D ' ' ' ' The high squeak in the first type of song is usually pitched on C ' ' ' ' or D ' ' ' ', the highest note of the piano or just above it, while the other phrases may begin anywhere from two tones to an octave lower.

"Singing on the spring migration is to be heard in Connecticut in March or April. My average date for the first song heard is March 19 and for the last April 16. The earliest song heard was on March 2, 1930, and the latest May 2, 1939. Three times, in my experience, I have heard rusty blackbirds sing in the fall: October 13, 1935; October 31, 1937; and October 12, 1945.

"Call notes I have heard are a short *kick*, not so loud as the *chack* of the redwing, and a rattle like *turururo*."

Francis H. Allen has sent me the following study: "The *chuck* note of this species, as I hear it, is rougher than that of the redwing, though much less rough than that of the grackle, as well as higher pitched than the latter.

"On April 17, 1938, in West Roxbury, Mass., I took rather careful notes on the song of the rusty blackbird. I watched one for a long time at close range. It sang pretty constantly in a willow over a brook and used the two phrases I have been familiar with, but not always in regular alternation as is commonly the case. The more familiar phrase I syllabify as *wisslter-ee*. This phrase would be repeated over and over, but frequently a phrase with the final *ee* on a lower pitch would be interpolated. This latter phrase was never repeated until at least one of the former had intervened. It was always followed immediately by the phrase first mentioned, with a shorter interval than between the repetitions of that phrase or between that and a following low-pitched one. The phrase with the low-pitched final note began with a higher pitched *wisslter* than that of the other. The 'shuffling' notes, always present in the rusty blackbird's song, seemed more liquid and less rustling, heard at this close range, than I have before considered them. For my own immediate purposes I syllabified the two phrases roughly as *oodle-a-wee, eedle-a-woo*. The order, however, should probably be reversed, so that a continuous performance might go like this: high-low low-high, high-low low-high, etc. The commas indicate a longer rest than the blanks. If I numbered the *eedle-a-woo* phrase as 1 and the *oodle-a wee* phrase as 2, the succession would then be: 1–2, 1–2, 1–2, 2, 2, 2, 2, 2, 1–2, 1–2.

"What is of special interest is the fact, which I observed many times, that the tail was spread with many phrases, but was spread wider with No. 1 than with No. 2. The width of the spread was relative, not absolute."

Field marks.—The rusty blackbird is not always an easy bird to identify in the field. In spring the migrating flocks may easily be confused with the early flocks of male redwings, for at that time the latter often show little or no red on the wings when perched, and might be mistaken for rusties. The females of the two species are not at all alike, and their habits are different.

In the fall, the rusty blackbirds deserve their name, as the black plumage of the males and the dark plumage of the females are both more or less veiled with the rusty edgings, and this is much more conspicuous in the younger birds.

In the Central-Western States, this species is even more difficult to distinguish from Brewer's blackbird. The latter has a thicker bill at the base and a purplish black head, which the rusty does not have. In the fall, the rusty blackbird is much more extensively rusty than is the Brewer's.

Enemies.—The narrow escape of a terrified rusty blackbird from a sharp-shinned hawk, as related above, shows that these blackbirds recognize the accipitrine hawks and probably the larger falcons as deadly enemies.

As the rusty blackbird breeds mainly north of the area where cowbirds are abundant, it is seldom imposed upon by these birds, and being larger, would probably not be a very satisfactory foster parent to this parasite.

Friedmann (1934) reports: "Mr. T. E. Randall found two nests of this bird in Alberta, each with eggs of the Nevada Cowbird. Mr. A. D. Henderson writes me that he found the species victimized in Alberta. These are the first records for this bird as a molothrine victim."

Harold S. Peters (1936) recorded one louse, *Myrsidea incerta* (Kell.), as an external parasite on this blackbird.

Fall.—Although not early migrants, the rusty blackbirds desert their breeding haunts as soon as the young are able to fly and to feed themselves. According to Kennard (1920) this occurs about the middle of July in northern New England; they are no longer seen in solitary pairs, but "again become gregarious, and are seen in small flocks, flying high overhead, between the lakes, or feeding along their shores, getting ready for their southern migration."

The fall migration begins early in September, but is not in full swing until October, when the birds are pouring through the northern States in immense flocks; the flight continues through November in diminishing numbers, and a few birds linger into December. Tufts tells me that the average date, over a 4-year period, when the species was last seen at Wolfville, Nova Scotia, is October 17. He has one record for early winter, December 16, 1921, "when a bright-colored male was seen feeding on the main highway in company with three blue jays. There was snow on the ground at the time."

When the flight is well underway it is sometimes quite spectacular. Wendell Taber tells me that, at Lynnfield, Mass., on October 10, 1937, he saw a migrating flock that extended for at least a mile, or as far as he could see; the birds were flying southward on a broad front extending from east to west; the wave was from 20 to 40 birds deep from vanguard to rearguard, and only one bird deep vertically.

Brewster (1906) writes:

In autumn Rusty Blackbirds are most numerous in the Cambridge Region during the month of October, when roving flocks may be found quite as often in upland fields and pastures as in the lowlands. Wherever they find a field of ripening corn—whether of the yellow, or the sweet, variety—they are sure to visit it almost daily, from the time of their first arrival to that when the last stalks are harvested by the farmer. Early in the season they puncture the kernels and suck out the pasty contents, and after the corn has hardened they sometimes swallow it whole. During the greater part of October they may be seen associating with Robins in "cedar pastures" or even with Blue Jays in oak and chestnut woods. Indeed there are few places in our country districts which they do not visit occasionally at this season. At evening the scattered flocks all fly to the swamps, sometimes congregating in considerable numbers to spend the night together.

During the fall migration, in October, these birds sometimes gather in large numbers in the tall deciduous trees, oaks, walnuts, maples, and elms that form a dense grove of thick foliage along a stream that flows past my back yard, close to the center of the city and within a stone's throw of brick buildings. Scores of them pour in after sunset in loose, scattering flocks, and move about chattering in the trees, or drop down to the banks of the stream to feed or drink. But they never spend the night here; they are always restless and active, and they move away before darkness comes, to find some other roosting place for the night.

Winter.—Most of the rusty blackbirds spend the winter in the Southern States, but there are several records of individuals, or even small flocks, surviving the rigors of our northern winters. A Nova Scotia record has been mentioned. John C. Phillips (1912) gives several winter records for Massachusetts and tells of seeing a flock of 18 that spent the whole of a severe winter in Essex County. "They were getting most of their food, apparently, from a large pile of horse manure."

From Alberta, Frank L. Farley (1932) writes: "Eleven Rusty Blackbirds spent the entire winter of 1919–20 in the stockyards in Camrose. On November 6th, 1919, the thermometer registered 24 below zero. Towards the end of January the cold was intense, the mercury on several occasions dropping to 55 below zero, yet the blackbirds appeared to get along just as well as the snow-bunting with which they fed."

At Buckeye Lake, Ohio, according to Trautman (1940), "wintering individuals fed about the water as long as it was free of ice, but whenever the lakes, ponds, and streams were ice-covered, they were to be found in fields of uncut corn or of rank weeds near brushy thickets. Wintering birds roosted in cattail marshes and in the denser and more brushy inland swamps."

Skinner (1928) says that, in North Carolina, "during the winter from November to February there was a flock of fifty Rusty Blackbirds almost constantly about the fields near the Pinehurst Dairy. This flock was composed of both sexes, but began to split up and scatter about the first of March. Although these birds were usually on the ground, they often alighted on low trees—oaks, pines, gums, dogwoods and sycamores—and on board fences and the wires and posts of wire fences. Occasionally they are seen on race-courses or golf links, and often about streams or the thickets over streams. Still it is quite noticeable that these birds prefer the uplands with other blackbirds more than any other locality."

In South Carolina, "great numbers of Rusty Blackbirds frequent the rice plantations in winter, associating with Florida Grackles (*Quiscalus quiscula aglaeus*) and Boat-tailed Grackles (*Megaquiscalus major*) where stacks of rice have been left in the fields," according to Arthur T. Wayne (1910).

DISTRIBUTION

Range.—Alaska and Canada, south to Texas and the Gulf coast.

Breeding range.—The continental rusty blackbird breeds from northern Alaska (Kotzebue Sound, Barrow, Fort Yukon), northern Yukon (Porcupine River at Alaska boundary, King Point), northwestern and central Mackenzie (mouth of Peel River, Pikes Portage), northern Manitoba (Churchill, York Factory), northern Ontario (Fort Severn, Lake River Post), northern Quebec (Fort Chimo), and central Labrador (Nain, Makkovik); south to south-central Alaska (Bethel, Fort Egbert), central and northeastern British Columbia (between the Rocky Mountains and coastal ranges: Atlin, Nulki Lake), south-central Alberta (Calgary, Red Deer), central Saskatchewan (Big River, Emma Lake), central Manitoba (probably Oxford Lake), western and southern Ontario (Savanne, Bruce County, Algonquin Park), and southern Quebec (Inlet), through the northern Appalachians to northeastern New York (Raquette Lake, Long Lake), northern Vermont (Franklin, Saint Johnsbury), northern New Hampshire (Averill, Lake Umbagog), central-western, central, and southeastern Maine (Oxford, to Washington counties), and southern New Brunswick (Scotch Lake).

Winter range.—Winters casually north to southern British Columbia (Okanagan Landing), central Alberta (Camrose), southern Saskatchewan (Eastend), southern Manitoba (Portage la Prairie), central Minnesota (Fosston, Elk River), southern Wisconsin (Madison, Waukesha), southern Michigan (Kalamazoo, East Lansing), southern Ontario (Kitchener, Reaboro), central and southeastern New York (Geneva, Rhinebeck), central New Hampshire and southern Maine (Falmouth, Calais); south casually to central Colorado (Loveland, Colorado Springs), central and southeastern Texas (Abilene, Seabrook), the Gulf Coast, and northern Florida (Cedar Keys, New Smyrna).

Casual records.—Casual in southwestern and southeastern Alaska (Nushagak, Kodiak Island, Wrangell), California (Amador County, Santa Rosa and San Clemente Islands, Jamacha), Idaho (Potlatch), Arizona (Grand Canyon, Tucson), and western Texas (Alpine).

Accidental in Siberia (Indian Point), Alaskan islands in Bering Sea (Saint Paul, Saint Lawrence), Baja California (Valladeres) and Greenland (Fiskenaes, Fredrikshaab).

Migration.—The data deal with the species as a whole.

Early dates of spring arrival are: North Carolina—Raleigh, February 10 (average of 10 years, March 2). Virginia—Naruna, February 14. West Virginia—Bluefield, February 8. Maryland—Laurel, Feb-

ruary 25 (median of 7 years, March 18). Pennsylvania—Berks
County, February 10. New Jersey—Milltown, February 23. New
York—Mastic, February 16; Geneva, February 21. Connecticut—
Fairfield, March 2. Rhode Island—Providence, March 17. Massa-
chusetts—Belmont and Concord, February 20. Vermont—Rutland,
March 11 (average of 13 years, April 3). New Hampshire—East
Westmoreland, March 5. Maine—Hebron, March 9; Ellsworth,
March 10. Quebec—Quebec, March 31. New Brunswick—Scotch
Lake, March 22 (median of 26 years, April 7). Nova Scotia—Wolf-
ville, March 20 (median of 8 years, March 25). Prince Edward
Island—North River, March 31. Newfoundland—St. Anthony,
April 28. Greenland—southwest Greenland, March 8. Arkansas—
Winslow, February 21. Tennessee—Knoxville, March 1. Ken-
tucky—Russelville, March 11. Missouri—Jasper City, February 20.
Illinois—Toulon, February 22; Chicago, February 25 (average,
March 15). Indiana—Worthington, February 25. Ohio—Toledo,
February 19; Buckeye Lake, February 22 (median, February 28).
Michigan—Three Rivers, February 27; Blaney Park, March 18.
Ontario—London, March 14 (average of 10 years, March 29); Ottawa,
March 19 (average of 28 years, April 20). Iowa—Winthrop, Febru-
ary 29. Wisconsin—North Freedom, March 7. Minnesota—
Owatonna and Wilder, March 10 (average of 14 years for southern
Minnesota, March 23); Fosston, March 16 (average of 10 years for
northern Minnesota, March 24). Kansas—Topeka, February 12.
Nebraska—Red Cloud, February 12 (median of 8 years, March 1).
South Dakota—Aberdeen, February 20. North Dakota—Fargo,
March 21 (average for Cass County, March 29). Manitoba—Trees-
bank, March 19 (median of 53 years, April 6). Saskatchewan—
Densmore, April 4. Alberta—Glenevis, April 3. British Columbia—
Atlin, April 14. Yukon—west of Dawson, May 2. Alaska—Kalskag,
April 10; Fairbanks, May 1.

Late dates of spring departure are: Florida—Gainesville, April 14.
Alabama—Decatur, May 15. Georgia—Athens, May 7. South Carolina—Greenwood, April 29. North Carolina—Raleigh, May 9
(average of 6 years, April 17). Virginia—Naruna, May 6. District of Columbia—May 14 (average of 27 years, April 18). Maryland—Laurel, May 10 (median of 7 years, April 26). Pennsylvania—
Renovo, May 22. New Jersey—Morristown, May 18. New York—
Long Island, June 3; Rochester, May 23. Connecticut—Norwalk,
May 15. Massachusetts—Northampton, May 28. Vermont—St.
Johnsbury, May 24. Maine—Ellsworth, May 21. Quebec—Montreal, May 28. Louisiana—New Orleans, May 10. Mississippi—Bay
St. Louis, April 25. Arkansas—Winslow, May 1. Kentucky—Dan-
ville, May 1. Missouri—St. Louis, May 1. Illinois—Chicago region,

May 16. Indiana—Bloomington, May 16. Ohio—central Ohio,
May 31; Youngstown, May 27. Michigan—Sault Ste. Marie, May
26; Detroit area, May 12. Ontario—Ottawa, May 24. Iowa—
Sigourney, May 17. Wisconsin—Winneconne, May 12. Minnesota—
Cloquet, May 25; Minneapolis, May 17 (average, May 8). Texas—
Austin, May 4; Dallas and Houston, April 28. Oklahoma—Tulsa
County, April 29. Kansas—Blue Rapids, May 5. South Dakota—
Aberdeen, May 16. North Dakota—St. Thomas, May 20.

Early dates of fall arrival are: North Dakota—Fargo, September 5.
South Dakota—Arlington, October 5. Kansas—Lawrence, September 13. Oklahoma—Kenton, September 18. Texas—Decatur, October 12. Minnesota—Iron Junction, August 20 (average of 6 years
for northern Minnesota, September 12); Minneapolis, September 8
(average of 12 years for southern Minnesota, September 21). Wisconsin—Ladysmith, September 10. Iowa—Marshalltown, August 27;
Grinnell, September 13. Ontario—Lake Nipissing, August 18–19;
Hamilton, September 13. Michigan—Blaney Park, August 8; Charity Islands, September 13. Ohio—Lucas County, August 23; Oberlin,
September 10. Indiana—Richmond, September 15. Illinois—Chicago region, September 5 (average, October 1). Tennessee—Athens,
September 29. Arkansas—Winslow, October 14. Mississippi—Saucier, November 8. Louisiana—Covington, November 17. New
Brunswick—Grand Manan, August 23. Quebec—Anticosti Island,
September 14. Maine—Livermore Falls, September 10. New Hampshire—New Hampton, September 8. Vermont—St. Johnsbury, Sep-
tember 2. Massachusetts—Springfield, September 10. Rhode Island—Providence, September 22. Connecticut—New Haven, Sep-
tember 11. New York—Brooklyn, September 1; Geneva, September
5 (average of 7 years, September 28). New Jersey—Passaic, Septem-
ber 29. Pennsylvania—Renovo, September 14. Maryland—Laurel,
October 1 (median of 5 years, October 20). District of Columbia—
September 16 (average of 18 years, October 22). West Virginia—
Bluefield, September 21. Virginia—Blacksburg, October 11. North
Carolina—Raleigh, October 14 (average of 11 years, October 26).
South Carolina—Clemson College, October 19. Georgia—Atlanta,
September 20; Fitzgerald, October 18. Alabama—Autaugaville,
October 19. Florida—St. Marks, October 17; Everglades National
Park, October 20.

Late dates of fall departure are: Alaska—Wrangell, November 30;
Point Barrow, October 24. Yukon—west of Dawson, September 17.
British Columbia—Okanagan Landing, December 5; Metlakatla,
November 26. Alberta—Glenevis, December 2. Mackenzie—Fort
Simpson, October 13. Saskatchewan—Camrose, December 10; East-
end, November 26. Manitoba—Treesbank November 28 (median

of 46 years, November 2). North Dakota—Jamestown, November 17. South Dakota—Sioux Falls, December 8. Nebraska—Lincoln, November 25. Kansas—Newton, November 19. Minnesota—Minneapolis, December 9 (average, November 14); St Vincent, November 29 (average of 9 years for northern Minnesota, November 18). Wisconsin—Green Bay, November 25. Iowa—Sigourney, December 9. Ontario—Plover Mills, November 30; Ottawa, November 5 (average of 25 years, October 17). Michigan—Isle Royale, November 30; Kalamazoo, November 22. Ohio—Canton, November 30; Buckeye Lake, November 24 (median, November 21). Indiana—Sedan, November 25. Illinois—Chicago, November 28. Missouri—Concordia and Jasper City, November 26. Kentucky—Versailles, November 20. Tennessee—Athens, November 21. Newfoundland—Tompkins, October 4. Prince Edward Island—Mount Hubert, October 17. Nova Scotia—Yarmouth, October 27. New Brunswick—St. John, November 10; Scotch Lake, November 1 (median 16 years, October 16). Quebec—Anticosti Island, December 4; Montreal, November 8 (average of 9 years, October 24). Maine—near Portland, November 5; Ellsworth, November 4. New Hampshire—Winchester, November 20. Vermont—Burlington, November 24. Massachusetts—Martha's Vineyard, December 2. Rhode Island—Providence, November 25. Connecticut—New Haven, December 13. New York—Orient, December 8; Schenectady, November 23. New Jersey—Englewood, December 19. Pennsylvania—Jeffersonville, December 9; Pittsburg, November 26. Maryland—Laurel, December 28 (median of 4 years, December 4). West Virginia—Bluefield, December 6. Virginia—Blacksburg, November 28. North Carolina—Raleigh, December 16 (average of 8 years, November 17).

Egg dates.—Alberta: 53 records, May 15, to June 30; 39 records, May 21 to June 6.

Alaska: 10 records, May 25 to June 26; 5 records, June 3 to June 19.

Maine: 17 records, May 18 to June 16; 10 records, May 24 to May 29.

New York: 8 records, May 7 to May 27.

Nova Scotia: 7 records, May 10 to May 21.

EUPHAGUS CAROLINUS NIGRANS Burleigh and Peters

Newfoundland Rusty Blackbird

HABITS

Similar in its habits to the continental race, this bird shows a preference for the vicinity of water in choosing its breeding spots. Thus, C. J. Maynard (1896) writes, "There are spots on the Magdalen Islands which might rightly be termed sloughs, for they are perfectly inaccessible as the surface, although apparently solid, is in reality so thin that it will not bear the weight of a dog. This floating mass of vegetation, however, supports bushes and in some cases small trees, all of which grow very thickly together. I had observed blackbirds about them on several occasions, but as they kept well in the center of the large tracts, I could not make out at first what they were but after a time found that a large colony of Rusty Grackles were evidently building in one of the above described places." Peters and Burleigh (1951) found the rusty blackbird in Newfoundland stayed about boggy areas of stunted spruce, around woodland pools, margins of ponds and streams and in wet lowlands with heavy underbrush. They migrate in flocks, and nest in small groups of several pairs. When they are disturbed they all fly up into a tree, facing the same direction."

Spring.—Robie W. Tufts writes that the average date of arrival in Nova Scotia over a 10-year period is March 24.

Nesting.—My personal experience with the nesting habits of the Newfoundland rusty blackbird has been limited to two northeastern localities, both of which were quite typical of the species as a whole. On June 18, 1904, near East Point in the Magdalen Islands, we found a colony of these birds nesting among the boggy pond holes and treacherous floating bogs such as those described by Maynard. In the spruce thickets along the edges of these bogs the blackbirds were abundant. Their nests were well concealed at moderate heights in the thickest spruces. The young were by that time all out of the nests and mostly able to fly; their anxious parents were very noisy and solicitous, flying about us, scolding and chirping in great distress.

On a later date, June 19, 1921, my companion, Herbert K. Job, returned to the same general locality and collected for me a nice set of four fresh eggs, which was probably a second laying; the nest was 10 feet up, at the top of a broken-off spruce in a damp pasture thickly overgrown with young spruces.

On June 17, 1912, while we were exploring some extensive marshes along the Sandy River in central Newfoundland, my guide found a rusty blackbird's nest containing four young birds about 2 or 3 days old; the nest was only 3 feet up in a small, bushy red spruce in a bog, where there were other small spruces scattered about; the nest was made externally of fir and spruce twigs, internally of dry grasses, and neatly lined with fine grasses. I have four sets of eggs in my collection, taken for me by J. R. Whitaker near Grand Lake, Newfoundland, at dates ranging from May 3 to June 10; the nests were all placed in spruces at heights ranging from 5 to 9 feet.

Robie W. Tufts has sent me the following nesting data for Nova Scotia: "My earliest record for fresh eggs is May 12, 1905, on which dates two females were found sitting on their respective nests, which contained four fresh eggs each. These were collected. Next day both these birds were seen building new nests nearby. On May 23 and 24, respectively, five eggs were taken from each of their nests. On May 12, 1921, five eggs far advanced in incubation were collected; this would suggest that the nest contained fresh eggs about May 1. The average date, however, for fresh eggs down the years has been about May 14."

Eggs.—The eggs are similar to those of the mainland race, light bluish green, spotted with brown and gray; a set consists of four or five.

Plumages.—The molts and plumages follow the pattern of the mainland race.

Food.—The food of the Newfoundland rusty blackbird is in every way the same as that of its mainland form. Peters and Burleigh (1951) noted, in Newfoundland, that the birds fed along the shores of ponds and bogs, even wading at times in the shallow water. "They feed upon many kinds of insects, worms, crustaceous and other small animal life, and also upon seeds of weeds and grains."

Voice.—Of its voice, which is similar to that of the better known mainland form, Peters and Burleigh (1951) write: "Their so-called song resembles nothing more than several rusty hinges being opened and closed, and it is far from musical."

Field marks.—"A rather short-tailed, black bird, slightly smaller than a robin," according to Peters and Burleigh (1951), "in fall it becomes rusty above and brownish below."

Winter.—The Newfoundland rusty blackbird has been found in winter in South Carolina (at Mount Pleasant, January 16, and at Huger, February 13 and 26), North Carolina (Asheville, March 18 and April 7), Georgia (Sherwood Plantation, December 25) and Virginia (near Fairfax, November 19).

DISTRIBUTION

Range.—The Newfoundland rusty blackbird breeds in the Magdalen Islands, Nova Scotia (Halifax, Barrington), and Newfoundland. It has been recorded in winter in North Carolina (Asheville) and Georgia (Grady County).

EUPHAGUS CYANOCEPHALUS (Wagler)

Brewer's Blackbird

Plates 22 and 23

Contributed by LAIDLAW WILLIAMS

HABITS

At least two proposals have been made to divide this species, using the names *E. c. minusculus* and *E. c. aliastus*, but neither of these subspecies has been accepted by the A. O. U. committee on nomenclature.

As Brewer's blackbird is found in large, conspicuous flocks in open places, often close to human habitations, it is a familiar bird throughout a considerable part of western North America. Although it occurs throughout the year in areas such as farming districts and even in villages and towns, this species also resorts to higher elevations where it nests remote from man.

The Brewer's blackbird has profited by human alteration of the environment. A large part of its time is spent perching on electric wires, where it rests, preens, calls, displays, and uses the wire as a guard perch during breeding activities. This bird forages extensively on lands that have been converted from brush or forest to pasturage, and on freshly plowed soil; it eats some grain (usually waste); and frequents golf courses, lawns, and irrigated areas. Such advantageous conditions possibly contribute to the increase of this species. Dawson and Bowles (1909) say that in Washington it has profited by human settlement of the land and by the spread of cattle; and Kennedy (1914) says that in the Yakima Valley the bird has "prospered greatly" due to irrigation. Grinnell and Miller (1944) state that, in some areas in California, it "apparently has increased as a result of human occupation of the land." The Brewer's blackbird seems to have been extending its range eastward in recent years, and it has now been recorded as a breeding species in Ontario, eastern Minnesota, Wisconsin, and Illinois. What seems to be the first published record for Ontario, of both occurrence and breeding, was made by Allin and Dear (1947); on June 14, 1945, a male was collected and a nest with young found in a cleared area near Port Arthur. The male was taken in a colony of eight birds, including a brown-eyed female, that occupied 8 acres. Concerning the bird's eastward extension in Minnesota, Roberts (1932) says that it is "one of several birds that have extended their ranges eastward across the state in comparatively recent years." It has been abundant in the Red River Valley "since the earliest records for that region; the first nesting colonies in the eastern part of the state were discovered at Minneapolis in 1914.* * *

Previous to that time it was either not present or so rare as to have escaped observation. Now it is a common summer resident, breeding in colonies throughout the state; absent or rare as a nesting bird in most of the southern counties."

"The presence of the Brewer's Blackbird in Wisconsin prior to the years 1926 and 1927 was rare," writes Schorger (1934), who remarks that "the recent extension of its range is quite remarkable." The year 1926 marked the beginning of the influx, and Schorger says that "it is now possible to state that Brewer's Blackbird is at present a common summer resident, breeding in a narrow area extending from Polk County in the northwest, to Walworth County in the southeast."

I (1952) carried on a behavior study of this species principally at a breeding colony, "the river-mouth colony," at the mouth of the Carmel River, Monterey County, on the central coast of California, for six breeding seasons from 1942 through 1947, with check observations in 1948. The colony is situated at the edge of a marsh. The birds nest in Monterey pines (*Pinus radiata*) which, although native to the region, are planted along the streets of a subdivision adjacent to the marsh, on what was originally chaparral land. Although the birds forage on lawns, streets, and food-trays, they spend a large part of their time on the marsh area, undisturbed during the years of this study. Tules (*Scirpus*), which grow in patches on the edge of the marsh, as well as the pines, are used for roosting and daytime resting places. Electric light wires and poles along the streets are used for perching by the flock as well as for display and guard perches by individual birds.

In the study, 318 Brewer's blackbirds, 158 males and 160 females, were color banded. Over the period of study, 117 marked birds bred in the colony; in addition to these there were 8 birds of each sex entering into the breeding activities of the colony that I was unable to band. Many of the remaining 201 banded birds were found at other colonies in the region at various seasons, and some of them were found breeding at those places. The banding station was maintained throughout the year near a house in the center of the colony.

The breeding period, which extends roughly from the end of January into July, may be divided into the following phases of activity: 1. segregation and assortment into pairs—pair formation; 2. nest-building, copulation, and egg-laying; 3. incubation; 4. nestling care; 5. fledgling care.

Spring and courtship.—In phase 1 old pairs (that is, pairs returning from the previous year) reassociate and new pairs are formed. By "new" is meant some combination involving either young birds in their first year, birds banded during the season (ages could not be determined after the fall molt), or, rarely, a recombination of individuals that had previously bred in the colony.

My study at the river-mouth colony revealed that there is no period of male isolation as in a typical territorial species. The flocking behavior of fall and winter gradually gives way as the birds associate more and more in pairs. The activities of the pair are not confined to any territory (except that later on there is a focus of attention at the nest site) and the birds may at first carry on pairing activities while grouped together in a flock.

A number of displays and accompanying calls are used throughout the breeding season.

The RUFF-OUT is employed by both sexes, but more frequently by males. The bird holds the bill nearly horizontal, or pointed somewhat upward and ruffs out many of the contour feathers, especially those of the head and neck, the breast, and upper tail coverts (the rump feathers remain flat); at the same time it partially spreads the wings downward and fans and depresses the tail. As the ruffing and spreading reaches a climax either a *squeee* or *schl-r-r-r-up* is uttered and the display immediately subsides. The notes are never uttered without the accompanying display, though the latter varies greatly in the extent of ruffing and spreading. When used by the female it is less developed and the utterance is more subdued. The whole display lasts only a second or two. It functions as a threat, but is also used in mutual display between pairs, as described below. As a threat it is much more frequently used by the male.

The MALE PRE-COITIONAL DISPLAY, employed by the male immediately preceding his mounting the female, is a more exaggerated form of RUFF-OUT, but the bill is pointed downward and the display held longer. If it is performed on the ground the wings and tail may actually scrape the earth as the male approaches the female. Sometimes, in this display, the male struts in a half-circle in front of the female before mounting.

In the FEMALE GENERALIZED DISPLAY, the bill is held upward at a slight angle; no feathers are ruffed out; the wings are held somewhat out from the body, drooped, and vibrated; the tail is cocked but not spread. This display is always accompanied by a series of *kit* notes. The display has a definite attracting effect on the male and also is part of the female's response to his advances.

The FEMALE PRE-COITIONAL DISPLAY, or "copulatory invitation," should probably be considered a more fully developed female generalized display, which it resembles, except that the body is tilted forward and the tail cocked at a steeper angle. It is accompanied by a specific series of soft, low, tapping notes. Before the male mounts the wing quivering ceases and the female's body becomes rigid.

The MALE ELEVATED TAIL DISPLAY is similar to the female generalized display in body appearance, wing action, and tail cocking, but with the tail somewhat spread and the wings possibly held out a bit wider. This is accompanied by the series utterances *chug-chug-chug* (see p. 325). It is possible that this display may function as an invitation to the female or an indication of the male's receptive state. It is never addressed to another male and has no significance as a threat.

The HEAD-UP DISPLAY is used regularly by males, rarely by females. With the bill pointing nearly vertical the body is drawn upward without ruffing the feathers. In its fullest development the bird has a slim, drawn-out appearance. This display is held for an appreciable length of time, unlike the momentary ruff-out, and there is no accompanying call. It functions as a threat.

All these displays except the pre-coitional ones have been observed in more or less rudimentary form in the flock during the non-breeding season. The threat displays (ruff-out and head-up) are used in disputes over food and in other aggressive situations at all times of year.

Activities that I have termed "pairing behavior" are as follows:

Pairs WALKING TOGETHER segregate from the flock and forage together, usually keeping within a few feet of each other.

Pairs, either isolated or in the flock, perch on wires or on the tips of pine boughs; if on the wires the distance between male and female is quite regularly about 18 inches. There they indulge in MUTUAL DISPLAY, exchanging the ruff-out, with *squeee* and *schl-r-r-r-up* notes. They may keep this up for several minutes. It usually ends by their flying off together, the female usually taking flight first and the male immediately following; sometimes the takeoff is nearly simultaneous.

When perched on the wires the male may hop at, or dart toward, the female. This action, THE DART, is often preceded by the female assuming the generalized display, or it may cause her to assume it; or the female may respond with the ruff-out.

Frequently, in response to THE DART, the female may fly off, the male in close pursuit, in a more or less circular flight. In this action, THE CHASE, a third bird or even a second pair, often joins.

At the river-mouth colony, Phase 1 usually begins toward the end of January or in February, when a change from winter behavior can be detected. Instead of spending a large part of the day in long foraging expeditions away from the colony area, the flock remains longer in the vicinity of the colony. Pairs sort themselves out; a pair may perch, isolated, on the wires or on the tips of pine limbs and engage in mutual display, then fly to the ground to forage together. Other pairs may be feeding there also. Although the flock may be all together, pairs can be detected within the group walking together and maintaining a fairly constant interval between the male and the female, which is less than the distance to the next pair. Suddenly they all flush and the pairs bunch together into a flock and fly off. Often the flock circles about and alights on the wires. If the members of a pair do not happen to light together individuals shift their positions on the wire until the flock is sorted pair by pair. Thereupon mutual display starts again.

Individual pairs vary as to the date when they commence to act as a pair. Also there is a gradual increase in the time that pairs spend in segregation and a corresponding decrease in the time spent in the flock. Although pairing behavior may start as early as the end of January, there are brief intervals in the day when the birds revert to flock formation even as late as April.

Constancy in pairing behavior with the same mates is also arrived at gradually. Members of old pairs are less frequently involved in pairing behavior with individuals other than the "proper" mate and thus they tend to be constant from the beginning. New pairs, however, perform pairing behavior with other birds in many instances until finally the "true" pair forms and remains constant through the remainder of Phase 1. This is not to say, however, that after pairs are formed, mated males do not respond to the displays of a neighbor's female, as indeed they do. Females perform the generalized display more and more frequently as Phase 1 progresses, even though copulation is in the future. The intruding male responds to this display by approaching in the ruff-out or even pre-coitional display, or he may just walk or fly toward the female. But the female is constantly guarded by her mate, who drives off the intruder, either by flying directly at him, or walking deliberately head-up, between the female and the intruder. The latter responds with the head-up display and both, still holding this posture, walk stiffly abreast of each other away from the female. Sometimes both males shift to the ruff-out and remain facing each other, exchanging this display. When they do, this the action looks very like a pair in mutual display (there is no head-up in mutual display, however). Rarely, a fight occurs when both males flutter up together and peck and claw at each other.

After the pair is formed the members are almost always together, becoming separated only for brief intervals. The male guards his mate from the approaches of other males with increasing constancy.

In Phase 1 there is some toying with nesting material by both sexes and even carrying it to a site. The male of the pair is sometimes the first to hold nesting material in the bill, but he rarely places it at a nest site. Actual nest construction is accomplished almost entirely by the female, and not until Phase 2 commences. Bendire (1895) says that "both sexes assist" in nest construction, but at the river-mouth colony the male's activity with nesting material is almost entirely functionless as far as the actual construction is concerned.

In Phase 1 there is considerable aggressive behavior and even fighting for the possession of nest sites, even though actual nest con-struction is still to come. This fighting is largely between females. A fight between females at a nest site usually brings a response from their mates, who alight nearby but do not always act belligerently at first. The approach of the males may cause one or both females to assume the generalized display, and the males then tend to guard their mates.

Pairs acting as a team will defend a nest site, but in such cases it is more often the female rather than the male who initiates the attack. Males will, however, defend the nest site without the female being present.

Although Phase 1 may start as early as the third week of January, actual nest construction, copulation, and egg-laying for the first brood (Phase 2) does not commence earlier than April, usually not until the second or third week (the earliest observed copulation: April 6, 1945). Thus Phase 1 of the first cycle may be stretched out for as long as 12 weeks. Phase 1 of subsequent cycles is exceedingly brief and can possibly be considered absent.

Males were both monogamous and polygynous; an individual might be polygynous one year and monogamous the next. The number of breeding males in the colony varied from 13 to 31; females from 14 to 36. Polygyny varied from one polygynous male in 1943, when the population was 13 males to 14 females, to 12 polygynous males in 1947 when the population was 18 males to 36 females. The number of mates per polygynous male was generally two; but in 1946 six males has two females each and one had three. In 1947 seven males had two, four had three, and one had four females.

When polygyny occurs it usually comes about in the following fashion: When the female is incubating (Phase 3) the male does not guard her constantly as he did formerly; and he takes no part in incubation. Consequently he pays more attention to other females. If an unmated female, or, more rarely, a female whose mate does not seem to be aggressive enough to guard her, is present, the unoccupied male may take this female polygynously as a "secondary" female. More than one secondary female may be acquired successively in this fashion.

In most cases the male guards the secondary female as assiduously as he had his primary female, and I did not become aware of the new attachment until the secondary female had already started nest construction and copulation was being performed. As in second cycles, Phase 1 may be extremely short, a matter of only a few days, or may be passed over entirely.

At about the same time that the secondary female starts to incubate, the primary's eggs have usually hatched and the male assists in feeding the nestlings. The periods that the young are in the two nests of a polygynous male may overlap and the male generally feeds at both nests. . This attention by the male to nestlings at two nests has been observed on the same day. Where simultaneous observation of both nests was possible, the male was sometimes seen to make feeding trips to each nest, dividing his time irregularly between the two. Likewise, attention to nestlings in one nest and fledglings from the other has occurred on the same day.

In the exceptional year, 1947, when the ratio of males to females was one to two, polygyny was at its height. There were certain cases of polygyny in which it was difficult to determine whether the usual attention was paid to the secondary female. There were also certain cases in which a polygynous male acquired mates almost simultaneously, and the timing of the cycles of these females were more or less parallel.

Males were observed in 1947 feeding nestlings at more than the usual two nests; one fed at the nests of his three females; another brought food to the nests of his four mates. However, since never more than two nestling periods of any one polygynous male were known to overlap, two was the maximum number of nestling broods fed on the same day.

Concurrently with this state of polygyny there was a remarkable faithfulness in the remating of primary pairs. Of the 45 cases in which both members of a primary pair were present the following year, 42 remated and only 3 were "divorced." In addition, among the returns there were six birds that had been mated to unbanded birds the previous year. Some of these unbanded birds may, of course, have returned the following year and might have added to the total of either faithful pairs or divorces. In some cases primary pairs remated for a number of consecutive seasons. Of the total of 70 primary pairs, 44 were maintained for 1 year, 15 for 2 years, 7 for 3 years, 3 for 4 years, and 1 for 5 years. No male had the same female in secondary status more than once.

A monogamous male's mate and the first seasonal mate of a polygynous male are considered primary females; the mate, or mates, of a polygynous male which are subsequently acquired in the same season are considered secondary females. (In a few cases in the exceptional year 1947, designations as to primary and secondary status were made with possibly some arbitrariness.) Thirty-four banded females were always primary throughout their years of occurrence; 20 banded females were secondary and 15 changed their status. Less than half the females in the 1-year group of survival were primary but only three banded secondaries survived for two or more years, whereas the primary, and those of changing status, showed survival periods extending into the fifth and sixth year. Because 15 females changed their status over the years of their survival, it is believed that had the 20 banded females which were in the "always secondary" group survived longer (only two survived for 2 years and one for 3) they, too, might have changed and become primaries for part of their years as breeders in the colony.

Nesting.—Writing of the Brewer's blackbird in California, Grinnell and Miller (1944) say that its habitat "in the spring season [is] grassland, meadows, or moist lake and stream margins, with trees or tall bushes in the vicinity which may be used for lookouts, roosting and nesting."

C. W. Lockerbie (MS.) says of the bird in Utah: "The large open mountain valleys along the eastern slope of the Wasatch mountains are favorite summer habitats for these birds, e. g., Parley's Park, Summit County, 1 to 4 miles wide and 10 miles long and about 6,500 feet elevation, with willow clumps along all water courses, a few tall cottonwood trees, and much of the land in wild hay. Dairying and stock raising are the only pursuits. No less than 50 pairs of blackbirds breed in this area; 200 birds, more or less, after July 15 are about the usual number observed. My earliest observations has been, May 25 and my latest September 7, though their residence period doubtless extends beyond these dates."

In Colorado "they range from 4,000 to 10,000 feet in altitude and seem to prefer the open meadow along streams and adjacent to evergreen forests," says R. J. Niedrach (MS.).

In Nevada the species "uses a wide variety of situations for nesting sites," writes Linsdale (1936a). This statement may well be applied to its nesting adaptability over its whole range. The nests may be placed on the ground or up to 150 feet above the ground; in the sedges of a marsh; in bushes of wet or dry areas; in many kinds of living trees and in the broken tops of stubs; in windbreak hedges at ranches; in ornamental trees and shrubs in parks and gardens or along the streets of towns; near plowed fields in agricultural areas; in semiarid situations; or along streams in mountain meadows at high altitudes.

Bendire (1895) writes that at Camp Harney in southeastern Oregon the nests were frequently placed on the ground "or rather in the ground, the rim of the nest being flush with the surface." Quite a number were found in this situation, even when suitable trees and bushes were available. These ground nests were located on the edges of perpendicular banks. Ground nesting has also been reported for California by Grinnell, Dixon, and Linsdale (1930), who also report nesting in "a clump of sedge" and in "drowned brush-clumps out in the water." Linsdale (1938) reports ground nesting in Nevada; Cameron (1907) in Montana, and Schorger (1934) in Wisconsin.

Among the writers who mentioned nest situations in willows fringing a swamp or stream, in cottonwoods, and in various bushes on river and creek banks are Dawson and Bowles (1909) for Washington and Cameron (1907) for Montana.

Nesting in the broken tops of dead trees in the Lassen Peak region of California is reported by Grinnell, Dixon, and Linsdale (1930). Dawson and Bowles (1909) describe nests "in cavities near the tops of some giant fir stubs none of them less than 150 feet from the ground." But the "most favored nest sites" in California, according to Grinnell and Miller (1944), "are in dense masses of foliage, especially of conifers."

La Rivers (1944) found a large variety of trees and shrubs used for nesting in Nevada; sagebrush (*Artemisia*), the most prevalent shrub, was most often used.

Although usually nesting in groups, the Brewer's blackbird does not nest in such dense colonies as some other icterids, notably the tricolored redwing (*Agelaius tricolor*). This may be due to wider tolerance in habitat requirements, causing less concentration. Some observers report the nesting pairs as "somewhat scattered" and other pairs nesting "singly." Bailey (1902) writes that "it nests in much smaller colonies than many of the blackbirds, five to ten pairs being the common number."

In California R. M. Bond (MS.) found a colony "in a row of small *Eucalyptus ficifolia* near Carpenteria, three adjacent trees contained 21 nests (spring of 1935) with none in the remaining half mile or so of row. There were five or six nests in a yard tree (*Acacia melanoxylon*) about 30 feet from the nests in the eucalyptus * * *. The next nearest nesting colony was about a mile away in a windbreak of Monterey cypress."

Linsdale (1938) writing of the Toyabe region of central Nevada says that Brewer's blackbirds were found "in small colonies" and that the colonies varied in size from 3 or 4 to about 20 pairs. Within each colony "pairs tended to select similar nest sites." Ridgway (1877) found a large colony in a group of piñon pines at the south end of Pyramid Lake, Nev., on June 3, 1867. There were more than 100 nests, nearly every tree containing at least one. Several trees had two or three nests. Each nest was on a horizontal limb, usually near the top of the tree, well concealed in a tuft of foliage, and the majority of nests contained young.

La Rivers (1944) found that on a 15-acre tract 14 miles northwest of Reno, Nev., during the period May 17 to June 16, 1934, there were 107 nests of this species, a density of "slightly more than 7 nests per acre." This indicated, he adds, a "heavy infestation for the region."

In the area within about a 12-mile diameter around the river-mouth colony I found nine other colonies and no pairs nesting singly, although it is possible that a single pair might have been overlooked. Some nests on the peripheries of the colonies were considerably more isolated than the majority of nests toward the centers. The river-mouth colony was a quarter mile west of the nearest other colony, the second nearest being a mile to the north.

Monterey pines were used for nesting, at some of the nearby colonies, and also live oaks, Monterey cypress, and *Baccharis*. One colony was at a golf course, another at a dairy farm, and another in trees in the business district of Carmel. All these colonies were adjacent to favored foraging areas.

In no year did the river-mouth colony exceed an area covering 9 acres. Every year the greatest density of nests was confined to the center area of 1 acre. Considering only the first nesting for the season of each female in the years 1944–46, the density for the whole colony varied from 4.1 nests per acre in 1944 to 6.7 in 1946, whereas in the center acre alone the density varied from 14 in 1944 to 23 in 1945. Three or four pines in the center acre were particularly attractive to the birds. One pair of these trees, with trunks 4 feet apart and branches intermingling, but separated from the other trees, had a height of 45 feet and a combined spread of about 48 feet. This pair of trees contained 7 nests in simultaneous use in 1945. The nests varied from 21 to 42 feet above the ground; no two nests were closer together than about 9 feet nor farther apart than about 37. This represents the maximum crowding in the colony. Possibly such crowding was partly due to the fact that the trees were not evenly distributed over the 9 acres.

The arrangement of nests of polygynous males does not suggest a territory embracing them all, as is the case in some polygynous species, such as the yellow-headed blackbird (*Xanthocephalus xanthocephalus*). Although the nests of a polygynous male may be in the same tree, one of the nests may be as near or even nearer to the nest of another male than to its own second nest. But frequently the nests of a polygynous male were in different trees, often considerably farther separated than the nests of different males, with one or more other nests in between. The distance between two nests of one male has been as much as 282 feet.

The height of nests above the ground at the river-mouth colony varied from 7½ to 42½ feet. Thirty-five nests in 1945 averaged 27.2

feet and 37 in 1947 averaged 22.4 feet above the ground. (Many of the nesting trees, being in gardens or along the streets, had their lower branches trimmed.) Nests were ordinarily placed near the ends of branches in thick tufts of needles, and often partly supported by bunches of cones. The birds occasionally used planted Monterey cypress also.

The nest was described by Dawson (1923) as "a sturdy, tidy structure of interlaced twigs and grasses, strengthened by a matrix of mud or dried cowdung, and carefully lined with coiled rootlets or horsehair." Some writers report mud used in the nest and others make no mention of it. A "mud cup" is mentioned by I. McT. Cowan (MS.) at Vancouver, British Columbia. Schorger (1934), in his description of a nest in Wisconsin, does not mention mud. Goss (1891) says that the nests he found on the ground at Chama, N. Mex. "were all without a trace of mud."

At my colony grasses, pine needles, etc., were seen in the bills of nest-building females; many females frequently gathered horse manure and mud, a combination that, when dry, makes a firm, plasterlike cup.

The dimensions are given by Macoun (Macoun and Macoun, 1909): "In size it averages over 6 inches across, with a cup over 3 inches and a depth of at least 1½ inches."

In Phase 2 (nest-building, copulation, and egg-laying) the male, although he takes no part in the actual construction of the nest, usually accompanies the female on each trip as she gathers material and carries it to the nest. At this time they make a long, continuous series of trips in contrast to the toying with and dropping of nest materials, or occasional trips to a nest-site, of Phase 1. When the female enters the tuft to place the material and mold the nest, the male perches nearby and displays the ruff-out, uttering *schl-r-r-r-up* and *squeee*. He uses one of several habitual guard perches, a wire, a pole or a branch tip. From this perch he may drive other males if they come near the nest site or into the tree. In this respect there is a certain amount of localized aggression by the male. But such localization does not take place until after the pairs have been formed in Phase 1. The greatest portion of the male's aggressive activity is directly concerned with guarding his female against the approaches of other males. This guarding, or aggression concerned with a sexual situation which reaches its height in Phase 2, is not localized; i. e., it has little, or only incidental, connection with any particular area or territory and may occur at any point on the wires, along the streets, on the edge of the creek, or several hundred yards out on the adjacent marsh.

In Phase 2 the female exhibits the pre-coitional display frequently, and other males as well as her mate respond; among the males, rushes and threat displays are frequent. These advances by intruding males and the guarding actions by the mate, which begin in Phase 1, reach their height of occurrence in Phase 2, which, from the beginning of actual nest construction to the completion of the clutch, lasts about 10 days or 2 weeks, at the most.

Eggs.—The number of eggs in a clutch varies from 3 to 7; Bendire (1895) and Hoffmann (1927) are the only writers I know who have given a number as high as 8. Cowan (MS.) says that in British Columbia "5 eggs is the most frequent number." Dawson and Bowles (1909), referring to the species in Washington, and Dawson (1923), referring to California, give the number per set as "4–7, usually 5 or 6." La Rivers (1944) in a month of field study near Reno, Nev., found 107 nests containing a total of 521 eggs. The clutches were divided as follows: 23 with 3, 21 with 4, 22 with 5, 29 with 6, and 12 with 7 eggs each.

Ray (1909) found numerous nests of this species in the region about Lake Tahoe, Calif., in May and June 1909, nearly all of which were in small "tamarack pines, often mere saplings, from four to fifteen feet up, and but poorly concealed" (a notable exception was one nest placed on a wharf piling which was standing in water 3 feet deep). He says that "five was the usual complement of eggs, tho often four or six, and sometimes only three." The eggs he examined "showed great variation in size, shape and coloring."

Dawson (1923) also notes variation in pattern and coloration. He writes that the eggs present—

Two divergent types of coloration, with endless variations and intermediate phases. Light type: ground color light gray or greenish gray, spotted and blotched with grayish brown or, more sharply, with sepia. Eggs of this type rehearse relationships, now with the Quiscaline Grackles, and now with the Yellowheads (*Xanthocephalus xanthocephalus*), or the Cowbirds (*Molothrus ater*). An egg in the M. C. O. collection has a background of pale niagara green sharply spotted with a blackish pigment which tones out to dusky drab, and is thus indistinguishable from the egg of an Agelaiine Blackbird. Dark type: Ground color completely obscured by overlay of fine brown dots, or else by confluent blotches of Rood's brown, walnut brown, or cameo brown.

Bendire (1895) writes: "The average measurement of two hundred and forty-five specimens in the United States National Museaum collection is 25.49 by 18.60 millimetres, or about 1 by 0.73 inch. The largest egg in this series measures 27.94 by 20.07 millimetres, or 1.10 by 0.79 inches; the smallest, 20.83 by 15.49 millimetres, or 0.82 by 0.61 inch."

Incubation.—Bendire (1895) states that the incubation period is 14 days. Saunders (1914) says that the eggs hatch in 12 days in Montana.

Because of the inaccessibility of most nests (high up and at the end of limber branches) that entered into my study of the species, it was possible to look into only a small number of them. Of these it was feasible to make daily inspection at only 5, of which 3 sets had 4 eggs, and 2 had 5 eggs each. One set of 4 eggs hatched in 12 days and all the others in 13 days, reckoning the incubation period from the day the last egg was laid until all were hatched. No thorough study of incubation rhythms was made, but watching females at inspectable nests revealed that they spend time on the eggs before the full set is laid. This correlates with the fact that hatching of the young (except in the set of 4, in which all hatched on the 12th day) was spread over as much of 3 days and indicates that incubation may start before the clutch is complete. The tangible facts ascertained from the few accessible nests coincided with those inferred from parental behavior at other nests.

Although the male takes no part in incubation and may even form an attachment with another female at this time, his attention does not in all cases leave the incubating female entirely. Monogamous males may spend much time on the guard perches near the nest. Polygynous males have been known to guard at two nests if the incubation periods overlap; but if one female is in Phase 1 or 2 while the other is incubating the male gives much more attention to the former (if the nest of either incubating female is destroyed she very quickly reverts to Phase 2). In a few cases the male has been seen to feed the incubating female on the nest.

Young.—Data on the length of the nestling period is extremely scanty in the literature. In my study I was able to ascertain definitely the nestling period in only three instances. In each case it was 13 days, calculating the period from the day the last egg had hatched until all the young had left the nest under natural conditions. As in the determination of the incubation period the information obtained at these three nests corroborated the observations at many inaccessible nests. The male Brewer's blackbird at the river-mouth colony was found to assist the female in feeding the young, both in and out of the nest, in 72 out of 99 monogamous nestings and 76 out of 109 polygynous ones. In the remaining cases in each category the male disappeared, or the eggs did not hatch, or the nestlings died before male attention could be determined, or there were not sufficient observations to prove or disprove male attention.

Although I was unable to carry out extended periods of watching at any one nest, periods varying from 1 to 3 hours at various nests throughout the fledging period indicate that although the female usually exceeds the male in the number of trips per hour with food, the male sometimes equals and even exceeds the female in such trips, especially in the early part of the period when the female is brooding the young. When two sets of nestlings were being fed by a polygynous male during the same hour period the combined rate for both nests might equal the maximum rate for a male feeding at only one. But males feeding at two nests were not observed to exceed this rate.

Verna L. Johnston (MS.) writes concerning a nest with five young at Live Oak, Calif. The nest was 15 feet above the ground in a deodar tree in school grounds. Both male and female fed the young at 2- to 4-minute intervals most of the time during the 9 days (May 3 to 12, 1945) that she watched them. "The male often fed the female, sometimes on a fence from which she then flew to the nest and fed the young, sometimes on the nest after feeding the young."

At my colony nestlings that died were sometimes removed by the parents. On six occasions nestlings too undeveloped to have left the nest by their own exertions were found on the ground 50 or more feet from the nearest nest. These bore no apparent marks of having been carried by a predator. Two of these dead nestlings were actually seen being carried by the parents in flight and deposited. One of these young, which was seen being carried by the male, was newly hatched and weighed 8.29 grams. Others that were found on the ground were larger. The places of deposition were those regularly used to drop excreta taken from the nest: a pathway, pavement of the street, and the edge of the creek.

At the river-mouth colony, fledglings were observed to take initial flight of 3, 4, and 7 feet from the nest; and juveniles were fed by the male up to the 26th, and by the female to the 25th day after leaving the nest.

Females usually attempt second broods if the first is unsuccessful. There have been as many as three attempts in one season. Second broods have sometimes been raised even in cases when a first brood has also been fledged. Two broods are frequently raised in Oregon, according to Gabrielson and Jewett (1940).

Two females at the river-mouth colony were seen carrying nesting material for a second brood nest on the same day that they were still feeding fledglings. One female even fed a fledgling 3 days after the day on which she was first noted placing material for a second nest.

At this colony no young were observed leaving the nest later than July 7 (1943). At Bridalveil Campground, at an altitude of 7,200 feet, in Yosemite National Park, Calif., a pair were seen by Marshall (MS.) feeding nestlings as late as July 22 (1946).

Plumages.—Linsdale (1936b) writes that in Nevada the down of the nestlings he examined was "nearly black, contrasting with the whitish down of red-wings." According to Ridgway (1902) the "young" [i. e., in juvenile plumage] are "very similar in coloration to winter females, but texture of plumage very different and feathers without gloss". The immature male in first winter plumage is "similar to the adult male, but feathers of head, neck, back, scapulars, chest, and sides narrowly tipped with grayish brown (paler and more buffy on underparts)."

Food.—The Brewer's blackbird feeds both on animal matter, principally insects, and vegetable matter, principally seeds.

Analysis of the stomach contents of six mature birds collected in an alfalfa and wheat area on the outskirts of Meadow, Millard County, Utah, on June 10, 1943, by Knowlton and Telford (1946) are reported as follows:

One stomach held 2 adult and 63 nymphal treehoppers, *Campylenchia latipes* (Say), besides other insects. Another stomach contained 8 adult and 22 larval alfalfa weevils, a clover leaf weevil, a histerid beetle and an elaterid beetle, etc. Total recognizable contents consisted of: 17 nymphal grasshoppers; the 18 Hemiptera included 1 pentatomid, 3 lygaeids and 1 mirid; of the 84 Homoptera, 65 were membracids, 15 were aphids including 8 pea aphids, and 2 leafhoppers; 57 Coleoptera, among them 19 adult and 16 larval alfalfa weevils, 2 clickbeetles, 3 white grubs, a buprestid and histerid; 1 adult Trichopteron; 40 larval Lepidoptera; 2 larval Diptera; 10 of the 15 Hymenoptera present were ants. Three spiders also were present.

This interesting blackbird is sufficiently abundant in many parts of Utah to be of importance in the control of cutworms, grasshoppers and certain other insect pests.

The termite *Zootermopsis angusticollis* has been observed by Cowan (1942) as a food of the Brewer's blackbird in British Columbia. La Rivers (1941) while working on a program for the control of the Mormon cricket (*Anabrus simplex*) in northern Nevada during the summer of 1939, made the following observations:

This bird [the Brewer's blackbird], in company with the Sage Thrasher [*Oreoscoptes montanus*] and Western Meadowlark [*Sturnella neglecta*], is one of the destructive "Big Three" of the northern Nevada cricket fields. It has been known to destroy entire bands of adult crickets, but has never been reported as working on the egg-beds. It can safely be said that each of these three species of birds is responsible for more destruction of the Mormon cricket than all the other species together. * * * However, while the blackbirds feed extensively on the crickets in lean areas, they may almost ignore them adjacent to fields where they can obtain abundant seed. In one region south of Whiterock I observed a band of approximately 200 blackbirds working on a hillside which bore a cricket population of five per square foot. After an hour's observation I investigated their work and found, at the spot, only one attacked cricket to the square yard. Females [the birds ate only female crickets] constituted fifty percent of the cricket population, and, on this basis, the kill ratio amounted to 1 out of 22.5, a very low figure.

Knowlton and Harmston (1943), working on grasshoppers and crickets eaten by birds in Utah, examined the stomach contents of 105 *Euphagus cyanocephalus*. They report "40 contained Orthoptera, including 51 adult and 9 nymphal grasshoppers in 30 stomachs; the other ten stomachs held 16 field crickets." Bryant (1912) lists the Brewer's blackbird as a feeder on grasshoppers during an outbreak of these insects in California, but says that it does not rank among the most important predators, judged either by the number of insects per day each bird eats or by the number of birds eating the grasshoppers. He concludes that the value of birds in controlling such insects is greater during the periods of normal insect numbers than at times of extraordinary abundance. Bryant (1911) found the Brewer's blackbird to be an efficient destroyer of the butterfly *Eugonia californica* during an outbreak in northern California in 1911.

Several authors have reported caterpillars in the diet of this species. Munro (1929) observed them feeding on "forest tent caterpillars" at Rollins Lake, British Columbia. McAtee (1922) reports the taking of canker-worms by this blackbird at three places in California where the worms were threatening prune crops.

The corn earworm (*Heliothis obsoleta*) has been accused of being "the most destructive insect enemy of corn in the United States" and one of the most important of the 17 species of bird to feed on this pest is the Brewer's blackbird, according to Phillips and King (1923).

This species was seen by Murie and Bruce (1935) associating with western sandpipers (*Ereunetes mauri*) which were feeding on the brine fly *Ephydra millbrae* along a road traversing the mud flats on San Francisco Bay. The blackbirds also were "almost certainly feeding on the flies." Bond (MS.) found *Ephydra hians* as an item in the diet of this bird at Moss Landing, Monterey County, Calif., in 1931 and *Ephydra sp.* at Owens Lake, Calif., in 1938.

The Brewer's blackbird is considered by Kalmbach (1914) to be an effective enemy of the alfalfa weevil. Howell (1906) says: "Four species of blackbirds are known to consume boll weevils [in Texas], the most important of which seems to be the Brewer's blackbird."

Emlen (1937) says: "Blackbirds have frequently been accused of stealing almonds; but although three species, Brewer (*Euphagus cyanocephalus*), Bicolored (*Agelaius phoeniceus*), and Tricolored (*Agelaius tricolor*) were all common in the orchards, there is no definite evidence that they were feeding on almonds during the preharvest months." Beal (1948) writes:

During the cherry season in California the birds [Brewer's blackbird] are much in the orchards. In one case they were observed feeding on cherries, but when a neighboring fruit grower began to plow his orchard almost every blackbird in the vicinity was upon the newly opened ground close after the plowman's heels in its eagerness to secure the insects turned up.

The laboratory investigation of this bird's food covered 312 stomachs, collected in every month and representing especially the fruit and grain sections of southern

California. The animal portion of the food was 32 percent and the vegetable 68 percent.

Caterpillars and their pupae amounted to 12 percent of the whole food and were eaten every month. They include many of those pests known as cutworms. The cotton-boll worm, or corn-ear worm, was identified in at least 10 stomachs, and in 11 were found pupae of the codling moth. The animal food also included other insects, and spiders, sow bugs, snails, and egg shells.

The vegetable food may be divided into fruit, grain, and weed seeds. Fruit was eaten in May, June, and July, not a trace appearing in any other month, and was composed of cherries, or what was thought to be such, strawberries, blackberries or raspberries, and fruit pulp or skins not further identified. However, the amount, a little more than 4 percent for the year, was too small to make a bad showing, and if the bird does no greater harm than is involved in its fruit eating it is well worth protecting. Grain amounts to 54 percent of the yearly food and forms a considerable percentage in each month; oats are the favorite and were the sole contents of 14 stomachs, and wheat of 2, but no stomach was completely filled with any other grain. Weed seeds, eaten in every month to the extent of 9 percent of the food, were found in rather small quantities and irregularly, and appear to have been merely a makeshift.

Stomachs of nestlings, varying in age from 24 hours to some that were nearly fledged, were found to contain 89 percent animal to 11 percent vegetable matter. The largest items in the former were caterpillars, grasshoppers, and spiders. In the latter the largest items were fruit, probably cherries; grain, mostly oats; and rubbish.

The results of an investigation of the food habits of the redwing (*Agelaius phoeniceus*) and Brewer's blackbird in California has been presented by Soriano (1931). The stomach contents of 285 Brewer's blackbirds, taken in all the months of the year and from 15 counties in various sections of the State, were examined. The animal food taken included Coleoptera (represented by at least 13 families), Diptera, Hemiptera, Homoptera, Hymenoptera, Lepidoptera, Orthoptera, Chilopoda, and Arachnida. Cereals, mainly wheat, were the vegetable foods most taken. Other seeds found included *Amaranthus*, *Amsinckia*, *Stellaria*, *Sorghum*, *Erodium*, *Polygonum*, and *Ribes*. "In the months when insects and vegetable food, especially cereal in the harvest season, are both abundant," writes Soriano, "vegetable food is taken less, showing that these birds are primarily more insectivorous in food habits than vegetarian. Most of the insects taken belong to the destructive families of insects from the point of view of the farmer and fruit grower. Very few beneficial insects are taken. The destruction of these harmful insects means a great help to the farmers and growers in particular, and to consumers in general." Most of the cereal taken is not "from newly planted seeds, for the birds take sown seeds uncovered, as also grain from pastures, barnyards, orchards and grain fields. Economically this cereal is not important and it can be considered waste grain." The rest of the vegetable food is almost all weed seeds. "Economically, in the widest human interests," concludes Soriano, the redwing and the Brewer's blackbird "are beneficial, being more insectivorous than vegetarian in food habits. However, being gregarious birds, they can now and then inflict such great damage on crops that to give them full protection is not fair to the farmer whose crops are immediately threatened."

The species as I have observed it in northern Monterey County, Calif., forages in a wide variety of open or grass covered situations. Parent birds have been seen gathering food for nestlings on the beach where the receding tide has left bits of kelp; the birds have been seen feeding on the dry parts of the beach above high tide mark, on wet sand bars at the mouth of the Carmel River, on mud flats near the river mouth, and on the grassy areas of the adjacent marsh where they mingle with pintails (*Anas acuta*), shovellers (*Spatula clypeata*), Wilson's snipe (*Capella gallinago*), killdeer (*Charadrius vociferus*), and other shore birds. They feed in both green and dry pastures and grasslands (where they may or may not feed near cattle), on recently burnt-over grass areas, along the bare shoulders of highways in open country, on freshly plowed fields, on lawns and golf courses, on sidewalks, and in the gutters of streets of towns.

I have never seen them foraging in dense brushland or forested areas, although the flocks fly over such areas. During the nesting season, at least, some foraging is done among the needles of the pines in the colony. For example, four times in three seasons I have seen birds poking into the insect spittle in the pines. One bird was seen carrying insects with spittle to nestlings. Dr. Kathleen C. Doering, of the Department of Entomology of the University of Kansas, identified specimens collected from similar spittle masses in the same tree as *Aphrophora*.

I have also seen flycatching tactics on numerous occasions when the birds took short flights from the ground to capture a flying insect. Such sallies are also made from telephone wires.

Grinnell, Dixon, and Linsdale (1930) mention the distance from its nest the bird will forage to feed nestlings. Parent birds gathered food from the ground on a chaparral-covered slope half a mile away from their nest on the edge of a lake. At the river-mouth colony parents range some distance for food. The maximum recorded was on May 17, 1944, when birds went to a feeding tray just half a mile away to get food for nestlings 4 and 5 days old and for fledglings out of the nest.

The species seems capable of shifting quickly from one foraging situation to another, e. g., the instance mentioned by Beal (1948) of a flock leaving cherries to follow the plow. This trait has been commented on by Winton Wedemeyer (MS.) and others. The bird is commonly observed following various tillage operations. R. M. Bond (MS.) reports that in 1939, 1940, and 1941 Brewer's blackbirds "followed the cultivating equipment near Somis, Ventura County,

Calif., often in hundreds. Available insect specimens turned up by the plow were few, and the great majority of individuals were larvae of the wireworms *Limonius californicus* and *Melanotus longulus*, both serious agricultural pests."

Observers have reported seeing the bird turn over chips of dry cow-dung in pastures. Linsdale (1938) writes of a male that "turned over pieces of cow manure with its head and looked beneath them for food." I have watched this action a number of times. It is usually accomplished by the bird putting its bill beneath the chip and flipping it over; sometimes the bird nudges the dung forward as it lifts, thus over-turning it. The bird inspects both the newly exposed ground and the underside of the chip and takes food from both places. One piece of manure turned over measured 98 by 70 by 30 millimeters and weighed 33.85 grams. Sometimes the birds poke vigorously into horse manure, flipping pieces aside with the bill. They turn over other objects such as small pieces of wood, clods of earth, and even small stones. One piece of wood overturned by a blackbird measured 95 by 45 by 13 millimeters. Occasionally the bird thrusts its bill underneath and then opens the bill to pry the object up, overturning it with a forward and upward motion. One bird enlarged a hole in soft mud by inserting the closed bill and then opening the bill. It then picked something out of the hole. Once a male bird was seen digging vigorously into the turf of a golf course, pulling out bits of dirt which it flicked off the bill; finally it pulled out a whitish object about 50 millimeters long, possibly an insect larva.

La Rivers (1941) describes the method of eating Mormon crickets as follows:

The Brewer's assault upon the cricket is confined entirely to the females, which the birds covet for their eggs. These they take by splitting the dorsum of the abdomen transversely along the soft membranous tissue between the sclerites, a feat accomplished by grasping one end of the body in the bill, the other in a claw, and tugging; some go to less trouble and merely tear the head off, pulling with it the entire abdominal, and much of the thoracic, contents, which are all consumed. An unexplained habit of these birds is their snipping off of the female cricket's ovipositor, something they quite frequently do.

I have observed blackbirds feeding in water that reached as high as the belly feathers. Semiaquatic feeding has been very well described by Richardson (1947):

Manzanita Lake on the campus of the University of Nevada has extensive growths of the water-weed *Anacharis canadensis*. Each year by the end of May the new growth of this plant forms a dense mat an inch or less below the water surface. For several years now both Red-winged (*Agelaius phoeniceus*) and Brewer (*Euphagus cyanocephalus*) blackbirds that nest in the vicinity of the lake have been observed feeding on insects associated with the waterweed. The blackbirds alight on the plants, the water usually coming to the middle or upper part of the birds' tarsometatarsi. Typically, the wings are then fluttered as the bird hops two or three feet to new vantage points. Less often a bird will walk, even a distance of thirty feet, without moving the wings. The tail, as appeared to be the habit in one individual especially, may be submerged and possibly pressed against the underwater vegetation for support.

The most readily visible food obtained, and certainly the major item for a period of weeks in the early summer, is recently emerged damselflies. The naiads of this insect crawl to the surface of the waterweed and metamorphose on projections just above the water. The blackbirds have been seen repeatedly catching these newly emerged and still pale and flightless adults. * * *

Brewer Blackbirds of both sexes have been seen several times walking and feeding on pad-lily (*Nymphaea*) leaves, even one leaf serving to hold up a bird. On two occasions, once on the Truckee River and once on the Carson River, Brewer Blackbirds have been seen hovering over open water and snapping food from the surface. A male of this species was seen similarly to obtain a large piece of bread in Manzanita Lake and carry it to shore to be eaten.

Both Ken Stott, Jr., and A. D. DuBois (MSS.) describe Brewer's blackbirds soaking popcorn in water before eating it. I have often seen them soak bread in a bird bath or, at the river-mouth colony, in the water of the marsh.

When mixed flocks of redwings and Brewer's are seen feeding on the ground a difference can be noted in the angle at which the tails are cocked; those of the redwing are held obviously higher. In a large mixed assemblage the difference is quite noticeable.

The movements of a foraging flock of Brewer's blackbirds has been well analyzed by Mulford (1936) as a combination of the walking of individuals in zigzag movement but all progressing slowly in the same general direction, and the flying up and realighting in front of the flock by those which had been left walking in the rear. Dawson and Bowles (1909) have commented that this flying up to the front of the foraging flock is sometimes a constant motion and when a large assemblage is present it creates a rolling, or surging, effect, as the mass of the birds moves over a large field.

Neff and Meanley (1957) have analyzed the winter food in Arkansas.

Behavior.—The gait of the Brewer's blackbird is usually a walk, accompanied by short forward jerks of the head. When the bird runs there is no jerk. Mulford (1936) outlines the process of taking two walking steps as follows: "1. Head is thrust forward as one leg and foot are lifted. This moves the center of balance forward. 2. Leg and foot are brought forward. 3. Head is pulled back as body is brought forward by step and as foot is set down on ground. 4. The sequence of movements is repeated with other foot. The result of this is that the bird moves forward in a series of movements rather than in one continuous movement. The walk is jerky."

Mulford also comments that there is no characteristic formation to the flock when in flight but it is an "amorphous mass," either compact or spread out; of rounded form or with irregular margins; in one big group, or more or less divided into subgroups.

The speed of flight has been measured by Rathbun (1934). A flock was paced from a car for one mile and was found to fly at 27, 35, and 38 miles per hour. There was no wind.

In the region of Carmel, Calif., the Brewer's blackbird uses principally two types of growth for roosting, patches of tules in the marsh, where they associate with the redwing, and thick tufts of foliage of the Monterey pine, a situation similar to that which they use for nesting.

In late summer, fall, and winter I have noted late afternoon flight lines from feeding areas up Carmel Valley toward roosts at the mouth of the Carmel River and in the Carmel business district. The birds frequently gather at a dairy farm 3½ miles east of the business-district and river-mouth roosts. From this point, as well as from further up the Valley, a number of flocks of Brewer's and mixed flocks of Brewer's, redwings, and tricolored redwings fly westward down the Valley. The number of Brewer's in either mixed or pure flocks has been found to range from 40 or 50 to 300 birds. The westward flight line has also been noted from a point 2 miles further down the Valley than the dairy farm, and a corresponding eastward movement in the morning has been seen there.

At both the business-district and river-mouth areas the flocks of Brewer's blackbirds gather on electric wires before roosting time and occasionally fly down to forage. The actual flight to the roosting places in the pines is made individually and at irregular intervals, the birds lighting first on the branch tips then working their way into the thick parts of the foliage. There is often considerable moving about in the foliage and bickering for positions, and considerable utterance of various calls, especially the *kit* notes.

On November 17, 1943, a female at the river-mouth colony roosted in a tuft very close to her nest situation of the preceding spring. On October 30, and December 28, 1944, the same female roosted in a similar tuft of the same tree. Her nest of 1944 was also within a few feet of this roost.

In the morning the birds fly out individually and at irregular intervals from the roosting tufts in rapid, zigzag flight and gather on the wires. Those leaving the tules sometimes fly directly eastward up the valley. Some leave the pines at the river-mouth and fly toward the business district. Morning observations in the business district show large gatherings on the wires, some birds coming from tufts of the street pines, and others, possibly, from the river mouth. After an interval on the wires, a large flock has been seen (December 12, 1947) taking flight eastward and passing out of sight. This line of flight would carry them up the Valley along the flight lines mentioned above. Some of the birds in the business district, as well as some at the river mouth, remain to forage locally.

Indications are that in this region the Brewer's blackbird goes to roost earlier and leaves the roost correspondingly later than some other species in the area. The redwing has been heard calling up to 10 and 12 minutes after the last Brewer's notes were heard. (The female noted above went to her roosting tuft, silently, 13 minutes before the last redwing note was heard from the adjoining marsh, October 30, 1944, and 14 minutes before the last bush-tit (*Psaltriparus minimus*) notes were heard on November 17, 1943.) In the morning, also, several species of passerines have been heard before the first calls of the Brewer's were detected. Golden-crowned sparrows (*Zonotrichia coronata*) have been heard making their first call 21 minutes before the first Brewer's blackbird notes.

Similar diurnal rhythms between pine roosting places at Berkeley, Calif., and distant foraging places have been described by Mulford (1936). Stott (MS.) furnishes the following notes from San Diego, Calif., where he watched a wild flock at the zoo:

"While their roosting place changes from time to time (bamboo clumps, palm crowns, *Grevillea* trees, etc.) they follow a fairly stable pattern in their diurnal activities about the pool. Each morning they appear first on a telephone line which stretches across a canyon north of the Mirror Pool area. Suddenly a small group at the south end of the line drops to the ground on a macademized road; there it is shortly joined by another group and another, until the entire flock has forsaken the wire. Its members strut up and down the road in search of scattered popcorn. In the late afternoon the blackbirds begin to congregate near the pool until they form a more or less compact flock. Subsequently, they fly in small groups back to the telephone wire on which they had congregated earlier in the day. Their next move is a unified one and takes them to their current roosting place."

Lockerbie (MS.) mentions the use of cattails for roosting in the Salt Lake City, Utah, region. Bassett (1931) found them using floating duck-hunting blinds, covered with eucalyptus boughs, that were anchored nearly a mile from the high tide line in San Pablo Bay, Calif.

Voice.—In the belief that there is no really satisfactory way to syllabify the notes of this species, I have merely attempted a tentative and, it is hoped, suggestive rendering of the utterances. In some of the cases the associated activity of the bird is mentioned, together with the function of the call, when this is known, in order to aid the reader in identifying the notes. Most calls are subject to considerable individual variation but are always identifiable as one of 13 types, that can be grouped in two broad categories—single utterances and series utterances.

The single utterances are either short or long and drawn out. Among the short, incisive, single-syllable utterances are the following sounds: *Tschup*—a "scolding" note uttered when there is a disturbance near the nest or young, or in other situations of excitement. A "flocking" note—uttered when groups take off and fly in flock formation, that is shorter, higher, less loud, and with less *s* sound and more *t* sound than *tschup*. *Tup*—uttered by adults when approaching the nest with food; more frequently used when approaching the place where a young fledgling is; but also used by adults when flying with older fledglings following; it is shorter, softer, lower, more muffled than the *tschup*. A "location" note—uttered by the fledgling during intervals between food-bearing visits of the parents; like *tschup*, but weaker, it has a shorter vowel sound and a more nasal quality. The "squawk"—a low, scratchy note uttered when one bird makes contact with another, it is used when one bird grabs another in fighting, when a blackbird is caught by a hawk, and also used when a blackbird dives at a hawk; Mulford (1936) writes it *chaw*.

Long drawn-out, single utterances, of one or more syllables, include the following: *Squeee*—a very loud hoarse whistle with decided upward inflection (this note and *tee-uuu*, described below, carry farther than any of the other calls). *Schl-r-r-r-up*—a comparatively subdued, toneless, whirring gurgle, aptly described by Mulford (1936) as "a rush of air without vocal accompaniment"; this call and the preceeding one are subject to considerable individual variations (for the associated activities see under "Spring" and "Courtship"). *Tee-uuu*—a loud, clear, thin whistle with decided downward inflection, that sometimes becomes *pit-eee* or *tsee-eur*, but in all cases with the second syllable lower in pitch than the first; it is the only clear whistled note in the repertory and functions as a "warning" note (see under "Enemies").

The second major category, series utterances, includes short notes uttered in series, sometimes with a definite rythm, as follows: *Kit-tit-tit-tit*, etc.—sometimes regular in delivery, sometimes irregular, varying in quality of tone, intensity and rapidity of utterance; at times it sounds more like *kit-r-r kit-r-r*, when it has a decided rythmic effect; it is used both in the female generalized display and in belligerent encounters, by males (rarely) when bickering over food, and is frequently heard accompanying bickering at the roost (when the sex of the bird cannot always be identified). The female copulatory note—a soft, low, steady series of tapping notes, very different from any of the other utterances—is used in the female pre-coitional display (see under "Spring" and "Courtship"). *Chug-chug-chug*, or *tucker-*

tucker-tucker, or *tit-tit-tit*—used by males accompanying the elevated tail display (see under "Spring" and "Courtship"); sometimes these utterances resemble the *kit* notes.   Peeping sounds—made by young nestlings.   *Tut-utz-utz*—a low, hoarse, scratchy series of rhythmic notes, the "begging" notes of older nestlings and of fledglings; when the parents are away foraging the fledgling utters the "location" note, but when the adult arrives the young bird changes to this "begging" note in anticipation of being fed.

Field marks.—The species most easily confused with the Brewer's blackbird in the field are the redwing and tricolored redwing, the rusty blackbird (*Euphagus carolinus*), the bronzed grackle (*Quiscalus quiscula*), and the cowbird (*Molothrus ater*).   Peterson (1941) says that the male rusty blackbird in summer plumage has "dull greenish instead of purplish head reflections [as in Brewer's].   The iridescence is almost lacking [and is] not noticeable as in the Brewer's Blackbird or the Bronzed Grackle."   The rusty and Brewer's blackbirds are about the same size but the bronzed grackle is noticeably larger and has a "longer tail, which is somewhat wedge-shaped."   The female Brewer's has a dark brown iris in contrast to the pale yellow iris of the female rusty blackbird.   In winter the female Brewer's blackbird does not have the rusty wash, as do rusty blackbirds of both sexes.   But the immature male Brewer's in the first winter plumage, according to Ridgeway (1902), has the "feathers of head, neck, back, scapulars, chest, and sides narrowly tipped with grayish brown."   There is the possibility that in this plumage, also, the male might be confused with the male rusty blackbird in winter plumage.   But the tipping is more grayish and less rusty brown.   The male redwing and tricolored redwing are obviously distinct from the Brewer's.   Hoffman (1927) points out the distinctive marks for females: "The unstreaked breasts distinguish the Brewer Blackbirds from female Redwings [and also, it may be added, tricolored redwings], and the greater size, darker plumage and long, sharp-pointed bills distinguish them from the female Cowbirds.   The yellow (apparently white) eye and long sharp-pointed bill distinguish the male Brewer from the male Cowbird."

Enemies.—La Rivers (1944), reporting on one nesting season of Brewer's blackbirds near Reno, Nev., found that 107 nests held 521 eggs, 205 of which resulted in fledglings that left the nest safely, a mortality of 60.65 percent.   "Twenty-three eggs were known to be sterile, but other sterile eggs were obviously among those which disappeared without known cause.   The total number of eggs and nestlings unaccounted for because of unknown predation, amounted to 83, but the remaining 233 could nearly all be ascribed with certainty to one of 23 known factors, 21 of which were biologic, the other two climatic [i. e., wind and hail]."

Among the biological factors which accounted for 86.07 percent of the mortality, predation was preeminent. Those predators that were "persistent bird and egg feeders" were: Scrub jay (*Aphelocoma coerulescens*), American magpie (*Pica pica*), crow (*Corvus brachyrhynchos*), bridled weasel (*Mustela frenata*), western ringtail (*Bassariscus astutus*), desert bullsnake (*Pituophis catenifer*), striped racer (*Coluber taeniatus*), and blue racer (*Coluber constrictor*). Those predators that could only be "classed as occasional opportunists" were: Sharp-shinned hawk (*Accipiter striatus*), horned owl (*Bubo virginianus*), Steller's jay (*Cyanocitta stelleri*), Piute ground squirrel (*Citellus mollis*), Beechey ground squirrel (*Citellus grammurus*), Sierra mantled ground squirrel (*Callospermophilus chrysodeirus*), Douglas chickaree (*Sciurus douglasi*), striped skunk (*Mephitis mephitis*), and Great Basin rattlesnake (*Crotalus viridis*).

"Protective factors" included the floral elements; some types of plants used afforded better nest protection than others. Also the height of the nest emplacement was important; 57.94 percent of the nests were placed 5 feet or less above the ground, 32.71 percent from 5 to 10 feet, and 9.34 percent above 10 feet. It was found that "mortality progressively increased" from the highest to the lowest. Still another protective factor was the tactics of parent birds in driving off predators. However, in some cases, La Rivers believes, "noisy, quarrelsome, conspicuous birds" attracted predators to their nests.

Bond (1939) in an examination of the remains of prey items from beneath five nesting sites of the prairie falcon (*Falco mexicanus*) in the region of the Lava Beds National Monument, Calif., found two Brewer's blackbirds at one eyrie, none at the other four. He found remains of eight Brewer's blackbirds beneath the nesting site of a duck hawk (*Falco peregrinus*). A large number of pellets of the horned owl and barn owl (*Tyto alba*) were also collected by this author from beneath roosts in the same region. Two collections were made, one on August 12, 1937, at which time only those pellets "were taken that seemed, on the basis of state of preservation, to have been cast later than the preceding winter." At the time of the second collection November 5, 1937, only those were taken which seemed certainly to have been deposited since the August collection. In the first collection the remains of 12 Brewer's blackbirds were found among a total of 3,391 items of bird, reptile, mammal and insect remains, which amounted to 0.0035 percent of the total. In the second collection two Brewer's blackbirds were found out of a total of 994 bird and mammal remains, or 0.002 percent. Considering the effect of predation on birds by both hawks and owls in this area, Bond states that the Brewer's blackbird was among the five species of birds most often killed by hawks. In the combined owl pellet collections the total of

14 Brewer's blackbird items out of a total of 106 bird items puts it among the 7 bird species that were represented by more than 5 percent. But most of the birds taken by both hawks and owls were common to exceedingly abundant in the area (the Brewer's blackbird was in the latter class). Bond concludes "it is quite clear that none of the species is endangered, or probably appreciably reduced in numbers, by either the hawks or owls."

Sumner (1928) reports finding a headless young Brewer's blackbird on two different days in a tree cavity occupied by young screech owls (*Otus asio*) at Claremont, Calif. W. H. Behle (MS.) found remains of this blackbird near the nest of a short-eared owl (*Asio flammeus*) in Utah.

Bond (MS.) writes that he has "seen both Cooper's hawk [*Accipiter cooperi*] and the pigeon hawk [*Falco columbarius*] catch a Brewer's blackbird, the former on the ground and the latter in the air."

In Carmel Valley, Calif., on December 15, 1942, I witnessed the capture of a Brewer's blackbird by a sharp-shinned hawk. A mixed flock of about 500 redwing, tricolored, and Brewer's blackbirds were alternately perching on power wires and flying down in small groups to feed in straw stubble near some horses. The hawk was first noticed pursuing the blackbird, which it seemed to have singled out from a group of about 300 birds flushed from the stubble. The hawk followed the blackbird in a twisting and turning flight; there was a "squawk" from the blackbird as it was caught. The raptor then flew to a nearby thicket, seeming to labor in flight with the blackbird in its claws. The whole chase and capture and removal to the thicket took place not more than 10 to 15 feet above the ground and lasted only a few seconds.

The usual warning note, *tee-uuu*, is given when a hawk flies over the colony. It is uttered by other members of the flock when a hawk is pursuing one of them. It may also be used when some large bird is merely passing by, even high overhead. The chorus of *schlr-r-r-up, squee* and *kit-tit-tit* of the nesting colony suddenly stops as one or two birds utter the warning note and others join in. Looking up one may see a hawk, a crow, a night heron, or even a large gull passing over. When it has gone by the *tee-uuu* ceases and the chorus recommences. This alarm has been noted on two occasions when airplanes have passed over. On at least four occasions I have heard this note when a duck hawk has soared overhead or flown by; I have also heard it when sharpshins have attacked or have merely flown by and on numerous occasions when crows have flown past.

R. M. Bond (MS.) observed the reactions of a flock of Brewer's blackbirds to the attacks of a marsh hawk (*Circus cyaneus*) as follows: "For about 15 minutes, during an extremely cold spell in February

1936, I watched about 25 Brewer's blackbirds feeding on spilled grain on the snow. I was seated in a parked car a few feet away. During the whole of this time, an adult male marsh hawk tried to catch one, and at each strike the blackbirds in danger would shift a couple of feet, easily dodging the raptor, pick up a few more grains and dodge the next blow. The blackbirds rose only a few inches from the ground each time. None was ever caught, though there were some near misses, and I suppose the hawk succeeded eventually." Bond (1947) has found the remains of a Brewer's blackbird at a marsh hawk's nest near Watsonville, Calif.

"It is a well known trait of the Brewer Blackbird to badger large birds such as hawks and crows," writes Grinnell and Storer (1924). The list of those animals harried or mobbed by this species is extensive and includes the great blue heron (*Ardea herodias*) (Bond, MS.), snowy egret (*Egretta thula*), white-tailed kite (*Elanus leucurus*), horned owl, pygmy owl (*Glaucidium gnoma*), weasel, gray squirrel (*Sciurus griseus*), cats, dogs, and humans. Attacks by the blackbird may be made singly or in groups. An incident of an attack on a sparrow hawk (*Falco sparverius*) taken from my notes will suffice to illustrate the manner of single attack on a flying bird: "Monterey County, June 13, 1948—A male Brewer's swooping at a sparrow hawk, seemingly making contacts on the lower back or tail. Would fly rapidly from behind and a little above, catch up to the hawk, sail down on it and seem to make contact uttering the squawk at the same time. At this moment the hawk would twist in flight, seemingly in order to evade the blackbird." These attacks, although seen several times on that day, did not seem to extend very far beyond the limits of the nesting colony. La Rivers (1944) mentions a sparrow hawk killing a Brewer's blackbird that was harrying it. Allan R. Phillips (MS.) says that on June 14, 1936, he "took a young male from a sparrow hawk that was carrying it away, pursued by adult blackbirds, so *Falco sparverius* is an occasional enemy of the young."

The sharp-shinned hawk has been observed to be mobbed by blackbirds. The warning note was uttered by several members of the river-mouth colony on May 23, 1945, as several other Brewer's blackbirds pursued a sharp-shinned hawk flying away and at some distance from the colony. A group of about five redwing and/or Brewer's pursued a pigeon hawk, on September 25, 1945, at the river-mouth colony. At first a much larger group of about 25 of both species of blackbirds hovered about a pine where the hawk was perched. When the hawk took flight the five took after it, pursuing from a little above the hawk. Once the hawk turned and swooped upward toward its pursuers, who immediately turned back for a short way, but when the hawk resumed the general direction of its flight, the blackbirds again took up the chase until they were out of sight.

When a subject for mobbing, as a gray squirrel, for instance, appears in a tree near a nest at the colony the effect on a large part of the colony is almost instantaneous; there is a sudden outburst of rapidly repeated *tschup* notes and many birds gather nearby. Some of them swoop at the animal and even strike it. Even though the attack is strenuous and prolonged it is not certain that it has any effect in routing the squirrel. I have no proof that the squirrel preys upon the eggs or nestlings, but since the nests are often built resting on one or more cones, the nests could, of course, be destroyed by the squirrel in taking cones.

On two occasions, at the river-mouth colony the birds were seen harassing weasels. One of these was on May 13, 1947, when 35 redwing and Brewer's blackbirds were seen hovering about 5 or 6 feet above the weasel and following it as it ran over the marsh in which the redwing colony was situated, adjacent to the Brewer's colony. The Brewer's uttered loud, excited *tschup* notes, sometimes lighting on the tule stems or the grass as close as 5 feet behind the mammal. The cloud of noisy birds thus followed the weasel for about 150 feet, until it disappeared.

Death along highways has been considered by several writers. Robertson (1930) covered the same 30.3-mile route over paved and unpaved roads 287 times during one year. He found 136 dead birds of at least 27 species, 9 of which were Brewer's blackbirds. R. M. Bond (MS.) remarks that he has been impressed by the fact that, in his experience, Brewer's blackbirds almost invariably fly clear of approaching cars on the highways, in contrast to the frequency with which redwings and tricolored redwings are struck. He counted 156 traffic casualties for the three species on roadsides in Oregon, California, and Nevada during his travels in those States in the years 1935 to 1937. Only four of these casualties were Brewer's blackbirds. The proportion of *Agelaius* would have been even higher if counts had been made where redwings and tricolored redwings were crossing a highway from nesting colony to feeding grounds."

Baumgartner (1934) records only one Brewer's blackbird among the 353 specimens of 42 species of birds recorded as automobile casualties during two trips by car to the Western States, totaling 16,700 miles, in the summers of 1927 and 1929.

Other accidental causes of death are recorded by Linsdale (1931, 1932), who found this bird to be an indirect victim of "pest control" programs in California, where it eats poisoned grain put out for ground squirrels; and by Lincoln (1931), who mentions two banding returns of Brewer's that were killed by flying into structures.

The Brewer's blackbird is parasitized by the cowbird.   Friedmann
(1929) in discussing the host species of the cowbird writes: "Bendire
thought this bird was only occasionally imposed upon, but subsequent
observations have shown it to be a common host of the Cowbird in
the plains and prairies of the west.   A. A. Saunders (Auk, XXVIII,
no. 1, Jan., 1911, p. 40), writing of the Cowbird in Gallatin County,
Mont., says, 'I have found their eggs more often in the nests of
Brewer's blackbird than any other species,' and, in the same paper,
says of the Brewer's blackbird, 'a large percentage of their nests
contain Cowbird's eggs.' "

Cowan (MS.) writes that on two occasions he has found nestlings
of Brewer's blackbird infested with *Protocalliphora*.   Dr. Carlton M.
Herman, parasitologist for the California Division of Fish and Game,
identified a flea from an abandoned nest at the river-mouth colony
in 1945 as *Dasypsyllus gallinulae*.

Fall and winter.—In the Salt Lake region of Utah the bird is
most abundant during the winter months, according to Lockerbie
(MS.), and Behle (MS.) says, "it congregates in winter in the valleys
along the Jordan River, in Utah, at ranches where livestock is fed
and at dumps and feed yards; there it occurs in great flocks numbering
100 to 1,000 individuals."   In the Rockport region of Texas, writes
Mrs. Jack Hagar (MS.), "many of this species winter in cut-over fields
with redwings, cowbirds, and great-tailed grackles," but she believes
that "great numbers go on south" of Rockport to winter, as the flocks
are larger in spring and fall.   In the Houston region, according to
G. G. Williams (MS.), "it arrives, usually, within a few days of November 1, and disappears in the last half of April.   It frequents the grazed-
over areas of our wide, flat, treeless coastal plain, as well as the stubble
fields of the great rice farms that occupy huge areas here.   It is
almost never seen in localities where trees predominate.   In the
northern parts of Harris County (of which Houston is the county
seat), where the coastal plain gives way to forest, the species becomes
less and less common, and is replaced by the rusty blackbird."
Lowery (MS.) writes: "I would regard the Brewer's blackbird as a
regular and fairly common winter visitor to southern Louisiana.
They seem to be more common in the vicinity of ponds and sloughs
which border cane fields and other open situations.   I do not believe
they are quite as numerous as the rusty blackbird, but they are
nevertheless a very prevalent Louisiana winter bird."

In the fall and winter, the Brewer's blackbird is especially gregarious,
associating in flocks composed not only of its own kind but also of
other icterids.   However, since I have always found it nesting in
some sort of colony organization, it might be considered more or less
gregarious the year around, in the Carmel region, although the size

of the foraging and roosting flocks is much greater in fall and winter.

The young, when first beginning to fly freely but while still depending on the parents for at least some of their food, flock together with other adults. These flocks soon associate with redwings and fall-sojourning tricolored redwings. In large aggregations of the three species the Brewer's are generally outnumbered by the other two combined.

Color-banded Brewer's were found to wander to places as far as 6 miles north, 4½ miles east, 6 miles south, and 4 miles northwest of the banding station at the mouth of the Carmel River. During the fall and winter the diurnal rhythm mentioned under "Behavior" is noticed, although it seems that not all birds follow this pattern rigidly.

In fall and winter the color-banded birds of the river-mouth breeding colony can be found at other colony areas, and birds that bred at these other areas may also mix with the flock at the river mouth. But there is always a nucleus of the locally breeding birds to be found at the colony.

In the months of September and October, and occasionally in November, at the river-mouth colony, there is a mild recrudescence of what I have called pairing behavior (see "Spring" and "Courtship," above). There are instances of temporary segregation into pairs, walking and flying about together (the female generally leading in flights); mutual display; occasional displacement by the male of another male alighting near the female; and occasional darts and chases. Three females which had bred in the colony the preceding spring made one trip each with nesting material to the vicinity of their former nest sites (September 25, and October 5, 1945, and September 14, 1947). With but a few exceptions these cases of pairing behavior involved individuals that had been paired the previous spring, and some that had remated for two or more of the preceeding breeding seasons. However, most of these pairing performances were not of long duration, and on many days of observation none at all was seen. As compared to spring the activities never seemed as fully developed, and on days that they were observed fewer pairs were engaged in them.

For much of the fall and winter season pairs of the previous spring, or those which had remated for one or more of the preceeding seasons, might not even be seen in the same flock. When they did appear in the same flock, they did not behave as a pair, except on the occasions of "recrudescence," as mentioned. Probably, therefore, it should be stated that even though the pairs remate on successive seasons, they do not maintain a pair bond continuously through the nonbreeding season. However, since the re-pairing of the male and his primary female in successive breeding seasons is more than a matter of chance (42 out of 45 possible cases), it might be stated that a true pair bond exists over a period of years in the Brewer's blackbird and that this breaking of the bond in the nonbreeding season should be considered as only an interruption.

No migration trends were indicated from the 318 individuals that I color-banded at Carmel, on the central coast of California. Most of those birds that survived a full year or more were found in the region in both winter and summer. However many of the banded birds that did not breed in the river-mouth colony could not be kept track of, and their exact status in all seasons was not always known. The wandering of the blackbirds over a larger area (about 12 miles in diameter) in the nonbreeding season made careful checks difficult.

There may be indications of a differential sex migration in this species. Bendire (1895) states that the "birds wintering along our northern border appear to be nearly all adult males." Gabrielson and Jewett (1940) say that in Oregon "the bird is present through the winter in small flocks composed mostly of males."

DISTRIBUTION

Range.—Western and south-central Canada to Mexico and the Gulf coast.

Breeding range.—The Brewer's blackbird breeds from southwestern, central, and southeastern British Columbia (Comox, Fernie), central Alberta (Grimshaw, Lesser Slave Lake), central Saskatchewan (Carlton, forks of the Saskatchewan), southern Manitoba (Duck Mountain, Shoal Lake), northern Minnesota (Crookston, Hibbing), western Ontario (Port Arthur) and northern Wisconsin (Hayward, Oconto, Green Bay); south to northwestern Baja California (La Grulla, San Rafael), central-southern and central-eastern California (Kenworthy, Saline Valley), southern Nevada (Lincoln County), southwestern and central Utah (Meadow, Pine Valley, Parleys Park), central Arizona (Flagstaff, Marsh Lake), western and central-southern New Mexico (Fort Wingate, Manhill), northern Texas (Canyon, Vernon), Oklahoma (Gate; casually Creek County) Kansas, northern Iowa, southern Wisconsin (Belleville, Walworth County), northeastern Illinois (Wauconda, Northfield), northwestern Indiana, and southwestern Michigan (Kalamazoo County). Summer specimens have been taken farther north in British Columbia (Kathlyn Lake, François Lake), Alberta (Banff, Deer Mount), and Saskatchewan (Prince Albert).

Winter range.—Winters from southwestern British Columbia (Vancouver), northern Washington (Bellingham Bay), central Alberta (casually, Camrose), central-eastern Montana, central Oklahoma, Arkansas (Fayetteville, Stuttgart), southwestern Tennessee (Mem-

phis), southern Mississippi (Saucier, Gulfport, Tupelo), Alabama,
Georgia (casually, Atlanta, Athens, Augusta), western North Carolina
(casually, Asheville) and western South Carolina (casually Clemson,
Chester); south to southern Baja California, Michoacán (Pátzcuaro),
Oaxaca, central Veracruz (Orizaba, Las Vigas), and the Gulf Coast,
casually east to western Florida (Panama City).

Casual records.—Casual in northern Ontario (Lake Attawapiskat),
northeastern Indiana (Ligonier) and northwestern Ohio (Spencer,
Jerusalem Township).

Accidental in Keewatin (Baker Lake).

Migration.—Early dates of spring arrival are: Indiana—Elkhart,
March 19. Ohio—Lucas County, April 3. Iowa—La Porte, March 8.
Michigan—McMillan, April 29. Wisconsin—New Richmond, March
24; Milwaukee, March 27. Minnesota—Elk River, March 15 (average
of 8 years for southern Minnesota, April 13); Fosston, March 24
(average of 7 years for northern Minnesota, April 13). Oklahoma—
Caddo, February 26. Kansas—Onaga, February 20 (average of
23 years, April 10). Nebraska—Alexandria, February 26. South
Dakota—Vermillion, March 9; Sioux Falls, March 26 (average of 4
years, April 4). North Dakota—Cass County, March 24 (average,
March 29); McKenzie County, April 12 (average of 9 years, April 18).
Manitoba—Treesbank, March 20 (median of 50 years, April 9).
Saskatchewan—Sovereign, March 8; South Qu'Appelle, March 26.
New Mexico—Clayton, March 7. Arizona—Tombstone, February
13. Colorado—Boulder, March 25; Fort Morgan, April 8. Utah—
Ogden, March 23. Wyoming—Wheatland, March 10; Cheyenne,
April 2. Idaho—Rupert, March 29. Montana—Fortine, March 20;
Alberta—Glenevis, April 8; Camrose, April 9 (median of 12 years,
April 20). California—Big Creek, March 2. Oregon—Weston, March
1; Pinehurst, March 5. Washington—Pullman, March 17. British
Columbia—Vancouver, March 18; Okanagan Landing, March 22
(median of 13 years, April 2).

Late dates of spring departure are: Veracruz—Las Vigas, April 5.
Durango—Río Sestin, April 14. Coahuila—southeastern Coahuila,
April 25. Baja California—Santa Catarina Landing, May 14. Flor-
ida—Vernon, April 8. Georgia—Athens, April 9. South Carolina—
Clemson College, April 17. North Carolina—near Asheville, April 12.
Mississippi—Gulfport, April 5. Tennessee—Johnson City, April 20.
Kentucky—Woodford County, May 12. Missouri—St. Charles,
April 26. Indiana—Ligonier, May 18. Ohio—Toledo, May 2.
Iowa—Sioux City, May 19. Michigan—McMillan, May 21. Texas—
Commerce, May 7. Oklahoma—Oklahoma City, April 25. New
Mexico—Silver City, May 15. Arizona—Fort Huachuca, May 8.
California—Fresno, April 30; Death Valley, April 29.

Early dates of fall arrival are: California—Fresno, September 7. Colorado—Fort Morgan, September 9. Arizona—Tucson, September 18. New Mexico—Koehler Junction, August 20; Silver City, September 1. Nebraska—Belvidere, August 31. Kansas—Onaga, September 10. Oklahoma—Norman, September 25. Texas—Somerset, October 4; El Paso, October 13. Michigan—McMillan, July 23. Iowa—Sigourney, August 20. Illinois—Chicago, September 13. Missouri—Freistatt, November 7. Tennessee—Memphis, November 5. Mississippi—Saucier, November 9. Delaware—Bombay Hook, October 22. North Carolina—Swannanoa, November 14, South Carolina—Dale, November 25. Georgia—Atlanta, November 14. Sonora—San Pedro, September 15.

Late dates of fall departure are: British Columbia—Cranbrook, November 26; Okanagan Lake, November 22. Washington—Blaine, November 27. Oregon—Weston, December 14; Pinehurst, October 25. California—Big Creek, November 23. Alberta—Morrin, December 1; Glenevis—October 26 (median of 17 years, August 28). Montana—Chouteau County, October 19. Idaho—Rupert, November 24. Wyoming—Laramie, December 9; Yellowstone Park, November 20. Utah—Ogden, November 16. Colorado—Fort Morgan, November 30; Beulah, November 3. Arizona—Tombstone, November 28. New Mexico—Aztec, December 9. Saskatchewan—Wiseton, November 27; McLean, November 21. Manitoba—Treesbank, November 22 (median of 47 years, October 31). North Dakota—McKenzie County, December 6; Cass County, November 28 (average, November 1). South Dakota—Sioux Falls, November 28 (average of 7 years, November 14). Nebraska—Valentine, November 3. Kansas—Onaga, November 27 (average of 22 years, November 13). Oklahoma—Oklahoma City, November 29. Minnesota—Minneapolis, November 30 (average, November 3); St. Vincent, October 20. Wisconsin—St. Croix County, November 22. Michigan—Blaney Park, October 30. Iowa—Wall Lake, November 18. Illinois—Chicago, October 27.

Egg dates.—Alberta: 16 records, May 20 to July 2; 8 records, May 26 to June 5.

California: 288 records, March 29 to July 10; 160 records, April 29 to May 30.

Nevada: 18 records, May 12 to June 10; 9 records, May 20 to May 26.

North Dakota: 14 records, May 25 to June 4.

Oregon: 29 records, May 12 to July 8; 16 records, May 20 to June 5.

CASSIDIX MEXICANUS MEXICANUS (Gmelin)

Boat-Tailed Grackle

Contributed by ALEXANDER F. SKUTCH

HABITS

[AUTHOR'S NOTE: In recent years, according to the A. O. U. Check-List (1957) the races of *Cassidix mexicanus* have been spreading slowly northward. Phillips (1950) reported a specimen of the typical race from Cameron County, Tex., and we can suppose that in time others will reach southern Texas.]

In Costa Rica the boat-tailed grackle appears to be confined to the Pacific coast, where it forages among the mangrove swamps, and is quite unknown in the interior. But in northern Central America and southern México it spreads over most of the country, and to the local inhabitants is one of the best-known of feathered creatures. Most other birds of the region are given only general or family names; CHORCHA must suffice for many kinds of orioles, and CARPINTERO does service for a great variety of woodpeckers. The familiar grackle not only bears a specific name, but male and female are honored with distinct titles. The big handsome, yellow-eyed males, clad in sleek black plumage glossed with violet and blue, are called CLARINEROS (trumpeters); the much smaller females, soberly attired in shades of brown, are known to everyone as SANATES. And well may the boat-tailed grackles have two names, for more than any other bird of northern Central America, they seek the neighborhood of man. The palm trees of the town plaza are their favorite nesting place; in the evening one sees them streaming in noisy flocks from the surrounding fields, where they forage during the day, to the village shade trees, where they roost. They abound in the coastal towns; and the stirring whistled screech of the clarinero at once recalls to my memory some palm-shaded Caribbean port; but they are scarcely less numerous in the interior, and in Guatemala frequent the towns of the central plateau, up to at least 7,000 feet above the level of the sea. They are equally at home in the most humid districts of the Atlantic littoral and amid the cacti and thorny scrub of the scorching, semi-desert regions of the interior and of the Pacific plain. But they are never found amid the forest.

But I have never remained longer than necessary in the towns, and only at Alsacia Plantation did I live on intimate terms with the grackles. The plantation house stood on the upturned end of a sharp spur jutting out from the mountains which form the boundary between Guatemala and Honduras into the level Valley of the Río

Morjá, a tributary of the Motagua. Here on the hilltop, several hundred feet above the Valley floor, a numerous company of grackles established their headquarters in the tall coconut palms that shaded the house. From my arrival in February until the following July, I awoke every morning with their voices in my ears. In the earliest dawn the clarineros repeated over and over again, in a calm, subdued voice, a long-drawn note between a screech and a whistle, which sounded very pleasant and contented, and reminded me of one running up the entire scale on some stringed instrument with one deft stroke. How different from the shrill calls they uttered later in the day, at the height of their amorous passion!

Then, as the morning grew lighter, with much commotion and clucking on the part of the females and excited calling by the males, they left their sleeping places among the coconut fronds and flew down to seek their breakfast. Many alighted on the *Conostegia*, a melastomaceous shrub with small pink flowers that grew abundantly on the grassy hillside below the house, to eat the small, black, sweetish berries. Others settled in the cowpen and on the road, where they walked about seeking small, creeping things on the bare ground, or on the lawn to forage in the grass. One morning I watched four sanates perform an office of kindness to a gaunt old cow who stood alone in the pen. One bird alighted on her back and pecked at vermin among the hair. After a slight show of resistance, she allowed a second to settle beside her and share the feast. Two more sanates moved about on the bare ground at the beast's feet, and at intervals jumped up to pluck something from her flanks or belly. They clambered over her legs and tail, performing the same service, while the cow stood patiently still.

Many of the grackles, upon leaving their roost, flew directly down into the Valley. As the morning wore on the rest melted away, singly or in small flocks, to the banks of the Río Morjá, which wound through the banana plantations half a mile away. Here they foraged along the moist shore or in the shallows, or searched among the piles of driftwood and washed-out banana plants stranded in the shoals. The clarineros walked sedately along the shingly beach and flicked small stones aside with their bills, to see what edible morsels might be lurking beneath them. On hot afternoons they delighted to bathe in the shoals at the margin of the steam, shaking wings and tail so vigorously that they sent up a shower of crystal drops which sparkled in the sunlight. One afternoon I saw a sanate approach a clarinero that was bathing and stand as close to him as she could, although there was an abundance of room elsewhere, seeming to enjoy the shower he was creating. She used him as the boat-tailed grackles of the towns sometimes employ the lawn sprinklers. Finally all the bathers flew up to the boughs of the willow and cecropia trees on the banks, vigorously shook the water from their plumage, and carefully preened their feathers with their slender bills.

As the sun sank low and the air grew cooler, the grackles flew up the hill in small flocks, sometimes cackling like a company of purple grackles, to congregate again in the coconut palms. On the way many would settle again in the *Conostegia* bushes for a dessert of berries before going to roost. From the time of their arrival until it was nearly dark, our hilltop presented a lively scene. The varied calls and squeaks of the males mingled with the constant chatter of the more numerous females. Many of the birds would settle upon the fronds of a single tree, but seemed unable to make themselves comfortable, and so flew out to alight upon another. Often they shifted back and forth a dozen times before at length adjusting themselves for the night. The fresh breeze that generally blew up from the valley at about sunset and tossed the great fronds of the coconuts made it more difficult for the birds to settle down. The long tails of the clarineros flagged back and forth as they perched on the leaves, causing them evident inconvenience. It was a delight to watch their graceful maneuvers in the wind, when they hovered, soared, and poised with dangling legs above the treetops, as sea gulls play above a windy shore.

On some particularly breezy evenings the clarineros engaged in spectacular if inconsequential sparring matches, meeting face to face and rising well above the treetops, until the wind took hold of them and twisted them around, and they were obliged to forget their opponents and devote all their attention to the maintenance of their own equilibrium. There seemed to be no point to these encounters, which were probably entered in a spirit of fun, the more to enjoy the wind by their vigorous exercise in it, as boys engage in sham battles in the water. The sun hung well above the western hills when the grackles began to congregate among the coconut trees; the last red glow was fading from the sky when finally they had all ensconced themselves out of sight among the inner fronds of the palms, and their final sleepy notes gave way to the awakening calls of the pauraque. But the clarineros, especially at the outset of the breeding season, were light sleepers, and often awoke during the night to shatter the monotonous humming of insects with their shrill calls.

At first I was happy to have such active, spirited birds as close neighbors, but at length I began to wish them elsewhere; for like their northern relatives, the purple grackles, they ate the eggs of other birds. A number of pairs of tanagers, flycatchers, thrushes, and other small birds built their nests on our hilltop, yet few succeeded in rearing their young. The grackles kept all other large birds so well at a distance, that I strongly suspect that they themselves were responsible for most of the depredations, especially since I once surprised a clarinero standing over the nest of a tiny Bonaparte's euphonia (*Tanagra lauta lauta*) which he had just torn to pieces in order to remove the eggs.

Courtship.—At "Alsacia" the sanates began to build their nests during the last week of February, and at this time the noise and excitement of the clarineros reached their highest pitch. The colony contained about a hundred birds, and there were at least two or three females for each male. The clarineros did not appear to have any particular mates, but formed merely random and temporary unions with the sanates. Yet although they shared the same territory and were idle, they never seriously quarreled. Sometimes two would stand side by side on the same perch, calling peacefully for several minutes, when of a sudden one would rush at the other and drive him away; but the bird thus threatened never turned to fight and the other forgot his animosity in a moment, so there never resulted any disagreeable encounters. The case was quite different with the sanates, who often came to grips in their disputes over their nest sites.

The clarineros were ardent in courtship. Often one flew down beside a sanate which was feeding or gathering material for her nest on the ground. He addressed her with wings half-raised and quivering, his great tail held level with his body and his head depressed, his contour feathers all fluffed out, making him appear larger than he was, while with half-opened bill he uttered pleading calls. Sometimes his voice was shrill and insistent, sometimes soft and appealing as the peeps of a little chick lost from its mother in the grass; but no matter what language he used, the ardent suitor was sure to be ignored by the busy sanate, who went resolutely about her work. At other times he perched beside her in a tree and paid his court in much the same manner. So long as the sanate ignored him, his passion would die away almost as suddenly as it began.

The nuptial flights of the grackles were aerial displays of the most thrilling sort. They began when a sanate fled the attentions of a clarinero who addressed her on the ground or on a coconut frond, or when he tried to overtake her as she flew about her usual business, whether to find food or to gather material for her nest. As she fled from him he uttered his shrill nuptial calls and increased his speed to overtake her. She doubled and twisted and dodged and used every stratagem to escape him. Far out over the valley they went, until they were high above the tallest ceiba trees. Closely as he pressed her, she always managed to elude him; and I never saw one of these breathless pursuits end in a capture. The wild chase over, the twain doubled back to the hilltop separately or together, or continued their flight to the river.

Although the clarinero was so spirited in courtship, it was the sanate who decided when she desired his attentions, and this was usually about the middle of the afternoon. Then she vibrated her wings and called with pleading peeps much weaker than his. Sometimes she might continue this for a considerable period without attracting a clarinero, although several were in sight. When a clarinero responded, he flew to her with shrill, ear-piercing cries and quivering wings, and their union was completed in a moment. Then they separated, perched not far apart with wings still violently vibrating, and continued their calls; but their notes were weaker than before and soon died away.

Nesting.—The nests of the boat-tailed grackles are usually built in colonies and are often placed near water, in willow trees or bushes along the banks of lakes or rivers, or among rushes and reeds at the marshy borders of lagoons. But often they are situated in the shade trees or clumps of bamboos about human habitations, sometimes at a considerable distance from water. At "Alsacia" all the grackles built well up on the hill, far above the river. A few of the sanates in this colony placed their nests in orange or lemon trees, or only 8 or 10 feet above the ground in low, thornless bushes growing in the pastures that surrounded the house. But the majority preferred to nest high up in the coconut palms where they roosted.

I wanted very much to see the nests, but at first was timid about climbing so high above the ground among the giant fronds of the coconut trees; for despite their tremendous size, they are only exaggerated leaves, and those of us who grow up in the temperate zones develop a prejudice against supporting our weight on leaves. At the beginning I sent up a slender lad to look into the nests and report their contents to me. But after I had watched the boy clambering in perfect safety among the fronds, I overcame my prejudices and ventured up myself. A man may climb by the aid of these giant leaves as though they were branches, provided of course that he keeps his weight fairly close to the trunk, and ascends to the top of the palm tree. The lowest dying leaf must be avoided, for it is on the point of becoming detached and may fall at the slightest touch. Above this the fronds are strong and safe. In their broad, cuplike bases fallen flowers and blasted fruits, shreds of decaying sheaths and miscellaneous debris, have accumulated and turned to mold, in which graceful pendent ferns, as well as grasses and various other plants, strike root and form a veritable aerial garden. Among these air plants hang the green clusters of ripe and ripening fruits, each larger than a man's head, and the whole bunch of some two dozen coconuts, weighing considerably more than a strong man can lift. Among these lower fronds ants establish their colonies, spiders spin their webs, and one expects to encounter, amid all this debris and decaying vegetable matter, scorpions, cockroaches, and other unpleasant creatures. Some trees swarm so with stinging ants that it is unhealthful to climb them. Few of the grackles nested among these lowest fronds.

Higher, where hang the young and the half-grown fruits, and from this point to the summit, the trunk and the bases of the fronds are enswathed by their sheaths, which soon dry to form a coarse fabric of loosely netted brown fibers, of much the texture and aspect of burlap. These sheaths tear and decay away while the fronds to which they belong remain green, with the result that the oldest are devoid of them. Here, in the axils of the younger fronds, against the coarse fabric of the sheaths, many grackles built their nests, among the white palm flowers covering the stiff upright branches of the spadix, each standing in front of its hooded spathe, fluted on the outer side.

But the place most favored by the sanates for their nests was in the very center of the palm tree's crown, between the two youngest of the expanded leaves, which stood upright face to face, providing between their broad green surfaces a cozy nook where the structures could be supported. Here the birds were in a verdant realm of their own, whence, through the narrow interstices of the fretwork made by the broad ribbons of the leaflets crossing at varying angles, they caught only imperfect glimpses of the outer world of plain and mountain that spread in a vast panorama about the lofty hilltop. The wind sent ripples along the pleated surfaces of these youngest leaves and tossed the older fronds below, the sun at high noon poured down its rays between the upright young leaves; but affairs on the ground below passed unseen and unregarded. A more attractive site for a bird's nest could scarcely be conceived. The first sanate to build in the coconut palm invariably selected this choice location, and sometimes allowed a friend to place her nest between these same fronds, but on the opposite side of the rachises. If any mishap befell one of these nests and left the position vacant, it was most likely to be occupied again so long as the grackles continued to build. Aside from its sequestered position, the nook between the youngest fronds offered many advantages, not the least of which was its cleanness, for here amid the fresh green leaves the birds were above the ragged sheaths and the vermin they harbored—above everything unclean save the droppings of the roosting grackles themselves.

It was only after these most favored sites had all been claimed that the sanates were content, perforce, to build among the mature fronds lower down. On the broad bases of the latter they found a firm and secure foundation for their nests. Five or six sanates frequently nested at one time in the same coconut palm.

As the sanates built, frequent quarrels arose among them, usually between birds which desired the same nest-site, or between those that had begun their nests too close together and were in each other's way as they worked. They menaced each other with open bills and high-pitched, irritated cries until at length one flew at the other, and the two sparred face to face as they fluttered toward the ground. Then they would separate and fly off to forage or to gather more material for their nests. These quarrels never resulted in injuries to the contestants, but caused the birds to scatter their nests in different parts of the tree rather than crowd them all in the same place.

Like oropendolas, the sanates often attempted to steal nest material. Often one bird would grasp the end of a long fiber that dangled from a neighbor's bill. Perching side by side on a coconut frond, the two tugged at the coveted prize, until at last one or the other tore it from her opponent's grasp and flew to her nest with it. Sometimes one or even two birds would pursue a third who had found a particularly desirable piece of material. The sanate who had her fibres rudely torn from her bill never manifested resentment, but soon went cheerfully off to search for more. The clarineros took no part in building the nests and viewed with indifference the quarrels between the females from whatever cause they might arise.

The completed nests were large and bulky open cups composed of a variety of ingredients. The foundation was sometimes prepared by piling in the chosen site a quantity of coarse materials such as weed stalks, small grasses torn up by their roots, and miscellaneous vegetable material. Above this the bird wove a roomy cup of coarsely fibrous stuffs picked up from the ground, chief among which were uncleaned strips from the decaying outer leaf sheaths of the banana plants; but grass stems, bits of rag and string, and fibrous weed stalks made flexible by partial decay, were also employed. The nests built in bushes and dicotyledonous trees were suspended among the finer twigs by fibers twisted firmly around them and woven back into the walls of the cup. In the coconut palms, the nests between the youngest leaves were attached by fibers woven around the leaflets. Those lower in the crown of the palm were attached to the branches of an inflorescence if they happened to touch it; but most of these nests merely rested upon the broad bases of the leaf stalks, for usually the builders found nothing suitable to which they might bind them. The completed cup was plastered on the interior, to within an inch or so of the rim, with a substantial thickness of fresh cow dung or mud, and this in turn was lined with finely fibrous material. Those sanates which worked hardest finished their nests in 5 days, but others less hurried took twice as long. The ample cups measured from 4 to 5 inches in internal diameter and from 2¾ to 4 inches in depth.

Eggs.—The female began to lay 3 or 4 days after finishing her nest. The earliest egg in the colony appeared on March 3, and during the next week many birds started to lay. Usually an egg was deposited each day until the set was completed, rarely 2 days elapsed between the laying of the first and second eggs in sets of two. Of 49 nests which we were able to reach at "Alsacia," 33 contained 3 eggs each, 15 contained 2 eggs, and there was a single set of 4.

The big, glossy eggs of the boat-tailed grackle are strikingly marked and usually very beautiful. The ground color varies from bright blue to very pale bluish gray, on which are dots, blotches and intricate scrawls of brown and black. The blue ground color of some eggs is locally washed with shades of brown or pale lilac. It would be tedious to describe all the diversities of pattern that fall within this general scheme; for the variation is so great that, if all the eggs in a populous colony were mixed together, each bird might conceivably be able to recognize her own by its distinctive markings. The measurements of 62 eggs temporarily removed from the nests at "Alsacia" average 33.6 by 23.0 millimeters. The eggs showing the four extremes measured 36.5 by 23.4, 32.9 by 24.6, 31.0 by 22.2 and 34.1 by 21.4 millimeters.

Incubation.—As she had built alone, so each sanate incubated alone, without help from a clarinero. But long before the first egg hatched, calamities began to occur. The earliest builders, who had seized upon the most coveted nest sites between the youngest fronds, found to their sorrow that this supreme and most desired position had one disadvantage which to the sober-minded would have outweighed all of its manifold attractions. It was inevitably unstable; for here the nests were supported between two fronds which still grew and bent outward in opposite directions as new leaves pushed up between them at the apex of the palm. The coarse fibers of the outer wall of the nest were, as we have seen, wrapped around the ribbonlike segments of one or both of the supporting fronds. But these formed an entirely inadequate foundation; the slender ribbons sank down under the weight of the heavy, dung-lined structures, and the eggs rolled out even when the whole nest did not fall. The swaying of the fronds in the wind hastened the undoing of the nests. We attempted to save many by tying them securely with cord as close as possible to the original position; but even with this help it was difficult to make them remain in their precarious situations, and most came to disaster. Of the many birds which had built between the youngest fronds, only one to my knowledge succeeded in bringing out her nestlings alive, and then only because we tied up her nest when it began to lean. Yet despite the terrible example constantly before her, with infinite faith a sanate would begin a new nest in the top of the palm as soon as the expansion of a fresh frond had prepared another of these deceptive sites.

The sanates which, from necessity rather than by preference, placed their nests on the broad bases of the mature fronds, fared somewhat better; yet even with them the loss of eggs and nestlings was enormous. This was largely because the birds continued to roost in the same trees where they nested—an extremely unsatisfactory arrangement. As the clarineros and the sanates not actually engaged in incubation—they were always in the majority—settled in the palms for the night, the excitement and disorder which prevailed there was so great that I wondered whether the incubating females managed to remain on their eggs. The angry cries which at this time emanated from birds unseen in the crown of the tree were doubtless from sanates trying to protect their nests from intrusion. As the nesting season advanced, the number of grackles which went to roost in the orange and grapefruit trees growing beside the coconut palms increased, possibly as a result of the protests made by the females nesting in the palms. Not only was the safety of the nests jeopardized by the disorder so prevalent each evening, but they and their immediate surroundings were defiled by the droppings of the roosting birds. Some queer things happened as a result of the grackles' disorderly habits. In one of the tallest of the palms was a nest which sheltered two nestlings. When they were nearly ready to take wing, one was found dead among the leaf bases in the vicinity of its nest, while the other in some mysterious manner made its way to a neighboring nest, where there was a single nestling 2 days younger than itself.

With the boat-tailed grackle the habit of colonial nesting is imperfectly developed, perhaps of comparatively recent origin; for conditions such as existed in the colony at "Alsacia" are a tremendous handicap to the reproduction of the species and therefore not likely to survive a long period of evolution. Oropéndolas and caciques, birds of the same family which like the boat-tailed grackles nest in colonies that contain more females than males, arrange the matter much better. At nightfall, all the males, and all the females who do not remain in their nests, retire to roost at a considerable distance, leaving the incubating females to pass the night free from unnecessary disturbances. After witnessing the disadvantages with which the sanates must contend while attempting to rear a family in a crowded colony, one understands better why so many kinds of birds select a nesting territory which they zealously defend from the intrusion of all others of their own species.

Only rarely, in nests with two eggs, did both hatch on the same day. More often, one egg hatched each day, so that in sets of three the hatching of all the eggs required three days; or two might hatch on one day and the third on another day. In a few sets the eggs were marked as laid, and these hatched in the order of laying. The incubation period was measured from the laying of the last egg to the hatching of this egg. At four nests the incubation period was 13 days; at two other nests, 14 days.

Young.—The newly hatched grackles had pale salmon-colored skin and bore a sparse but long gray down on the head, back, wings and legs. Their eyes were of course tightly closed, but they could already peep weakly and their bills when opened for food revealed a bright red interior. Their calls of hunger were heeded only by their mothers, for the clarineros were indifferent to this as to every other domestic claim. The only responsibility they assumed was that of guarding the nests. Whenever they espied a man approaching the coconut trees, their sharp *tlick lick, tlick lick* warned the females to flee from their nests, with the result that it was almost impossible to catch sight of them as they incubated or brooded. If we climbed into the crown of a tree which sheltered young grackles, the noise and excitement were immense. Clarineros and sanates, even those whose nests were safe in neighboring trees, circled around and filled the air with excited clucks. There was one particular clarinero, guardian of an isolated palm growing in the corral, who was bolder than all the others. While I rested in the crown of this tree to look at the nestlings under his tutelage, he ventured closer than any of the sanates dared to come, and often alighted near the end of the frond against which I leaned, bending it perceptibly under his weight and making me instinctively clutch another support. He interrupted his clucks with a little tinkling note rapidly repeated, and at times in his anger uttered an indescribably harsh, agonized call, which set the sanates, who all the while had been flying in circles around the tree and complaining in voices weaker than his, into faster movement and louder calling. A single female, mother of nestlings in this tree, perched on a frond and relieved her distressed feelings by giving it angry pecks.

No hawk or other large bird dared to fly close to our hilltop. Both clarineros and sanates joined in harrying the vultures, both the red- and the black-headed species, which circled too near the palms that sheltered the nests or attempted to alight upon them. They pursued the carrion feeders far down the hillside, striking them repeatedly on the back until they retreated to a satisfactory distance. I am not sure whether they had a natural aversion to birds so unclean, mistook them for hawks, or whether the vultures would actually have eaten the nestlings if given the opportunity. But the grackles even attacked a curassow (*Crax globicera*), probably the first they ever in their lives saw, and certainly not a natural enemy; for these big gallinaceous birds come into the clearings as seldom as the grackles enter the forest where the curassow is at home. One morning, while I was in the Valley, my attention was drawn by a harsh cry to a male curassow flying heavily, with labored wingbeats, high above the hillside in front of the house. As he approached the top, two clarineros and several sanates flew out from the palm trees to buffet him. Flying "near his ceiling" and doubtless already fatigued by his unwonted journey, the big bird wavered in his course and lost altitude as his assailants beat down upon him, but managed to remain in the air until he rounded the brow of the hill and was lost from view.

Just as the historian must record both the pleasant and unpleasant events—alas! too often the latter—in the history of nations, so must the bird watcher reveal the disagreeable as well as the lovable traits of the birds which come under his notice; thus, I must record the following episode in the history of the grackles at "Alsacia." There was a nest, in the coveted position between the youngest fronds of one of the smaller palm trees, which as usual with such nests I found it necessary to support by tying. It had also been considerably damaged by a high wind, but the two nestlings that it cradled continued to thrive. One morning in this tree I saw a fight between two sanates, who clinched and fluttered to the ground; but I did not give much attention to their quarrel, for such flurries were of frequent occurrence. The next day, when I climbed into the crown of this palm tree, I found that grass and weed stems had been piled on top of the dilapidated structure which only yesterday had been the home of two healthy, 10-day-old grackles. Removing the new accumulation, I found the cold, dead bodies of the nestlings interred beneath it. The intruder had apparently won the fight and must have begun her nest above the living nestlings, trampling or smothering them to death, for they were too fine and vigorous to have died during the night if they had been left unharmed.

The sanates found most of their nestlings' food on the ground and often bore it a long distance to the nest. Sometimes they flew from the river, half a mile away, bringing a morsel that they had found along the shore. The nestlings received grubs from among the grass roots, green caterpillars, and sometimes small lizards. Their eyes opened between their third and fifth days, but they continued to be very ugly little creatures until they were feathered at the age of 2 weeks or a little more. When from 16 to 19 days old, they would try to crawl from the nest if disturbed by one of my visits, but they could not yet fly. The rasping cry of distress which, at this stage of development, they uttered when touched, drove the adults to a frenzy. Those young birds which forsook the nest at the age of 19 days could not yet fly, but remained climbing around among the broad bases of the coconut fronds for 2 or 3 days longer. The full nestling period was from 20 to 23 days. At the time of quitting the nest the young birds of both sexes resembled the adult females, but their breasts were more grayish, their irides brown instead of bright yellow as in the adults, and their faces and foreheads still bare of feathers.

The destruction, during the course of cleaning the pasture below the house, of one of the isolated nests built 10 feet above the ground in an *Inga* tree, gave me the opportunity to make an experiment. Although the nest tree had been cut down, the two vigorous week-old nestlings were picked up unhurt from the ground. I placed one of these in a nest in a coconut palm which already held three 10-day-old nestlings; it was attended by their mother along with her own off-spring. The other fallen nestling was deposited in another nest in the same palm tree, from which the original occupant had vanished a few days earlier. Apparently none of the four females which at the time were building or attending nestlings in this tree, nor any of the other grackles which frequented it, took notice of the foundling, for it died of neglect after a day or two. Each female appears to attend strictly to her own nest, and to ignore the nest and offspring of her neighbors.

Soon after leaving the nest, the young grackles began to follow their mother afield as she foraged, before long going even as far as the river, where they perched on a banana leaf arching above the bank while awaiting her return from her search along the shore; or else they pursued her along the sandy margin of the stream, begging for food with vibrating wings. In May and June, the young birds became an increasingly conspicuous element in the flock—for despite their numerous failures, the sanates succeeded by persistent efforts in rearing a goodly number of offspring—and the youngsters' half-pleading, half-imperious call, *witit witit*, mingled with the whistles and clucks of the older birds. The young males continued to solicit food from mothers larger than themselves. Once I watched two youngsters, a clarinero and a sanate, alternately beg for and receive food from their mother and help themselves to the ripe banana which she was eating. Sometimes, as the young birds waited for food to be brought them in the hibiscus hedge beneath the coconut trees, they picked off the leaves and bright red flowers, or pecked at the un-opened buds, seeming to try to find food for themselves before they could distinguish what was edible.

By the first week of July the nesting season was drawing to a close.

Since the grackles had begun to nest at the end of February, they had had time for rearing two broods. One sanate, who in some unknown manner had lost her tail and got a piece of red tape entangled around her right leg, making it easy to recognize her, built a second nest and hatched a second brood after her first had been successfully fledged; but how many birds actually succeeded in raising two broods to the point where the young could shift for themselves, I was not able to determine.

During the night of July 6 the grackles which roosted in the coconut trees were restless, shifting their positions and often crying out in the dark. After this the great majority of them withdrew from the hilltop which had so long been their home. There remained only a few sanates who still had young in the nest, one whose two eggs were just hatching, and two faithful clarineros. The early mornings were strangely silent after the grackles departed.

On the Pacific side of Guatemala, where the dry season begins in mid-October or early November, 2 or 3 months earlier than in the Caribbean region, the boat-tailed grackles begin to breed correspondingly earlier. At an altitude of 3,300 feet on the Finca Mocá, a great coffee plantation situated on the Pacific slope, I found grackles feeding nestlings as early as the first week of January. These birds must have begun to build no later than the middle of December, more than 2 months before those at "Alsacia," which began to build in late February and had no nestlings before mid-March.

Food.—Few birds, I imagine, subsist upon a greater variety of food than the boat-tailed grackle, or display greater ingenuity in procuring nourishment. "Everything is grist for their mill." Their diet includes both animal and vegetable products. Much of their food is picked up from the ground, where they extract the larvae of beetles and other insects from among the roots of the grasses, and capture small lizards. They are said to hunt in freshly plowed land, following close behind the plowman. They pluck ticks and other vermin from cattle, often alighting upon the animals' backs for this purpose. They spend much time foraging in the vicinity of water. On bare shingly flats along the shores they turn over small stones by inserting the tip of the bill beneath the nearer edge and pushing forward, then devour the small crustacea, insects, worms, or the like that they find lurking beneath. It is chiefly the more powerful males that hunt in this fashion. Often the grackles wade into shallow water, where apparently they capture tadpoles and small fish. Or if the water be deep, they may adopt other modes of fishing; A. W. Anthony (Griscom, 1932) tells how at Lake Atitlán in Guatemala the grackles caught fish as they flew low over the surface of the water, seizing their prey by means of a quick snap and hardly wetting their plumage in the process. At other times, however, these grackles plunged boldly into the lake, like a tern or a kingfisher, immersing themselves to a depth of not more than 3 or 4 inches. Sturgis (1928) records that the great-tailed grackles frequent the most isolated rocks in Panamá Bay, where doubtless they devour a variety of small marine creatures. In Costa Rica, Carriker (1910) found the bird common among the mangroves of the brackish estuaries so numerous along the Pacific coast.

Like other grackles, this species pillages the nests of other birds, devouring their eggs or nestlings. In Guatemala, I surprised a male boat-tailed grackle resting upon a fence-post where a pair of Bonaparte's euphonias (*Tanagra lauta lauta*) had built a nest, well concealed in a cranny caused by decay. The roof had been torn from the little domed nest and the newly laid eggs had vanished. Although I arrived too late to catch the grackle in the act, the circumstantial evidence pointed strongly toward him as the despoiler of the nest and-devourer of the eggs. In México, Chester C. Lamb (1944) saw a male grackle seize a female yellow warbler which had dashed into the face of the bigger bird in a vain attempt to save her eggs. The warbler was killed, her skull crushed by the grackle's powerful bill.

Of vegetable food, the grackles are fond of ripe bananas and of small, sweet berries, especially those of the melastomaceous shrub *Conostegia*. They greedily eat maize, tearing up the germinating grains from newly planted fields. One Guatemalan farmer told me that his efforts to start a cornfield were frustrated by the grackles until he adopted the expedient of scattering a considerable quantity of grain about the edges of his field. This kept the hungry birds occupied until the planted maize had grown large enough to withstand their attacks. Yet this same farmer considered that the grackles, by destroying grubs and other insect pests, did on the whole more good than harm on his estate. Later, as the maize crop nears maturity, the grackles renew their depredations upon the milpas, tearing open the husks to reach the tender, milky grains, which the females at this season feed to their fledglings.

Behavior.—The big male grackle glides downward with wings set, the tips of the primaries separated from each other and distinctly curved upward by the weight of his heavy body, and with his long tail folded together upward so that the feathers lie in a vertical plane, like that of the purple grackle, and vibrating from side to side in the breeze. Usually he flies upward with heavy, resonant wing beats, like those of the male oropéndola; but at times he may fly silently. The flight of the female grackle is almost silent; but when laboring upward with long fibers for the nest streaming from her bill, her wing beats may be sonorous like the male's, but not so loud. Rarely she folds her tail feathers together in the manner of the male, but not completely. In sustained flights on a horizontal or ascending course, both sexes move with perfectly regular and rapid wing beats, neither folding their wings intermittently nor spreading them for gliding. On the ground, the grackles walk rather than hop.

Although I never witnessed a serious dispute between the male boat-tailed grackles in the colony at "Alsacia," in other regions these birds may be more pugnacious. While traveling by rail through southern México, I saw from the train window two male grackles fighting in good earnest. They clinched and rolled on the ground, continuing their battle as long as I could keep them in view.

Voice.—The range and power of the male grackle's voice are wonderful—he lacks only a set song. At one extreme, he utters a little tinkling note, rapidly repeated and very pretty, at the other, his calls are so loud that they are best heard at a distance. If one may say that a bird with so varied a language has one call which is most characteristic, that call is a single, long-drawn utterance, something between a squeak and a whistle, which rises through the scale. Then there is a resonant *tlick tlick tlick*, delivered while the bird is either in flight or at rest, and a spirited, rollicking *tlick-a-lick tlick-a-lick* which seems the outpouring of rare good spirits. There is also a rolling or yodeling call, very vigorous, and quite in contrast with the lazy, screeching note, like the slow swinging of a gate with rusty hinges, which is also a part of his varied vocabulary. Sometimes while perching the male grackle puffs himself up, swelling out all his feathers, and half opening his bill, slowly expels the air with a low, undulatory sound, such as can be made by whistling through the teeth.

As musicians, the grackles display a good deal of originality. They often invent new calls, and when they hit upon one which takes their fancy, repeat it over and over again. One bird fell in love with a pretty phrase, which sounded like *wheet-tóck*, and uttered it constantly for a week or more, until at last, like a popular song, it grew stale and was forgotten. Another clarinero was much taken with a buglelike call that went *ta-dee ta-dee ta-dee* and was really very martial and stirring, especially when heard at a little distance, for it had great carrying power. After delivering a call, the grackles frequently perch with their long, sharp, black bills pointing straight upward, a pose which displays to good advantage the sleek glossiness of their purple necks. This position is also assumed on other occasions, and sometimes two splendid birds, perching side by side on a coconut frond, point upward at the same time and hold the pose for a good fraction of a minute, looking very self-conscious. The females, too, sometimes assume this attitude, but the trait is not nearly so strongly developed among them as among the males.

The female grackles are not only smaller but quieter than the males. Their most characteristic utterance is a rapid, clicking sound, a *tlick tlick tlick* sharper and less sonorous than the corresponding note of the males. They use this while building their nests, quarreling with their neighbors, or flying. Single throaty clucks are also uttered by both sexes. Sometimes a female attempts to deliver the note that I ventured to call the most typical of the male, but hers is a weak, squeaky imitation.

Field marks.—It is scarcely possible to confuse the boat-tailed grackle with any other bird of southern México or Central America. The larger size of the male, his bright yellow eye, and his long fan-shaped tail, easily serve to distinguish him from the other wholly black or blackish members of the troupial family that inhabit the region. Perhaps, at a distance, the giant cowbird (*Psomocolax oryzivorus*) might be mistaken for a male boat-tailed grackle. But in flight, these cowbirds with red eyes close their wings momentarily after each five or six beats, while the grackles fly with regular, uninterrupted strokes—peculiarities which will serve to distinguish the two species almost as far away as they are visible.

The members of this race are somewhat larger and darker, especially the female, than those of the other races inhabiting the South Atlantic and Gulf Coast States.

DISTRIBUTION

Resident in the lowlands and tableland of México from eastern Jalisco, southern Nuevo Léon, and southern Tamaulipas to Guatemala, British Honduras, El Salvador, and northern Nicaragua; extending northward in recent years (a specimen from Cameron County, Tex., has been identified as of this race; this record has not yet been acted on by the A. O. U. Committee on Classification and Nomenclature of North American Birds).

CASSIDIX MEXICANUS NELSONI (Ridgway)

Sonoran Boat-Tailed Grackle

HABITS

Primarily a denizen of the coastal district of Sonora, and the interior of that province from Rancho Costa Rica southward, this race of the boat-tailed grackle has recently been reported from central-southern Arizona (Tucson). From Monson's boat-tail it may be distinguished by its smaller size, shorter tail, and the paler color of the female plumage.

Little or nothing has been recorded of its habits, but these are probably similar to those of the other races of this boat-tailed grackle.

CASSIDIX MEXICANUS MONSONI Phillips

Monson's Boat-Tailed Grackle

HABITS

This recently described race occurs in the plateau of northern México, in Chihuahua, and in recent years has spread northward to adjacent parts of the United States. It resembles the boat-tailed grackle, but has a less massive bill, a more slender tarsus, and the plumage of the female is noticeably darker in tone.

In its habits, as far as known, it does not appear to differ appreciably from the boat-tailed grackle.

It breeds from southeastern Arizona (Benson, Randolph), north-central New Mexico, and western Texas (Brewster County), south to Chihuahua. It has been recorded sparingly in winter in Pinal and Graham Counties, Ariz., the Bosque del Apache Refuge, near San Antonio, N. Mex., and along the Río Grande at Juarez, Chihuahua; it is presumed to winter mainly in Chihuahua, but has extended its winter range in the United States northward in recent years.

CASSIDIX MEXICANUS PROSOPIDICOLA Lowery

Mesquite Boat-Tailed Grackle

HABITS

George H. Lowery, Jr. (1938), has given the above name to the large grackles of this species that are found in the "Gulf Coast region of central southern Texas, north to at least Port Lavaca, and south into northeastern Mexico in the states of Tamaulipas, Nuevo León, and Coahuila. In Texas it is closely associated with the range of the mesquite (*Prosopis glandulosa* Torrey)." He gives as its subspecific characters: "Resembling *Cassidix mexicanus mexicanus* (Gmelin) more closely than any other form, but wing, tail, exposed culmen, and tarsus shorter; male in color almost indistinguishable, but female conspicuously different from *C. m. mexicanus*, the under parts being decidedly lighter, ranging from Light Brownish Olive to Buffy Olive; also the pileum, sides of head and neck much lighter, tending toward olive rather than brown."

The separation of this subspecies removes the type race, *Cassidix mexicanus mexicanus*, from our list, and the bird that has for so long stood in our literature as the great-tailed grackle must now be called the mesquite grackle, and restricts it to eastern and southern México, Central America and northwestern South America. However, papers by Allan R. Phillips (1940), Laurence M. Huey (1942) and Lawrence V. Compton (1947), to which the reader is referred for details, show that boat-tailed grackles of the *Cassidix mexicanus* species have been extending their ranges in the upper Rio Grande valley into New Mexico and into southern Arizona. From correspondence in 1946 and 1947 with Phillips and Lowery, I infer that there is much still to be learned as to the subspecies involved and their ranges. Roger T. Peterson (1939) also reported grackles of this species breeding in New Mexico. The species had been reported previously, as breeding in New Mexico, by J. Stokley Ligon (1926).

When I was in southern Texas, in 1923, I found the mesquite grackle to be astonishingly abundant from Matagorda Bay to the Rio Grande. It was unquestionably the most abundant bird all along the coast, as well as the noisiest and most conspicuous, almost a nuisance at times, especially in the heron rookeries.

Dr. T. Gilbert Pearson (1921) says: "One of the most noticeable, noisy, and abundant species of birds along the lower Texas coast is the Great-tailed Grackle. It possesses an astonishing repertoire of whistles, calls, and guttural sounds and one sees or hears them everywhere. On islands surrounded by salt-water it is found and one may see it also about fresh-water ponds, or in the towns and on the high prairie or chaparral lands if water of any kind is in the vicinity."

Courtship.—Mrs. Bailey (1902) gives the following account of this grotesque performance:

Seated on an oak top, where his humble spouse could see him to the best advantage, an old male would begin by spreading his wings and tail to their fullest breadth and making a crackling "breaking brush" sound which he evidently considered a striking prelude. This done he would quiver his wings frantically and opening wide his bill emit a high falsetto squeal, *quee-ee, quee-ee, quee-ee, quee-ee,* perhaps attuned to the feminine blackbird ear. But his *coup d'état,* which should have wrung admiration from the most unappreciative mate, consisted in striking an attitude, his long bill pointed as nearly straight to the sky as his neck would permit. Poised in this way he would sit like a statute, with the most ludicrous air of greatness. Incredible as it may appear, instead of standing spellbound before him, his spouse, practical housewife that she was, whatever her secret admiration may have been, through all his lordship's play calmly went about gathering sticks.

Dr. Arthur A. Allen (1944) describes it as follows:

In the lone pine the grackles were executing their courtships, accompanied by such sounds as shatter an adult's nerves, but delight children when you draw your fingers over a toy balloon and let the air out at various speeds; first low squeals and then high squeals, followed by a crashing sound as if the bird were beating its wings on dry twigs. All this accompanied a display of plumage that was equally ridiculous, for the bird first threw his head back on his shoulders and inflated himself until he appeared twice his natural size, his feathers standing on end and his enormous tail spreading. In the bright sun the brilliant iridescence of otherwise black feathers shot out gleams of purple and green. Next he threw his head forward and, as he collapsed, he rapidly fanned the air with his wings, producing the crashing sound already mentioned.

Nesting.—On the southern Texas coastal plain, from Matagorda Bay to Brownsville, we found the mesquite grackle nesting in enormous numbers in practically all of the heron colonies where there were trees or shrubs; there were sometimes a score or more nests in a single tree; nests were often built in the lower portions of the nests of Ward's herons, some were in prickly pear cactus, yuccas, and even in long grass. On May 9 and 10, 1923, we explored the heron colonies around Karankawa Bay, near Port Lavaca; the largest of these, at Wolf Point, was a densely populated colony of Louisiana, Ward's, and black-crowned night herons and reddish egrets, a few black vultures and, as I wrote in my notes at the time, "countless thousands of great-tailed grackles."

The willows, huisache, and other small trees and bushes were full of the nests of the grackles; the dense colony seemed to be much overcrowded. Many nests in the huisache trees were 10 or 12 feet from the ground, but many others in the bushes were only from 3 to 6 feet up. The nests were rather bulky structures, made of dry and

green weed stalks, and grasses.   We found nests with eggs and others
with young during May.

Major Bendire (1895) writes:

The Great-tailed Grackles are more or less gregarious at all times, and generally
breed in companies, often in considerable colonies, among the willow thickets and
chaparrals bordering the streams and irrigation ditches, or in the tops of mesquite,
ebony and colima trees, so common a feature in the lower Rio Grande Valley;
they nest less often in hackberry, prickly ash, and oak trees, as well as in the extensive canebrakes bordering the numerous lagoons and fresh-water lakes and
in the rushes in the salt marshes near the Gulf coast. * * *

According to Mr. Sennett, when breeding in swamps their nests are frequently
placed within 2 feet of the water, and from 4 to 30 feet from the ground when in
trees.   Their nests, of which I have several before me, resemble those of the rest
of our eastern Grackles in size, construction, and materials; some of them are
almost entirely composed of Spanish moss, while others are mainly built of small,
round stems of creeping plants which are flexible enough to admit of their being
securely woven together.   Mud is often used to bind the materials together, and
the upper rim of the nest is generally securely fastened to the surrounding branches
or reed stalks among which it is placed.   Some nests show no traces of mud in
their composition, but the materials forming the outer walls appear to have been
quite wet when gathered.   The lining usually consists of dry grass and fine roots,
and when near towns bits of cotton cloth, feathers, paper, etc., are often found
mixed among the other materials.

Nidification usually begins during the latter part of April; it is at its height in
the first half of May and lasts through June.   One and sometimes two broods
are reared in a season.   Young birds of various sizes and fresh eggs may frequently be found in the same colony.

Dr. Pearson (1921) says: "Near the main buildings on the Wolf
Point Ranch in Calhoun County, the prairie is decorated by two
'motts.'   In local usage the word 'mott' means a thick growth of
slender live-oak trees.   The combined area of these two motts is
certainly not over an acre and a half in extent, yet they held on May
29, not less than 1,000 nests of the Great-tailed Grackle.   The noise
produced by the birds could be heard from the deck of the yacht
where we lay at anchor half a mile distant."

Eggs.—The mesquite grackle ordinarily lays three or four eggs, but
sometimes five.   Bendire (1895) describes them as follows:

The ground color is usually pale greenish blue, and is often more or less clouded
over with purple vinaceous and smoky pale umber tints, which are usually heaviest
and most pronounced about the smaller end of the egg.   The markings consist
mainly of coarse, irregularly shaped lines and tracings of different shades of dark
brown, black, and smoky gray, and less-defined tints of plumbeous.   In rare
instances an egg is found which is only faintly marked with a few indistinct lines
of lavender gray about the small end, the rest of the egg being immaculate.
They are mostly elongate ovate in shape; a few are blunt ovate, while others
approach a cylindrical ovate.

The average measurement of 93 eggs in the U. S. National Museum
collection is 32.18 by 21.75 millimetres, or about 1.27 by 0.86 inches.

The largest egg in the series measures 36.58 by 22.61 millimetres, or 1.44 by 0.89 inches; the smallest, 28.19 by 20.57 millimetres, or 1.11 by 0.81 inches.

Young.—There is no evidence that this race differs, in the care of the young, from the eastern or Florida races.

Plumages.—The molts and plumages of the mesquite grackle are similar to those of the boat-tailed grackle, with due allowance for subspecific characters.

Food.—The food and feeding habits of this grackle are evidently similar to those of the boat-tailed. It seems to be equally omnivorous, feeding mainly on the ground or in shallow water on various forms of insects and their larvae, small crustaceans, little fishes, and whatever small aquatic animals, dead or alive, it can pick up around the shores. Some grain and small fruits are eaten, but no great damage to human interests is done. However, these grackles destroy large quantities of eggs of other birds in the colonies where they breed.

Behavior.—George B. Sennett (1878) writes:

When I think of this bird, it is always with a smile. It is everywhere as abundant on the Rio Grande as is *Passer domesticus*, English Sparrow, in our northern cities, and, when about the habitations, equally as tame. This bird is as much a part of the life of Brownsville as the barrelero rolling along his cask of water or the mounted beggar going his daily rounds. In the towns or about the ranches, he knows no fear; is always noisy, never at rest, and in all places and positions; now making friends with the horses in the barns or the cattle in the fields, then in some tree pouring forth his notes, which I can liken only to the scrapings of a "cornstalk fiddle"; now stealing from porch or open window some ribbon for his nest, then following close behind the planter, quick to see the dropping corn. With all his boldness and curiosity, the boys of the streets say they cannot trap or catch him in a snare. He will take every bait or grain but the right one; he will put his feet among all sorts of rags but the right ones; and the boys are completely outwitted by a bird. He performs all sorts of antics. The most curious and laughable performance is a common one with him. Two males will take position facing each other on the ground or upon some shed, then together begin slowly raising their heads and twisting them most comically from side to side, all the time steadily eyeing each other, until their bills not only stand perpendicular to their bodies, but sometimes are thrown over nearly to their backs. After maintaining this awkward position for a time, they will gradually bring back their bills to their natural position, and the performance ends. It is somewhat after the fashion of clowns' doings in a circus, who slowly bend backward until their heads touch their heels, then proceed to straighten up again. It is a most amusing thing to see, and seems to be mere fun for the bird, for nothing serious grows out of it.

With all their familiarity, I have seen these birds in the open chaparral as wild and wary as other birds, knowing very well when out of gunshot range. Their flight is rather slow, and when they make an ascent it is labored; but once up, with their great tails and expanse of wing they make graceful descents.

Baird, Brewer, and Ridgway (1874) tell the following story:

Captain McCown states that he observed these Blackbirds building in large communities at Fort Brown, Texas.　Upon a tree standing near the centre of the parade-ground at that fort, a pair of the birds had built their nest.　Just before the young were able to fly, one of them fell to the ground.　A boy about ten years old discovered and seized the bird, which resisted stoutly, and uttered loud cries. These soon brought to its rescue a legion of old birds, which vigorously attacked the boy, till he was glad to drop the bird and take to flight.　Captain McCown then went and picked up the young bird, when they turned their fury upon him, passing close to his head and uttering their sharp *caw*.　He placed it upon a tree, and there left it, to the evident satisfaction of his assailants.

Voice.—The vocal performances of the mesquite grackle are similar to those of its eastern relative, equally noisy and equally unattractive.

Captain McCown (Baird, Brewer, and Ridgeway, 1874) says that "these birds have a peculiar cry, something like tearing the dry husk from an ear of corn.　From this the soldiers called them corn-huskers." Friedmann (1925) states that "the notes are very harsh and suggest the sound of the crackling of twigs."

Pearson (1921) remarks that this grackle "possesses an astonishing repertoire of whistles, calls, and guttural sounds."　Charles W. Townsend (1927) was evidently better pleased with the voice of the mesquite grackle, for he says: "I had excellent opportunities to watch this bird and was struck with the great variety of its clear and at times musical notes and songs mixed with others that were not so pleasing, all so different from the songs of the Boat-tailed Grackle.　I have recorded them as a clear almost Flicker-like *week-it, week-it*, and *see, see, see;* also a clear and pleasing *wheet, whit-a, whit-a, whit*, followed by *whee-ee-ee*, the last vibratory and pleasing."

Field marks.—The large, glossy black males are unmistakable, with their enormous tails which distinguish them from the smaller grackles and other blackbirds.　The females are much smaller brown birds, with more normal tails; they are lighter brown than the females of other grackles, and decidedly lighter-colored than the females of the boat-tailed grackle, the sides of the head and neck and the under parts being light brownish olive or buffy olive.

DISTRIBUTION

Range.—The mesquite boat-tailed grackle breeds and is mainly resident from southeastern New Mexico (Carlsbad) and western, south-central, and east-central Texas (Toyahvale, Eagle Lake); south to southern Coahuila (Las Delicias, Saltillo), Nuevo León (Monterrey, Montemorelos) and southern Tamaulipas (Gómez Farías).

Casual records.—Casual in winter on Gulf coast of Louisiana (Avery Island).

Florida Boat-Tailed Grackle

Contributed by ALEXANDER SPRUNT, JR.

HABITS

With the recognition of the Atlantic coast population of the boat-tailed grackle as racially distinct, only a remnant of what was formerly considered the range of the Florida race is left to it—Florida and the Gulf coast west to Galveston and Port Arthur, Tex. Along the Gulf coast its distribution is at times discontinuous. A. H. Howell (1932) found it common at only one locality between Pensacola and Cedar Keys, i. e., at St. Marks, which lies on the Gulf, south of Tallahassee. F. M. Weston writes me (MS.) that "The boat-tailed grackle is so rare in the Pensacola region that I had been located here for ten years before I saw one. Having once found them, I was able to establish the fact that the species is resident here in very small numbers, for I have seen them in every month of the year at some time during the past 18 years." He further states that, in contrast to the scarcity about Pensacola, the bird is "enormously abundant in the vast fresh marshes at the head of Mobile Bay, about 60 miles west, and common all the way down both sides of the Bay to the restricted salt marshes just behind the Gulf beaches."

While collecting about Choctawhatchie Bay, about 70 miles east of Pensacola, Worthington and Todd (1926) found that "this species was not detected on the north side of the bay * * * but * * * on the south," where a flock of "about twenty birds, mostly females, was encountered on May 4th * * * and two specimens were secured." The Pensacola region shows some surprising ornithological gaps, and the occurrence of the boat-tail in that area is illustrative.

Westward from Mobile along the coast it is abundant, and no difficulty is experienced in observing it almost anywhere. Curiously enough, it does not appear to winter very commonly on the Mississippi coast although a common breeder. T. D. Burleigh (1944) says: "Despite the comparatively mild winters and no apparent scarcity of food, very few of these birds remain on this part of the coast during the winter months. The last small flocks are usually seen late in October, and it is the last of February or even later before they reappear again. On Deer Island [Miss.] I noted the Boat-tailed Grackle only once during the winter months. * * * It is possible that these grackles winter more commonly on the outer islands. * * * On the mainland and Deer Island, February 22 is the average date of arrival in the spring."

Courtship.—The courtship antics of this race are similar to those described for the eastern form (see p. 366). A great variety of locations may be used by the displaying or singing bird; E. A. McIlhenny, (1937), writing about the bird in Louisiana, gives a clear picture of such proceedings as follows:

Their favorite station for plumage exhibition is the top of a small bush or low tree. If these are not available, they will alight on the ground or on a muskrat house or pile of debris. Here they stay quietly for some minutes, with their feathers compressed and beak and neck pointing skyward, then suddenly one of them will give a series of squeaking, chuckling, raucous cries, during which all the feathers are fluffed, tail spread, wings half opened and vibrated rapidly, making a loud, rattling sound [see Voice]. The others of the group immediately follow the leader's example, and for a minute or two each individual is animated and noisy, only to drop back to the compressed statuelike pose. This noisy exhibition takes place either while at rest or on the wing. * * *

If, over such a group of males, flies a female seeking a mate, all of the males at once take flight on loudly flapping wings and with rattling quills, squeaking and calling in their most seductive manner, begin chasing her. Should none of this group of males attract her, she quickly outflies them and proceeds to look over other groups until she finds her choice. When a mate is selected she flies in front of and near him, leading him off to one side, until the other males in the group drop out of the chase. The pair then alights on the ground and mating is accomplished.

This race, like the eastern one, is more or less polygamous, as may be inferred from a statement by C. J. Pennock (1931): "Observe a glistening old male atop a buttonbush, in a sawgrass marsh, his seraglio close under his view." Brooks (1932) commenting on this, said, "the implication of polygamy in Mr. Pennock's concluding paragraph also calls for investigation. Seraglios are always interesting. Is it possible that we have at our very doors an Icterine with the fascinating habits of an *Oropendola?*" It is indeed possible.

E. A. McIlhenny (1937), however, remarks that, "the boat-tailed grackles are not monogamous; neither are they polygamous. They seem to be promiscuous. The female chooses her mate, who is decidedly temporary, and as soon as sexual mating is accomplished, she leaves him, and he does not attempt to follow." On the other hand, S. A. Grimes writes (MS.) that in his opinion, "the Florida birds * * * are polygamous rather than promiscuous in breeding habits." Thus, even experienced observers differ about polygamy and promiscuity.

Courtship takes place in early February in Louisiana, and in south Florida activity among the males usually starts about the middle of that month, though forward and backward seasons may vary the time. Courtship has been noted at Lake Okeechobee in late January, but a month later is more normal.

Nesting.—The nesting habits of the Florida boat-tail are similar to those of the eastern race (see p. 367). In the Lake Okeechobee area of Florida, the birds often nest in willows.

Eggs.—Unlike the eastern grackle, for which four or even five eggs have been reported in many nests, three is the usual number of eggs for the Florida race. E. A. McIlhenny (1937) states that in Louisiana the complement of eggs is "invariably three," although he once did find four in a nest. The eggs are identical with those of the eastern boat-tailed grackle.

Incubation.—E. A. McIlhenny (1937) says: "The male pays not the slightest attention to the female after copulation is accomplished; neither does he visit the nesting location in the early part of the nesting season with any regularity, nor does he assist in the building of the nest or in the care of the young." He prefaces these observations by stating that in its courtship and lack of attention to the young, "the Boat-tailed Grackle differs from any other American bird I have ever observed."

S. A. Grimes of Jacksonville, Fla., writes (MS.) that he has "never seen a male *mexicanus* of any race lend a hand in any manner to assist in nest-building, incubation, or care of the young," and Ivan R. Tompkins of Savannah, Ga., tells me (MS.) that no male he ever collected had "worn incubation patches."

As might be expected, the peculiar breeding habits of the boat-tailed grackle is reflected in the sex ratio of the young. In a polygamous species one would expect a preponderance of females, and such is normally the case with the boat-tail. Illustrative of specific figures in this regard, McIlhenny (1940), who checked 89 nests at Avery Island, La., and found that the hatch comprised 70 males and 145 females, rather more than a 2-to-1 majority. In his extensive banding operations McIlhenny found this ratio consistently carried out in trapped birds. In 1935 and 1936 he banded 1,848 boat-tails, of which 609 were males and 1,239 females, practically the same proportion. He adds the interesting observation that banding has proved that "while the females of the previous year nest as yearlings, the males do not reach the breeding age until the second year."

Another characteristic at which I have often wondered is the unusual percentage of infertile eggs in nests of the boat-tail. On many an occasion, when investigating the home life of this bird and examining nests of young, I have found at the bottom an unhatched egg or even two; and now and then a search of the nests after the season has revealed these lonely reminders of an unborn progeny. I have not heretofore mentioned this in print, nor have I ever made any systematic count of the occurrence of this peculiarity. The only author who ever has, as far as I know, is E. A. McIlhenny (1937), who found in one Louisiana colony that" twelve out of nineteen nests examined contained one egg each that did not hatch, and three out of nineteen * * * contained two." He also found that in the first nesting there were no infertile eggs, in the second an occasional one, while in the third, "the majority of nests contained one or more infertile eggs." It may be that the unique breeding habits of the male are reflected in this manner, or perhaps these are examples of lowered vitality, decreased virility, and the like.

Plumages.—The comments given in our account of the eastern race of this species apply to this one as well (see p. 368).

Food.—In his full and interesting study of the boat-tail in Louisiana, E. A. McIlhenny (1937) notes that the food supplied to the young "varies considerably. On some days it is almost exclusively small fish; on other days it may be spiders, and on still other days almost entirely crickets, grasshoppers or other insects. * * * Then again, when a batch of dragon-flies (either *Libellula* or *Diplax*) is coming off, the food supply consists entirely of dragonfly nymphs. On other days, if tadpoles or small frogs are especially abundant, these will constitute the food for the young. The Cricket Frog (*Acris crepitans*) is the one most used. I have not seen seed or grain or plant food fed to the young." He adds that caterpillars are taken now and then by boat-tails, which in so doing perform a distinct service to agriculture. He says: "Frequently, in the autumn, fields of soybeans may be infested by great numbers of caterpillars which sometimes destroy the entire bean crop. When the Boat-tails find such an infestation they flock to these fields in enormous numbers and do not leave them until all caterpillars are eaten." Thus, while some of the bird's food habits may not seem to be of marked benefit to man's interests, this, at least, certainly is, and so is the bird's destruction of such insects as crickets and grasshoppers.

Behavior.—Like the eastern race, this boat-tailed grackle stays fairly close to water. It goes further inland from the coasts, wherever streams or ponds occur, but the only area where it really penetrates far into the interior seems to be in Florida. Occurring on both coasts, the birds are scattered across and through the Peninsula, and one finds it wherever there is any swampy or river-lake habitat. However, it is very much of a city bird as well as rural, and is found breeding in Jacksonville, as well as at practically every small farm on the east coast of Florida. In its tendency to feed and to spend most of its time on the ground this bird is also like its eastern counterpart. It is at times predatory, not only attacking other birds but even on occasion practicing cannibalism.

Interesting instances of preying on other birds are given by E. A. McIlhenny (1937), who states that the first instance he witnessed occurred in 1911, while he was in the company of George Bird Grinnell, on the coast of Louisiana. "We were in my big launch," he says, "anchored * * * off the mouth of Bayou Michow. Near the boat was a stake * * * on which a swallow had alighted. A male Boat-tailed Grackle flew out from land, coming to the stake to alight. The swallow did not move until the Boat-tail was almost upon it, when it spread its wings, but the Boat-tail gave a quick snap and killed it * * *. The grackle sat on the stake a half minute or so looking at its victim floating on the water, then swooped down, picked it up and went ashore with it." He adds that he has "frequently seen male Boat-tailed Grackles feasting on ducks that had been killed and drifted to shore." The muskrat trappers of Louisiana complain that these birds ruin the pelts of caught animals by pecking into them and eating the flesh. The writer heard many such complaints from trappers in Cameron Parish when investigating muskrat trapping there in 1934, and was shown some animals which had been thus disfigured.

Injured birds are taken by the boat-tailed grackle when opportunity offers, and McIlhenny lists broken-winged red-backed sandpipers (*Pelidna alpina sakhalina*) as such victims. Even its neighboring redwings are not safe from it, as the same author has seen grackles devour young in the nest while the impotent parents strove vainly to drive away the marauder. It is also a confirmed egg thief and seeks out heron, egret, and other such nests to indulge this appetite. On the Texas nesting islands I have often seen the boat-tailed grackle despoiling nests of the reddish egret. In such instances it is not so much the predation of the grackle as the attitude of the egret which is interesting, for the latter simply stands by with a most fascinated expression and calmly watches the proceedings, making not the slightest effort to interfere with them!

Much to my surprise, I discovered osprey-jaeger tactics among these grackles in the Lake Okeechobee area, in Florida, while conducting Audubon wildlife tours there in 1941. One of the tour routes lay along the road which skirts the northern shore of the Lake and a great attraction of it was the fact that flocks of feeding eastern glossy ibis (*Plegadis guarauna*) were to be seen every trip. These flocks, often associated with snowy egrets (*Egretta thula thula*), were attended by numbers of boat-tails as they probed about for crayfish.

When an ibis secured a crayfish it would, instead of gulping its prey at once, spring into the air and fly upward. Instantly, it would be beset by grackles which almost invariably either snatched the crayfish from the ibis's beak or forced it to be dropped, whereupon another of the tormentors would seize it. This victimization, not simply an isolated occurrence, was indulged in regularly and we saw it many times. Always, from the observer's standpoint, it was a spectacular performance, and at some little distance the birds looked like great grains of dark corn popping from a giant popper.

At times the boat-tail bedevils species much larger than itself, setting upon them and driving them off with vituperative and vociferous energy. Audubon relates that he has watched "seven or eight of them teasing a Fish Hawk for nearly an hour, before they gave up the enterprise." It is not unusual to see them converge on turkey and black vultures (*Cathartes aura septentrionalis*) and (*Coragyps atratus*), these slow-moving unfortunates having no protection from their nimble and persistent tormentors but flight.

The boat-tail often feeds in close proximity to cattle, both in barnyards and on the open range, principally to secure the insects disturbed by the animals' feet. Whether they actually take ticks from the hides of cattle I am not sure, but I have, on many occasions, seen them alight on the backs of cows. The Florida crow (*Corvus brachyrhynchos pascuus*) definitely secures ticks in this manner and is highly regarded in the cattle sections of that State as an aid in controlling the screwworm. About Lake Okeechobee the cattle are fond of entering the drainage canals, which are choked with water-hyacinth (*Eichhornia crassipes*); from knee to shoulder depth, there, standing in water they feed to repletion on this plant. At such times one can see numbers of boat-tails about them, walking about on the floating vegetation or actually perched on the animals' backs, snapping up insects stirred up by their movements.

Albinism occurs in this form as in the eastern race (see p. 371). During February and March 1946, such an individual was observed on six occasions on the north shore of Lake Okeechobee, Fla., always within one hundred yards of the same spot.

Voice.—The voice of this bird is similar to that of the eastern race (see p. 371), including the characteristic rolling or rattling sound, as may be seen from the following accounts. A. H. Howell (1932) describes a bird he heard near Jupiter Inlet, Fla., as ending its song "by a peculiar, guttural, clattering sound that seemed to be of vocal origin, though accompanied by a fluttering of the wings." E. S. Dingle (1932) remarks: "Besides the great number of sounds that issue from its throat, one frequently hears a curious rolling noise, made by the wings when the bird is perching, but occasionally during flight." F. M. Chapman (1912) has likened this "singular rolling call" to the sound produced by a coot in pattering over the water.

That it is instrumental, indeed, is the first impression experienced by all who have written about it, but to those who have followed it subsequently it is plain that this is not the case. The two most concise and detailed accounts of recent years are those of Francis Harper (1920) and C. W. Townsend (1927).

Harper says that while studying this grackle's voice in Florida he—

Began to pay close attention * * * particularly to that part of it which Chapman describes as a "singular rolling call, which bears a close resemblance to the sound produced by a Coot in pattering over the water". * * * I noticed that the bird * * * vibrated or slightly fluttered its wings, so that their tips appeared to strike either together or against the upper side of the tail. At the same time the bill had the appearance of partly closing. I therefore concluded that the sound was not vocal, but wing-made; a number of subsequent observations strongly confirmed me in this opinion.

It was not until my last morning in Florida * * * that I was undeceived. I then had an excellent view of a bird * * * and saw that its wing-tips did not touch during the final part of the song, though they vibrated a little. A little later another bird * * * did not appear to vibrate its wings at all * * *. I could plainly see the bill in a sort of rattling motion, however, and finally realized that it was the rapid striking together of the mandibles that produced the sound.

Almost identical with Harper's first impresions and following conclusions, are those of Bradford Torrey (1894) who says, in writing of this sound, "that the sounds were wing-made I had no thought of questioning. Two days afterward nevertheless, I began to doubt. I heard a grackle 'sing' in this manner * * * wing-beats and all, while flying * * * and later still, I more than once saw them produce the sounds in question without any perceptible movement of the wings, and furthermore, their mandibles could be seen moving in time with the beats. * * * My own * * * conjecture is that the sounds are produced by snappings and gratings of the big mandibles." Townsend (1927), who quotes both Harper and Torrey on the matter studied the sound both in Florida and South Carolina, and he too was at first under the impression that the noise was wing made, for he says in his field notes that "they flutter their wings slightly, making instrumental music in the form of a rattle." Later observations, however, caused him to change his mind, for he says:

On several occasions I noticed that during the rattle the wings were sometimes moved but little, or were motionless. Once or twice I saw one wing slightly elevated but not vibrated. I also heard the rattle many times given in flight, and there was no perceptible modification of the actions of the wings at the time. I think it can be definitely stated therefore, that the evidence eliminates the wings from any causative action of the rattle, although the vibrating movement is generally present and exactly synchronous with it. * * * But my observations lead me to think that the rattle is vocal, modified by throat vibration and not made with the bill.

Enemies.—S. A. Grimes of Jacksonville, Fla., writes that at times nests of the Florida boat-tail are invaded by black ants. He says that while examining a colony near Jacksonville Beach, "the northernmost outpost of the brown-eyed birds," he found that "black ants had taken possession of all the nests, filling the interstices with their larvae and pupal cocoons. Only one grackle had held out against the ants. The eggs in her nest were pipped, but it was evident the young stood little chance of survival."

These specific reasons why many young do not survive hardly

suffice to explain the high mortality among nestling boat-tails. Facts on this phase of the life history are almost totally absent from the literature. I must confess never having mentioned it, and in this respect am as much at fault as any. McIlhenny (1937), who has spent much time in observing the bird and its habits in Louisiana, found in a detailed study of 74 nests on Avery Island, La., that success in raising young was only 54 percent; only 20 nests raised 3 young from that number of eggs; 26 raised 2 young; and 5 nests, 1 young. In 23 nests the entire setting was lost; nor could he discover the reason.

One possible cause for the loss of eggs is suggested by Grimes' notes. He states that while investigating nesting boat-tails in the Amelia River marshes of Florida he found "six or eight nests, some with and others without, fragments of eggshells. * * * Worthington's Marsh Wrens (*Telmatodytes palustris griseus*) were numerous in the marsh and may have been guilty of the egg puncturing."

McIlhenny (1936) lists the purple gallinule (*Ionornis martinica*), as well as "many other species," as preying on young boat-tails, and gives the following interesting example:

On Sunday, May 10, Stanley Solar and I were observing a large colony of nesting Boat-tailed Grackles. * * * We had already remarked the large number of empty nests, that the Sunday before, had contained small birds. We heard a young grackle crying in distress, and on going toward the place from where the noise came, saw a Purple Gallinule standing on the edge of the nest holding with one foot a half-grown grackle while it deliberately tore at its back with its beak. On our nearer approach, the gallinule took the still living young grackle in its beak and flew with it about 75 yards to the pond's bank, where we watched it tear it to pieces and eat it. It first tore a hole in the back of its victim, and pulling out the viscera in sections, swallowed the pieces as they came free. It then tore bits of tender flesh from the body, paying no attention to my approach in a boat to within about sixty feet of it.

Field marks.—Essentially similar to its eastern counterpart (see p. 373), this race may be distinguished in the field by the iris, which is dark brown, whereas it is yellow in the eastern one.

DISTRIBUTION

Range.—The Florida boat-tailed grackle breeds and is mainly resident, but wandering in winter, along the shores of the Gulf of Mexico from southeastern Texas (Galveston, Port Arthur), southern Louisiana (Ged, Madisonville), southern Mississippi (Bay Saint Louis, Deer Island), southern Alabama (Chuckvee Bay, Alabama Port) to Florida (Bay County); south to the Florida Keys.

Egg dates.—Florida: 41 records, March 3 to June 4; 24 records, March 20 to April 16.

Texas: 103 records, April 3 to June 9; 54 records, May 1 to May 22.

CASSIDIX MEXICANUS TORREYI Harper

Eastern Boat-Tailed Grackle

Contributed by ALEXANDER SPRUNT, JR.

HABITS

During my boyhood I was accustomed to spend each summer on Sullivan's Island, a beach resort across the harbor from Charleston, S. C. This stretch of sea sand, bearing little vegetation other than bushes, small trees, and grass, has bulked largely in history for here was the palmetto-log fort which, commanded by General William Moultrie, saved Charleston from British invasion in 1776 by beating off the fleet of Sir Peter Parker. Again in the 1860's Fort Sumter, a few hundred yards off the eastern end of the Island and directly in the bottle-necked harbor entrance, withstood for four years the attacks of the Federal fleet.

It was along the beaches of this Island, front and back, that I made my first field studies of the birds of the Carolina Low Country. As a boy I roamed Sullivan's Island from end to end and across, haunting its inlets, its myrtle thickets, and its grassy flats. There I began my life list and there I started, as what boy has not, my first collection of eggs. The first "cabinet" for this collection was a deeply cupped nest of what we called the "Jackdaw," a name by which many southern coastal dwellers still know the boat-tailed grackle. In it were treasured specimens (one of each, blown with a hole at each end) of eggs of the nesting birds of the Island. Thus it was that this grackle was literally one of the first avian species I came to know, and this association continues today, for it is a daily sight about my home. Long contact has not diminished my interest for there is much about this fine bird to attract and hold the attention of any student of ornithology. Its handsome plumage, remarkable vocal efforts, and peculiar breeding habits all combine to make it an object of unusual interest.

Spring.—The boat-tailed grackle is not much of a migrant. Just what volume of movement may take place from the south Atlantic area to the northern limit of its range in southern Delaware is uncertain. It is generally resident wherever found from Tidewater Virginia southward. It may appear to be more common in winter in many localities because of its gregarious habits, but I have never noted any appreciable seasonal change in population numbers in coastal South Carolina, and this seems to be true in North Carolina and in Georgia.

In Virginia it appears in spring, according to H. H. Bailey (1913) "early in April" and nests as far north as Accomac County, with Hog Island supporting the largest concentration of breeding in the Tidewater area. "A few" he continues, "may be found as far north as Cedar and Chincoteague Islands," and he concludes with the observation that the species is "extending northward each season." This last has definitely proved true, and recent years have seen the boat-tail nesting in Maryland and Delaware. Accurate arrival dates are not available but Cottam and Uhler (1935) found it "obviously nesting" at Sinepuxent Beach near Ocean City, Md., on May 22. A definite breeding date for Delaware is illustrated by the discovery on May 5, 1933, of a nest with eggs near Milford, by Herbert Buckalew (1934). He states that the birds nested in the same area in 1934.

Recent observations have shown the boat-tail now to be resident in Virginia. J. J. Murray writes me that he would "sum up the present status of the bird in Virginia as follows: common at Back Bay at all seasons; fairly common on lower Eastern Shore (Northampton County)."

Courtship.—It would almost seem that the boat-tail is conscious of his good looks, for few birds display such elaborate posturings and grotesque antics before the female. Indeed, it is not necessary for the male to have an audience of prospective consorts; often he is seen performing with no female nearby. Spreading his wings and tail in a wide variety of poses, he bows, bobs, sidesteps, and jumps about in a great flurry of excited movements. The undoubted beauty of his glossy plumage, and the brilliantly metallic reflections of his feathers appear to wonderful advantage under such circumstances.

One particular posture, frequently indulged in, is highly characteristic. It is accompanied by no movement whatever and is for that reason perhaps even more striking. Often, in the midst of great activity, the bird will become quite still, then raise its head high, with the beak pointing straight up, neck stretched vertically, and remain so in statuesque immobility for many seconds, sometimes minutes. When several are performing in this way at a time, they present a ludicrous appearance, the wings drooping slightly, the huge tail rigid and every head pointed upward as if they were intent on watching something hundreds of feet above in the sky. Then suddenly, the pose is broken and they return to a vociferous and active pursuit of other antics.

Vocal accompaniment of practically all other poses is invariable, and the din resulting when numbers are engaging in courtship is astonishing. The ground, bushes, trees, and telephone poles are used in these performances; where the bird is at the moment seems to make little difference.

As noted in connection with the Florida race (see p. 358), the male by no means confines himself to one mate. The mention made by writers of the gregarious nature of the bird, and the use of such terms as "loose colonies" or "small groups" in describing its domestic habits do not convey an accurate picture of the real state of affairs. F. M. Chapman (1922) states: "It is unknown whether the Boat-tail has more than one mate," and A. H. Howell and H. C. Oberholser, both of whom have written extensively about southern birds, say nothing of this matter in their accounts. The fact is, as T. G. Pearson and the Brimleys (1942) state, "the Jackdaw is decidedly polygamous," and all the evidence I have been able to secure personally convinces me that this is correct, even though some experienced observers suggest that the bird is promiscuous, rather than polygamous.

In Georgia and South Carolina the average dates for courtship activities occur in mid-March, with eggs laid by early or mid April. North Carolina egg dates occur in late April, while in Virginia, Maryland, and Delaware, they average from May 5 to 20.

Nesting.—The nest is constructed by the female alone and is composed of grasses and mud, rather bulky and very firm and compact. Semidecayed rushes, flags, or marshgrass is usually the foundation; when this material dries and hardens, the result is an exceedingly durable structure that is deep and basket-shaped. It is placed in various aquatic growths such as sawgrass (*Cladium effusum*), flags (*Typha latifolia*) and bullrushes (*Spartina alterniflora*), all being typical overwater locations, the growth varying with the locality. On the south Atlantic coast many colonies are over dry land but always near water. A favorite nesting shrub in the Charleston area is the wax myrtle (*Myrica carolinensis*), very like the northern bayberry. Now and then the live oak (*Quercus virginianus*) is used, and in such cases, of course, the nests are at much greater elevations, at times between 40 and 50 feet. In the great majority of situations, elevation varies from 3 or 4 feet to about 10 or 12.

Eggs.—The eggs vary in number from three to five. Apparently, any excess of three is peculiar to the eastern boat-tail and not to the Florida race. I have found four on scores of occasions on the South Carolina coast and sometimes five. The latter number is unusual, the former all but the rule. Audubon gives "four or five" as the set number, C. A. Reed (1904) puts it at "three to five." Two and sometimes three broods are raised.

Bendire (1895) describes the eggs as follows:

The eggs of the Boat-tailed Grackle resemble those of the preceding species [great-tailed grackle], both in shape and coloration, excepting that the cloudy purple vinaceous and pale umber tints are generally more evenly distributed over the entire shell, when present, and are not so noticeable at the small end of the egg. In some instances the lines and tracings with which they are marked are also perceptibly finer as well as more profuse, being more like the markings found in the eggs of the Baltimore and Bullock's Oriole. They also average somewhat less in size.

The average measurement of 98 eggs in the U. S. National Museum collection is 31.60 by 22.49 millimetres, or about 1.24 by 0.89 inches. The largest egg measures 34.29 by 24.64 millimetres, or 1.35 by 0.97 inches; the smallest, 27.94 by 21.59 millimetres, or 1.10 by 0.85 inches.

Incubation.—Incubation consumes 14 days and is accomplished entirely by the female. Here again, confusion exists among writers. H. C. Oberholser (1938) states; "It takes about 15 days to hatch the young, in which performance the male seldom assists, although he does aid in taking care of the young." However, it is the universal and confirmed experience of those who have studied the bird on its nesting grounds that the male never assists in incubation and does not aid in the care of the young.

Audubon (1834) gives a curiously mixed account of the nesting behavior, in that he intimated that both birds build the nest, which is an error, and that when this is done, the male departs and shows no further interest in the domestic proceedings, which is correct! He states further that the male "places implicit reliance on the fidelity of his mate * * * many *pairs* now resort to a place previously known to them, and in the greatest harmony construct *their* mansions. * * * Each pair choose their branch of smilax." Also that the birds repair last year's nest if any of it still exists, but if not, "they quickly form a new one from the abundant materials around." The reader certainly gathers the impression that both birds engage in nest building, which is not the case. However, Audubon then observes that after the eggs are laid "all of the male birds fly off together and leave their mates to rear their offspring." Rev. John Bachman's observations, so frequently of value to Audubon, who quotes him at length, bear out this practice; they can be summarized by Bachman's statements that he "never found the males in the vicinity of the nests from the time the eggs were laid," and that "the females alone take charge of their nests and young." The experience of present-day observers confirm these observations, and although contemporary writers say remarkably little about the apparent refusal of the male to take any part in nesting activities, the statements given in the account of the Florida race (see p. 359) should be conclusive.

Plumages.—Dr. Chapman (1922) writes:

The difference between the sexes is more pronounced in the Boat-tailed than in the Purple Grackle, the female of the former being a generally brownish bird with small trace of the glossy plumage of her mate. Furthermore, she has a much shorter tail. Young birds of both sexes resemble their mother. The post-juvenile molt is complete. The female acquires a plumage essentially like that of the adult, but that of the male is much duller than that of the mature bird. There

is no spring molt and the shining fully adult plumage is not donned until the
first post-nuptial, that is, second fall molt, after which there is no further change in
color.

Food.—If any bird exhibits catholic tastes in its diet, it is the
boat-tailed grackle. Practically anything is fish which comes to its
net, and literally, fish, flesh, and fowl, as well as grain go to make up
its food. Generally speaking, it might be said to be a grain eater in
fall and winter and a flesh eater the rest of the year.

When indulging its highly gregarious habits in the fall, considerable
waste grain is consumed, predominantly corn and rice. Audubon
(1842) noted the rice-eating propensities by saying that the boat-tail
"commits serious depredations in such green fields." Some damage
results to these crops, particularly on the gulf coast, when grackles
descend on both standing and stacked grain. Similar damage was
once widespread on the South Carolina coast when rice was such a
golden crop there. These birds sometimes follow spring planting and
uncover grain as it is sowed. H. C. Oberholser (1938) points out,
however, that "not all the consumption of grain should be considered
injurious, since a considerable portion of this obtained is probably
waste, gleaned from the fields after harvest."

F. E. L. Beal (1900) writes that an examination of 116 stomachs
revealed that the food was made up of 40 percent animal and 60
percent vegetable matter. "Crustaceans amounted to about two-
fifths of the animal food in the stomachs examined, and comprised
crawfishes, crabs and shrimp. Grasshoppers are eaten in July and
August, but few in other months. Beetles and various other insects
are taken in small quantities. Grain, chiefly corn, constitutes 46.8
percent of the total food, and is taken in every month of the year,
and as part of this is corn 'in the milk,' some damage must result to
this crop."

During the spring and summer the food consists largely of a wide
range of aquatic life—fish, frogs, insects, crustacea, and spiders.
The boat-tail's ability as a fisherman is considerable, and it is often to
be seen wading in pools or marshy creeks, up to its belly, making
accurate stabs of the beak at minnows of various sorts. In some of
these maneuvers it immerses the entire head, in others it hovers like
a petrel. The boat-tail seems very fond of the crayfish, and often
searches this creature out on its own; but, as related under "Behavior,"
it sometimes seizes them from other birds, notably the eastern glossy
ibis and probably some of the herons. The bird is no mean performer
as a flycatcher, and secures various insects on the wing with apparent
ease.

In that part of the range where the cabbage palm (*Sabal palmetto*)
occurs, and this is a large part, too, the boat-tail resorts to this tree to

feed upon the small, blackish berries which are borne in great clusters on its pendant stalks.

Behavior.—The boat-tailed grackle is essentially a coast dweller. Showing such a decided preference for salt water that it is seldom seen anywhere else, it frequents at all seasons the barrier and sea islands, the marshes, and the shoreline. Occasionally it follows up some of the tidal rivers for a short distance inland; but Audubon (1834) noted that it "seldom goes further inland than forty or fifty miles, and even then follows the margin of large rivers as the Mississippi, the Santee, the St. John's and the Savannah." For coastal South Carolina, 40 miles inland would be liberal and such distances are more apt to be characteristic of the Florida race on the gulf coast. Ivan R. Tompkins writes me (MS.) that he once saw "a number of boat-tails at Nahunta, Brantley County, Ga., on March 14, 1938. Nahunta is about 29 miles west of Brunswick, on the coast, and the birds were busy around a gum-swamp habitat. All were definitely yellow-eyed birds." In the range of the species in South Carolina, the writer cannot recall having seen the species more than about 20 miles inland. It is very much of an urban, as well as a rural, bird; it abounds in many coastal towns and cities, occurring as a breeder in such seaboard cities as Wilmington, N. C., Charleston, S. C., and Savannah, Ga. Brunswick, Ga., has a very large population, and even the hurried tourist can hardly fail to be impressed by the number of these birds in the many live oaks which add so much to the attractiveness of that community. At The Cloister, a resort hotel on Sea Island, near Brunswick, the birds are semidomesticated in and about the patio, largely through the efforts of the genial hostess, Mrs. G. V. Cate, and will take food and pose for photographs for visitors with remarkable tameness.

Conspicuous as this bird is at any season, it attracts perhaps more attention in the fall and winter, for at these seasons it is particularly gregarious, going about in great flocks. The term "darkening the sky" seems still applicable to grackles, redwings, and cowbirds, such veritable clouds of them are to be seen frequenting the grain fields in the south in winter.

The boat-tail is markedly terrestrial. It spends much time on the ground searching for food, both in dry fields and the mud of marshes or extensive flats, where the rather long legs result in its being a good walker. Its attitude is at all times alert and vigorous, with the huge tail held high and the gait firm and sure-footed, though one gets the impression of a waddle at times. In windy weather the tail appears to be a positive encumbrance. Its great area catches the wind like a sail and at times turns the owner completely around. Sometimes the bird is all but upset, and is often obliged to sidestep ludicrously in order to turn broadside to a brisk breeze.

In flight, the wings are moved rapidly, and here again the tail seems to get so much in the way as to constitute a handicap. The flight is somewhat labored in appearance, particularly into the wind, and one is strongly reminded of the slow progress of a blimp in a headwind when watching the boat-tail in like circumstances. The wings often make a pounding noise as the bird passes close overhead, making evident the effort being put forth.

Like the Florida race (see p. 360), the eastern boat-tail tends at times to be a predator. It is also a highly proficient fisherman, often wading into pools and streams to belly depth, or stalking about the shallows with tail held high, making occasional and accurate jabs at minnows.

Slightly wounded specimens are exceedingly agile and lead one an exhausting chase, at the termination of which they bite and scratch the collector's hands vigorously, often to bloody effect.

This species is occasionally subject to albinism and, as might be supposed, the effect is invariably striking. While I have never seen a totally albinistic specimen, on two occasions I have observed it in the partial state. In January 1944, in company with E. B. Chamberlain of the Charleston (S. C.) Museum, we unsuccessfully pursued such a specimen on James Island, which was very wild. A day or two later however, the bird was secured and brought to the museum, where it is now preserved. The body is white, the wings and tail black.

Voice.—Of the several characters which make this bird conspicuous, its vocal accomplishments are in the very front rank. There may be more noisy birds but if so, I have yet to hear them! One or another may be noted for vociferous effort but the boat-tail is without equal. While some of its productions are not unmusical, most can hardly be described as anything but raucous, harsh, guttural and rasping. Translation into words of even approximate equality is impossible; at any rate, none of the "chips," "churs," or "kwees" I might invent would go far toward interpreting its astonishing medley of what might as well be groans, grunts, clacks, and shrieks.

In that characteristic style of the time, Nuttall (1832) dignifies the boat-tail's vocal attainments by saying that "their concert, though inclining toward melancholy, is not altogether disagreeable." Audubon renders the calls into "*crick, crick, cree*" with a variation in more pleasing vein during the "love season" of "*tirit, tirit, titiri,* rising from low to high with great regularity and emphasis." There is little point in giving other verbal renditions of the voice. Suffice it to say that, during spring and early summer it is all but incessant and no one can be within range of the birds' voice without being abundantly aware of their presence throughout the entire day.

The boat-tailed grackle produces one sound, however, which has attracted the study and conjecture of many ornithologists, and without mention of which, no account of the voice would be complete. I speak of the remarkable rolling, or rattling sound so thoroughly characteristic of the "jackdaw." Many have noted it, some have commented on it, but no one who watches or listens to this grackle very long can fail to be impressed by it. Curiously enough, Audubon does not mention it at all, although he could hardly have failed to notice it, and his great friend and collaborator, John Bachman, also omits reference to it. Nuttall however, while not stressing this sound, at least recognized its existence, though he intimates that word of it came to him second-hand for he says (1832) that "some of its jarring tones are said to bear a resemblance to the noise of a watchman's rattle." That this refers to the sound in question is not to be doubted.

Contemporary writers have used the words "rolling" and "rattling" to describe the sound, but the point in controversy is whether it is instrumental or vocal. A. T. Wayne (1910) says: "A peculiar habit of the male of this species is to perch upon a limb of some tree and with their wings make a loud rolling sound. This peculiar noise is also frequently made while the birds are flying."

From his observations of the Florida race (see p. 363), Townsend (1927) concluded that the rattle was vocal, not mechanical, and my notes of March 24, 1926, made at Charleston, S. C., explain this: "The bird was seen on a tree in a favorable light within twenty yards and studied with eight power prismatic glasses. After three or four wheezy trills with bill wide open, he would partly close it and appear to gulp and the feathers of the throat vibrated as the guttural rattle was produced. I could see the bill vibrating also, but it did not occur to me then, nor does it seem probable to me now, that the bill made the sounds. The vibration of the throat would seem to point to its vocal origin. Certain parts of the song of the purple martin are very similar to this guttural rattle, and the throat of the bird may in the same way be seen to vibrate. I observed this at Mr. Wayne's home."

I have watched literally hundreds of boat-tails make this rattling sound and have studied them at very close range, with and without binoculars, in South Carolina, Georgia, and Florida, particularly the latter. While for some time I could not decide whether the sound came from the throat or the clacking of the mandibles, it was perfectly clear that it was most certainly not produced by the wings.

On dozens of occasions the rattle sounded with the wings absolutely motionless, not even the slightest vibration of their tips taking place.

However, I often noted that immediately after the rattle had sounded the wings were fluttered strongly. At times it appeared that the mandibles produced it, for they were definitely moved, but with this movement much less pronounced, the rattle sounded just as loud. Still not completely satisfied in my mind, I have come to the provisional conclusion that the sound is produced by a combination of the rattle made in the throat and vibrations of the mandibles. It may take slow-motion moving pictures to prove the source of this very interesting noise, as it did in the long controversy over the drumming of the grouse.

Enemies.—One might assume that such a large and assertive species as the boat-tail would be about as free from natural enemies as any passerine bird. This may be generally true of the adults, but not of the young. In addition to the abnormally large loss in nestlings already commented on, other agencies actively militate against them.

Audubon (1834) has written that the alligator (*Alligator mississippiensis*) is frequently attracted by the "cries of the young when they are nearly fledged" and that, on hearing such notes, "well knowing the excellence of these birds as articles of food, swims gently toward the nest and suddenly thrashing the reeds with his tail, jerks out the poor nestlings and immediately devours them," but predation from this cause today is rare; at least I have never observed or heard of it.

Parasites cause some mortality among the young. Audubon (1834) stated that "My friend Dr. Samuel Wilson of Charleston, attempted to raise some from the nest * * * and for some weeks fed them on fresh meat but they became so infested with insects that not withstanding all his care they died." That similar circumstances are often present in the nest is well known. T. G. Pearson and the Brimleys (1942) say: "This is one of the species whose nests at times unfortunately are infested with parasites which, if they do not bring death to the young * * * certainly add nothing to the comfort of the household."

Man, too must be listed among the enemies of the boat-tail, because of the bird's tendency to despoil grain crops. Numbers are shot in various localities, and in past years it must have been also the practice to use the young as food; Audubon (1834) quotes the Rev. John Bachman as saying that grackles "are excellent eating whilst squabs."

Field marks.—It is hardly possible to confuse the eastern boat-tailed grackle with any other grackle except its larger relative, the type species, *Cassidix mexicanus mexicanus*. The completely dark plumage and huge, keeled tail and the fact that it seldom strays far from a maritime habitat will always distinguish it. The female is very much smaller than the male, of a uniform dark brown above,

and ochraceous-buffy below. The startling difference between the sexes often astonishes those not familiar with the bird, as I have many times noted while conducting Audubon wildlife tours in Florida and South Carolina.

The bright yellow eye of *torreyi*, another character that distinguishes both sexes of the eastern boat-tailed grackle, though not as apparent as either plumage or tail, is none the less invariable and easily visible at some distance.

When seen in bright light and close at hand, the eastern boat-tailed grackle is a strikingly handsome bird. The brilliant metallic reflections of the plumage, the intense, glowing color, and the trim alertness of the carriage, all combine to command enthusiastic admiration.

DISTRIBUTION

Range.—The eastern boat-tailed grackle breeds along the Atlantic coast from southern New Jersey (Fortescue) south to Georgia. It winters from Cape Henry, Virginia (in mild winters north along the Eastern Shore of Virginia) south to Florida.

Egg dates.—South Carolina: 25 records, April 26 to June 12; 20 records, May 9 to May 23.

QUISCALUS QUISCULA STONEI Chapman

Purple Grackle

HABITS

Frank M. Chapman (1935a) proposed the above scientific name for the bird that we have always called the purple grackle (*Quiscalus quiscula quiscula*), naming it in honor of Witmer Stone. He apparently restricts this name to the grackles in which "the head varies from greenish to purplish blue and rarely violet, the back and sides are bronzy purple with more or less concealed iridescent bars, the rump is purplish bronze, sometimes with bluish spots." In the same article he advances theories to show how the forms of the genus *Quiscalus*, as we now know them, probably originated and spread.

Far too much has been published on the relationship, nomenclature, and distribution of the races of this genus of grackles to be even summarized here. The reader who wishes to follow the discussion is referred to nine important papers on the subject: Dr. Chapman's preliminary study in the Bulletin of the American Museum of Natural History (1892); his other articles were published in The Auk (1935a, 1935b, 1936, 1939a, 1939b, 1940); Arthur T. Wayne's articles on the status of the species in South Carolina, also in The Auk, (1918); and

Dr. Harry C. Oberholser's study of the subspecies published in The Auk (1919b).

As far as can be gathered from a study of these papers, map and tables, the breeding range of the purple grackle (*Quiscalus quiscula stonei*), extends from northern South Carolina and Georgia through the Atlantic States, east of the Alleghenies, to southern New York and southern Connecticut; I should extend this race eastward to include extreme southeastern Massachusetts; and there seems to be a westward extension as far as south-central Louisiana, between the ranges of the bronzed grackle on the north and the Florida grackle on the south. In Rhode Island and southeastern Massachusetts I have collected quite a number of our breeding grackles and have observed many others at short range and in good light; although our birds here are somewhat intermediate in their characters, I believe that they are nearer to the purple grackle (*Quiscalus quiscula stonei*) than to the bronzed grackle (*Quiscalus quiscula versicolor*); I shall therefore treat our local records as applying to the former race. Farther north in Massachusetts, the bronzed grackle seems to be the commoner form, though the purple grackle has been recorded much farther north. Robie W. Tufts writes to me: "On or about Nov. 20, 1931, a specimen was taken at Grand Manan by Allen L. Moses, who mounted the bird. This specimen was taken to P. A. Taverner at Ottawa, who supported Mr. Moses in his identification. Mr. Moses shot the bird thinking it was a bronzed grackle and was about to toss it into his fox pen when he noticed the transverse markings on the back."

Spring.—Crow blackbirds, as they are often called, start migrating northward from their not far distant winter range during the latter part of February and reach their breeding grounds in southern New England around the middle of March. St. Patrick's Day, March 17, has always been associated in my mind with the arrival of the grackles about my home; then we may expect to hear the creaking notes of the males and see the glossy black birds posturing in the leafless treetops or exploring the tops of the tallest pines and spruces for possible nesting sites, preparatory for the coming of the females a week or two later. If weather conditions are favorable, they may remain, but a late snow storm or severe cold spell may cause them to retreat.

Courtship.—On April 8, 1946, two grackles, apparently both males, were moving about in the branches of a big ash tree close to my study window. One was evidently following the other as he traveled along the branches or hopped from one branch to another. Every few seconds one would stop, crouch down on the branch, lower his head, puff out his body plumage, spread his wings downward, and lower and spread his tail, at the same time giving voice to his unmusical notes. The other male went through the same motions at intervals, alternating with the first one. Eventually they separated and flew away in different directions. Apparently, it was a competitive display for the benefit of some hidden female, of which there were several in the yard.

Mating is èvidently earlier at Cape May, N. J., for Witmer Stone (1937) writes:

As early as March 13, many of the Grackles are flying in pairs, the male just behind the female and at a slightly lower level. They are noisy, too, about the nest trees and there is a constant chorus of harsh alarm calls; *chuck; chuck; chuck;* like the sound produced by drawing the side of the tongue away from the teeth, interspersed with an occasional long-drawn, *seeek*, these calls being uttered by birds on the wing as well as those that are perching. Then at intervals from a perching male comes the explosive rasping "song" *chu-séeeek* accompanied by the characteristic lifting of the shoulders, spreading of the wings and tail, and swelling up of the entire plumage.

As early as March 5 I have seen evidence of mating and sometimes two males have been in pursuit of a single female, resting near her in the tree tops, where they adopted a curious posture with neck stretched up and bill held vertically.

Nesting.—At the extreme northeastern end of their breeding range, near my home, we have found purple grackles nesting in a variety of situations. Many years ago, in eastern Rhode Island, a colony of a dozen or more pairs nested for several years in a hillside grove of red cedars (*Juniperus virginiana*). The nests were placed in the cedars, 10 or 12 feet from the ground, and were made of dried grasses and weed stems, lined with fine dry grass. In the extensive cattail marshes surrounding Squibnocket Pond on Martha's Vineyard Island, we found two well-hidden grackles' nests in the tall, dense, green flags, firmly attached to these cattails, and placed from 2 to 3 feet above the water. In that same vicinity there was a colony of eight or ten nests of these birds, 7 or 8 feet up, in a swampy thicket of large bushes.

On May 29, 1904, at Chatham, Mass., while passing through an apple orchard in full bloom, we noticed a pair of grackles making quite a fuss; their nest was soon located in an upright crotch near the top of one of the apple trees, about 12 feet from the ground; the nest, made of seaweed and coarse grasses and lined with fine grass and horsehair, contained five fresh eggs.

By contrast, our local purple grackles sometimes select much more inaccessible nesting sites. Within sight of my former residence is a row of tall white pines (*Pinus strobus*), along the banks of the Taunton River; every year several pairs of grackles have nested near the tops of the these trees, where the nests must have been between 50 and 60 feet from the ground; the nests were never disturbed by egg-collecting boys. We found another safe nesting site in a cedar swamp on Cape Cod. The swamp had been flooded as a reservoir and the white cedars (*Chamaecyparis thyoides*) were standing in water from 4 to 5 feet deep; it was a very large colony and there were evidently many nests in the cedars, but we did not care to make any accurate count of the nests, nor could we even estimate the number of the birds that were flying about over the swamp.

Bendire (1895) gives the following description of the nests: "The nests are rather loosely constructed and bulky. The materials used vary greatly according to locality; the outer walls are usually composed of coarse grass, weed stalks, eelgrass or seaweed, sometimes with a foundation of mud, and again without it. The inner cup of the nest is composed of similar but finer materials, and is generally lined with dry grass, among which occasionally a few feathers, bits of paper, strings, and rags may be scattered; in fact anything suitable and readily obtained is liable to be utilized. Exteriorly the nests vary from 5 to 8 inches in height, and from 7 to 9 inches in diameter, according to location. They are ordinarily about 3 inches deep by 4 inches wide inside." After describing nesting sites, similar to those mentioned above, he adds:

Sometimes natural cavities in trees or hollow stubs, as well as the excavations of the larger Woodpeckers, are also used, and along the seashore, where the Fishhawk is common, they often place their nests in the interstices of these bulky structures, notably so on Plum Island, New York. Speaking of this locality, the late Dr. Charles S. Allen [1892] says: "In every Fishhawk's nest, except those on the ground, I always found from two to eight or ten nests of the Purple Grackle. They were situated in crevices among the sticks under the edges of the nest, or even beneath the nest itself, so as to secure protection from rain and bad weather. They were very bold in collecting fragments from the table of their powerful neighbors."

Mr. J. H. Pleasant, Jr., of Baltimore, Maryland, writes as follows: "On May 19, 1888, I discovered a colony of Purple Grackles nesting under the eaves and rafters of a hay barn. In some instances the entrance to the nest was so small that it was extremely difficult to obtain the eggs. The crevices in which the nests were built were very much of the same character as those frequently chosen by the English Sparrow, and were situated at an average height of 25 feet from the ground; over a dozen nests were observed."

T. E. McMullen has sent me the data for 20 New Jersey nests: 9 of these were in grapevines or ivy vines climbing over various deciduous trees; 9 others were in red cedars; one was 20 feet up in a gum tree, the highest one was 45 feet from the ground in a large pine, and the lowest nests were 6 or 8 feet up in vines.

Eggs.—The purple grackle lays ordinarily four or five eggs to a set, very rarely seven; sets of six are not especially rare; the only set of seven that I have found contained two eggs that were quite different from the other five. The eggs are generally ovate in shape and are slightly glossy. Bendire (1895) describes them as follows:

The ground color of the Purple Grackle's eggs varies from a pale greenish white to a light rusty brown; they are generally blotched or streaked with irregular lines and dashes of various shades of dark brown, and in an occasional set different tints of lavender markings are also noticeable. Only in rare instances are these markings so profuse and evenly distributed over the entire egg as to hide the ground color. They vary greatly in style and character in different sets.

The average measurement of 85 eggs is 28.53 by 20.89 millimetres, or about 1.12 by 0.82 inches. The largest egg in the series measures 32.76 by 23.11 millimetres, or 1.29 by 0.91 inches; the smallest 25.65 by 20.57 millimetres, or 1.01 by 0.81 inches.

Young.—Of the young, Bendire (1895) says: "Incubation, in which both parents assist, lasts about two weeks, and they are equally solicitous in the defense of their eggs or young; the latter are able to leave the nest in about eighteen days, and sometimes a second brood is raised. They are fed almost entirely on insects while in the nest." Eighteen days seems a long time for the young to remain in the nest; 12 or 14 days would seem to be the usual time. It seems strange that so little has been published on the care and development of the young of such a common bird as the purple grackle.

Plumages.—The plumage changes of the purple grackle are very simple and hardly noticeable after the young bird's first summer. Dwight (1900) calls the color of the natal down pale sepia-brown. The whole juvenal plumage is "dull clove-brown, the body feathers often very faintly edged with paler brown. Tail darker with purplish tints." A complete postjuvenal molt takes place early in August, at which the iridescent black plumage of the male is acquired, and old and young birds become indistinguishable. The nuptial plumage is "acquired by wear which produces no noticeable effect as is regularly the case with iridescent plumages." Adults have one complete annual molt, the postnuptial, beginning early in August.

Of the plumages of the female, he says: "In juvenal dress the female is perhaps paler below than is the male and usually indistinctly streaked. There is a complete postjuvenal moult and later plumages differ from the male only in being much duller and browner with few metallic reflections. They also show more wear."

Witmer Stone (1937) makes the following interesting observation: "The progress of the molt in Grackles can easily be noted by the appearance of the wings and tail as the birds fly overhead, although the new and old body plumage of the adults are the same. They show gaps in the flight feathers as early as July 18 and some are still molting as late as September 8, 11 and 16 in different years. When the tail molt begins the long central feathers drop out first so that the tail appears split or forked, this gap becomes wider as successive pairs of feathers are lost, but by the time the outer pair is dropped the new central feathers have grown out and the outline of the tail is pointed or wedge-shaped."

Harold B. Wood has sent me the following notes on the colors of the iris in the purple grackle: "The young have brown irides, which by the absorption of the pigment, change to gray and lemon, ivory or white. The young of the year have uniformly dark brown irides until fall. Early spring birds have gray, lemon, ivory, or white irides. No bird which I trapped and banded with brown or gray eyes ever returned to the traps." As the iris in the adult is pale lemon color, or almost white, it appears that the brown iris is confined to the youngest birds and that the gray iris marks a transition stage of adolescence.

Food.—Beal's (1900) report on the contents of 2,346 stomachs of crow blackbirds includes the food of both the purple and the bronzed grackles, and will be considered under the latter subspecies. It seems proper to discuss here only such reports as refer especially to the purple grackle.

In his report on the birds of Pennsylvania, B. H. Warren (1890) gives the following list of the contents of several series of stomachs, collected in various months:

March—Twenty-nine examined. They showed chiefly insects and seed; in five corn was present, and in four wheat and oats were found. All of these grains, however, were in connection with an excess of insect food.

April—Thirty-three examined. They revealed chiefly insects, with but a small amount of vegetable matter.

May—Eighty-two examined. Almost entirely insects, cut-worms being especially frequent.

June—Forty-three examined. Showed generally insects, cut-worms in abundance; fruits and berries present, but to very small extent.

July—Twenty-four examined. Showed mainly insects; berries present in limited amount.

August—Twenty-three examined. Showed chiefly insects, berries, and corn.

September—Eighteen examined. Showed insects, berries, corn and seeds.

October—During this month (1882), the writer made repeated visits to roosting-resorts, where these birds were collected in great numbers, and shot 378, which were examined. Of this number the following is the result of examinations, in detail, of 111 stomachs:

Thirty, corn and *coleoptera* (beetles); twenty-seven, corn only; fifteen, *orthoptera* (grasshoppers); eleven, corn and seeds; eleven, corn and *orthoptera;* seven, *coleoptera;* three, *coleoptera* and *orthoptera;* three, wheat and *coleoptera;* two, wheat and corn; one, *diptera* (flies).

The remaining 267 birds were taken from the 10th to the 31st of the month, and their food was found to consist almost entirely of corn.

These examinations show that late in the fall, when insect food is scarce, corn is especially preyed upon by these birds, but during the previous periods of their residence with us, insects form a large portion of their diet.

Bendire (1895) makes the general statement that—

Their food consists largely of animal matter, such as grasshoppers, caterpillars, spiders, beetles, cutworms, larvae of different insects, remains of small mammals, frogs, newts, crawfish, small mollusks and fish. While it must be admitted that Indian corn, oats, and wheat are also eaten to some extent, much of the vegetable matter found in their stomachs consists of the seeds of noxious weeds, such as the ragweed (*Ambrosia*), smartweed (*Polygonum*), and others. Fruit is used but sparingly, and consists usually of mulberries, blackberries, and occasionally of cherries. One of the gravest charges against them is the destruction of the young and eggs of smaller birds, especially those of the Robin. * * *

They spend much of their time on the ground, being essentially ground feeders; they walk along close to the heels of the farmer while plowing, picking up beetles, grubs, etc., as they are turned up by the plow, or search the meadows and pastures for worms, grasshoppers, and other insects suitable for food.

The purple grackle eats the Japanese beetle, that imported pest that does so much damage to lawns, fruit trees, and flower gardens. I constantly see grackles and starlings feeding on my lawns, and like to think that they are probing for the grubs of this beetle: but I have never seen them feeding on the adult beetles in my rose garden. However, Japanese beetles were found in all the stomachs of purple grackles, meadowlarks, starlings, cardinals, English sparrows, wood thrushes, catbirds and robins, that were taken in the heavily infested areas in New Jersey and eastern Pennsylvania. Smith and Hadley (1926) say: "The purple grackle accounts for more of the beetles than any other bird. * * * Several were completely gorged with them. * * * The percentage of beetles eaten by the more important birds is as follows: Purple grackle, 66.3; meadowlark, 50.7; starling, 42.3; cardinal, 38.6; catbird, 14.8."

About our city parks these grackles are scavengers, picking anything edible from the rubbish cans, or eating any crumbs or bits of food dropped from the lunch baskets of visitors. Frank R. Smith sends me a story illustrating the sagacity of the bird: "This morning, as I passed through the park back of the National Museum, I noticed a grackle that had found a dry, hard crust, left from a lunch. The bird made several attempts to eat the crust, but its hardness resisted his efforts. Picking it up, he flew across the walk and alighted near a hydrant, beneath which a bird-bath was sunk to the level of the ground. Soaking in the water sat a pigeon; and the grackle, while evidently wanting to enter, feared to trust his prize so near the larger bird. After several false starts, he waded boldly into the water and turned his back on the pigeon, so that his own body was between the bread and the bird he feared. He dropped the bread into the water, waited a few seconds, picked it up and walked out to the grass, where he ate the softened bread. During this time the pigeon sat watching him curiously."

Hervey Brackbill writes to me: "Acorns are a prominent fall food.

Flocks as large as a couple of hundred birds come into the oak-wooded suburbs of Baltimore in late September and October, and feed both in the trees and on the ground beneath. The grackles, incidentally, do not open the acorns as blue jays do, by holding them down with their feet and hammering them with their bills; they grip them back in the angle of their mandibles and crack them by direct pressure."

Clarence Cottam (1943) observed an unusual feeding habit of grackles and crows at the outlet of a reservoir where—

About 12,000 cubic feet of water per second was passing through the electric turbines, "boiling up" to form the headwater of the Cooper River. Apparently the turbines were cutting up or otherwise killing large numbers of gizzard shad and other small fishes. These, brought to the surface by the churning water, attracted Ring-billed, Herring, Laughing, and Bonaparte's Gulls, as well as crows, Purple Grackles, and even a solitary Red-wing. * * * The grackles and crows fed over the turbulent water, picking up morsels of food with the skill and dexterity of the typical water birds. The feet and even the breast feathers of many of the crows and grackles were seen to touch the surface of the water momentarily as the birds hovered over this (for them) uncharacteristic feeding place. * * * Purple Grackles * * * use a wide variety of foods, and we have occasionally observed them feeding in shallow water on stranded insects and even small fishes. To see several dozens of these birds feeding in deep and turbulent water after the manner of gulls and terns, however, was indeed a surprise.

Economic status.—The grackle's reputation among farmers is almost as black as its plumage, for its faults, and it has plenty, are more conspicuous than its good deeds. Nor is it any more popular among its bird neighbors, as can be seen by the hostility they show toward it, for many a robin's or other small bird's nest has been robbed of its eggs or callow young to satisfy the appetites of young grackles. Analysis of stomach contents does not show any large percentage of such food, but it must be remembered that the yolks of eggs and the soft parts of small young are quickly digested and thus not easily detected; and the egg shells are not always swallowed.

The grackles are condemned by farmers on account of the considerable damage done by them to the grain crops during the planting season and until after harvesting has been completed. They are accused of pulling up the sprouting corn and wheat in the spring, but much of this is done to obtain the cutworms that are attacking the seedlings. Warren (1890) says on this point: "Some four years ago I was visiting a friend who had thirty odd acres of corn (maize) planted. Quite a number of 'blackies,' as he styled them, were plying themselves with great activity about the growing cereal. We shot thirty-one of these birds feeding in the cornfield. Of this number nineteen showed only *cut worms* in their stomachs. The number of cut worms in each, of course, varied, but as many as twenty-two were taken from one stomach. In seven some corn was found, in connection with a very large excess of insects, to wit: Beetles, earthworms, and cut worms. The remaining five showed chiefly beetles."

Perhaps the chief damage to the corn crop is done when the grain is in the milky stage in the summer; the grackles are flocking at that season and, where they are abundant, they swoop down in great black clouds into the standing corn; they strip the husks off the ears and eat the tender kernels, taking perhaps only a few from each ear, but rendering many unfit for the market. Sometimes as much as a quarter of the crop is thus damaged. The farmer is nearly helpless to protect a large field, for shooting only drives the birds from one portion of the field to another. All that can be said in favor of the grackle here is that it is a persistent enemy of the destructive corn borer.

Later in the season, after the corn is harvested and shocked, the grackles do some damage to the ripened ears by extracting the hard kernels; and Nuttall (1832) says that "in the Southern States, in winter, they hover round the corn-cribs in swarms, and boldly peck the hard grain from the cob through the air openings of the magazine."

Referring to the attacks on sprouting winter wheat, Judd (1902) writes: "During November 1900, a flock of from 2,000 to 3,000 pulled wheat on the Bryan farm, and only continual use of the shotgun saved the crop. At each report they would fly to the oak woods bordering lot 5, where they fed on acorns. Nine birds collected had eaten acorns and wheat in about equal proportions. The flock must have taken daily at least half an ounce of food apiece, and therefore, if the specimens examined were representative, must in a week have made away with 217 pounds of sprouting wheat, a loss that would entail at harvest time a shortage of at least ten times as much."

Although grain forms nearly half (47 percent) of the food for the year it is not all a loss to the farmer, as much of it is waste grain dropped during harvesting or left on the ground after that. Some slight damage is done to green peas, cherries, strawberries, blackberries, and other small fruits, but less than is done by some other birds.

All this damage may seem considerable, but it is largely offset by the good done in the destruction of those insects, harmful to the interests of the farmer, which make up over 50 percent of the food for the year. Consequently, where grackles are overabundant, they should be controlled or the crops be protected, otherwise they are fully as useful as harmful.

Behavior.—While feeding on my lawn the grackle walks with a slow, dignified gait, head held high and tail somewhat elevated, or runs nimbly over the ground, nervously flirting its long tail up and down and occasionally making long, high hops in pursuit of some insect. Occasionally it jumps or flies up a foot or two to catch a flying insect in the air. It forages also in the shrubbery or trees, evidently after insects, but for the most part it finds its food on the ground, picking something off the grass or probing in the earth for grubs or worms. When robins are feeding on the lawn at the same time, the grackles watch them and follow them about; as soon as a robin is seen pulling up a fat worm, the grackle rushes in and seizes the worm, driving away the gentler bird; the robin seems to be unable to defend itself and must yield its prize to the more aggressive robber. I have often seen a grackle, while foraging on my lawn on a warm sunny day in spring, stop and squat close down on the ground, remaining there for several minutes with its body pressed close to the warm earth, as if it enjoyed the warmth or perhaps just taking a sunbath. It may have been "anting," as other birds do in order to anoint their plumage with formic acid.

In this connection, the following observation by Mary Emma Groff and Hervey Brackbill (1946) is of interest:

The recent discussions of anting and supposedly substitute activities by birds makes it seem worth while to describe the behavior of Purple Grackles (*Quiscalus quiscula stonei*) in anointing themselves with a juice, apparently an acid, from the hulls of English walnuts (*Juglans regia*). * * * The walnuts grow in clusters of as many as five or six, at the ends of branches. The grackles would alight upon these clusters—just one bird to each—and begin pecking a hole in the sticky hull of one of the nuts, usually throwing away the pieces of hull they gouged out, occasionally seeming to swallow a piece. When a good-sized hole had been made, the birds would dip their bills into it, undoubtedly wetting them against the pulpy interior, and then thrust their bills over and into their plumage. A great part of the body was thus anointed—the breast, the under and upper surfaces of the wings, the back, and very often apparently the rump at the base of the tail. * * * Particularly birds that were watched worked as long as 10 to 15 minutes at a stretch. Many males sang at intervals, with display, and there was also much noise because of commotion among the birds, two or three of which would often contest for the same cluster of nuts. * * * The indication that it was an acid the birds were using was obtained when one of the English walnut hulls was cut open and litmus paper quickly placed against it; the paper instantly gave a strong acid reaction."

In the air the purple grackle flies in a direct line, not undulating like redwings, and generally at a considerable height, with strong steady wing beats; its flight is well sustained but not especially rapid. Witmer Stone (1937) says that when they descend from a height to alight in the trees "they sail down on set wings which form a triangular, kite-like outline, with the long tails of the males deeply depressed into the characteristic boat or keel." As fly-catchers the grackles are not experts. Stone saw one "pursuing a flying beetle on the street, an unusual performance; the bird was exceedingly clumsy in turning on the wing and after following its erratic prey for several minutes without result it gave up the chase. On August 31, several Grackles were observed darting up into the air from the tree tops in pursuit of flying ants in which activity they also proved very clumsy."

In its relations with other species the grackle not only indulges in the well-known habit of stealing eggs or young birds from the nests of its neighbors, but sometimes attacks and kills other birds in open places. In the National Zoological Park, in Washington, Malcolm Davis (1944) saw a purple grackle kill an English sparrow, which it had been stalking in almost catlike manner. * * * The sparrow was not long out of the nest, but was able to fly and take care of itself. A few days later I walked along the same area, and saw the kill. The grackle approached the sparrow and as the smaller bird flew away, the attacker seized its prey in its beak and gave it several hard shakes, with the body of the sparrow hitting the hard concrete pavement. At this moment passersby frightened the grackle away, but later the bird returned to eat the viscera of the sparrow."

Frank B. Foster (1927) reports: "At my Game Farm on the Pickering Creek, in Chester County, Pa., we lost in the Pheasant field, almost three hundred little Pheasants (*Phasianus*), a few days old, which were destroyed by Purple Grackles (*Quiscalus q. quiscala* [sic]). The male Grackles were the ones that did the damage. They came into the enclosure and simply took the heads off the little birds, leaving the bodies."

The purple grackle is highly gregarious at all seasons; even during the nesting season the birds often breed in sizable communities; and those that are not incubating resort to communal roosts at night. In the larger roosts they are often associated with starlings, redwings, or cowbirds.

Several roosts in eastern Pennsylvania have been studied, of which the Overbrook roost, described by C. J. Peck (1905), is typical: "The Overbrook Grackle Roost is situated upon the property of Mr. David L. Hess at the corner of Sixty-third street and Lansdowne avenue, Philadelphia. The estate comprises about ten acres, is rolling and wooded and has an artificial lake of about an acre in extent. The trees are deciduous with a goodly sprinkling of conifers and are of fair size. The roost has been in constant use for more than twenty years—how much more I have been unable to ascertain." This roost was used by varying numbers of birds during every month in the year, the smallest numbers being found in December and January. He gives a short account month by month showing the fluctuations in the population of the roost. In January, fewer birds use the roost than at any other time of the year. "On a few very severe nights the roost may be deserted, but such nights are rare and usually four or five hundred birds remain throughout the month." Conditions are about the same until the last week in February, when the migration begins. "Probably five thousand birds use the roost during the last few days in February." In March the "number of birds rapidly increases throughout the month until from twenty to twenty-five thousand are using the roost nightly." In April and May, the nesting months, the numbers fall off, "but the number never seems to fall below two or three thousand—birds which have not mated as yet or else males which have nests near by, probably both." June is very much like May, except that a very few females and the first of the early young begin to come in. But all this is changed soon after August first.

The birds have for the most part completed their domestic cares and family groups are rapidly consolidated into large flocks which come to the roost from considerable distances. The numbers are very greatly increased and the birds in flying to and from the roost follow much more closely a regular well-defined route.

During September and October the greatest numbers are reached and the birds come in at night in great flights, one flock following another so closely as to give the impression of a single long-drawn-out flock. The flight begins about 5:30 p.m. and lasts for about twenty or twenty-five minutes, but scattered birds and small flocks continue to come in until dark. I believe that from fifty to seventy-five thousand birds visit the roost every night during these two months. * * * Robins use the roost to the number of one thousand or more, their numbers being hard to judge with any degree of accuracy on account of the way they mix with the Grackles.

By 6:30, on September 17, the noise from the birds had begun to subside; and by 6:45 darkness and silence had come.

When grackles and starlings select a roost in a thickly settled community, or in the trees of a city street, as they sometimes do, they create a decided nuisance. Lewis W. Ripley (1914) tells how such a roost was established in one of the finest residential streets in Hartford, Conn., and what was done about it: "The birds, numbering probably several thousand, began to come in just before dark, and by seven o'clock all had arrived, and from this time until about six in the morning constituted a first-class nuisance, whistling and chattering until about 8 p.m., and beginning about 4 a.m., making a tremendous racket so that it was difficult to sleep. Not less annoying was the filthy condition of the walks and lawns, and the damage to the clothing of those passing along the street was not inconsiderable."

Several plans were discussed for getting rid of them and some were tried without much success; ordinary roman candles had no permanent effect, even when fired by men in the trees; but finally it was learned that the persistent use of high-powered, 10-ball candles, weighing 56 pounds to the gross, would produce the desired result. "As a net final result, about eight dozen candles were used at a total expense of about $10 and, at the end of a week, only a couple of dozen birds are to be found where there were thousands."

Voice.—The unattractive voice of the purple grackle is described in the following notes sent to me by Aretas A. Saunders: "While the sounds produced by grackles are far from musical, nevertheless some of them are largely confined to a definite season, including the period of nesting, and therefore may be considered to be songs. The commonest of these is a grating, metallic sound that might be written *kuwaaaa*. The main note is pitched about F ´ ´, and the short note at the beginning is a tone to a tone and a half lower. The matter of pitch, however, is more difficult to determine definitely in sounds that are not of musical quality. This is particularly true in determining the octave. The pitch of this note is near F, but whether F ´, F ´ ´, or F ´ ´ ´ I do not feel entirely sure. This particular sound is to be heard from the first arrival of the birds in March to the end of the breeding season in late June. It is sometimes also heard in late September and October from individuals in the flocks that congregate at that season.

"In the time of courtship in late April or early May, grackles produce another songlike sound that is accompanied by spreading of wings and tail. This is a series of four or five notes, each higher in pitch than the former one. The lower notes are rather harsh, while the higher ones are squeaky. These sounds are something like *kogubaleek* or *koochokaweekee*. The pitch begins on C ´ ´ or D ´ ´ and rises to B flat ´ ´ or C ´ ´ ´ at the end. The common call-note of the grackle is a loud *chak*, very similar to that of the redwings, but louder and somewhat lower in pitch."

To the nonmusical ear the squeaky notes of the grackles sound like the creaking of a rusty hinge and are decidedly unpleasant, but when heard in chorus from a migrating flock the effect is rather pleasing. During the courtship display the contortions of body, wings, and tail seem to indicate that the notes are produced with considerable effort.

Field marks.—The grackles are the largest of our northern blackbirds and have the longest tails; these are wedge-shaped and rounded or graduated at the end; and the male often carries his tail keeled, the middle feathers lower than the others. Grackles differ from redwings in having a straighter, more level, less undulating flight. They can be distinguished from rusty blackbirds by the longer tails. The sharply defined bronze back of the bronzed grackle cannot be distinguished from the more variegated back of the purple grackle, except at short range and in favorable light. There are, of course, many intermediates to be seen near the borders of the ranges; these are very difficult to identify as to race.

Fall.—The migrations of purple grackles are not long ones. They leave the northern portions of their breeding range in November, but even here a few remain occasionally in mild winters, though they are rare north of Washington, D. C., in winter.

As soon as the breeding season is over and the young birds are well grown, they begin to gather in the summer roosts, the family parties joining to form immense flocks. During October and November, these great flocks wander about over the country, often joined by starlings, cowbirds, and other blackbirds, seeking suitable feeding places in the grain fields, grasslands, and swamps. Stone (1937) describes one of these large feeding flocks "which contained many thousand birds. They covered the ground in great black sheets, the rear ranks constantly arising and flying over to take their place in the van which gave the impression of rolling over the ground. When they took wing in force the long procession streamed past shutting off from view all that lay beyond and when they alighted in the trees the bare branches appeared to be clothed with a dense black foliage."

Winter.—The main winter range of the purple grackle seems to extend from the Carolinas southward to the Gulf coast, though Skinner (1928) says that it occurs mainly as a migrant in the sandhill region of North Carolina, and Wayne (1910) considers it rare in coastal South Carolina. Probably most of these grackles spend the winter farther south in the Gulf States.

Wilson (1832) gives the following graphic account of a large wintering flock:

A few miles from the banks of the Roanoke, on the 20th of January, I met with one of these prodigious armies of Grackles. They rose from the surrounding fields with a noise like thunder, and, descending on the length of road before me, covered it and the fences completely with black; and when they again rose, and, after a few evolutions, descended on the skirts of the high timbered woods, at that time destitute of leaves, they produced a most singular and striking effect; the whole trees for a considerable extent, from the top to the lowest branches, seeming as if hung in mourning; their notes and screaming the meanwhile resembling the distant sound of a great cataract, but in more musical cadence, swelling and dying away on the ear, according to the fluctuation of the breeze.

DISTRIBUTION

Range.—Central Louisiana, Tennessee, Pennsylvania, and Connecticut to Florida and Georgia.

Breeding range.—The purple grackle breeds from central and southeastern Louisiana (Lake Arthur, East Baton Rouge), central and northeastern Mississippi (Shubata, Lucedale), southern and northeastern Tennessee (Selmer, Shady Valley), eastern West Virginia (Franklin, Leetown), central and northeastern Pennsylvania (State

College, Scranton), central-southern and southeastern New York (Binghamton, Hempstead), and southwestern Connecticut (Bethel, Portland); south to Central Alabama (Greensboro, Auburn), northern Georgia (Kirkwood, Athens), western South Carolina (Greenwood), east-central North Carolina (Raleigh), and southeastern Virginia (Petersburg).

Winter range.—Winters within breeding range rarely north to southeastern Pennsylvania (Doylestown, Holmesburg) and Rhode Island (Newport); south to the Gulf coast, northern Florida (Cedar Keys, Gainesville), and southeastern Georgia (Riceboro).

Casual records.—Casual in Texas (Sour Lake), Kentucky (Barboursville), western Pennsylvania (Wilkinsburg), New Hampshire (Tilton), and New Brunswick (Kent Island).

Migration.—The data deal with the species as a whole.

Early dates of spring arrival are: South Carolina—Spartanburg, January 27. North Carolina—Raleigh, January 29; Charlotte, February 4. Virginia—Lexington, February 15. West Virginia—Bluefield, February 14. District of Columbia—January 21 (average of 38 years, February 23). Maryland—Baltimore County, January 17; Laurel, January 28 (median of 7 years, February 16). Pennsylvania—Doylestown, February 1; Beaver, February 17 (average of 19 years, March 8). New Jersey—Princeton, February 6. New York—Shelter Island, February 12 (average of 16 years, March 7); Geneva, February 24 (average of 12 years, March 13). Connecticut—Fairfield, February 16. Rhode Island—Providence, February 22 (average of 23 years, March 9). Massachusetts—Harvard, February 23 (average of 7 years, March 14). Vermont—Bennington, February 28 (median of 29 years, March 25). New Hampshire—Exeter, March 6. Maine—Orono, March 1. Quebec—Montreal, March 12 (average of 16 years, April 9); Kamouraska, March 24. New Brunswick—Memramcook, March 5; Scotch Lake, March 19 (median of 35 years, April 7). Nova Scotia—Shulee, March 12; Prince Edward Island—North River and Mount Herbert, April 4. Mississippi—Saucier, February 13. Tennessee—Elizabethton, January 28; Athens, February 6 (median of 8 years, March 1). Kentucky—Bowling Green, February 4. Missouri—Kansas City, February 1. Illinois—Murphysboro, February 2; Chicago region, February 22 (average, March 10). Indiana—Worthington and Richmond, February 5. Ohio—Toledo, February 1. Michigan—Three Rivers and Ann Arbor, February 20; Germfask, March 17. Ontario—Toronto, February 14 (average of 17 years, March 21); Ottawa, March 8 (average of 38 years, March 28). Iowa—McGregor, February 20. Wisconsin—Madison, February 26 (average of 21 years, March 21). Minnesota—Minneapolis, February 28 (average of 9 years, March 16); Fergus

Falls, March 14. Texas—Dallas, February 9. Kansas—Wilsey, January 29. Nebraska—Omaha, February 10; Red Cloud, February 12 (median of 24 years, February 28). South Dakota—Aberdeen, March 4; Sioux Falls, March 19 (average of 8 years, March 25). North Dakota—Cass County, March 21 (average, April 1). Manitoba—Treesbank, March 24 (median of 55 years, April 14). Saskatchewan—McLean, March 29. Colorado—Fort Morgan, February 27. Wyoming—Wheatland, April 1; Laramie, April 13 (average of 8 years, April 23). Montana—Billings, March 21. Alberta—Alliance, April 1.

Late dates of spring departure are: South Carolina—Charleston, April 3. North Carolina—Raleigh, May 8 (average of 7 years, April 15). District of Columbia, April 17. Maryland—Baltimore County, April 20; Laurel, April 14 (median of 6 years, March 31). New York—New York City region, May 17. Connecticut—New Haven, April 24. Ohio—Buckeye Lake, median, April 10. Texas—San Angelo and Cove, May 1.

Early dates of fall arrival are: Texas—Pecos, September 9. Ohio—Buckeye Lake, median, August 9. Connecticut—New Haven, October 6. New York—New York City region, October 5. North Carolina—Weaverville, October 25; Raleigh, October 26 (average of 12 years, November 1). South Carolina—Chester County, November 1.

Late dates of fall departure are: Alberta—Belvedere, November 12. Montana—Kirby, October 20. Idaho—Sandpoint, November 19. Wyoming—Douglas, December 18; Careyhurst, November 2. Colorado—Fort Morgan, November 20. Saskatchewan—Indian Head, November 7. Manitoba—Brandon, November 27; Treesbank, November 9 (median of 52 years, October 25). North Dakota—Grafton. November 14; Cass County, November 3 (average, October 20). South Dakota—Vermillion, December 26; Sioux Falls, November 28 (average of 5 years, November 9). Nebraska—Blue Springs, November 19. Kansas—Clearwater, December 10. Oklahoma—Oklahoma City, November 17. Texas—Dallas, November 30. Minnesota—Minneapolis, December 11 (average of 8 years for southern Minnesota, November 8); Isanti County, November 4 (average of 10 years for northern Minnesota, November 1). Wisconsin—Oshkosh, December 13. Iowa—Marble Rock, December 11; Hudson, December 1. Ontario—North Bay, November 20; Ottawa, November 12 (average of 26 years, October 11). Michigan—Detroit, December 6; McMillan, December 1. Ohio—Leetonia, December 7; Toledo, December 2. Indiana—Elkhart, December 4. Illinois—Urbana, December 13; Chicago region, November 18 (average, October 30). Missouri—Bolivar, November 26. Kentucky—Danville, December 2. Tennessee—Athens, November 12 (average of 6 years, October 29).

Prince Edward Island—Tignish, November 6. Nova Scotia—
Wolfville, December 30; Halifax, November 9. New Brunswick—
Memramcook, December 20; Scotch Lake, November 12 (median of
15 years, October 24). Quebec—Montreal, October 23 (average of
9 years, October 6). Maine—South Portland, December 8; Avon,
November 24. New Hampshire—Ossipee, November 13. Vermont—
Woodstock, November 25. Massachusetts—Belmont, December 2;
Harvard, November 24 (average of 6 years, October 29). Rhode
Island—South⁻ Auburn, November 27. Connecticut—Fairfield, De-
cember 15. New York—Dutchess County, November 30. New
Jersey—Milltown, December 19. Pennsylvania—Chester County,
December 24 (average of 32 years, November 5). Maryland—
Laurel, December 28 (median of 7 years, November 20). District of
Columbia—average of 8 years, November 16. West Virginia—
Bluefield, December 16. Virginia—Charlottesville, December 11.
North Carolina—Weaverville, December 17.

Egg dates.—Alberta: 6 records, May 12 to June 2; 3 records,
May 18 to May 24.

Florida: 20 records, March 30 to June 12; 10 records, April 12 to
April 25.

Illinois: 46 records, April 21 to June 5; 25 records, May 6 to May 22.

Massachusetts: 56 records, May 4 to June 17; 39 records, May 14
to May 21.

Maine: 42 records, May 12 to June 23; 22 records, May 25 to
June 6.

New Jersey: 24 records, April 15 to May 12; 14 records, April 22 to
April 29.

North Dakota: 10 records, May 10 to June 16: 5 records, May 19
to May 31.

Ontario: 13 records, May 3 to June 15; 7 records, May 16 to May 29.

QUISCALUS QUISCULA QUISCULA (Linnaeus)

Florida Grackle

HABITS

The above scientific name, which for so many years was used for the purple grackle of the Middle Atlantic States, is now restricted to the southern race, which formerly bore the subspecific name *aglaeus*. The reason for this change is that the Linnaean name *quiscula* is based on Catesby's (1731) description of the "Purple Jack Daw," which was evidently collected in South Carolina, probably in the coastal region. As Arthur T. Wayne (1910) has shown, the purple grackle is very rare in that region, the Florida grackle being the abundant resident form there, and since it is almost certain that Catesby's bird was this form, the name *quiscula* must be applied to the Florida grackle, the former name of which, *aglaeus*, must be relegated to synonymy. For a further study of the relationships and nonmenclature of the grackles of this genus the reader is referred to the papers mentioned under the preceding race (p. 374–5).

The best description of the Florida grackle is given by Ridgway (1902), who says that it is similar to the purple grackle—

but decidedly smaller (except bill and feet), and coloration far less variable; adult male with color of head, neck, and chest varying from dark purplish bronze to violet (the head usually more bluish); back, scapulars, and sides of breast dark olive-green or dull bottle green, often nearly uniform, but always with at least concealed bars of other metallic hues; rump varying from purplish bronze to violet, usually more or less spotted with steel blue, bronze, etc.; abdomen and under tail-coverts dark violet, sometimes mixed with dark blue; prevailing color of wings varying from violet purple to steel blue (the color most pronounced on greater coverts and secondaries), the middle and lesser coverts more or less barred with various metallic hues.

The range of the Florida grackle, where it is practically a permanent resident, includes the whole of peninsular Florida and extends westward along the Gulf coast, south of the range of the purple grackle, as far as southeastern Louisiana, and northward throughout the lowlands of Georgia and South Carolina.

Wayne (1910) says of the Florida grackle in coastal South Carolina: "This form of the Purple Grackle is a permanent resident in the coast region, being found at all seasons in great numbers. It is, however, a freshwater bird, rarely, if ever, visiting the salt marshes. In winter I have seen countless thousands of these beautiful birds on the rice plantations in company with the Boat-tailed Grackle, feeding upon rice which was left in the fields."

Eugene E. Murphey (1937) reports the Florida grackle as an abundant permanent resident in the middle Savanna Valley of Georgia, and says of its haunts: "Many of the fish ponds in this region have a dense growth of young cypress trees around their margins and in their shallower portions, the trees average fifteen to twenty feet in height and with their lower branches overhanging the water, and here the Florida Grackle breeds regularly. Its favorite breeding spot is, however, some old fish or mill pond where the dam has broken and the entire bed grown up into a thicket of young trees and bushes. Here it breeds in considerable colonies."

In Florida, this grackle is an abundant resident over the entire State, including the Keys as far south as Key West, according to Arthur H. Howell (1932), who says: "The Florida Grackle inhabits a variety of situations and adapts itself to very diverse conditions. The birds are usually abundant around the towns and villages, nesting in orange groves, in pines or live oaks in dooryards, or along roadsides. In the wilderness, they often nest in the smaller cypress swamps, or open pine forests, palmetto hammocks, or in bushes growing in or near a pond or stream."

Thomas D. Burleigh (1925) found about a dozen pairs living on Billy's Island in Okefenokee Swamp, in southern Georgia near the Florida line. This island "is merely a bit of solid land in the middle of seemingly endless miles of swamp, and is characterized, as are the other scattered islands, by what was once a fine virgin stand of long-leaf pine (*Pinus palustris*)." The birds seemed to show a decided preference for the remaining trees near the logging camp.

H. H. Kopman (1915) says of it in Louisiana: "This is the only form of the common Crow Blackbird that occurs in the swampy coastal section of the State, so far as I have been able to learn. It is abundant and occurs in practically all situations except the open marsh. It is often found in great flocks in the wet woods in winter and early spring. It nests chiefly in the neighborhood of habitation, especially in groves of live oaks, and water oaks."

Nesting.—Burleigh (1925) says of the nests found in the tall longleaf pines on Billy's Island:

The nests, never more than one to a tree, ranged from twenty-five to fully a hundred feet from the ground, some of them being at the outer end of the upper branches where they were quite inaccessible. The average height was fifty feet, and they were usually in a crotch of one of the limbs eight or ten feet from the trunk. I managed to reach three of them, and found in two five eggs and in the third four, all of them half incubated. The nests proved very similar in construction, being well built of gray usnea moss intermixed with dry pine needles and grasses, coated on the inside with mud and then well lined with fine grasses. In each case the female was incubating but flushed quietly and showed practically no concern over the nest, disappearing and not being seen again.

Referring to the nesting habits of this grackle in Florida, Bendire (1895) writes: "Most of the nests found by Dr. Ralph were placed in low bushes, from 2 to 7 feet above the water in cypress swamps; others were found in orange trees and small pines, at no great distance from the ground. One nest, containing four eggs, in which incubation was about one-fourth advanced, taken by him March 30, had been placed directly under an occupied nest of the Green Heron, with an interval of about 6 inches between them. * * *"

He says, that the nests vary somewhat in composition:

Some are made of coarse grass, leaves, etc., taken from the ground in swamps, pressed firmly together, and thickly covered on the outside with Spanish moss, with which a few pieces of grass, twigs, etc., are mixed, and they are lined with finer dry grass. In other nests the outer walls are mainly composed of coarse grass, weeds, and but little Spanish moss; these materials are cemented together with cow manure and mud, and the nests are lined with wire grass (Aristida); again flags, wet sphagnum moss, pine needles, and small twigs are used to a considerable extent in these structures. * * *

A nest now before me, built in an orange tree, about 8 feet from the ground, measures 5½ inches in height and 8 inches in outer diameter. The inner cup of the nest is 3¾ inches deep by 4¼ inches in diameter.

Howell (1932) says that the nests of the Florida grackle are sometimes found "in bunches of pendant Spanish moss, and not infrequently in hollow trees or broken-off stubs.

Pigeon Key, near Key Largo, is typical of many small Keys bordering the Bay of Florida; the dry, or nearly dry, land in the center supports a growth of fair-sized black mangroves, while a dense fringe of red mangroves forms an almost impassable barrier around its shores. Here on May 8, 1903, we found a small breeding colony of Florida grackles nesting in the black mangroves. I shot two of the birds for identification and collected a set of two fresh eggs from a nest about 10 feet up in a black mangrove sapling; this bulky nest, which I still have, seems to have been loosely constructed with a mass of seaweed, very coarse weed stems, small dead twigs, with a lot of moss and other rubbish in the foundation and sides; the cup is built up with somewhat finer weeds and grasses and lined with still finer grass, but it is far from being a neat structure. There were a number of other nests higher up in the larger trees; those that we examined contained young birds.

Earle R. Greene (1946) mentions two nesting sites at Key West: "A 'sandbox' tree, standing in the courtyard of the Key West postoffice has long been a favorite nesting place and a number of nests are annually built among its branches. The custodian of the building is kept busy during the season looking after young that fall to the ground, to the great concern of their parents. A 'Spanish laurel' tree on Simonton Street is another preferred nesting site; this tree is one of the finest of its kind in the area."

Eggs.—The four or five eggs usually laid by the Florida grackle are practically indistinguishable from those of the purple grackle.

The measurements of 40 eggs average 29.4 by 20.8 millimeters; the eggs showing the four extremes measure 33.0 by 20.0, 24.0 by 22.4, and 28.8 by 19.2 millimeters.

Food.—In a general way, the food of the Florida grackle is similar to that of the species elsewhere, but C. J. Maynard (1896) mentions the following items, some of which are peculiar to this race:

In early Winter large flocks may be seen on the tops of the palmettoes, feeding on the fruit, and they also eat berries in their season. Later small flocks are found on the margin of streams, frequently wading into them in search of little mollusks, crabs, etc., and it is not rare to meet with one or two scattering individuals in the thick hammocks, overturning the leaves in order to find insects or small reptiles which they devour. I once saw one catch a lizard which was crawling over the fan-like frond of a palmetto, and fly with it to the ground.

The reptile squirmed all the while in its frantic endeavors to escape, but the Blackbird held it firmly and, after beating it to death, removed the skin as adroitly as if accustomed to the operation, then swallowed the body.

Wayne (1910) says: "The Florida Grackle is a very destructive bird as it eats the eggs of all birds which breed in swamps, making a systematic search for nests which contain eggs, Swainson's Warbler (*Helinaia swainsonii*) being generally the victim. It also eats the eggs of the freshwater terrapin."

Winter.—The Florida grackle is generally regarded as a permanent resident throughout its breeding range, but Mr. Greene (1946) says that his records indicate that it is absent from the Florida Keys from September to February, inclusive. He thinks that they may join those farther north on the mainland.

DISTRIBUTION

Range.—The Florida grackle is resident from southeastern Louisiana (Isle Bonne, Chef Menteur) and southern Mississippi (Bay St. Louis, Agricola), to central-western and southeastern Alabama (Reform, Dothan), central Georgia (Montezuma, Augusta), eastern South Carolina (Anderson), eastern North Carolina (Lake Mattamuskeet, Kittyhawk), and southeastern Virginia (Newport News, Pungo); south to southern Florida (Key West, Grassy Key, Key Biscayne).

QUISCALUS QUISCULA VERSICOLOR Vieillott

Bronzed Grackle

Plate 24

Contributed by ALFRED OTTO GROSS

HABITS

The bronzed grackle is a bird that has well adapted itself to radical changes in environment brought about by civilization. More and more of them have come to accept conditions existing about our farms and many have even invaded our populous cities and towns to nest and to roost near human habitations. They have accepted every advantage thus afforded and have thrived on the food provided by man in the waste of his door and farmyards, and especially on his bountiful crops. These birds through their extreme resourcefulness have been eminently successful as a species in maintaining and increasing their numbers in spite of persecution.

The almost universal common name applied to the grackles as a group is crow blackbird. The name is well chosen, for many of its traits, as well as its dark coloration, suggest the crow; and it is a convenience to have a common name that applies to the purple and the Florida as well as to the bronzed grackle.

These three birds are difficult for the layman to differentiate in the field, and even the ornithologist has his troubles when it comes to identifying individuals in immature plumages. Most of the details given in this account of behavior and habits, the song, food, nesting, molts, immature plumages, etc., can be applied equally to either of the other two races. The bronzed grackle intergrades with the purple, the northernmost of the two southern forms where the ranges come in contact, nevertheless it is an exceptionally stable form and shows no geographic variation in color throughout its extensive range.

Spring.—A considerable number of bronzed grackles spend the winter in favorable places throughout southern New England. The first flocks, many of which are made up of a hundred or more individuals, appear during the first week of March to mark the beginning of the spring migration. They do not arrive in Maine until the middle of the month; at this time the snow is still on the ground and in the dense interiors of the coniferous forests it is still several feet deep. Usually the first arrivals I see at Brunswick are the individuals of a noisy, querulous band that land in my backyard to gobble up the food provided for the evening grosbeaks, tree sparrows, and other winter birds which are still enjoying the hospitality of my feeding stations. The grackles are audacious and greedy, but extremely restless and wary. If one individual becomes frightened the whole flock takes wing with a whirr and they are off to another section of the town, but in due time they return to repeat the raid on the feeding shelf, which meanwhile has been replenished.

The coming of few birds attract more general attention than do these conspicuous bands of noisy grackles. Their arrival creates mixed emotions. Most people have a greater thrill on seeing or hearing their first robin or bluebird. Later in the season when the great hordes of grackles have passed on and the summer residents settle down for the season, they are a more welcome sight on our lawns. The male especially is a trim and handsome fellow. His bright, piercing yellow eyes, his iridescent plumage flashing in the bright sun, his bold strides, and the swagger of his tail combine to form a personality well worth studying.

Otto Widmann (1907) gives an account of the arrival of the bronzed grackle in Missouri as follows:

Real migration begins in the latter part of February and in early March in the southeast; it reaches the central, and along the Mississippi River even the northern, in the second, less often in the third week of the month, very rarely later, as in 1906, when winter reigned to the end of March. The first-comers are probably mostly transients, bound for the north, keep in dense flocks and roost in the river bottoms. It is only after the bulk of the species has invaded the state during the latter half of March, that the first of our summer residents make their appearance on the breeding grounds and announce that they intend to occupy them again as soon as their mates have arrived. They return in the evening to the common roost and, should the weather turn bad, are not seen at their old stands again for days, but as soon as warm weather sets in they return, are joined by the first females, and mating begins with much chasing and noise making. The transit of tremendous flocks of migrants continues through the first two weeks of April, during which time the ranks of summer residents fill up, and nest-building begins. During all this time of mating and nest-building, and until incubation begins, the whole colony leave the breeding ground in the evening and go to the common roost, preferably willows in the bottoms, to which they come from all sides for miles to spend the night together.

The grackles destined to go further north proceed leisurely on their migration during March. They seem content to rove about the countryside in marauding bands in search of food, waiting for the further progress of spring. It is not until the first week of April that the first birds usually appear in New Brunswick, Nova Scotia, and Quebec; and in the midwestern Provinces of Canada, as well as in Colorado and Montana, it is well after the middle of April before the vanguard can be expected to arrive.

Banding and wholesale trapping of grackles has shown that the migrating flocks are not mixed but are usually made up of either males or females. The first birds that arrive in spring are males.

Courtship.—Courtship starts early in the season. It has been frequently observed long before nesting activities begin; early in March one may see the amorous but ludicrous males going through their curious gestures and paying ardent attention to the females.

Sometimes two or more males will be seen in rapid pursuit of a single female. When alighting in the tree tops with other birds they adopt peculiar postures, puff out their plumage, partly open their wings, spread their tails, stretch out their necks, and hold their heads in a vertical position. Intermittently they utter the hoarse raucous calls no doubt attractive to their intended mates, but not appreciated by human ears. If disturbed, they all fly off together but when the flock returns they again separate in pairs to continue the performance as before. Charles Wendell Townsend (1920), who has closely observed the courtship of many of our birds, gives the following account of the performance of the grackle:

> The courtship of the Bronzed Grackle is not inspiring. The male puffs out his feathers to twice his natural size, partly opens his wings, spreads his tail and, if he is on the ground, drags it rigidly as he walks. At the same time he sings his song—such as it is—with great vigor and abandon. * * *
> During the period of courtship the male in flight depresses the central feathers of its tail forming a V-shaped keel. I was first inclined to think that this was of use in flight like a rudder, but I am inclined to think that it is in the nature of courtship display, for this arrangement of tail feathers is not seen when a bird is actively engaged in flight for the purpose of obtaining food. Under these circumstances the tail is spread in the ordinary manner.

Francis H. Allen (M. S.) supplies these notes on courtship: "May 17, 1905, Boston Common: A male following a female about. He walked close behind her with the feathers of his shoulders erected into a ruff behind his head. It was evidently to exhibit the iridescence of the feathers. Meanwhile he repeatedly uttered the jarring note which Bendire renders as *tchch.* June 5, 1938, West Roxbury, (Massachusetts): A pair courting on the road. The male walked around the female displaying, while the female stood still with tail closed but held elevated at an angle of about 45°. They separated without any culmination of the affair."

Nesting.—The grackles are quite adaptable in their nesting habits; depending on the conditions at a particular locality, a diversity of nesting sites ranging from marshes and nests in holes of tree stumps to those near the tops of tall trees may be selected. There is little difference in the nesting habits of the races of the grackle. Some individuals nest alone in places apart from the nesting sites of their fellows, but more often flocks of a hundred pairs or more will nest close together in a grove of trees. I have seen as many as a dozen occupied nests in a single giant boxelder tree standing near my boyhood home in central Illinois. Some of the nests were saddled on large horizontal limbs, at points well over 40 feet above the ground. Two of the nests were not more than 2 feet apart. Milton B. Trautman (1940) found 28 pairs nesting in a large Norway spruce (*Picea abies*) at Buckeye Lake, Ohio.

These birds have readily adapted themselves to an environment created by man and have taken over orchards and shade trees near farms. The nests have been found in a variety of hardwood trees such as oaks, maples, elms, sycamores, willows, cottonwoods, etc., but the grackles always manifest a strong partiality to conifers. They have invaded not only the shade trees of our cities and towns but have built their nests in niches and suitable places on public buildings and homes, in direct competition with English sparrows and starlings.

A more primitive nesting site and perhaps one used long before the coming of white man is that of holes in large dead trees and stumps; a few are found in old nesting cavities excavated by large woodpeckers. This practice is still common in certain localities especially in the western and northern sections of the nesting range. Hartley H. T. Jackson (1923) in an account of the birds of Mamie Lake, Wis., writes: "Abundant in the vicinity of Mamie Lake, June 5 to 24, 1918, where they were nesting in the dead stumps and snags in overflows, usually at the mouth of creeks. The nests for the most part were two to four feet above the water, but were difficult of access in our canoe on account of logs, snags and fallen timber in the water."

E. S. Cameron (1907) in writing of the nesting of the bronzed grackle in Custer County, Mont., states:

These birds nest here in the holes, or hollows, of dead trees, so that their nests are generally invisible from the outside. However, on June 1, 1893, Mr. H. Tusler showed me a nest of this species placed in a hollow formed by the fork of the two main branches of a box elder. Although well protected on all sides by wood, it was possible to examine this nest, which was only six feet from the ground, and made entirely of slough grass, with a thick internal layer of mud. It contained six lovely eggs. * * *

In 1894 there was a small colony of grackles in the large cottonwoods on the south bank of the Yellowstone, below Terry ferry crossing. All the nesting holes were high and very difficult to reach, excepting one where the nest was in the top of a burnt cottonwood stump, about twelve feet from the ground. The birds had eggs on June 3, and young hatched out on June 11 which both parents were feeding on crane flies.

Robert Ridgway (1889) found many nests built inside of holes in large dead trees and in tree stumps along the river near Mount Carmel, Ill. Similar conditions are reported for southeastern Missouri where Otto Widmann (1907) states the birds "still nest in tree holes of deadenings." Many such reports seem to indicate that nesting in holes of trees is still a common practice. A modification of this habit

is a unique nesting site of a pair of bronzed grackles that built their
nest and reared their young in a squirrel box placed on the top of a
hackberry tree at Nashville, Tenn., reported by J. R. Tippens (1936).
They have also been found in birdhouses.

A departure from the usual habit of nesting in trees on the uplands
is illustrated by Bendire (1895). He quotes Mr. J. W. Preston, who
saw a large colony in a tract of bushy land at the northern extremity
of Heron Lake, Minn.: "Here the nests were placed in low shrubs
and wild-gooseberry bushes, some not more than 1 foot from the
ground. * * * I have seen an odd nest of this Grackle built in a
bunch of common reed (*Phragmites*), which looks like broom corn at
a distance and grows from 5 to 12 feet high. This nest resembled
that of a Yellow-headed Blackbird, the material being evenly woven
together."

Along the lower Mississippi and Illinois Rivers in Illinois I have
seen large numbers of grackles nesting in the willow swamps. The
nests were built in willow trees at various distances, some not more
than 3 or 4 feet above the water to other 30 feet high. Edmund S.
Currier (1904) found one nest in an open marsh in the midst of a
colony of red-winged blackbirds at Leech Lake, Minn. It is obvious
that the colonial instinct of this grackle was satisfied by the presence
of the redwings. "This nest was woven together in the top of a
clump of flags, and its weight had lowered it to within a few inches
of the water." William Brewster (1906) in writing of the nesting of
the bronzed grackle in the Cambridge region of Massachusetts states:

Most of the grackles frequenting this locality build their nests in dense thickets
of alders and other low bushes sometimes not more than a foot or two above the
ground or water; others breed in company with the redwings in beds of cattail
flags well out in the open marshes. Within the past ten years I have found a
few nests placed in button bushes or among cattails growing in shallow water, at
Great Meadow. This habit of nesting in swamps and marshes is unquestionably
of recent origin in our neighborhood, for during earlier years of my experience
the birds seldom or never resorted to very wet places excepting in autumn, when
they used to assemble in large numbers at evening in the maple woods bordering
on Little River where they roosted in company with Robins and Cowbirds.

On June 22, 1937 at Churchill, Manitoba, Frank L. Farley (1938)
found a bulky nest of the bronzed grackle in a dead spruce standing
in the water at the edge of a marsh. It was built under a thick
brushy branch about 3 feet above the water.

Several observers have reported finding nests inside buildings,
where the unusual associates of the grackles were barn swallows.
John and James Macoun (1909) quote W. H. Moore: "This species
nests in barns on islands and intervales along the St. John river, N.
B.; sometimes there being three and four nests in one barn. They
are usually built on beams or in the angle of a post and brace of the

framework." William Youngworth (1932) reports a similar situation as follows: "In July 1929, I watched several pairs of bronzed grackles attending to nesting duties at Scranton in southwestern North Dakota. The birds had built their nests on the steel beams inside of a large coal briquet plant which was not in operation at the time." In correspondence, Lyle Miller writes that a large colony of grackles nest on the girders of a large water tower at Youngstown, Ohio, and he has also found numbers nesting on the girders of a large steel bridge at Lake Milton, Ohio. Others have reported finding them in similar situations, sometimes in places shared by nesting phoebes. A most unusual nesting site is in the bulky masses of the osprey's nests. Apparently the grackles are not molested by the giant birds, and from the associations have derived protection as well as scraps from the osprey's dinner table.

The nesting season ranges from the first week of March to the latter part of June, depending on the latitude and various conditions of the locality. In the extreme southern sections of the range, nests are common early in March. The height of the nesting season in Massachusetts is reached about the middle of May; but in Maine, and also in the more northern sections of the range, most of the nests are built in the latter part of May and in June.

The structure of the nest of the bronzed grackle varies much less than do the nesting sites. It is always a substantially built, bulky affair of sticks, coarse grass, weeds, roots, leaves, and similar materials. In most nests a liberal supply of mud in the interior serves to plaster the loose nesting materials into a more permanent mass. Inside the mud layer is a lining of fine grasses and rootlets; sometimes hair and feathers are added. Many of the nests I have seen in the corn belt of the Midwest had foundations made up almost entirely of corn husks. Some of the nests, especially those near human habitations, had the foundation materials interwoven with string, paper, and rags. The nests are deeply cupped and serve well to hold the active young that are to follow. A typical nest has an outside diameter of 7 inches and a depth of 5 inches; the nesting cavity is 4 inches in diameter and 3½ inches deep.

The nests of the grackles are usually so well made that many of them remain in good condition even after being exposed to the buffeting of winter storms. H. Elliott McClure (1945) found many such nests, eight of which were used by mourning doves.

Eggs.—The eggs of the Florida, purple, and bronzed grackles are similar, and the reader is referred to Bendire's (1895) description under the account of the purple grackle. According to Bendire: "The average measurement of a series of one hundred and forty-eight specimens in the United States National Museum collection is 29.02

by 20.90 millimetres or about 1.14 by 0.82 inches. The largest egg measures 31.50 by 21.59 millimetres, or 1.24 by 0.85 inches; the smallest egg, 25.40 by 19.05 millimetres, or 1 by 0.75 inches."

The number of eggs in a set varies from three to six. Rarely have seven been found. The vast majority of the nests containing complete sets that I have examined have had four or five eggs, but six are not unusual. Ordinarily there is but one set of eggs in any one season, but if the first set of eggs or newly hatched young are destroyed, a second set will be laid.

The incubation period of the bronzed grackle is 14 days. The task of incubation is performed by the female, and I have never seen the male assist at any of the nests I have had under observation. However, the male is usually in evidence during this period and is quick to assist in defending the nest in the event an intruder appears. In fact, any unusual commotion about the nest brings the members of the entire colony to the scene after the alarm note of the male is sounded.

Young.—After the young appear the male shares with the female the work of feeding them, a task which increases in arduousness with the constant demands of the young as they grow older.

Ira N. Gabrielson (1922) in a study of a nest of young in a colony near Marshalltown, Iowa, made the following observations on the feeding behavior:

A blind was placed in position at a nest seven feet from the ground in a plum tree on May 30 at 11.00 a. m. At 1.00 p. m. I entered the blind and found the parents somewhat nervous so only remained about two hours. Only the female summoned up courage to feed during that time and fed both nestlings each trip but the last. Eleven minutes after entering the blind the female appeared carrying two earthworms and two or more unrecognized insects. After hopping nervously about from limb to limb above the nest she hurriedly fed both nestlings and left. At the sixth feeding she carried seven cutworms in her beak and fed them one at a time to the two nestlings. On the last feeding she came three times and thrust her bill into the nestling's mouth, apparently without feeding. On the fourth return she fed one nestling and the fifth time returned and gave the remainder of the food to the same one.

On May 31 I watched the nest from ten o'clock until three during which time the young were fed 26 times, the male feeding nine and the female seventeen times. On two occasions the parents arrived simultaneously to feed.

By the time the young are 16 days old they are fully feathered and by the 18th day they are ready to leave the nest, but if not disturbed, may remain a day or two longer. The adults continue to feed them, but as they gain strength and ability to fly they go on foraging parties and by the first week of July join the flocks at the common roost at night.

Amelia R. Laskey (1940) has written a very interesting account of a bronzed grackle obtained in May from its nest in a tree at Nashville, Tenn., after the parents had been shot. This bird was never caged. It was placed in a basket, which served as a nest; had absolute freedom, and was normal in its development. Much of its behavior was probably similar to birds brought up by their parents. During the first few days his hunger was expressed by characteristic squawking begging notes. About 2 weeks after his arrival the partially naked little bird had become fledged and was given the freedom of the out-of-doors.

This bird revealed many traits and characteristics that remind one of pet crows kept under similar circumstances. "From July until early fall he gradually molted his juvenal plumage and acquired the beautiful glossy black feathers of the adult bronzed grackle. In reflected light his plumage was rich in glistening purples, blues, and greens. His juvenal squawkings were replaced in mid-August by the characteristic squeaking, creaking songs of his kind." In late August this bird exhibited a distinct courtship behavior toward a hand-raised female cardinal. The pet grackle paid no attention to other grackles that visited the garden during the 4 months he was developing. In September he made long trips of a mile or more, even visiting blackbird roosts in the vicinity, but always returning to his foster home to be fed. After September 17 there was a marked change in his behavior, and from then on he seldom spent the night at home, but returned in the morning. During the first days of October he was frequently absent during the day but made trips back to be fed and to receive the attentions of his hostess. Finally, on October 6, after being fed "he flew to the peak of the porch, wagged his tail a bit and then flew to the west, singing. This is the first time he left singing and the last time he was seen." He probably left on the migration to the south with the other members of the roost he had been visiting.

Plumages.—According to Jonathan Dwight (1900), the plumages and molts of the bronzed grackle correspond to those of the purple, the descriptions of which follow:

Natal down. Pale sepia-brown.

Juvenal plumage acquired by a complete postnatal moult. Whole plumage dull clove-brown, the body feathers often very faintly edged with paler brown. Tail darker with purplish tints. Bill and feet sepia-brown, black when older.

First winter plumage acquired by a complete post juvenal moult early in August. The iridescent black dress is acquired, old and young becoming indistinguishable.

Some birds assume metallic green heads and some blue, while the backs are of all colors and patterns so that age can have nothing to do with the varied colors of this species.

First nuptial plumage acquired by wear which produces no noticeable effect as is regularly the case with iridescent plumages.

Adult winter plumage acquired by a complete post-nuptial moult beginning the first of August. Indistinguishable from first winter.

Adult nuptial plumage acquired by wear as in the young bird.

Female.—In juvenal dress the female is perhaps paler below than is the male and usually indistinctly streaked. There is a complete post juvenal moult and later plumages differ from the male in being much duller and browner with few metallic reflections. They also show more wear.

H. B. Wood (1945) made a study of the molt of 146 grackles which he trapped at Harrisburg, Pa., between March 19 and September 18, 1944:

Evidence of molting, with new feathers, first appeared on July 23. The molting period extended until mid-September and with other observed grackles until mid-October. The first feathers molted were those along the edge of the wing, the last were the central tail feathers. * * * the sequence of molting was determined to be in the following order of feather groups: lesser wing-coverts, greater coverts, secondaries, forehead, crown, nape, rump, primary-coverts, upper tail-coverts, cheeks, neck, back, belly, under tail coverts, scapulars, proximal primaries, breast, chin, and finally the distal remiges and then the median rectrices. The old axillars were retained by some birds until all but the primaries and rectrices were completed. * * * Practically all the birds exhibited great regularity in their molting areas. The proximal remiges were shed and regained quickly, but the distal four were lost in regular order and slowly redeveloped. * * * In nearly all the birds, the secondaries were either all old or all new. * * * The median body feathers were shed and grown before the laterals, both dorsal and ventral, as along the spine before the side areas.

Frank M. Chapman (1921b) discusses the plumages of the bronzed grackle as follows: "The nestling plumage of this species resembles that of the Purple Grackle, and, as in that species, the plumage of the adult is acquired in the fall (post-juvenal) molt. There is, however, a more pronounced difference between the color of the winter and summer plumage in the Bronzed, than in the Purple Grackle, the shining brassy back and abdomen of the fall and winter Bronzed Grackle becoming dull seal bronze in summer."

"The Bronzed may be known from the Purple and Florida Grackles by the absence of the iridescent bars which, whether exposed or concealed, are present in the back and abdomen of the other two birds."

The head and upper breast of the adult male bronzed grackle varies from greenish blue to purple, the neck and chest sometimes brassy green; rest of the plumage a uniform bronze or brassy-olive with more purplish on the wings and tail. The lesser and middle wing coverts are not marked with bars or metallic tints. The females are similar to the males but are very much smaller and duller in coloration.

L. L. Snyder (1937) in a study of 204 trapped grackles found the prismatic colors varied from a red-purple group at one end of the series to a metallic green at the other. Fourteen percent appeared in the first and 24 percent in the latter group; 62 percent were intermediates.

The average weight of 99 males was 131.4 grams and of 105 females was only 100.8 grams. The average of each of the five measurements made of the males were decidedly greater than the average of the same measurements of the females.

Mabel and John A. Gillespie (1932) have noted the eye color of immature bronzed grackles. "The youngest birds * * * possess a dark brown iris. With the acquisition of black to the feathers, the iris becomes correspondingly paler in shade. Late summer immatures often have eyes of greyish green. This color presumably precedes the straw yellow eye which we have always found in adult birds."

A considerable number of albinistic, chiefly partially albinistic, plumages of the bronzed grackle have been reported by various observers.

Longevity.—We do not have sufficient data to determine the life expectancy of the bronzed grackle, but a number of recoveries of banded birds are of interest in this respect. Christian J. Goetz (1938) recovered three birds at his station at Cincinnati, Ohio, in 1938 that were banded at the same station in 1931, an elapsed time of 7 years. Since these birds were adults, two males and a female, they were at least 8 years old. Mr. Goetz also recovered two birds that were at least 7 years old.

May Thacher Cooke (1943) reports a bird banded as an adult at Fort Smith, Ark., April 21, 1931, and recovered 11 years later at Endora, Ark., on March 12, 1942. This bird was at least 12 years of age. A bird banded by R. T. Gordon in South Dakota, August 17, 1924, was recovered in Minnesota in October 1940 (reported by Geoffrey Gill, 1946). If this bird was an adult when banded it would be at least 17 years old, a longevity record for the bronzed grackle, as far as I have been able to ascertain.

There have been a great many recoveries of birds from 5 to 6 years of age, which probably represent the average attained by the bronzed grackle.

Food.—The food of the bronzed grackle is so varied that it can be considered omnivorous. Its food, consisting of both animal and vegetable matter, varies so much with the season, the supply, and local conditions, that its economic status, like that of the crow, has aroused diverse and controversial opinions. On the credit side is the fact that much of the animal life eaten consists of destructive insects, but at times when, in late summer and autumn, this gregarious bird assembles in immense flocks, much grain, especially corn, is destroyed. It is the latter that accounts for the vast majority of complaints lodged against this bird.

The food of the whole year based on the examination of the stomach contents of 2,346 stomachs by F. E. L. Beal (1900), was 30.3 percent animal and 69.7 percent vegetable matter. In addition to insects the animal matter was composed of spiders, myriapods, crayfish, earthworms, sowbugs, snakes, snails, fish, frogs, toads, salamanders, lizards, birds, eggs, and mice.

Insect food constitutes 27 percent of the entire food for the year, and is the most interesting part of the bird's diet from an economic point of view. When it is examined month by month, the smallest quantity appears in February (less than 3 percent of the whole food)—In March it rises to one-sixth, and steadily increases till May when it reaches its maximum of five-eighths of the whole; it then decreases to one-sixth in October and appears to rise again in November. * * * The great number of insects eaten in May and June is due in part to the fact that the young are fed largely on this kind of food.

Analysis of the insect food presents many points of interest. Among the most important families of beetles are the scarabaeids, of which the common June bug or May-beetle and the rose bug are familiar examples. These insects are eaten, either as beetles or grubs, in every month except January and November; In May they constitute more than one-fifth and in June one-seventh of the entire food. The habit grackles have of following the plow to gather grubs is a matter of common observation which has been fully confirmed by stomach examinations. Many stomachs were found literally crammed with grubs.

Next in importance to beetles as an article of blackbird diet are the grasshoppers.—They constitute less than 1 percent of the total February food.—The proportion of grasshoppers in the stomachs increases each month up to August when it attains a maximum of 23.4 percent of all the food. After August the grasshopper diet falls off, but even in November it still constitutes 9 percent of the total for the month. The frequency with which these insects appear in the stomachs, the great numbers found in single stomachs (often more than thirty), and the fact that they are fed largely to the young, all point to the conclusion that they are preferred as an article of food and are eagerly sought at all times.

Caterpillars, including the army worm, averaged 2.3 percent in each month, but in May a maximum of more than 8 percent is reached.

A letter, from Benjamin J. Blincoe, tells of the bronzed grackle feeding on the larvae of the sphinx moth which were infesting a tobacco field: "A short time before sunset on the evening of July 21, 1932, while Mrs. Blincoe and I were motoring along a country road near Dayton, Ohio, we noticed a scattered flock of grackles, the individuals of which were alighting in a tobacco patch and in the road ahead of us. Stopping the car we soon saw dozens of the grackles alight in the road with large green larvae of the sphinx moth, that is so troublesome to the tobacco plant. Holding them securely in their mandibles, they beat the fat larvae against the ground with such force that the impact could actually be heard. We could also see many grackles picking the larvae from the rather small tobacco plants. We counted at least a hundred grackles with the larvae and I believe many more were helping with the good work of ridding the plants of the destructive larvae."

The Hymenoptera are represented mostly by ants, while flies are entirely absent. Spiders and myriapods are eaten to a small extent every month.   The spiders attained a maximum of more than 7 percent in May, not only the spiders but their cocoons full of eggs appear to be taken whenever found.

In the South, A. H. Howell (1907) has revealed that the large flocks of grackles in February and March feed on the destructive boll weevil.

W. J. Howard (1937) gives an account of the grackles among other birds that were feeding on the 17-year locusts at a time of an outbreak of these insects at Crown Hill Cemetery, Indianapolis, Ind.

It is at the times of major insect infestations that birds become important factors in controlling destructive insects, and at such times their work becomes obvious even to the casual observer.

A few insects eaten by the bronzed grackle beneficial to man's interests and among these are a considerable number of predaceous beetles belonging mainly to the family Carabidae.   These valuable destroyers of noxious insects are eaten, according to Beal (1900), in every month of the year in quantities varying from more than 7 percent of the food in January to 13 percent in June.

The comparatively few other forms of animal food eaten are of little economic importance, yet they serve to emphasize the grackle's omnivorous nature and also some of its characteristic feeding behavior.

There are numerous reports of the bronzed grackle feeding on crayfish especially from the Middle Western States.   Thomas S. Roberts (1932) has written an account of the habits of this bird in capturing and eating these crustaceans:

It patrols the water's edge, often wading body-deep, and is quick to seize any moving creature, be it insect, small fish, or crawfish.   It does not hesitate to plunge from a low, overhanging bush or tree-trunk, though it lacks the power and dexterity of a Kingfisher.   Of crawfish it is especially fond.   Dragging them, squirming and struggling, from under the stones and roots, it carries them ashore and onto a convenient, hard surface, where they are pounded and mauled until they cease struggling.   The next move is to open a large hole in the back just behind the carapace, through which the meat is extracted until nothing but the empty shell remains.   The writer has watched both males and females thus engaged along the shore of one of the park lakes of Minneapolis.   The dead crawfish was held firmly on the ground with one foot while the white meat was picked out bit by bit and piled in a heap near by until there was a good, sizable billful, when it was gathered up and conveyed to the waiting nestlings.   The ground for a quarter of a mile was strewn with discarded and fresh remains of many crawfish, showing that for days the Grackles had been supplying their young with this delectable viand.

Lorus J. Milne (1928) saw 20 bronzed grackles capturing specimens of the amphipod *Gammarus fasciatus* on the shallow sandy bank of a small stream flowing into Grenadier Pond, High Park, Toronto, Canada.   Each bird would gather several amphipods together into a pile on the sand before eating them.

Many reports have been made of the bronzed grackle catching

small fish. Mr. Frank C. Pellet (1926) observed bronzed grackles feeding on minnows at a Mississippi River power dam near Hamilton, Ill. The birds alighted in the shallow water running over the cement apron below the dam and watched for the passing minnows. When a fish was caught they flew to a nearby rock, or to the top of the dam, and hammered their victim to death. Mr. Pellet, who observed the performance for many days, is of the opinion that grackles living near water may depend upon fish to a considerable extent.

L. L. Snyder (1928) observed a bronzed grackle, perched on a stone in the center of his bird bath, spear a minnow, which was then laid on the grass at the border of the bath. The performance was repeated until the grackle had secured three minnows; these were then picked up and carried away. Upon examining the bath at a later date, he found that every one of two dozen minnows had disappeared. After several days had elapsed, the bath was restocked with fish but these likewise disappeared.

P. A. Taverner (1928) had a similar experience with grackles catching goldfish at a large pool located in his garden at Ottawa, Canada. Again and again one was seen to snatch up a fish, beat it to death on the concrete margin and then carry it away to its nest. When emptying the pool in the fall, Mr. Taverner usually took some 300 goldfish of varying sizes, but that year there were no young fish and the breeding stock was greatly reduced. Others have reported similar experiences at their fish pools.

Stanton Grant Ernst (1944) observed bronzed grackles catching, killing, and devouring small leopard frogs at a small pool located in a swampy woodlot near Olean, N. Y. Mr. Ernst describes their behavior as follows: "Circling the pool, they would suddenly run along the ground, fluttering their wings, and jab viciously at the small frogs which abound in the pool. I watched the birds kill three frogs, then frightened them away and examined the remains. Each frog was neatly pierced with a bill-sized gash in the soft throat or near the eyes." He observed them again two days later and reports: "I observed one bird eating a frog in a small oak above the pool and noted that the other was actually in the water and that the belly feathers were wet. This bird repeatedly stabbed at frogs, apparently without success, but I later found two dead frogs floating in the pool; both had been pierced through the head."

Joseph W. Hopkins, of Colo, Iowa, writes concerning the habit of the bronzed grackle in capturing mice: "In Iowa, the bronzed grackle nests in colonies in nearly every coniferous grove. They soon take notice of any farm work which involves stirring the soil and take full advantage of it. The disc harrow penetrates rather deeply, and frequently turns field mice uninjured from their burrows. I saw a male bronzed grackle pursue and stun an adult mouse. He had some difficulty in doing it, for the mouse seemed able to dodge his blows, but after a half minute of chasing and vicious pecking he was successful and flew off with the mouse in his bill. Judging from the ease with which he sprang into the air and the rate of ascent, a considerably heavier load could be carried."

One of the most serious complaints lodged by the bird lover against the bronzed grackle is its pernicious habit of destroying the eggs and young of other birds and its practice of killing small adult birds. J. Nelson Gowanlock (1914), Winnipeg, Canada, observed a bronzed grackle visit all of the homes of an entire block at regular intervals of every 4 or 5 days. The grackle "entered the nests of the English sparrows built in the corners, and, after eating the eggs or young, would emerge, stand a moment or two ignoring the throng of distracted sparrows, and then fly on to the next house where the scene would be repeated. * * * he was certainly the coolest, most methodical and heartless nest robber I have ever seen or heard of."

Charles W. Townsend (1920) writes: "Robins' nests in the vines of my house have been despoiled of their eggs and young by this bird, and I have known it to kill adult birds of moderate size. I once found a Grackle holding down the freshly killed body of a Bicknell's Thrush while it pecked out its brains."

J. M. Wheaton (1882) writes: "I have repeatedly seen them destroy the nest and eggs of the chipping sparrow, built in my own garden. This appeared to be from mere love of mischief, as they were not content with destroying the eggs but returned to demolish the nest, and again pulled to pieces the half finished nest which the birds rebuilt."

K. Christofferson (1927), Sault Ste. Marie, Mich., saw a bronzed grackle kill two pine siskins, benumbed by cold, by pecking the birds on the head. The brains were devoured, leaving only a part of the skull. He also saw a bronzed grackle kill a young barn swallow.

There are many cases on record of bronzed grackles killing English sparrows. E. H. Forbush (1927) relates the following incident:

I saw a Bronzed Grackle on Boston Common with a full-grown dead "English" sparrow which it tried to carry away, as I thought, in its claws, but it dropped the smaller bird after flying up a few feet from the ground. In a letter received from Dr. John W. Dewis he says that he and others saw a flock of sparrows on the wing, pursuing and apparently attacking a Bronzed Grackle, also in flight and carrying a live sparrow in its bill. When a few feet from the ground the grackle dropped his prey which was fluttering, and squatted over it, threatening the sparrows which soon gave up the fight. The grackle then pecked out the eyes of its victim, disembowled it, ate the muscle from its right breast and left it. The bird proved to be a full-fledged "English" sparrow.

In Maine, grackles are frequently seen at the edges of ponds, or on salt water mud flats, where they secure worms, small crustaceans, and other edible articles. Occasionally, they will feed on a large dead fish left by the receding tide, and along fresh water ponds they have been known to pick up dead frogs and snakes, as well as fish that have drifted ashore. At times grackles even visit garbage dumps, with starlings, to pick up miscellaneous waste food.

I have often seen them feeding among the animals in piggeries, where this sleek, well-groomed bird seemed decidedly out of place; and they frequent the door yards of homes in cities and towns, as well as of farms, where they obtain bits of bread and other food. Hard bread they may first soak in water, if a convenient pool or bird bath is near, until it is soft and easily swallowed. They have been seen to retrieve bits of food floating on the water. William Brewster (1937) gives an account of grackles taking bread and crackers from the water at Lake Umbagog, Maine, as follows: "Today I saw them dip their legs to the thighs in the water and repeatedly one immersed the lower half of its body, also apparently floating on the water for an instant. The food was invariably taken up in the bill however."

The vegetable food of the grackle is as variable and diversified as the animal food, showing plainly that when one article of diet is not available, this very adaptable bird turns to food more easily obtained. Of the various items of vegetable food, the chief interest centers about grain and fruit, and it is through the consumption of these that blackbirds inflict the greatest damage on man. Frequent complaints have been made against the grackle by the farmer because of the large quantities of grain eaten, especially when immense flocks descend on the grain fields. According to F. E. L. Beal (1900), "the stomach contents were found to contain corn, oats, wheat, rye and buckwheat and of these, corn is the favorite, having been found in 1,321 stomachs, or more than 56 percent of the whole number. It is eaten in all seasons of the year; and in every month except January, July, August and November amounts to more than one-half of the total vegetable food. * * * "In August corn amounts to one-seventh of the whole food, and this together with a part taken in September, is green corn 'in the milk'." The birds easily strip the green husks from the ears in order to reach the growing corn and what they do not eat is left exposed to the weather to dry or rot. The maximum amount of corn, 82 percent, is eaten in February, according to Beal. At this season it is waste grain and of no economic importance.

Oats, eaten in irregular quantities, in August, forms 26 percent of the total food, this being the only month of the year in which this grain reaches a higher percentage than corn. In the southern States, bronzed grackles prey upon rice in company with other blackbirds and bobolinks.

Fruit is eaten in every month from March to December, but it does not become important until "in June, July and August it reaches 7, 13, and 10 percent respectively." The fruits of economic interest are blackberries, raspberries, cherries, currants, grapes, and apples. Blackberries and raspberries, the favorites, make up the bulk of the fruit eaten. The vast majority of the fruits eaten are wild and hence are not important from the standpoint of man's interests.

Weed seeds are freely eaten, especially during the colder months reaching a maximum of 11 percent in October. Chestnuts, acorns, and beechnuts form an important item of the food in the fall and early spring months.

Mr. A. W. Schorger (1941) has found that the bronzed grackle feeds freely on acorns, aided by a special ridge or keel located on the palate. We found the birds successfully opened small acorns of the yellow, Hill's scarlet, bur and pin oaks, but the normal acorn of the white and northern oak were too large to be manipulated. Attempts to open small acorns of the white oak were seldom successful due to the toughness of the shell. No fragments of the shell are eaten but the entire kernel is swallowed. Miscellaneous mineral substances such as sand, gravel, pieces of brick, bits of mortar, plaster of Paris, coal, cinders, etc., are eaten by the grackle to assist in grinding the food.

According to Beal, (1900) the food of 456 young collected from May 22 to June 30, inclusive, was made up of 74.4 percent animal and 25.6 percent vegetable matter. The animal food of the young is chiefly insects, amounting to 70 percent of the total food. During the first few days the young are fed chiefly on spiders and soft bodied insects in the form of larvae or grubs. Grasshoppers and crickets are a common food of the young, and as they grow older, hard shelled insects such as beetles are included in their diet.

Ira N. Gabrielson (1922) in the course of a study of a nesting colony of bronzed grackles near Marshalltown, Iowa, found the parent birds were flying to a partially inundated pastureland to secure cutworms, earthworms, crickets, spiders, tumble bugs, ground beetles, and other insects that had migrated into the short grass on little knolls to escape the high water. There were 16 nests and each of the 32 parents made an average of 6 trips for food per hour. At one nest observed from a blind, Mr. Gabrielson witnessed 33 feedings during the course of 7 hours. He saw the adults deliver 12 earthworms, 9 crickets, 60 cutworms, 2 spiders, 2 kernels of corn, and 7 unknown insects which were taken from the bountiful source in the pastureland.

The vegetable food of the young consists chiefly of corn and fruit but the corn, comprising 15 percent of the total food, is fed only to the older birds. The nestling bronzed grackles, in eating insect pests such as cutworms, May beetles, weevils, and grasshoppers far outweigh the harm done by the consumption of corn.

Because of the variable nature of the food of the bronzed grackle, there is little wonder that a marked difference of opinion has arisen in regard to its economic status. However, not until we have a view of the entire picture, can we safely pass judgment. There seems to be no justification of a general control of this species but when thousands of these birds descend on a farmer's cornfields, he should be permitted to employ every reasonable means to protect his interests.

Behavior.—The bronzed grackle, a sleek, well-groomed bird, is striking in appearance when his iridescent plumage flashes its varied colors in the bright sun. A single bird may sometimes appear cowardly toward an adversary, but in a group the grackles are aggressive. I have seen a mass of 40 give chase to a large, powerful eagle that flew over the colony and continue to harass the intruder until it was well away from their nesting place. Louis B. Kalter (1932) saw six to eight hundred grackles perched in some oaks near a small lake. About a third of them, apparently without provocation, left and pursued an osprey, which, with the birds in pursuit, circled the lake twice, then climbed higher into the air where the wind was much stronger and colder. The grackles abandoned the chase.

Two males will fight fiercely in competition for a female, and they frequently battle in the colony in defense of their territories. Grackles are devoted to their young and seem fearless in defending them from danger. One only needs to climb a tree containing a nest of young to be impressed with the vigor and boldness of their attacks. I have had them strike me with blows sufficient to knock off my hat when I attempted to remove a squawking young for closer examination.

Sometimes, when subjected to unusual conditions or under stress of hunger, grackles resort to wholesale killing. Ruthven Deane (1895) gives an account, by Jesse N. Cummings, of the activities of crow blackbirds at Anahuac, Tex. On February 14–15, 1895, an unusual snowstorm lasting 30 hours covered the ground to a depth of 20 inches on the level and remained for 3 or 4 days. On a large piece of ground along the bay shore, kept free of snow by water flowing from an artesian well, about 200 jack snipe gathered in a space not over 100 feet square. There, Cummings saw rusty and bronzed grackles kill 10 or 12 of the birds and he counted 30 or 40 dead ones in other places. At this same time the blackbirds also attacked the robins about his house, and while he did not ascertain the numbers killed, he saw many lying on the snow about his place and along the shore of the bay. The blackbirds fed on the brains of their victims, leaving the remainder

of the body untouched.   Presumably this behavior was brought about
by the lack of other accessible food, as a result of the snowstorm.
While the killing of birds for food is not unusual, in a number of cases it
has seemed that the killing of birds and destruction of their nests and
eggs was purely an act of destruction.

In striking contrast to this type of behavior is a case reported by
Wilson Baillairge (1930) in which a bronzed grackle served as a foster
parent to chipping sparrows at St. Michel, Quebec, Canada.   He
writes:

I frequently noticed a pair of Bronzed Grackles about the house.   Whenever
we went on the gallery the female Grackle flew from branch to branch in a near-by
tree, scolding noisily.   I looked for her nest but could not find it, but did find a
Chipping Sparrow's nest containing three young, in a grape vine trained along
the gallery.   I was surprised not to see any sign of the parent Chipping Sparrows,
and watched the nest carefully.* * *   Finally, I saw the female Grackle go to the
nest and feed the young Chipping Sparrows; she fed them three or four times in
my presence, not more than a few feet from me.   That afternoon one of the young
Chipping Sparrows flew from the nest to a tree nearby, and was followed by the
female Bronzed Grackle, which showed every sign of maternal anxiety."

Francis H. Allen (MS.) reveals the resourcefulness of this bird under
unusual conditions: "For at least two years a male grackle that spent
its summers on or near the Boston Public Garden lived and throve
with a malformed bill that interfered with feeding in the normal way.
The upper mandible was about twice as long as the lower, which appeared to be normal length.   It was also decurved and flattened and
had a squarish tip.   When feeding on the ground the bird had to turn
its head to one side to pick up its food, though no such accommodation
was necessary when it picked insects from the top of the grass.   Probably no bird less hardy and less resourceful than a grackle could have
survived so long with such a handicap.

"On one occasion I saw a grackle get completely under a newspaper
lying on the ground in the Boston Public Garden, for the purpose of
feeding.   Each time the paper, or part of it, was raised considerably
from the ground.   This illustrates the enterprising character of the
grackle."

Grackles have been seen anointing their plumage with the juices
of certain fruits, and with acid or pungent substances derived from
the hulls of fruits and nuts.   Certain birds are well known to use ants
for this purpose, a behavior called anting.   This term has come to be
generally applied to cases where other substances are used.   The
purpose of this act is not clear although a number of theories, for
example, to repel parasites, have been advanced.

Mr. H. R. Ivor (1941) has observed the bronzed grackle going
through the performance of "anting" with choke cherries.   He has
seen none of the many other birds he has observed "anting" use this

fruit for that purpose. On July 3, 1945, G. Hapgood Parks (1945) saw a male bronzed grackle anointing its feathers with juices devived from the green fruits of the cucumber tree (*Magnolia accuminata* Linnaeus). The bird was seen to take pieces of the fruit, and, frequently, entire "cucumbers" in his bill and rub them vigorously against his breast and body feathers. The bird preened his feathers with unusual industry. The tail and wing feathers as well as the body, breast, and neck received energetic attention. It also frequently scratched the head and neck first with one foot and then the other. The bird was trapped and found to be in normal condition and had the usual brilliant iridescence of its feathers. No parasites were found. A half hour later after releasing the banded bird, Mr. Parks saw two other unbanded, adult male bronzed grackles go through the identical behavior of "anting" with the cucumber tree fruits. Judging from the number of reports, the practice of anointing the plumage with various substances is not a rare behavior among grackles.

Voice.—The notes of the bronzed grackle are not pleasing and beautiful, nor are they at all musical, but they are characteristic and easily recognized. The song consists of one or two short notes followed by a prolonged squawk. The quality is harsh and squeaky, with a peculiar metallic sound difficult to describe. It has been likened to a noise of a squeaky hinge on an iron gate. A. A. Saunders (1935) has interpreted the call by *Kŭchăkŭ wēē ēē k ēē, ku wăă ă,* saying: "The male often produces this sound with a spreading and fluttering of the wings which resemble similar actions of singing Red-winged Blackbirds and Cowbirds." F. Schuyler Mathews (1921) compares the queer noises uttered by the grackles with "rattling shutters, watchmen's rattles, ungreased cart wheels, vibrating wire springs, broken piano wires, the squeak of a chair moved on a hardwood floor, the chink of broken glass, the scrape of the bow on a fiddle string, and the rest of those discords which commonly play havoc with one's nerves!" When a large number of grackles are singing in chorus all of these discordant sounds are beyond description.

The ordinary call note is a hoarse loud *chuck* or harsh *clack*. When answering the call, a fellow grackle may utter a kind of subdued *cuk*.

Witmer Stone (1937) describes the notes uttered at a nesting colony of grackles as follows: "About the nest trees there is a constant chorus of harsh alarm calls; *chuck; chuck; chuck;* like the sound produced by drawing the side of the tongue away from the teeth, interspersed with an occasional long-drawn *seeek*, these calls being uttered by birds on the wing as well as by those that are perching. Then at intervals from a perching male comes the explosive rasping song *chu-séeeek* accompanied by the characteristic lifting of the shoulders, spreading of the wings and tail, and swelling up of the entire plumage."

Francis H. Allen has sent us the following observations: "Among the less common notes of the bronzed grackle is a low-pitched mellow whistle, rather short, with an *r* in it, which might be rendered as *pree*. The *r* is not prominent and the effect is sweet and pleasing and quite ungracklelike. Another note, probably a courtship note, heard April 9, 1934, consisted of a sort of *chi* or *shi;* it was given generally three times in succession, but sometimes only twice. It was uttered both when the birds were perched and when two or three were flying together in what looked like a courtship flight."

Robert Ridgway (1889) in comparing the notes of the bronzed and purple grackles writes: "From an almost equal familiarity with the two birds, we are able to say that their notes differ decidedly, especially those of the male during the breeding season, the song of the western bird being very much louder and more musical, or metallic, than of its eastern relative." However, Aretas A. Saunders who has studied the songs of both forms intensely, fails to find any difference between the songs of the bronzed and purple grackle.

Enemies.—Man can be considered one of the worst enemies of the bronzed grackle, for great numbers are killed and poisoned, especially at the large roosts by farmers and others in their efforts to protect their crops. And many are killed for food, especially in the southern sections of the range. In a willow growth along the Mississippi River near Cairo, Ill., I saw a group of hunters enter a populous roost of grackles with shotguns, at sunset. After firing several volleys they picked up over three hundred of the birds. When questioned the hunters stated the birds were to be used as food for themselves and neighbors.

Grackles like other passerine birds have their enemies among the larger hawks and owls. A. K. Fisher (1893) reported finding the remains of grackles in the stomach contents of the marsh, Cooper's and red-tailed hawks and the short-eared owl. I found in the nest of the horned owl the remains of a grackle which had been brought by the parent birds to feed the young. The behavior of the grackles when a hawk or an owl appears near their nesting places is evidence that they are considered enemies.

Squirrels have been known to destroy the eggs and young of the grackle. Robert Ridgway (1889) saw a fox squirrel emerge from a bronzed grackle's nest, built in a hole in a large tree, with a young grackle in its mouth. "The squirrel was attacked by a number of the blackbirds, who were greatly excited, but it paid no attention to their demonstrations, and, after descending scampered into the woods with its prey."

Bagg and Elliott (1937) report that live grackles have been found with sticks completely pierced through the body. One bird that was shot had a smooth twig somewhat smaller than a pencil protruding four inches from the abdomen. "The bird must have carried the twig for some time, as it was worn smooth and the skin had grown firmly about it." Such accidents may occur when the birds are disturbed and caused to dash about in wild confusion at the roosts. I have seen birds of other species that apparently had rammed themselves into the stiff dead twigs of spruces.

Kenyon and Uttal (1941) report an unusual case in which a young grackle about two weeks old had met its death by swallowing a string. "A double length of string passed through the esophagus terminating in a tightly packed wad of string in the proventriculus and ventriculus; thus making an exit through the pyloris impossible. The total length of the string, including some three or four inches which protruded from the mouth, was eleven feet, ten inches."

Although a hardy bird, the bronzed grackle may succumb to storms and sudden changes of temperature. H. Elliott McClure (1945) found nine bronzed grackles among other birds that had been killed in the city park at Portsmouth, Iowa, by a tornado of moderate velocity that had struck the city. In a winter roost of bronzed grackles, starlings, cowbirds and redwings at Urbana, Ill., Odum and Pitelka (1939) found 63 dead bronzed grackles, among the many other birds, killed by a driving wind and rain storm followed by a sharp dip in the temperature: "The proportion of Bronzed Grackles and Cowbirds to Starlings in the total storm mortality was certainly much greater than that in the total roosting flock."

The bronzed grackle is sometimes parasitized by the cowbird; Herbert Friedmann (1929) reports three nests in Illinois and one nest in Iowa which contained eggs of the eastern cowbird and one nest in North Dakota parasitized by the Nevada cowbird (*Molothrus ater artemisiae*). Later Friedmann (1931) reported a nest of the bronzed grackle found in Texas which contained an egg of the Eastern cowbird.

It is of interest to note, although not involving parasitism, that eggs of other birds have been found in nests of grackles. M. G. Vaiden of Rosedale, Miss., states that in a mixed colony of bronzed grackles and mourning doves he found a nest of the bronzed grackle containing three young grackles and an egg of the mourning dove.

Most birds have been found to have a number of external parasites and the bronzed grackle is not an exception. Harold Peters (1936) has found the two lice, *Degeeriella illustris* (Kell.), and *Menacanthus chrysophaeum* (Kell.), the fly, *Ornithoica confluenta* Say, and the tick, *Haemaphysalis leporis-palustris* Packard, on specimens of the bronzed grackle.

A new blood parasite *Haemoproteus quiscalus* obtained from the blood of the bronzed grackle has been described by Coatney and West (1938).

Fall.—It is in late summer and autumn when the gregarious bronzed grackles congregate by the thousands, and often in hundreds of thousands, that they become one of our most conspicuous forms of bird life. These birds attract unusual attention when the roosts are near human habitations in the midst of our cities and towns. Dr. Lynds Jones (1897) has written an excellent paper concerning such a roost that was located on the college campus at Oberlin, Ohio, during the summer and fall of 1896. He describes conditions typical of many similar roosts. The vanguard of the grackles, which reached Oberlin on March 9, was greatly increased by March 28. From this time on flocks of varying size visited the roost but none passed the night in it. On April 20 the first nest was found and by May 14, young birds. May 16 was the first day when considerable numbers began to spend the night at the roost. On May 21, 100 birds were counted leaving the roost in the morning and on May 23, 352, of which all were adult males except one young with tail feathers half grown. Since the birds did not go far, Jones assumed that most of them had nests in or near the village.

This small company was recruited from day to day by old males and a little later by the more forward young. About July 10 adult females and more young came to the roost as the nests were deserted. At this time the trees became so crowded with birds that other places were sought by the overflow. On July 17 the birds came in at the rate of 52 per minute for an hour, the flight terminating with the arrival of an uncountable company just at sunset. Approximately 5,000 birds were in the trees of the roost, and many others in neighboring trees.

During the early part of July the birds did not wander far from the roost at any time, but by August 1 none were seen in the town during the day. From this time on the birds arrived in greater companies, after considerable flights across the country. The gregarious instinct asserted itself more and more as the season advanced and the necessity of a wider feeding ground increased. The numerous small flocks joined together until there was but the one huge flock, with a few stragglers.

On September 7 the first note heard in the colony was at 4 a.m. By 4:30 many were singing and shifting about in the trees; and at 4:40, 300 were counted leaving the trees. At 5:04, the birds of the roost arose, not in one mass—

but in consecutive order from the south to the north edge of the group of trees, as though by previous arrangement, giving the impression that the foliage was melting away into that black stream. * * * As long as it could be seen, the flock remained intact, and did not stop to rest. The flight was near the ground, and followed the contour of the country closely, rising only to clear farm buildings and woods, then dipping again to the former level. The lowermost birds were scarcely more than twenty feet from the ground. While the birds were flying there was no singing and not much noise of any kind except that made by the wings. It was evident that the birds had some definite feeding ground selected, toward which they were hurrying in a straight line.

In the evening the first birds arrived at the roost at 5:14 p.m. Between 5:34 and 5:45 about 5,000 arrived, coming in companies of from 200 to 800, an almost continuous flight. The birds continued to come in until a few minutes after 6. By 6:15 practically all were out of sight in the foliage and a few minutes later all noise had stopped.

A study of the flocks at a point away from the roost revealed that the mass assumed definite patterns of narrowed and expanded parts. It became more drawn out and broken as it proceeded. The vanguard would stop at some treetop and rest until the others had passed over, at which time it arose and formed the rear guard. In this way the whole flock secured a short breathing time, part by part. Rarely, two flocks were formed during the flight.

"There was no diminution in the number occupying the roost up to September 21, but not one bird appeared at the old stand on the two succeeding days. On the 24th less than a hundred occupied the trees during the night and none visited it afterwards."

Charles R. Keyes (1888) describes the great blackbird flights at Burlington, Iowa, as follows:

During September and October the cornfields of Iowa are visited by countless numbers of these black marauders, which wander about in mixed flocks of several thousands, passing the day in the fields and the night in woodland or marshes. And it is during this period that so many thousands are poisoned and killed by the farmers. About the first of October the birds begin * * * to rise out of the swamps and radiate in all directions towards the inland cornfields, where they spend the day, returning again to the swamps before sunset. These flocks are often a quarter of a mile in width and are more than an hour in passing—a great black band slowly writhing like some mighty serpent across the heavens in either direction, its extremities lost to view in the dim and distant horizon. Not unfrequently, three or four such vast flocks are in sight at one time. How far away from their night resorts they go each day has not been observed; an hour and a half before sunset, twelve miles away from the river, the mighty armies of Blackbirds are still seen coming over the distant hills and directing their course toward the marshes. It is evident, however, that many miles are daily traversed in their journeys to and from the feeding grounds. Making liberal deductions for any possibility of over estimating, the numerical minimum of individuals in a single flock cannot be far from twenty millions.

It has been noted by many observers that the times when blackbirds arrive and leave the roost varies according to the length of day. Margaret M. Nice (1935) made observations of the bronzed grackles and starlings which roosted in the shade trees of a residential district of

Columbus, Ohio. For 9 days, October 6 through 15, 1934, she determined with a Weston photometer the light values in the morning and evening at the times the birds left and arrived at the roost. On seven clear mornings their first flights left from 7 to 9 minutes before sunrise at light values of 13 to 16 foot-candles (median 14). On one cloudy morning they left 3 minutes before sunrise at a light value of 13.5 foot-candle. The largest flocks left at light values of 20.5 to 29 foot-candles. In the evening the first flocks were seen about half an hour before sunset. Light values ranged from 114 to 40 foot-candles but the height usually occurred between 65 and 52 foot-candles. The flight ended just about sunest, from 1 minute before to 3 after. Mrs. Nice determined that leaving and returning to the roost was closely correlated with light, and that the grackles went to roost when the light was about three times as bright as it was when they left it.

E. H. Forbush (1907) gives a graphic account of bronzed grackles on their fall migration flight which he observed at Concord, Mass., on October 28, 1904, as follows:

From my post of observation, on a hilltop, an army of birds could be seen extending across the sky from one horizon to the other. As one of my companions remarked, it was a great "rainbow of birds;" as they passed overhead, the line appeared to be about three rods wide and about one hundred feet above the hilltop. This column of birds appeared as perfect in form as a platoon. The individual birds were not flying in the direction in which the column extended, but diagonally across it; and when one considers the difficulty of keeping a platoon of men in line when marching shoulder to shoulder, the precision with which this host of birds kept their line across the sky seems marvelous. As the line passed overhead, it extended nearly east and west. The birds seemed to be flying in a course considerably west of south, and thus the column was drifting southwest. As the left of the line passed over Concord meadows, its end was seen in the distance, but the other end of this mighty army extended beyond the western horizon. The flight was watched until it was out of sight, and then followed with a glass until it disappeared in the distance. It never faltered, broke, or wavered, but kept straight on into the gloom of night. The whole array presented no such appearance as the unformed flocks ordinarily seen earlier in the season, but was of finer formation than I have ever seen elsewhere, among either land birds or waterfowl. It seemed to be a migration of all the Crow Blackbirds in the region, and there appeared to be a few Rusty Blackbirds with them. After that date I saw but one Crow Blackbird. It was impossible to estimate the number of birds in this flight. My companions believed there were millions.

Dr. Charles Blake of Lincoln, Mass., has written us concerning a flock of 3,100 bronzed grackles he observed on their migratory flight, October 30, 1942. The flock, which required 15 minutes to pass, was about 300 feet wide but contained a fixed wave so that the flight was undulated over a width of fully three-tenths of a mile.

Recoveries of bronzed grackles banded by Charles B. Floyd (1926), Mabel Gillespie (1930), and others, clearly indicate that the fall migration is in a southwesterly direction along the Atlantic coast.

Many grackles banded in New England during the spring migration travel northeastward to Nova Scotia, New Brunswick, and eastern Quebec.

Samuel E. Perkins III (1932) in compiling the recoveries of bronzed grackles banded at a dozen stations in different parts of Indiana discovered those birds wintered in a narrow, restricted area between Louisiana and Alabama. Purple grackles banded by Horace D. McCann (1931) in Pennsylvania likewise have the same habit of keeping to a restricted east-to-west winter range, not more than 100 miles across. Many other birds, such as the robin and mourning dove, have shown a distinct tendency to spread out fanlike in their fall migration to the south and to winter in a very wide area.

Examples of other interesting recoveries of the bronzed grackle are: One banded at Ottawa, Canada, taken in North Carolina; one banded in Saskatchewan, taken in Louisiana; and T. E. Musselmann has written me that a grackle banded at Quincy, Ill., in the spring of 1947 was taken 3 months later at a point 3,000 miles north in Alberta, Canada. As many such records accumulate in the future, we shall gain a clearer picture of the migratory routes as well as the summer and winter distribution of specific populations.

The distribution of the bronzed grackle over the various types of crops and farm land in Illinois was determined by a statistical survey conducted in 1906–1907 and reported by Forbes (1907, 1908), and by Forbes and Gross (1923). The sight of large flocks in the grain fields, especially in the autumn, leads us to a natural misconception of their distribution as a whole, but when adequate samples are taken of all types of land and crops under all conditions of weather and all times of day, as was done on this survey, a truer picture is gained of their status in relation to the crops. The survey was conducted continuously throughout an entire year but the results of one trip taken from the Indiana line to Quincy on the Mississippi River, from August 28 to October 17, 1906, will serve to illustrate this point. The accumulated records, revealed that the bronzed grackle was the most abundant of the native birds, representing 11 percent of the total population of all the birds of the agricultural areas, an average of 94 grackles per square mile. The interesting fact, however, is that their numbers were greatest not on grain fields, but on pasturelands, where 90 percent of this species was found at a population density of 307 birds per square mile. Only 4 percent of the grackles, at a density of only 10 per square mile, were found in corn; and 4 percent, at a density of 21 birds per square mile, were present in stubble.

Winter.—The bronzed grackles which occupy the extreme northern parts of the nesting range migrate to the south in the fall. In New England the great mass of birds leave by the end of

October, but in this region, especially southern New England, many individuals remain throughout the winter. Likewise, in the Midwestern States south of the Great Lakes at least a few birds seem successful in combating the rigors of cold weather and snow. In these northern sections of the winter range, the birds are generally seen as individuals, or else in very small groups, but large flocks are sometimes reported as late as November.

Milton B. Trautman (1940) found bronzed grackles wintering at Buckeye Lake, Ohio. In some years not more than 12 individuals were noted, but in other winters the aggregate numbers of the small groups ranged from 100 to 300 birds. "The wintering birds remained chiefly about the barn yards, in fields where stock was fed and in the larger uncut cornfields. They roosted in spruces, in cattail marshes and in the brush of inland swamps."

Otto Widmann (1907) writes concerning the wintering of these birds in Missouri as follows:

As a winter visitant the bronzed grackle is rare except along the Mississippi River from St. Louis southward. Opposite St. Charles along the bank of the Missouri River there is a large swampy tract of willows used as a winter roost for innumerable red-wings and with them hundreds of bronzed grackles have been seen going even in the middle of January, in mild weather, but as their numbers change constantly, there are hardly two days alike, showing that they also use other roosts farther south, to which they fly when the weather is not inviting northward. Should weather conditions remain unfavorable, the roost may remain deserted or nearly so for weeks at a time, until a change sets in when they appear again. Away from the roost they are seldom met with, because they go far to favorite feeding grounds and scatter over a large territory.

In the southern part of the winter range, along the Gulf coast from Florida to southern Texas, the bronzed grackle mingles with the southern forms of grackles and other blackbirds at the roosts as well as on foraging expeditions for food during the day.

DISTRIBUTION

Range.—Western and southern Canada to Alabama and Georgia.

Breeding range.—The bronzed grackle breeds from northeastern British Columbia (Tupper Creek), central-southern Mackenzie (Fort Simpson, Fort Smith), central Saskatchewan (Flotten Lake, Cumberland House), central and northeastern Manitoba (Grand Rapids, Churchill), western, central, and northeastern Ontario (Favourable Lake, Rossport, Moose Factory), southern Quebec (Blue Sea Lake, Anticosti Island, La Tabatiérre), southwestern Newfoundland, and northern Nova Scotia (Baddeck, Sydney); south along the eastern slope of the Rockies to central-southern and southeastern Colorado (Denver, Beulah, Fort Lyon), central and southeastern Texas (Abilene, Galveston), southwestern and central Louisiana (Calcasieu, Vidalia),

western and northern Mississippi (Centerville, Baldwyn), northern
Tennessee (Nashville), Kentucky, central West Virginia (Nicholas
County, Franklin), central Pennsylvania (State College), central New
York (Ithaca, Troy), northern Connecticut (Litchfield), Rhode
Island, and southeastern Massachusetts (Martha's Vineyard, Dennis);
also on Shelter Island, at the eastern end of Long Island, New York.

Winter range.—Winters casually north to northern Minnesota
(Fosston, Grand Marais), southern Wisconsin (Racine), southern
Michigan (Vicksburg, Ann Arbor), southern Ontario (Kitchener,
Gananoque), and along the Atlantic coast to New Brunswick (Mem-
ramcook) and central Nova Scotia (Wolfville); south to southern
Texas (Mission), southern Mississippi (Biloxi), central Alabama
(Greensboro), southern Georgia (Fitzgerald), and South Carolina
(Aiken, Mount Pleasant).

Casual records.—Casual in eastern Washington (Whitman
County), Nevada (Fallon, Crystal Springs), central-southern Texas
(Fort Clark), northern Ontario (Fort Severn), and on Sable Island,
Nova Scotia.

MOLOTHRUS ATER ATER (Boddaert)

Eastern Brown-Headed Cowbird

Plates 25, 26, 27, 28, and 29

HABITS

The two most characteristic habits of this bird are indicated in the
above names. The Greek word *Molothros* signifies a vagabond,
tramp, or parasite, all of which terms might well be applied to this
shiftless vagabond and imposter. It deserves the common name
cowbird and its former name, buffalo-bird, for its well-known attachment to these domestic and wild cattle. The species is supposed to
have been derived from South America ancestors, to have entered
North America through Mexico, to have spread through the Central
Prairies and Plains with the roving herds of wild cattle, and to have
gradually extended its range eastward and westward to the coasts as
the forests disappeared, the open lands became cultivated, and
domestic cattle were introduced on suitable grazing lands.

The cowbird is unique in a family of nest-building birds; the blackbirds all build strong, well made nests, and the orioles show remarkable
nest-building ability; the bobolink builds only a flimsy nest of grass
on the ground, but the cowbird builds no nest at all, relying on other
species to hatch its eggs and rear its young. Whether the cowbird
ever knew how to build a nest, and, if it did, how it happened to lose

the art and become a parasite, probably never will be known, though some interesting theories on the subject have been advanced. Much light is thrown on this subject by Herbert Friedmann (1929) in his study of the South American cowbird, to which the reader is referred. For the benefit of the readers who do not own this interesting and comprehensive book, we shall quote from it freely.

In his chapter on the origin and evolution of the parasitic habit he writes:

The evidence points unmistakably to the view that the Cowbirds originally bred in normal fashion and that parasitism is a secondarily acquired habit. The reasons for making this statement are:

1. The instincts of nest-building and incubation are so universally present in all groups of birds in all parts of the world that it seems likely that this is the primitive condition of the Cowbirds.

2. All the Cowbird's close relatives are nest-builders; in fact, its family, the Icteridae, is known as a family in which the nest-building instincts reach their pinnacle of development. * * *

3. Within the genera *Agelaioides* and *Molothrus* we find several stages in the evolution of parasitism exhibited by different species. The Bay-winged Cowbird, *A. badius*, uses other birds' nests and lays its eggs in them but incubates and rears its own young. Sometimes it makes its own nest. The Shiny Cowbird, *M. bonariensis*, is parasitic but has the parasitic habit very poorly developed, wasting large numbers of its eggs. Rarely it attempts to build a nest but in this it is never successful. This indicates that originally it built a nest but no longer knows how. The North American Cowbird, *M. ater*, is entirely parasitic but is not wasteful of its eggs. * * *

4. The parasitic Cowbirds (*Molothrus*) have definite breeding territories and are more or less monogamous. Howard has shown that the territory precedes the nest in the evolution of the instincts of guarding associated with reproduction. If the Cowbirds were parasitic from the very beginning it would be very hard to explain their territorial instincts. * * * The facts that the Cowbirds are fairly monogamous indicates that they were monogamous originally and probably nested in normal fashion as all monogamous birds do.

5. The most primitive of the existing species of Cowbirds is, * * * the Bay-winged Cowbird. This species is the only one of its group that is not parasitic and doubtlessly represents the original condition of the Cowbird stock. * * *

From the above it seems safe to assume that parasitism is not the original condition in the history of the Cowbirds. The problem, then is not whether the Cowbirds were always parasitic or not, but how they lost their original habits and became parasitic. * * *

The best theory advanced as yet, and one which my studies tend to support in part, at least, is that of Prof. F. H. Herrick. This writer studied the cyclical instincts of birds and found that not infrequently different parts of the cycle are interrupted by various causes which result in a general lack of harmony between successive parts of the cycle. He suggested that the parasitic habit may have originated from a lack of attunement of the egg-laying and the nest-building instincts which resulted in the eggs being ready for deposition before a nest was ready for them. * * *

The first writer to see that one explanation would not serve for all the different groups of parasitic birds was G. M. Allen (1925). * * * Wisely refraining from offering an explanation of parasitism, he suggests several "possible ways of origin."

One of the possibilities is that parasitism may have arisen from the occasional laying of eggs in strange nests by birds that are very sensitive to the ovarian stimulus provided by the sight of a nest with eggs resembling their own. This is substantiated by experimental evidence collected by Craig who found that in doves ovulation could be induced by comparable stimuli.

Otto Widmann (1907) offered the following interesting theory to account for the origin of the parasitic habit:

We know that fossil remains of horses, not much unlike ours, are found abundantly in the deposits of the most recent geological age in many parts of America from Alaska to Patagonia.

It was probably at that period that the Cowbird acquired the habit of accompanying the grazing herds, which were wandering continually in search of good pasture, water and shelter, in their seasonal migrations and movements to escape their enemies. As the pastoral habit of the bird became stronger, it gave rise to the parasitic habit, simply because, in following the roving animals, the bird often strayed from home too far to reach its nest in time for the deposition of the egg, and, being hard pressed, had to look about for another bird's nest where-in to lay the egg. * * * By a combination of favorable circumstances this new way of reproduction proved successful, and the parasitic offspring became more and more numerous. In the course of time the art of building nests was lost, the desire to incubate entirely gone, paternal and conjugal affection deadened, and parasitism had become a fixed habit.

Dr. Friedmann (1929) disposes of this theory as "more interesting than suggestive," and adds: "It is somewhat surprising to find a naturalist of Mr. Widmann's ability advancing such a theory. Probably he meant it more as a suggestion to be taken for whatever it might be worth than as a real attempt at an explanation." The trouble with the theory is that we have no known facts on which to base it, there being no record of a cowbird leaving its nest to follow cattle, horses, or bison. Probably the parasitic habit was developed before the cowbirds invaded North America. And we do not know to what extent the primitive cowbirds, in South America, had developed the habit of following the wandering herds.

Dr. Coues (1874) makes the following suggestion:

Ages ago, it might be surmised, a female Cow-bird, in imminent danger of delivery without a nest prepared, was loth to lose her offspring, and deposited her burthen in an alien nest, perhaps of her own species, rather than on the ground. The convenience of this process may have struck her, and induced her to repeat the easy experiment. The foundlings duly hatched, throve, and came to maturity, stamped with their mother's individual traits—an impress deep and lasting enough to similarly affect them in turn. The adventitious birds increased by natural multiplication, till they outnumbered the true-born ones; what was engendered of necessity was perpetuated by unconscious volition, and finally became a fixed habit—the law of reproduction for the species. Much current reasoning on similar subjects is no better nor worse than this, and it all goes for what it is worth.

The weakness in this theory is that such cases of adventitious laying in alien nests must have been very rare at first, and the inherited tendency to repeat the experiment would soon disappear by cross-breeding with individuals of normal breeding habits, unless the habit proved to be beneficial to the species, and no such proof is evident. We frequently find fresh eggs of robins and other birds laid on the ground, but failure to reach their nests has never developed parasitic habits in these birds.

The North American cowbirds have been split into three recognized races; two other races have been described, but have not been admitted to the A. O. U. Check-List.

The eastern cowbird, the subject of this sketch, breeds in eastern North America from southern Ontario, southern Quebec, Nova Scotia, and New Brunswick south to central Virginia, southeastern Kentucky, central Tennessee, south central Arkansas, Louisiana and central Texas, and west to Minnesota, northeastern Iowa, southeasten Nebraska, southwestern Kansas and New Mexico.

It may breed, or at least lay eggs, casually farther south in the Atlantic States. In this connection, the reader is referred to an interesting paper by Thomas D. Burleigh (1936) suggesting that the cowbird may lay eggs during migration, of which he gives some evidence. This may account for some of the southern breeding records.

Spring.—The eastern cowbird has not far to go on its spring migration. It is one of the earlier migrants, leaving its winter range in the Southern States during March and reaching the northern parts of its breeding range during the first 2 weeks in April, or sometimes before the end of March.

According to Friedmann (1929):

The Cowbird migrates by day, early in the morning and late in the afternoon. I know of no data tending to show that this species indulges in nocturnal migration, but it may do so to some extent. * * * The Cowbird commonly migrates with the Red-winged and Rusty Blackbirds and the Grackles; in fact these three are usually found together. Other less common associates are Meadowlarks and Robins in the east, and Brewer's and Yellow-headed Blackbirds in the west.

Doubtless many Cowbirds succumb annually to the perils attendant upon migration but so far as I have been able to find there are no definite records of such happenings. Because they migrate chiefly by day no Cowbirds have been picked up dead around lighthouses or the bases of tall monuments and buildings." Bendire (1895), however, tells of one that was blown out to sea and came aboard a vessel, "fully 1,000 miles east of Newfoundland."

During his studies of the cowbird at Ithaca, N. Y., Friedmann (1929) divided the spring migration into six more or less separate phases, much like similar phases in the migration of the red-winged blackbird. The first to appear were the vagrants, at dates ranging from March 1, 1919, to March 14, 1922; these were wandering individuals coming before the true migration, consisting mostly of males, which were usually not in song and did not display. The second group to come were the migrant males, passing through on their way to points farther north; they were arriving and departing at various dates ranging from March 20 to April 27. "These birds usually are seen scattered among flocks of Red-wings, not forming any solid flocks of their own kind. They come to the marshes to pass the night with the Red-wings and the Rusty Blackbirds but during the day scatter over the fields on the uplands, where, in small groups, they forage for food."

The third phase is marked by the arrival of the resident males, which come on the average between March 23 and April 8.

The resident males on arrival at once establish themselves on their posts and remain in their territories for a few hours early in the morning every day, spending the greater part of the day in foraging around for food, often going to stubble fields, and to plowed areas a little later in the season. As the season advances they spend more and more time in their respective territories and at the time the resident females arrive the males are spending at least half of each day within the limits of their areas. * * * The testes of these resident birds are noticeably larger than those of the migrants, and as all the birds collected at dusk in the marshes averaged smaller testes than those shot from their singing trees, it seems that the two groups do not associate in the marsh, the resident birds sleeping in their territories.

The migrant females are the first of that sex to arrive, passing to more northern points between March 23 and April 28. At first they appear as lone individuals among the flocks of migrating redwings, grackles, and cowbirds, but later they pass through in flocks of a dozen or more.

The fifth group is made up of resident females, arriving between April 3 and 11. "The arrival of these resident females acts as a spark to set off the pent-up energies and passions of the males and arouses them to an unbelievable frenzy. The persistence and determination with which the resident males pursue these females makes one wonder if either ever rest. The females evidently recognize the limits of the territories of their respective pursuers as they usually fly in wide circles closely followed by one or more males, but do not leave the general vicinity of their territories. The ovaries of these resident females are considerably larger than those of migrant birds of the same date."

The sixth and last phase of the migration marks the arrival of what are apparently the immature males and females. "About the beginning of May or even the last of April all the resident birds are established and no more migrants are to be found coming in to the marshes. There is a decided slump in the migration lasting until about the second week of May when suddenly there appear numerous flocks of Cowbirds of mixed sexes in the upland fields, around the cattle, and

near barns. * * * The gonads of these birds are smaller than those
of the only other Cowbirds then present (resident birds)." The males
of this group do not seem to take much notice of the females, whereas
the resident birds do, and, to some extent, so do the last of the migrant
males. "These two facts", says Friedmann, "point to the conclusion
that these birds are really the immature individuals."

Territory.—Friedmann (1929) proved to his own satisfaction
that both the female and the male cowbird are confined to a definite
breeding territory:

Not only has the female a definitely marked off breeding area, but the male has
a definite post, entirely comparable to the "singing tree" that Mousley describes.

During the summer of 1921 numerous individual male birds were seen daily
on certain trees or on definite telegraph poles. From these perches they would
sing and display; they might fly off but would soon circle around and come back.
There was no question but that they were tied down to their respective singing
trees. In one case the identity of the male was made certain because of a peculiar
harshness of his song, and as this individual was to be found daily in the same tree,
it seemed safe to assume that each day the same bird was seen at a given perch.
Not only was a certain tree used by each male, but a certain part seemed to be
preferred, usually the higher branches. * * *

That the female has a definite territory is not so easily noticed as she has no
"singing tree," and is, as in most birds, less conspicuous and less often seen than
the male. * * *

At Ithaca in the late spring and summer of 1921 I found that certain females
(probably the same one in a given place each day) seemed to have definite territories. Just off the northeastern corner of the main quadrangle of the Cornell
campus is a small body of water called Beebe Lake. One pair of Cowbirds stayed
on the north shore of the lake, another pair on the south shore. I was sure that
the birds I saw on the north shore were not the same as those of the south side
because on several occasions I saw the pair on one side and simultaneously heard
or saw the birds on the opposite shore. All the Cowbird eggs found in each
territory were very similar to each other and uniformly different from those found
in the other. * * *

The size of the territories is very variable, some being a mile or more long and
comparatively narrow, others * * * much smaller. * * *

The Cowbirds do not make any very spirited attempts to defend their territories and consequently in regions of unusual abundance the territorial factor is
much less noticeable. I have never seen Cowbirds fight and their method of
defense is restricted to an intimidation display.

The females are probably not always confined to definite territories
for their egg laying, for eggs evidently laid by two different females
are often found in the same nest.

Mrs. Amelia R. Laskey, of Nashville, Tenn., has sent me notes on
her 3-years' study of cowbird behavior. She remarks on territory
and mating: "In the area about our home, in each of the three breeding seasons, one male and one female became dominant. This area
may be called a 'domain' rather than a territory. The dominant male
and the dominant female used this area in their pair formation and

mating. They did not drive others from the food in the domain, carried on no boundary line defense, but tolerated both sexes in social contacts, feeding and flying together. I believe these birds displayed vestigial territory behavior in intimidating others so as to keep the domain for their own use in pair formation and mating, and this behavior perhaps may function to some extent in keeping other females from utilizing host nests within the domain of the dominant female.

"All evidence indicated that pairing and monogamous mating generally prevail. Although both sexes on many occasions throughout the breeding season associate in trios or larger groups, there was no indication of polygamous or promiscuous mating."

Courtship.—Friedmann (1929) noted three types of courtship display, the terrestrial, the aerial and the arboreal displays. He describes the first of these as follows: "The male would run alongside of the female, and when slightly ahead of her would turn a little so as to be placed somewhat diagonally to her, and would then ruffle the feathers of his neck and the interscapular region. Then he would bow or bend down his head a little and emit his squeaky, shrill note—*pseeee*. The wings and tail are not involved in the terrestrial mode of courting." Charles W. Townsend sent Friedmann the following note on this performance: "April 9, 1922—Three males and one female busily engaged in eating in a field. Every now and then a male would look up, puff up feathers, spread wings and tail and fall on head. This is evidently the bowing, as in trees, where he does not fall. Since this time I have seen the performance several times and it always impressed me as a falling and being stopped by his head and breast striking the ground. * * * It seems to me that the tree act is a low bowing, while on the ground act is an actual fall, for the bird suddenly lets himself go and brings up against the earth, a comical procedure."

Of the aerial display, Friedmann (1929) says: "The display of the males in mid-air consisted of ruffling out the feathers of the neck, interscapulars and throat, bending down the head, and arching the wings more than in usual flight, and giving their squeaky song. During the instants when the wings were arched in display the flight seemed unsteady, a sort of half-hearted attempt at a glide during an unaccustomedly long interval between wing beats." Two males that he watched following a female "seemed to display and sing in unison. The two males and the female kept on flying back and forth, at an altitude of about two hundred feet, from 7:15 a. m., when they were first seen, until about 8:30 a. m., when they were last seen, without resting or alighting even for a second."

Of the arboreal display, which is the commonest and best known, Friedmann writes:

The display is often, but not always, begun by the bird pointing its bill toward the zenith. This is usually done whenever another bird, especially another male, is very close to the displaying bird. Next it fluffs out the feathers of its hind neck, breast and sides and flanks. * * * It is during this part of the display that the bubbling guttural notes are given. Wetmore has written it *bub ko lum* and I cannot improve on his description. These notes are quite low and not audible in the field at a distance of more than fifty feet. During this stage of the performance the bird ʻsometimes rises and falls gently on its legs in a vertical direction, the rise hardly ever amounting to as much as the length of the tarsus.

After this the bird begins the display proper by arching its neck and spreading its tail * * *. Then it begins to raise its wings and bend forward * * *. All this time the feathers of the back are fluffed out just as are those of the underparts. Then the wings are brought out to their full expanse * * *, and the toppling over proceeds from now on with accelerated rapidity, the tail being lifted before the body pivots and swings over. * * * The display ends when the wings are brought back to the body * * *. The bird then rights itself and is ready to repeat the whole performance. [See plates 28 and 29.]

The entire display lasts about three or four seconds and the *tseee* note [see under "Voice"] usually has a duration of about a second or a little over. The frequency of display is extremely variable. I have seen a male display with almost clock-like regularity at intervals of five seconds for several minutes when no female was in sight, and I have also watched a male display once and not do it again for over an hour. Display becomes less and less frequent as the season wears on and is usually not indulged in to any extent after the middle of June, while song continues until a month later. Those displays that are given after the middle of June are usually incomplete. This incomplete display consists of spreading the tail, hunching the back and slightly arching the wings, but the bird does not fall forward.

A male bird, observed by Wetmore (1920) in New Mexico, "would sit quietly for a few seconds, then expand the tail and draw the tip slightly forward, erect the feathers of the back and to a less extent those of breast and abdomen, and then sing *bub ko lum tsee*. In giving the first three notes he rose twice to the full extent of his legs and sank back quickly."

C. J. Maynard (1896) describes the courtship flight as follows:

Two or more males often pay their attentions to one female, singularly, without attempting to quarrel, when she will suddenly take wing and all will start in pursuit. The flight of a female at this time is exceedingly swift, for she will usually manage to keep ahead of her followers who ardently press on, giving a rather sharp, prolonged cry as they dart through the air. All the males within hearing join in, and it is not unusual to see a half dozen at a time after one of the other sex who will lead them a long chase, now darting upward to a considerable height, then doubling, will glide through the tangled branches of a clump of trees, emerging on the opposite side with great rapidity. This exciting race is evidently maintained merely as a matter of sport, for when the object of chase becomes weary she will quietly settle on the branch of a tree, and her admirers gather around her, calmly arranging their feathers. After resting for a time one will commence his gallantries once more, when the female darts into the air again and the males dash vehemently after her as before.

In this connection, it may be well to consider the sex ratio and the sex relations. The prevailing impression that the males far outnumber the females is probably more apparent than real, for the males are more conspicuous and less retiring; Friedmann (1929) says: "From my observations I would put it as about three males to two females." The sexual relations of the cowbirds may not be above criticism, but they are probably not as bad as they are often painted. Cowbirds have been called monogamous, polygamous, polyandrous, and even plain promiscuous; probably any one of these terms could be applied to certain individuals under certain circumstances; but there is much evidence to indicate that the cowbird was originally monogamous and is so by preference today in most cases.

Friedmann (1929) writes:

At Ithaca I have found that each male and each female has a definite territory * * * and that there is a more or less definite pairing between the birds. My experience has been that if the birds are not strictly monogamous, at least the tendency towards monogamy is very strong. My observations have been supported by those of Dr. Alexander Wetmore, of the Smithsonian Institution. He informs me that in Utah he had exceptionally favorable conditions for observing the sexual relations of the Cowbirds, and that in a relatively small area, (which was quite open and made observation of the birds an easy matter), he watched six pairs of Cowbirds, each pair having their own territory, and the birds remaining true to their mates. The male of pair A stayed with female A, and did not consort with any of the other five females.

At Lake Burford, N. Mex., Wetmore (1920) noted that a pair of cowbirds, mated on June 2, "remained constantly nearby for ten days or more. On June 5 and 6 a second female appeared and fed with the others. The male was seen running at them with his bill pointing straight in the air and then pausing to sing and display. The second female disappeared at once while the pair remained together until June 13."

Cowbirds are often seen in small flocks even during the breeding season, which might give the impression of loose sexual relations, and it is well known that, if one of a pair of mated birds is killed, the survivor secures a new mate in a surprisingly short time, showing that there is always an available supply of unmated birds ready to fill in the gap. These flocks are probably made up of such unmated, surplus individuals and are usually seen in places where there are few or no nests; they are not, therefore, breeding birds. Moreover, these flocks may consist of immature, 1-year-old birds, which cannot be distinguished in the field from adults, which have arrived later in the season than the adults and have not mated.

As evidence of polyandry, Friedmann (1929) relates the following experience:

A pair with whose territory I was fairly familiar was noted several times and each time there was just a single male and a single female. The male used to stay in his singing tree and so was easy to find. The female, when wanting her mate, would fly into the open and give her flight rattle. The male would quickly take off after her. One day it was noted that when the female called for her mate he came directly from his favorite perch as always but another male, new to the territory, also answered her summons. This interested me not a little and I went back there the next day and waited for the female to call for her mate. Again both males answered her summons, the original male coming as always from his singing tree and the new one from a tall tree near a railroad track. Some time later in the afternoon the original male was seen again in his singing tree and the second male was noted in the tree from which he had flown in answer to the call of the female. On the next two successive days this male was seen in or near this tree and it certainly looked as though he had established himself there. A week later the place was revisited and both males were found, each in his own tree and both again answered the summons of the female.

This apparently was a case of polyandry—an unmated male established himself in the territory of a mated pair. He was not there originally as the pair had been watched considerably before his advent. The original male seemed not to mind the presence of the other. However, no actual intercourse between any of the birds was observed.

Mrs. Nice (1937) writes: "With a small population of Cowbirds, this investigator found the species predominantly monogamous, with some tendency towards polyandry. But here on Interpont, with an abundance of Cowbirds, promiscuity prevails just as the older writers maintained. A banded male has been seen with three different banded females and one unbanded female, while banded females are seen with varying numbers of males from one to five." And Forbush (1927) says: "Cowbirds are free lovers. They are neither polygamous nor polyandrous—just promiscuous."

Nesting.—The remote ancestors of the cowbirds may have been, and probably were, nest builders, incubating their eggs and rearing their own young, as other birds do. It is difficult to imagine how they could have evolved otherwise. I once saw a poor apology for a nest that I thought might have been built by a cowbird. While driving across the North Dakota prairies, on June 14, 1901, we saw a crude bunch of straws and dried grasses lodged in a bush; it had the appearance of a roughly built nest, but it was too large and bulky and too loosely and carelessly put together to have been built by any other bird in that region; a hollow in the center held a single egg of a cowbird. As cowbirds were abundant in that section and other nesting birds were scarce, it occurred to me that perhaps a cowbird, being unable to find a suitable host, had made an attempt to build its own nest. It is more than likely, of course, that some small animal may have placed the material there and the cowbird had mistaken it for a bird's nest. We were too far away from any human habitation for any man or boy to have put it there, so I will let the reader decide how it got there; I offer it only as an interesting suggestion.

Regardless of what significance the above suggestion may have, our North American cowbirds, as we know them today, are all wholly parasitic, laying their eggs in the nests of other birds and leaving their young to be raised by their foster parents. In the long list of birds thus imposed upon, the vireos, the wood warblers, and the small sparrows figure most prominently. No attempt will be made to list here all of the victims of the cowbirds; this has been well done by Bendire (1895) and more thoroughly done by Friedmann (1929). The latter makes the following general statements as to the families afflicted:

Most of the victims of the Cowbird are contained in four families—the tyrant flycatchers, the finches, the vireos, and the warblers. Of the thirty-six species and subspecies of tyrant flycatchers in the North American fauna eleven are known to be parasitized, while of the remaining twenty-five, seven do not breed within the breeding range of the Cowbird. * * * The Cowbird is known to victimize sixty-two species and subspecies of finches. The total number of North American forms of this great family is one hundred and ninety-four, of which about a hundred are not known to breed within the range of the Cowbird. Yet the family is one of great importance to the parasite as some of its component species are very frequently victimized. The Cowbird is probably one of the chief factors in checking the increase in the smaller Sparrows and Finches.

The Vireos, while relatively few in species, are nevertheless a very important factor in the natural economy of the Cowbird, and the latter is undoubtedly the most serious single enemy of the birds of this family. No birds are more frequently affected, either absolutely or relatively, and none make less protest at the frequent impositions of the parasite. Of the twelve species of Vireos in the North American Check-List, nine are known to be victimized. Including subspecies, twelve of the twenty-five forms are included in the present list of victims. Of the remaining thirteen, five do not breed within the Cowbird's range and six others have ranges which only slightly coincide with that of *Molothrus*.

Of the fifty-four species of Warblers in the North American fauna thirty-six are known to be more or less imposed upon by the Cowbird, and, of the remaining eighteen, at least ten do not breed in any part of the Cowbird's range, or are, at most, so rare, that the absence of records means nothing. The other eight are still little known and very few of their nests have been found. Including subspecies, forty-four, of the seventy-three kinds of Warblers, are included in the list of the victims of the Cowbird.

Of other families of less importance in the economy of the cowbird, he says that five species of mockingbirds and thrashers, are rarely victimized by the Cowbird; with only one of the five has the Cowbird definitely known to be successful. He continues:

The Wrens are almost negligible factors in the ecology of the Cowbird, and the latter is of no great consequence in the life histories of these birds. Four of the fourteen species in the A.O.U. Check-List are known to be victimized by the Cowbird, all of them very infrequently.   Counting subspecies, six, of the thirty-six in the North American fauna, are included in the roll-call of the Cowbird's victims.

Paridae are of little importance in the economy of the Cowbird, and the latter plays an inconsequential role in the lives of the Titmice.   Five species of the eleven in the A.O.U. Check-List are recorded as victims of the Cowbird; all of them uncommonly.

The kinglets and gantcatchers are, with one exception, infrequently molested by the Cowbird.   They are interesting in that they are among the smallest birds definitely known to be affected by the parasite.   Three of the six species in the North American fauna are included in the list of hosts: one of the three being represented by two geographic races.

The Thrushes are of considerable importance in the natural history of the Cowbird, and are among the largest birds commonly and regularly parasitized.   Not only do we often find Cowbirds' eggs in the nests of some of these birds, but frequently they may be seen caring for the young Cowbirds.   Even here, where the rightful young are of approximately the same size as the young parasites, it is rather unusual to find any but the Cowbird surviving in a victimized nest. * * *

Seven of the fifteen North American species in this family are more or less imposed upon by the Cowbird, two of them being represented in the present list by two races each.

Bendire (1895) listed 91 species and subspecies victimzied by the eastern cowbird, including a few victims of the Nevada cowbird, which had not at that time been given separate status.   Friedmann (1929) listed 114 species and subspecies as victims of the eastern cowbird alone, adding three species, the ruby-throated hummingbird, Nelson's sparrow, and the brown creeper as hypothetical.   In subsequent papers (1934, 1938, 1943, and 1949), he increased the known hosts of the eastern cowbird to 149 forms.

Cowbirds' eggs are sometimes found in nests of birds that are wholly unfitted to become foster parents for the young, in which cases the eggs never hatch or the young never survive.   If the eggs of the owner of the nest are much larger than those of the parasite, the cowbirds' egg will not receive enough warmth from the body of the incubating bird to hatch.   If the young of the selected foster parents are fed on food unsuitable for the young cowbird, the latter cannot be expected to thrive on it; one can hardly conceive of a mourning dove, which is listed as a victim, feeding a young cowbird on "pigeon milk," or of a kingfisher feeding it on fish.   A swallow might hatch a cowbirds' egg and feed the young one in the nest, but not afterward, as young swallows are taught to feed on the wing.   It is quite important for the cowbird to select an open nest of some altricial bird that feeds its young in the nest until they are nearly able to fly; the young of precocial birds leave the nest soon after they are hatched and the young cowbird would be deserted; the egg of a cowbird has been found in a killdeer's nest, but, if the egg ever hatched, the young must have been left in the nest to starve or die of exposure.

The temperament of the host species is also of importance; the hawklike character of the shrikes makes them absoultely free from cowbird molestation; and the pugnacious kingbirds are seldom imposed upon. Birds nesting in holes are mostly free from cowbird interference; Friedmann (1929) says that woodpeckers, house wrens, nuthatches, chickadees and bluebirds are "very seldom molested, in fact the Bluebird is the only one of these birds for which I have found more than a very few records." Some birds are intolerant of cowbirds' eggs; He mentions the robin, catbird, and yellow-breasted chat as "examples of absolutely intolerant species. Others such as the Yellow Warbler are intolerant to a certain extent." I should except the catbird, as I once found a catbird sitting calmly on a nest that contained four eggs of the cowbird and only one of its own! Friedmann adds: "Birds react to Cowbirds' eggs in several ways. The great majority of species seem not to mind the strange eggs in the least and accept, incubate and hatch them. Of these birds, some occasionally cover over the parasitic eggs by building a new floor over them if they have no eggs of their own at the time. This is true of such birds as the Red-eyed, Warbling, Blue-headed and Yellow-throated Vireos, the Prothonotary, Yellow and Chestnut-sided Warblers and the Redstart. This has also been recorded in the following species—Meadowlark, White-crowned Sparrow, Cardinal and Indigo Bunting, but only a single time in each case."

He conducted experiments with a robin's nest and a catbird's nest to learn what the reaction of these birds to strange eggs would be. The robin threw out a song sparrow's egg, which was spotted much like a cowbird's, but accepted a chipping sparrow's, which was mainly blue like the robin's, though much smaller. The catbird ejected both the song sparrow's and the chipping sparrow's eggs. Of the yellow-breasted chat, he says: "The eggs of the Chat are very similar to those of the Cowbird, but nevertheless the nest is almost invariably deserted if a parasitic egg is laid in it. This is doubtless due to the extreme shyness and nervousness of the Chat, rather than to any superior ability to distinguish the strange eggs from those of its own. * * * Nevertheless on at least two occasions Chats have hatched and reared Cowbirds." Some other birds, perhaps more than we know about, eject the alien eggs from the nests, B. H. Warren (1890) says: "I have twice found broken eggs of Cowbirds on the ground near nests of the Yellow-breasted Chat, and on three occasions have discovered the shattered remains of these eggs directly beneath the pendant nests of Baltimore Orioles."

Among the birds that show their intolerance by burying the cowbird's egg in the bottom of the nest, or by building a second story over it, so that the alien egg fails to hatch, the yellow warbler is the star performer. Two-story nests of this warbler are fairly common, where cowbirds are numerous, * * * three-storied nests are not very rare, and as many as four or even five stories have been built. In addition to those birds mentioned above as addicted to this habit, R. M. Anderson (1907) reports a "Traill Flycatcher's nest with a Cowbird's egg imbedded." C. R. Keyes (1884) found a scarlet tanager's "nest with a Cowbird's egg embedded in the bottom." Amos W. Butler (1898) reports a nest of the Maryland yellowthroat, containing three stories. "Two additional nests were built upon the original structure, burying beneath each the egg of a Cowbird." E. A. Samuels (1883) claimed to have a double nest of the American goldfinch in his collection, but this seems open to question, as the goldfinch usually nests later in the season than the cowbird.

The female cowbird is an expert nest hunter; in fact, she has to be. Coues (1874) describes her nest hunting graphically:

It is interesting to observe the female Cow-bird ready to lay. She becomes disquieted; she betrays unwonted excitement, and ceases her busy search for food with her companions. At length she separates from the flock, and sallies forth to reconnoitre, anxiously indeed, for her case is urgent, and she has no home. How obtrusive is the sad analogy! She flies to some thicket, or hedge-row, or other common resort of birds, where, something teaches her—perhaps experience—nests will be found. Stealthily and in perfect silence she flits along, peering furtively, alternately elated or dejected, into the depths of the foliage. She espies a nest, but the owner's head peeps over the brim, and she must pass on. Now, however, comes her chance; there is the very nest she wishes, and no one at home. She disappears for a few minutes, and it is almost another bird that comes out of the bush. Her business done, and trouble over, she chuckles her self-gratulations, rustles her plumage to adjust it trimly, and flies back to her associates.

Russell T. Norris (1944) gives the following account of a cowbird laying in a song sparrow's nest, which was photographed by Hal H. Harrison (see pl. 25):

Just before 4:30 a.m., about 22 minutes before sunrise, we heard the sputtering note of a Cowbird, and a few seconds later a female Cowbird alighted on the camera. After looking around cautiously, she flew to the ground at the base of the tripod and began to walk nervously toward the nest. As she reached the rim of the nest, she paused and carefully surveyed the surrounding territory, then stepped into the nest, and turned about several times. Finally she settled down, and Harrison pressed the button on the battery. As the flash went off, the Cowbird flushed. She had been on the nest no more than 15 seconds and had not deposited her egg. * * *

At 4:38 a.m. I noticed a movement in the grass behind the nest, and after a few seconds the Cowbird appeared. She approached the nest warily, stepped up onto the rim, and paused there. Then she entered the nest and began to turn about as she had on her previous visit. After a few seconds, she stepped back onto the rim and looked around. She three times repeated this procedure of standing on the rim, then uneasily turning about in the nest. In one instance she mounted the rear rim and looked back into the grass. At approximately 4:40 a.m. she settled on the nest, and Harrison released the shutter. The Cowbird raised herself slightly but remained a few seconds before flying away. Upon examining the nest I found a fresh Cowbird egg. Undoubtedly the egg was being laid as the picture was taken.

He ends the story with the statement that: "Two sparrows and the Cowbird hatched and were reared successfully until the Cowbird was seven, the sparrows six days old, when the nest was destroyed by a predator."

Harry W. Hann (1941), in the studies of the cowbird at the nest of the oven-bird, came to the following conclusions:

1. The female Cowbird regularly finds the nest of the host by seeing the birds building.

2. She sometimes watches the building process intently and this doubtless stimulates the development of eggs, which are laid four or five days later. This theory, first suggested by Chance for the Cuckoo, accounts for the delicate synchronization of the egg-laying of the Cowbird with that of the host, and does not preclude the possibility of laying several eggs on successive days.

3. The eggs of the Cowbird are usually laid during the egg-laying time of the host, but exceptions are common. Extremes noted during the study were three days before the first Oven-bird's egg was laid, and three days after incubation began.

4. A Cowbird lays but one egg in a nest unless nests are scarce; in that case it lays more.

5. The female Cowbird makes regular trips of inspection to nests during the absence of the owners, between the times of discovery and laying, and knows in advance where she is going to lay.

6. Her regular time for laying is early in the morning before the host lays, and she will frighten the owner from the nest if she happens to be there first. * * *

7. The Cowbird is both alert and determined when she come to the nest to lay. She moves about in the vicinity of the nest and looks carefully for as much as three minutes before entering, but will return to the nest if she is frightened away.

8. She spends from a few seconds to a minute in the nest when laying and flies directly from the nest as soon as the egg is laid.

9. The Cowbird disturbs the nest of the Oven-bird but little when she enters to lay, and I have found no broken eggs which were attributable to her entering.

10. Parasitized nests regularly have one or more eggs removed by the female Cowbird. These are not removed at the time of laying, but during the forenoon of the previous day, or the day of laying, or rarely on the following day. * * *

11. Eggs removed are eaten by the Cowbird, but are not removed for that purpose along, or their disappearance would not be correlated so closely with the laying of her own eggs. The number of eggs removed from parasitized Oven-birds' nests was eighty-five per cent of the number of eggs laid [by the cowbird] and included four eggs of the Cowbird itself. From nonparasitized nests of the Oven-bird only a single egg disappeared during the study.

12. The statement by Borroughs that a Cowbird takes an egg from a nest only when two or more eggs are present is borne out by this investigation.

Mrs. Nice (1939) has twice seen a cowbird remove an egg from a song sparrow's nest, "the thief eating the egg and shell," in one case. And T. S. Roberts (1932) has seen one remove an egg from a scarlet tanager's nest and from the nest of a chipping sparrow. W. V. Crich, of Toronto, has sent me a photograph of a cowbird's egg in a last year's nest, showing that this clever nest hunter sometimes makes a mistake.

Eggs.—To determine the number of eggs laid by the cowbird during a season is a question that cannot be answered with certainty, but we have some data indicating that it lays no more eggs than many other passerine birds. Friedmann (1929) made a careful and thorough study of three well-known cowbird territories, in which all, or nearly all of the nests were located and in which the eggs of the three different females could be recognized. Of these he says:

Two of the birds laid five eggs each and the other laid four. In one of the cases where five eggs were laid (Terr. A.), I found no more after the fifth one although a great deal of time was spent in this breeding area. It was just because no egg was found on the sixth day that I kept very close watch of the bird and its territory on that and the following day. On the day this individual laid its fifth egg the other "five-egg" bird, (B), laid its first. For four days thereafter this bird (B) laid an egg daily and no more were found for individual A. On the day Cowbird B laid its fifth egg a heavy storm broke out and for a month and a half thereafter it rained more or less violently every day. As fast as nests were found, they were destroyed or washed away by the heavy rains, and, of course, it became impossible to keep any check on the actions of the Cowbirds. * * *

Above it was stated that no eggs were found in the territory of Cowbird A after that bird had laid its fifth. Of course the mere fact that none were found is no indication that none were laid. However, the second of the "five-egg" birds (individual B), was collected three days after it had laid its fifth egg. Only five discharged egg follicles were found in its ovaries and the oldest of these follicles was still very prominent so that if any more eggs had been laid, follicles would have betrayed the fact. This shows pretty conclusively that only five eggs were laid in the case of this bird. This, together with the fact that bird A was known to lay at least five eggs and, judging by the four-day rest (?) after the laying of the fifth egg, probably did not lay any more, suggests the idea that five eggs may possibly represent what in other birds would be called a clutch although this is doubtful. We cannot be certain that Cowbird A laid only five eggs although I feel that I would have found at least one more egg in the four days between the fifth egg and the stormy season, if the bird had kept on laying. * * *

A record of the eggs laid by each bird and the nests used may be of some interest. Cowbird A laid its first egg on May 23 in the nest of a Chestnut-sided Warbler; its second egg, May 24, in a Veery's nest, its third, May 25, in another Veery's nest, its fourth, May 26, in a nest of a Redstart, and its fifth, May 27, in the same Redstart's nest.

Cowbird B laid its first egg May 27 in a nest of Redstart, its second egg May 28, in the same nest, its third, fourth, and fifth, on May 29, 30, 31 respectively, all in one nest of a Red-eyed Vireo.

Cowbird C laid its first egg May 22 in a Veery's nest, its second, May 23, in a nest of a Redstart, its third, May 24, in the same nest, and its fourth and last recorded egg, May 25, in a Red-eyed Vireo's nest.

From the above it may be seen that the cowbird lays one egg each day, and that it is not specific in its choice of hosts. As evidence that the cowbird may sometimes lay more than five eggs, Friedmann quotes F. L. Rand, who had kept cowbirds in captivity, as saying that "a little hen Cowbird that had the liberty at all times in a suite of rooms, was tempted by me to enjoy as nearly as possible its natural bent in the direction of egg laying and the results obtained in the way of information were somewhat surprising. Eight or ten last year's nests were placed around the room, with dummy (candy) eggs in them; each morning about six o'clock the little hen would seek some nest or other in which it would drop her egg, but not always in the same nest; often times, the candy egg would be found on the floor; so, in fourteen successive days, the little hen had laid thirteen eggs; this would indicate apparently that the destructive nature of the bird is even greater than it has been thought to be."

Lawrence H. Walkinshaw (1949) concluded during his studies in Michigan, that 25 eggs were laid between May 15 and July 20, 1944, by a single cowbird "because (1) they were very similar in coloration (2) no two were laid on the same day (3) the length of 11 similarly colored eggs had significantly less variability than the length of 22 not-similarly colored eggs."

As to the number of eggs laid in any one nest, we have plenty of positive information; Friedmann (1929) writes on this subject:

In order to determine definitely whether or not the Cowbird normally lays but a single egg in a nest, data on approximately nine thousand victimized nests of one hundred and ninety-five kinds of birds were assembled, and it was found that in over two-thirds of the cases only one parasitic egg was found in a nest. This shows pretty clearly that the normal, the usual, the characteristic thing is for a Cowbird to deposit one egg in a nest.

Nevertheless about a third of the nests held more than a single Cowbird's egg apiece. This is no inconsiderable number or percentage of exceptions to the above rule, and calls for an explanation. It has been shown that under normal conditions each pair of Cowbirds has a more or less definite territory and that the female tends to restrict herself to nests within that particular area. However, the cyclical instincts of the female are so aborted that she is probably quite easily induced temporarily to forsake her territory long enough to deposit an egg in a near-by nest. * * * Again, in regions where the Cowbird is very common (and this applies to a great part of its range) territories are apt to overlap and in this way two Cowbirds may make use of the same nest. In this way I believe we can account for the fact that not infrequently eggs of two or more rarely even three different individuals are found in the same nest."

Furthermore, a cowbird may lay more than one of her own eggs in a nest, provided she can find at the proper time no nest in which she has not already laid.

Mrs. Nice (1939) says: "During seven years' study on Interpont, 98 of the 223 Song Sparrow nests located contained Cowbird eggs; 69 held one egg, 26 held two eggs and three held three. Only once did I find four Cowbird eggs in a single nest; this happened in June, 1928, and the nest belonged to a Maryland Yellowthroat." (See, also, her (1949) paper on the laying rhythm.)

One cowbird's egg in a nest is evidently the prevailing rule, but two eggs, often laid by two different females, are not a rare occurrence and we have numerous records of three or more in a nest. A. C. Reneau tells me that he has twice found three eggs in the nest of a phoebe and once in the nest of a Bell's vireo. A. D. Du Bois' list contains a record of three eggs in a cardinal's nest, three in a wood thrush's nest and five in another nest of a wood thrush. Frank R. Smith writes to me of three eggs in the nest of a wood thrush, deposited before the thrush had laid any eggs, and apparently laid by different females. T. S. Roberts (1932) says: "Three or four are uncommon, though Mr. Kilgore and the writer once found a Wood Thrush's nest containing two eggs of the Thrush and six of the Cowbird, the latter of two distinct patterns, suggesting that two Cowbirds had laid three eggs each in this nest. * * * At Mille Lacs on July 7, 1934, Mr. Marius Morse found a Willow Thrush's nest containing two eggs of the owner and *eight* of the Cowbird. * * * Apparently four different Cowbirds had laid one or more eggs each."

J. H. Langille (1884) writes: "I have frequently found more than one in the same nest; once not less than four in the nest of a Scarlet Tanager, which had only room enough left for two of her own. Mr. Trippe once found a Black-and-white Creeper's nest with five of the eggs of the interloper and three deposited by the owner." Isaac E. Hess (1910) mentions a scarlet tanager covering four eggs of the cowbird and an oven-bird's nest that contained seven eggs of the parasite. I have already mentioned the case of a catbird sitting on four cowbird's eggs and one of her own. F. A. E. Starr writes to me: "I once found a red-eyed vireo's nest with the vireo sitting on six cowbirds' eggs and none of her own. A nest of the Wilson thrush was found containing one egg of the thrush and four eggs of the cowbird." Sanborn and Goelitz (1915) report a towhee's nest that "contained one Towhee egg and eight Cowbird eggs."

Bendire (1895) describes the eggs as follows:

The shell of the Cowbird's egg is compact, granulated, moderately glossy, and relatively much stronger than that of its near allies, the *Icteridae*. The ground color varies from an almost pure white to grayish white, and less often to pale bluish or milky white, and the entire surface is usually covered with specks and blotches varying in color from chocolate to claret brown, tawny, and cinnamon rufous.

In an occasional specimen the markings are confluent and the ground color is almost entirely hidden by them; in the majority, however, it is distinctly visible. These markings are usually heaviest about the larger end of the egg, and in rare instances they form an irregular wreath. The eggs vary greatly in shape, ranging from ovate to short, rounded, and elongate ovate, the first predominating.

The average measurement of 127 eggs in the U. S. National Museum collection is 21.45 by 16.42 millimetres, or 0.84 by 0.65 inch; the largest egg measures 25.40 by 16.76 millimetres, or 1 by 0.66 inch; the smallest, 18.03 by 15.49 millimetres, or 0.71 by 0.61 inch.

Incubation.—Friedmann (1929) says that the incubation period of the cowbird "is ten days, about the shortest of any of our passerine birds." But Mrs. Nice (1939) says: "On Interpont with the Song Sparrow as host the Cowbird egg has never hatched in ten days. Sometimes it hatches in eleven days, sometimes in twelve, and occasionally in thirteen or even fourteen days. It requires about one day less of incubation than the Song Sparrow egg, hence it normally hatches first and the bird gets an advantage from the start. Some eggs have been laid after incubation has started; these have hatched from one to five days later than the Song Sparrows, and most of the little birds perished."

Hervey Brackbill gives me the following personal observation on the subject: "The incubation period that I observed for one egg at Baltimore was about 11⅓ days. This egg was laid on May 18 before 8:47 a. m. (studies by other investigators indicate that laying is usually done at about 5 a.m.), and after steady incubation by a Wood Thrush hatched on May 29 at 1:25 p.m.; that is, at that hour I found the bird enclosed in only half of the shell and, when I touched it, it wriggled free of that. The incubation period of the thrush's own eggs was 12 to 13 days."

Young.—It will be seen from the above that the young cowbird usually hatches at least 1 day ahead of the young of its foster parents. It does not, apparently, make any effort to oust its nest mates, as the European cuckoo does, but it is so much larger than the young of the smaller foster parents, and grows so much faster, that some of the smaller young are often crowded out of the nest; also, it gets more than its share of the food brought to the nest, a condition that sometimes proves fatal to its nest mates. Mrs. Nice (1939) says on this subject:

Many writers assert that each Cowbird is raised at the expense of a brood of young. This is not true with Song Sparrows. Sixty-six successful nests without Cowbirds on Interpont raised an average of 3.4 Song Sparrows, while twenty-eight successful nests with Cowbirds averaged 2.4 Song Sparrows. So, taken by and large, each Cowbird was reared at the expense of one Song Sparrow. * * *

Song Sparrows often raise all of their own young that hatch along with a pensioner, anywhere from one to five Song Sparrows having been fledged in such nests. With two Cowbirds of like age in the nest, the Song Sparrows have been able to bring up only one or two of their own children. Smaller birds undoubtedly suffer more than do Song Sparrows, but there is little information on this subject. Once I found a nest containing three young Maryland Yellowthroats and a Cowbird just ready to leave.

T. C. Stephens (1917) studied the feedings of a young cowbird and two young red-eyed vireos in a nest of the vireo; during a period of a day and a half, the young cowbird received 58 percent of all the food; the cowbird was both older and larger than the vireos; the older vireo received 27 percent of the food and the younger one only 15 percent.

Alexander F. Skutch writes to me: "On May 20, 1931, I found under a bridge near Ithaca, N. Y., a phoebe's nest containing four of its own eggs and one newly hatched cowbird, half of its shell still remaining in the nest. On May 22, there were two young phoebes, hatched since the preceding day. The cowbird appeared to be fully four or five times their size. Its eyes were partly open and the sheaths of its remiges were sprouting. By May 25, one phoebe nestling and one of the unhatched phoebe eggs had mysteriously vanished. By May 28, the cowbird was well feathered, could perch, and showed fear. The young phoebe, six days old, was still blind and was very weak and helpless; its pin feathers were just sprouting. On May 29, I found the cowbird on the rim of the nest. The young phoebe's eyes were just opening. By May 30, the cowbird had left the nest, aged ten days. By June 3, the phoebe was well feathered. By June 6, the young phoebe had departed, at the age of fifteen days. The parent phoebes were feeding the young cowbird nearby, and the female had already laid a new egg in the nest! Returning on June 12, I found her incubating five eggs of the second brood in the old nest.

"On July 29, also near Ithaca, I found a nest of the red-eyed vireo with two young vireos and one cowbird, all about a week old. By August 2, the cowbird had left the nest; the vireos, still inside, were in a thriving condition and seemed about ready to depart."

Friedmann (1929) made a careful study of the growth and development of the young cowbird in the nest; it is not essentially different in pattern from the development of other young passerine birds, except for its increase in weight—the really important factor in its survival: "This Cowbird, probably less than an hour old, weighed 2.5 grams. * * * The average weight of a day old Cowbird is 4.5 grams. * * * At the end of the second day the young Cowbird may weigh from 7.5 to 8.5 grams. * * * For the first two days, the daily increase is close to 100 per cent, but from then on the rate is slower, averaging about 50 per cent on the third day and gradually lessening until it comes to be about 10 per cent on the eighth day and only 5.5 per cent on the ninth. When the Cowbird leaves the nest it averages about 33 grams or approximately 13 times its weight on hatching."

The rate of growth varies considerably, depending largely on the kind and amount of food furnished by different species of foster parents.

A lone cowbird, in a nest by himself, grows faster than one that has competition from other nest mates.

Probably all the altricial species that successfully hatch a cowbird's egg feed the young imposter, or attempt to do so, while it is in the nest, but the larger species are not always successful in rearing it to a nest-leaving age.   The smaller flycatchers, the vireos, the wood warblers, and the smaller sparrows are the most successful in this and therefore make the best foster parents.   As Friedmann (1929) remarks: "Obviously only those species that serve as foster-parents of the young Cowbird are important in the economy of the parasite.   Of the 195 birds on the list, 91 have been definitely recorded as rearing young Cowbirds.   Of the remaining number, a large number doubtless could, and do, act in this capacity but are less commonly victimized and so have been less often recorded."

In order successfully to rear a young cowbird until it attains its growth and is able to shift for itself, the foster parent must feed it for some time after it leaves the nest.   The smaller flycatchers, the vireos, the wood warblers, and the smaller sparrows that have acted as hosts usually do this, and most of them have been definitely recorded as doing so, as have also the house wren and the Carolina wren. Milton B. Trautman (1940) lists, among the larger birds observed feeding fledgling cowbirds out of the nest, the catbird, eastern robin, wood thrush, starling, yellow-breasted chat, eastern cardinal and red-eyed towhee; most of these were observed in the act only once or twice, but the cardinal was seen more than 13 times.   Baird, Brewer, and Ridgway (1874) mention that J. A. Allen saw a brown thrasher feeding a nearly full grown cowbird.   Probably the above lists could be considerably enlarged.

In at least three cases, a female cowbird has been seen to feed a young cowbird that was supposed to be its own young, but in no case could the relationship be proven.   J. R. Bonwell (1895), of Nebraska City, Nebr., reported seeing a female cowbird feed a young cowbird in a nest with young rose-breasted grosbeaks.   "Nearly every evening she would come and feed the young Cowbird, but if the young Grosbeaks would open their mouths for food she would peck them on the head and refuse them food."   Forbush (1927) mentions two other cases.   He knew Mason A. Walton, of Gloucester, well enough to accept his report of seeing a female cowbird feed a young cowbird in a yellow warbler's nest, but he had no way of knowing that she was the parent.

The other case cited was based on a careful observation by Laurence B. Fletcher (1925), who trapped a female Cowbird with a young one, saw the female feed it, banded both birds and saw the two together afterward, the same adult still feeding the young.   The adult, which bore Biological Survey band No. 64782, continued to feed the banded fledgling and no other, although there were other young Cowbirds near.

None of these observations prove that the female cowbird recognizes its own young; but they do indicate a lingering vestige of the lost maternal instinct.

Since the above was written, Russell T. Norris (1947) has published the results of his extensive study of the cowbirds of Preston Frith, to which the reader is referred. The conclusions he arrived at are not far different from what is indicated above, but the following paragraphs are of special interest:

Of 19 Cowbird eggs, one hatched four days before the host; 4 hatched one day before; 10 hatched the same day as the host; 3 hatched one day later than the host; and one hatched five days later than the host. * * *

In the 237 observed nests, the hosts laid 668 eggs, of which 383 (57.3 per cent) hatched; the Cowbirds laid 108 eggs, of which 46 (42.6 per cent) hatched; 37.7 per cent of the host eggs, 26.8 per cent of the Cowbird eggs produced fledglings. Of the host eggs that hatched, 64 per cent produced fledglings; of the Cowbird eggs that hatched, 63 per cent produced fledglings.

With four exceptions all parasitized nests that produced young produced at least one host young.

The 35 non-parasitized (successful) nests produced 2.94 fledglings per nest; 19 parasitized (successful) nests fledged 2.05 host young per nest, indicating that each parasite was raised at the expense of about one host young.

Plumages.—The plumage changes of the cowbird are simple. The natal down is described by Dwight (1900) as olive-gray. He describes the juvenal plumage in which the sexes are alike, as "above, including sides of head and neck, wings and tail, dark olive-brown, the feathers edged with pale buff, whitish on the primaries. Below, dull white, buffy on throat, breast and flanks much streaked with olive brown. Chin white or yellowish."

A complete postnuptial molt occurs in August or early September, producing a first winter plumage which is indistinguishable from that of the adult. In the male, the head, throat, and nape are purplish "clove-brown," but all the rest of the plumage, including the wings and the tail, is clear lustrous black with green and purple reflections. The female assumes at this molt the "mouse-gray" plumage of maturity. The nuptial plumage in both sexes is acquired by wear, which is not conspicuous. Adults have a complete molt in September and no prenuptial molt. The seasonal changes are inconspicuous.

Food.—Beal (1900) reports on the contents of 544 stomachs of the cowbird, taken in 20 States during every month in the year, examined by the Biological Survey:

The total food in these stomachs was divided as follows: Animal matter, 22.3 percent; vegetable, 77.7 percent. * * * The animal food consists almost entirely of insects and spiders, a few snails forming the exceptions. The insects comprise wasps and ants (Hymenoptera), bugs (Hemiptera), a few flies (Diptera), beetles (Coleoptera), grasshoppers (Orthoptera), and caterpillars (Lepidoptera). * * * Grasshoppers appear to be the cowbird's favorite animal food, and compose almost half of the insect food, or 11 percent of the whole. * * *

The vegetable food of the cowbird exceeds the animal food, both in quantity and variety. When searching the ground about barnyards or roads the bird is evidently looking for scattered seeds rather than insects, though the latter are probably taken whenever found. Various other substances are also eaten, but they are mostly of the same general character, such as hard seeds of grasses or weeds, with but little indication of fruit pulp or other soft vegetable matter.

In his list of vegetable matter the following items are the most prominent: Corn was found in 56 stomachs, wheat in 20, oats in 102, and buckwheat in one, as against seeds of ragweed in 176 stomachs, barngrass in 265 and panicgrass in 133. Grain as a whole amounted to 16.5 percent, or about one-sixth of the total food for the year, and probably one-half of this was waste grain. "In summing up the results of the investigation," he says, "the following points may be considered as fairly established: (1) Twenty percent of the cowbird's food consists of insects, which are either harmful or annoying. (2) Sixteen percent is grain, the consumption of which may be considered a loss, though it is practically certain that half of this is waste. (3) More than 50 percent consists of the seeds of noxious weeds, whose destruction is a positive benefit to the farmer. (4) Fruit is practically not eaten."

Dr. B. H. Warren (1890) says that cowbirds eat blackberries, huckleberries, cedarberries, wild cherries, and summer grapes (*Vitis aestivalis*). E. R. Kalmbach (1914) includes the cowbird among the birds that eat the alfalfa weevil; from the first of May to the middle of July, the weevil forms more than half of the bird's food. A. H. Howell (1907) credits the bird with feeding on the cotton boll weevil. And Hervey Brackbill sends me the following note: "One afternoon I came upon a female cowbird eating dandelion seeds from a full-blown seed-head. It must have found feeding from the upright stem inconvenient, for after I had seen it take a few billfuls, it suddenly thrust out one foot and pinned the stem to the ground and finished its eating that way."

Economic status.—It appears that in its food habits the cowbird is decidedly more beneficial than harmful, doing very little damage to the farmer's crops and destroying many destructive insects. The chief cause of its unpopularity is the harm that it does through its parasitic habits for it undoubtedly interferes with the successful hatching and rearing of large numbers of small insectivorous birds.

Paul Harrington, of Toronto, has sent me his notes on the study of 100 nests of various birds that contained cowbird eggs or young. Most of these were nests of small birds, 48 of warblers, 37 of finches,

5 of vireos, 5 of flycatchers, and 5 of various other birds. Cowbirds had deposited 115 eggs in these nests, only a single egg in 80 of them. He estimated from his records that, approximately, for every cowbird raised to a self-sustaining age there is a loss of three and one-third birds of some smaller species.

Before we condemn the cowbird for its parasitic habits, however, it must be shown that the young birds sacrificed for the cowbirds have more economic value than the parasites. Beal (1900) comments on this point: "When a single young cowbird replaces a brood of four other birds, each of which has food habits as good as its own, there is, of course, a distinct loss; but, as already shown, the cowbird must be rated high in the economic scale on account of its food habits, and it must be remembered that in most cases the birds destroyed are much smaller than the intruder, and so of less effect in their feeding, and that two or three cowbird eggs are often deposited in one nest."

Behavior.—Cowbirds are highly gregarious at all seasons: although the females and mated males scatter out in their breeding territories, in the vicinity there are generally to be found flocks of unmated or promiscuous birds with which the breeding birds associate more or less. They are sociably inclined toward each other and there seems to be no jealousy among them. Friedmann (1929) never saw them fighting, but Mrs. Nice (1937) has seen it five times, "the occasion being disagreements between males during communal courting parties."

The outstanding features of the cowbird's behavior is its well-known fondness for, or association with, grazing cattle from which it derives its appropriate common name. In its association with these animals it is quite fearless, searching for food about the heads of the grazing animals or even between their feet, sometimes even alighting upon their backs, where they are supposed to relieve the animals of annoying insects. The movements of the cattle undoubtedly stir up grasshoppers and other insects, making them more easily available for the birds. The statements by some earlier writers that cowbirds search the droppings of cattle to feed on intestinal worms is not substantiated by stomach analysis.

On the ground the cowbird walks or runs, but seldom hops; while feeding it often holds its tail erected high in the air, with the wings drooping below it. Its flight in the air seems rather unsteady, much like that of the red-winged blackbird, from which it can be distinguished in the mixed flocks by its smaller size. Cowbirds seem to be on good terms with other blackbirds and starlings, associating with them in enormous mixed flocks on their feeding grounds or roosting with them at night. They also, sometimes, join with swallows or martins in their night roosts. Forbush (1927), however, says that "in New England Cowbirds usually roost by themselves; often they choose thick coniferous trees or other thickets in the shelter of which they pass the night in great numbers. Another favorite roosting place is in the grass and reeds far out on wide meadows."

John E. Galley has sent me some notes on a large winter roost of cowbirds and starlings at Midland, Tex. At the height of their abundance, he estimated that the roost contained "between 10,500 and 11,000 individuals, of which 2,000 to 2,500 were cowbirds." They were roosting in Chinese elms around the courthouse. "The starlings were grouped in the topmost branches, the cowbirds below them."

Voice.—Aretas A. Saunders contributes the following study: "Some of the sounds produced by the cowbird are distinctly seasonal, produced mainly, if not entirely, by the male, and therefore should be considered songs, though they are not particularly pleasing musically. The commonest of these consists of a prolonged, high-pitched, squeaky note, followed by two or three shorter, lower-pitched, usually sibilant notes. This song may be written *wheeeee tsitsitsi*. My records are all somewhat different in details, but the first note is pitched from C'''' to F'''', and the notes that follow are from one and a half to two and a half tones lower, the lowest note in all the records pitched on G'''. The first note may be all on one pitch or slurred slightly upward. A not uncommon variation of this has the first note followed by a downward slur which is explosive and sibilant at its beginning; this sounds like *wheeeee tseeya* and is strongly suggestive of a sneeze.

"During courtship, when the male is going through his bowing and his wing and tail spreading, another kind of song is produced, something like that described for the grackle under similar circumstances, but it usually goes from low to high pitch abruptly—two or three low notes and then a few high, squeaky ones. The low notes are not harsh, but gurgly. It sounds like *glub-glub-kee-he-heek*. The interval between them is from three and a half tones to an octave, and the pitch between them varies from C''' to C''''.

"The seasons for these songs last from the arrival of the birds to early July, when the egg laying is over and the birds gather in flocks to feed in fields for the rest of the summer. Call notes are a short *chuck*, a slurred *preeah*, and a loud, harsh rattle."

Eugene P. Bicknell (1884) says: "There seems to be no regularity about singing in the fall; but I have heard imperfect songs and half-songs at different times within a month after the middle of September. Sometimes, in the autumn, when Cowbirds are assembled in small flocks, they become garrulous, when their commingled utterance of low notes produces a sound as of subdued warbling."

Field marks.—The cowbird is the smallest of our blackbirds and can generally be recognized in the mixed flocks by size alone, even if the brown head and glossy black body of the male and the plain dark-gray coat of the female cannot be distinguished.

Enemies.—Some of the host species are hostile to the cowbird, but few succeed in driving it away. Some remove the eggs of the parasite and some bury them; many eggs are removed by human observers, and the total egg loss must be considerable. An interesting demonstration of hostility is mentioned by Dr. George M. Sutton (1928). He noted that in Pymatuning Swamp the swamp-nesting small species were nearly immune from cowbird parasitism, because the red-winged blackbirds ganged up against the cowbirds and drove them out of the swamp. He "saw a flock of Red-wings once pursue a female Cowbird until she was utterly exhausted and plunged into the water to escape. Her pursuers chased her to the edge of the Swamp then headed her off and forced her back to the opposite bank."

While roosting in the swamps, cowbirds are preyed upon by mink and weasels and perhaps owls; during the day, they are subject to attack by hawks and falcons.

Harold S. Peters (1936) lists one louse, two flies, two mites, and one tick as external parasites on the eastern cowbird.

Fall.—As soon as the egg-laying season is over, the cowbirds begin to gather into large flocks and wander about over the country, feeding in the fields and pastures. The young birds join these flocks as soon as they are able to fly. While molting, in August, the young males are quite conspicuous, the glossy black feathers of the new plumage being scattered among the old brown feathers of the juvenal dress and giving them a curious, mottled appearance.

The enormous flocks are often quite spectacular as the great, black clouds of birds swoop down into the fields to feed or pour into their roosts at night, sometimes in association with other blackbirds or starlings. Elon H. Eaton (1914) says: "The flocks of cowbirds found during September in the grain fields and pastures are so large that on one occasion after discharging my gun into a flock which was passing I picked up 64 birds from the two discharges of the gun, which will indicate the density of the flock. My estimate of the flock referred to was that there were between 7,000 and 10,000 birds. The usual flock in the fall, however, consists of from 50 to 200 birds."

The fall migration gets under way in September, but is mainly conducted during October, some individuals lingering well into November.

Winter.—A few cowbirds sometimes remain to spend the winter as far north as Massachusetts and southern Ontario. During mild winters considerable flocks sometimes spend the winter on Cape Cod, which is usually free from snow, and along the coastal marshes of southern New England. On February 12, 1935, I was surprised to see seven male cowbirds on my window feeding shelf in Taunton, Mass., all fighting for the food. The weather, below freezing, was clear and cold, and the ground was covered with deep, hard-frozen snow, as it had been for the past few weeks of unusually cold weather. The males continued to visit the feeding shelf all through that month, and on the 28th three females appeared.

Thomas McIlwraith (1894) says: "In Southern Ontario nearly all the Cowbirds are migratory, but on two occasions I have seen them located here in winter. There were in each instance ten or a dozen birds which stayed by the farm-house they had selected for their winter residence, and roosted on the beams above the cattle in the cow-house."

Milton B. Trautman (1940) noted wintering cowbirds in Ohio for nine winters. "Usually, only a few individuals or a few small flocks totaling less than 20 birds were noted in winter. In 2 years 50 to 300 wintered. The birds remained throughout the day about barnyards and adjacent fields where cattle were kept. Some roosted at night in brushy inland marshes or in cattail swamps, and when only a solitary individual or a few were present, they most frequently roosted and associated with English Sparrows."

The regular winter range is south of the Potomac and Ohio River Valleys and extends to Florida and the Gulf coast. Here they join in large mixed flocks with redwings, rusty blackbirds, starlings, grackles, and meadowlarks, feeding in the stubble fields and ricefields.

DISTRIBUTION

Range.—Southeastern Canada and central and eastern United States to México and Florida.

Breeding range.—The eastern brown-headed cowbird breeds from southeastern Colorado, northwestern Kansas (Decatur County), eastern Nebraska, central Iowa (Polk County, Clayton County), eastern Minnesota, northern Michigan, central Ontario (Biscotasing, Ottawa), south-western and central-eastern Quebec (Blue Sea Lake, Capstan Island), New Brunswick (Tabucintac), and southern Nova Scotia (Digby, Yarmouth); south to central Texas (San Angelo, Waco, Caddo), south-central Louisiana, southern Mississippi (Saucier, Gulfport), central Alabama (Tuscaloosa, Birmingham), central Georgia (Augusta, Athens), western South Carolina (Clemson), western North Carolina (Asheville, Weaverville), and central and southeastern Virginia (Naruna, Virginia Beach).

Winter range.—Winters from central Oklahoma (Canadian County, Tulsa), central Missouri (Mount Carmel, St. Louis), southern

Michigan (Kalamazoo County, Jackson County), southern Ontario (Chatham, Ottawa), New York (Rochester, Utica), and Connecticut (North Haven), rarely north to northern Maine (Presque Isle); south to Chihuahua (Chihuahua), Morelos (Cuernavaca), central Veracruz (Tlacotalpam), the Gulf coast, and southern Florida (Fort Myers, Key West).

Casual records.—Casual in Bermuda.

Migration.—The data deal with the species as a whole. Early dates of spring arrival are: North Carolina—Raleigh, January 29. Virginia—Blacksburg, January 19; Naruna, January 24. West Virginia—Wheeling and French Creek, March 8. Maryland— Laurel, January 25 (median of 6 years, February 17). Pennsylvania— Harrisburg, February 12. New Jersey—Milltown, February 20. New York—Orient and Smithtown Branch, February 22. Connecticut—Fairfield, February 26. Massachusetts—Manchester, March 3. Vermont—Rutland, March 11. New Hampshire—Hanover, March 6. Maine—Topsham, March 15. Quebec—Montreal, March 20 (average of 12 years, April 11); Quebec, March 26; Kamouraska, April 2. New Brunswick—Fredericton, March 22. Arkansas— Fayetteville, February 20. Tennessee—Nashville, February 12. Kentucky—Eubank, February 10. Missouri—Bolivar, February 10. Illinois—Rantoul, February 13; Chicago region, February 19 (average, March 25). Indiana—Hobart and Frankfort, February 20. Ohio— Oberlin, February 13; Buckeye Lake, February 22 (median, February 28). Michigan—Vicksburg, February 23; McMillan, March 22 (median of 23 years, April 6). Ontario—Harrow, March 8; Ottawa, March 21 (average, April 6). Iowa—Sioux City, February 21. Wisconsin—Dane County, March 7. Minnesota—Hamel, February 28 (average of 16 years for southern Minnesota, April 13); Lake of the Woods County, April 13 (average of 16 years for northern Minnesota, April 24). Texas—Boerne, January 20; Dension, January 25. Oklahoma—Caddo, January 19. Kansas—Clearwater and Wichita, February 12. Nebraska—Red Cloud, February 12 (median of 25 years, April 25). South Dakota—Vermillion, March 12 (average of 6 years, April 5). North Dakota—Marstonmoor, March 28; Cass County, April 6 (average, April 29). Manitoba—Oak Point, April 14; Treesbank, April 22 (median of 57 years, May 3). Saskatchewan— McLean and Qu'Appelle, April 6. Mackenzie—Fort Simpson, May 14. New Mexico—Clayton, April 11. Arizona—Phoenix, February 10. Colorado—Pueblo, March 9. Utah—Saint George, April 26. Wyoming—Laramie, April 26 (average of 8 years, May 1). Idaho—Rupert, May 8. Montana—Fortine, March 22. Alberta— Veteran, April 7. California—Gilroy, February 20; Berkeley,

March 24 (median of 13 years, April 13). Nevada—Carson City, April 28. Oregon—Albany, February 28; Sauvie Island, March 28. Washington—Pullman, May 4. British Columbia—Mirror Lake, April 20. Okanagan Landing, May 10 (median of 15 years, May 19).

Late dates of spring departure are: Tamaulipas—Gómez Farías region, April 29. Bermuda—Hamilton, April 11. Florida—Pensacola, April 20 (median of 12 years, April 3). Alabama—Fairhope, April 17. Georgia—Athens, April 23 (median of 4 years, April 14). South Carolina—Meriwether, April 30. North Carolina—Weaverville, May 10; Raleigh, April 29 (average of 9 years, April 4). Virginia—Naruna, May 12. District of Columbia—Washington, May 10. Maryland—Laurel, May 7 (median of 4 years, April 15). Ohio—Buckeye Lake, median, April 18. California—Yermo, June 7; Farallon Islands, June 2.

Early dates of fall arrival are: Oklahoma—Fort Sill, August 13. Texas—El Paso, August 7. Iowa—Osage, July 3. Ohio—Buckeye Lake, median, August 18. Arkansas—Delight, August 10. Louisiana—Gueydan, September 7. Connecticut—Hartford, August 10. New Jersey—Cape May, July 5. Maryland—Cambridge, August 10. West Virginia—Jefferson County, August 10. Georgia—Athens, July 13 (median of 5 years, July 19); Augusta, July 16. Alabama—Greensboro, July 17. Florida—Pensacola, July 14 (median of 11 years, October 21); Jacksonville, July 23. Baja California—Los Coronados Islands, September 5. Sonora—Rancho La Arizona, August 10. Coahuila—Las Delicias, August 15. Jalisco—Autlán August 4.

Late dates of fall departure are: British Columbia—Port Hardy, October 10; Atlin, September 4; Okanagan Landing, August 30 (median of 5 years, August 26). Washington—Callam Bay, November 18. Oregon—Klamath County, September 28. Nevada—Yucca Pass, October 13. California—Berkeley, July 18 (median of 10 years, July 13). Alberta—Ferintosh, December 31; Warner, November 16. Montana-Fortine, November 1. Idaho—Rupert, September 17. Wyoming—Yellowstone Park, October 19; Laramie, September 5 (average of 8 years, August 13). Colorado—El Paso County, October 29. New Mexico—Clovis, December 12. Saskatchewan—Wiseton, October 21. Manitoba—Killarney, October 30; Treesbank, October 2 (median of 30 years, August 31). North Dakota—Inkster, October 18; Cass County, October 4 (average, August 26). South Dakota—Forestburg, November 20. Nebraska—Cortland, November 23. Kansas—Harper, December 9; Oklahoma—Tulsa, November 25. Minnesota—Faribault, November 17 (average of 4 years for southern Minnesota, August 17; Otter Tail

County, November 1. Wisconsin—Superior, Burlington, and Prairie du Sac, November 10. Iowa—Newton, December 3. Ontario— Port Dover, November 22; Ottawa, November 17 (average, October 5). Michigan—Schoolcraft, December 15; McMillan, November 17 (median of 20 years, August 30). Ohio—Toledo, December 21; Buckeye Lake, November 30 (median, November 23). Indiana— Richmond, November 28. Illinois—Chicago region, December 9 (average, October 1); Odin, November 27. Missouri—Bolivar, December 3; Concordia, November 20. Kentucky—Versailles, November 15. Tennessee—Nashville, December 8; Athens, November 21. Arkansas—Rogers, November 20. New Brunswick—St. Andrews, October 18. Quebec—Montreal, October 23. Maine—Portland, December 5. New Hampshire—Hanover, December 15. Vermont—West Barnet, November 25. Massachusetts—Belmont, December 8. Connecticut—Meriden, November 23. New York— Phelps, November 27. New Jersey—Milltown, December 13. Pennsylvania—Berwyn, November 23; Renovo, November 16 (average of 5 years, October 15). Maryland—Laurel, December 23 (median of 4 years, December 0). West Virginia—Bluefield, November 19. Virginia—Naruna, December 16. North Carolina— Raleigh, November 27.

Egg dates.—Alberta: 51 records, May 24 to July 1; 40 records, June 1 to June 15.

Arizona: 37 records, May 2 to August 2; 19 records, June 17 to July 16.

California: 130 records, April 3 to July 21; 66 records, June 7 to June 29.

Illinois: 162 records, April 26 to July 11; 86 records, May 21 to June 6.

Massachusetts: 68 records, May 14 to June 29; 34 records, May 30 to June 12.

Michigan: 39 records, April 30 to July 7; 23 records, May 29 to June 14.

North Dakota: 28 records, May 23 to July 15; 14 records, June 7 to June 18.

Oklahoma: 21 records, April 29 to June 26; 12 records, May 7 to May 29.

Ontario: 15 records, May 15 to July 1; 8 records, June 4 to June 20.

Texas: 40 records, April 7 to July 2; 20 records, May 10 to May 25.

MOLOTHRUS ATER ARTEMISIAE Grinnell

Nevada Brown-Headed Cowbird

HABITS

This large race of the species breeds in western Canada and in the northern part of western United States and winters south to southern Texas and México. It was originally described by Joseph Grinnell (1909) as similar to the eastern cowbird, "but somewhat larger, with proportionally longer and more slender bill; similar to *M. a. obscurus* (Gmelin), of the lower Sonoran zone in Arizona and southeastern California, but larger." In its plumage changes, feeding, and general habits it does not differ materially from its better-known eastern relative. It is reported to eat the Mormon cricket and many other harmful insects, an action greatly to its credit.

Spring.—Dr. Ian McT. Cowan (1939) writes of the arrival of these cowbirds in the Peace River District of northeastern British Columbia: "Ten cowbirds were seen on May 6 at Tupper Creek but not until May 10 did they become numerous. On that date a flock of about forty males and four females and another containing fifty-five males and one female appeared and fed for some time in the pasture on Austin's ranch. Later on May 14 the proportion of females increased to about a quarter of the aggregate of birds in each flock and flocks up to fifty birds were common. * * * At its maximum the sex ratio was approximately one female to three males. In May the main hosts of the cowbirds were hermit thrushes and juncos."

Nesting.—Friedmann (1929) gives a list of 52 birds known to have been imposed upon by the Nevada cowbird; they belong mainly to the same classes of birds that are hosts of the eastern cowbird and including some of the larger birds, such as the blackbird, towhee, grosbeak, catbird, brown thrasher, and robin. Eggs have been found in a nest of the California gull and of the ferruginous rough-legged hawk, which were, of course, wasted eggs.

Eggs.—The eggs of the Nevada cowbird are similar to those of the eastern cowbird, and the measurements have not been separated from those given by Bendire (1895) for the eastern form; they average slightly larger. The measurements of 40 eggs average 21.8 by 16.8; the eggs showing the four extremes measure 25.4 by 16.8, 23.4 by 18.0, 19.8 by 17.0, and 20.1 by 15.2 millimeters.

Behavior.—Coues' (1874) account of the behavior of this cowbird in the west is worth quoting, as follows:

Every wagon-train passing over the prairie in summer is attended by flocks of the birds; every camp and stock-corral, permanent or temporary, is besieged by the busy birds, eager to glean sustenance from the wasted forage. Their familiarity under these circumstances is surprising. Perpetually wandering about the feet of the draught animals, or perching upon their backs, they become so accustomed to man's presence that they will hardly get out of the way. I have even known a young bird to suffer itself to be taken in hand, and it is no uncommon thing to have the birds fluttering within a few feet of one's head. The animals appear to rather like the birds, and suffer them to perch in a row upon their backbones, doubtless finding the scratching of their feet a comfortable sensation, to say nothing of the riddance from insect parasites.

A singular point in the history of this species is its unexplained disappearance, generally in July, from many or most localities in which it breeds. Where it goes, and for what purpose, are unknown; but the fact is attested by numerous observers. Sometimes it reappears in September in the same places, sometimes not. Thus, in Northern Dakota, I saw none after early in August.

This disappearance, which occurs also with the eastern cowbird is evidently for concealment during the molting season. Seton (1891) states: "I noticed that on the Big Plain the cowbirds disappear for a time, apparently joining the rusty grackles and other species among the swamps and wet lands until after the attainment of the fall plumage, when for a time they again become conspicuous, and continue about the pastures until October."

DISTRIBUTION

Range.—Western Canada and western United States (except the southwestern portion) east to southern Louisiana south to México.

Breeding range.—The Nevada brown-headed cowbird breeds from central and northeastern British Columbia (Nulki Lake, Swan Lake, Peace River District), central-southern Mackenzie (Fort Simpson, Fort Resolution), northeastern Alberta (Athabaska Delta), central Saskatchewan (Flotten Lake, Emma Lake), southern Manitoba (Lake St. Martin, Hillside Beach), and western Ontario (Rainy River; intergrades); south through central and eastern Washington (rarely west to Tacoma) and eastern Oregon (Klamath County) to northeastern and central-eastern California (Alturas, Independence), southern Nevada (except Colorado River Valley), Utah (except extreme southwestern section), northeastern and central-eastern Arizona (Kayenta, Springerville), western New Mexico, Colorado (Fort Lyon), western Nebraska, and through western Minnesota to northwestern Iowa (Sioux City).

Winter range.—Winters from western and southern California, southeastern Arizona (Tucson), northeastern Texas (Dallas), and southeastern Louisiana (New Orleans, Pearl River); south to southern Baja California (Miraflores), Michoacán (Morelia), Guerrero (Chilpancingo), and Veracruz (Córdoba). Rarely east to eastern Iowa (Linn County, Johnston County), and eastern Kansas (Lawrence, Neosho Falls).

Casual records.—Casually in northern and coastal British Columbia (Massett, Atlin, Calvert Islands), and northeastern Ontario (Moose Factory); apparently only casual west of the Cascades in Washington (Cape Flattery), Oregon (Mercer Lake; one breeding record from Medford, and California (Farallon Islands).

MOLOTHRUS ATER OBSCURUS (Gmelin)

Dwarf Brown-Headed Cowbird

HABITS

This small cowbird is resident in México and the southwestern United States, north to southern Louisiana, southern Texas, southwestern New Mexico, southern Arizona, and southern California.

The molts and plumages of the dwarf cowbird are like those of the eastern bird, and its coloration is similar, but it is decidedly smaller. It feeds on similar food.

Bendire (1895) writes:

It can only be considered a summer resident in southern Arizona, although a few appear in winter there, as I shot an adult male on Rillito Creek, near Tucson, on January 24, 1873. It usually arrives from its winter home in southern Mexico about the middle of March, and is then found associating with different species of Blackbirds, especially Brewer's Blackbird, frequenting the vicinity of cattle ranches, roads, and cultivated fields. By April 15 the flocks have scattered, and small parties of from five to twelve may now be seen in suitable localities, such as the shrubbery along water courses, springs, etc., where other small birds are abundant. The character of its food, and its general habits as well, are similar to those of the common Cowbird, which it closely resembles, being only a trifle smaller. In middle Texas the two races intergrade to some extent, and it is claimed both breed there. In the lower Rio Grande Valley, Texas, the typical Dwarf Cowbird is common.

This cowbird is not supposed to breed above the Lower Austral Zone, but J. Stuart Rowley tells me that he found an egg in the nest of a Cassin's vireo near Lake Arrowhead in San Bernardino County, Calif., at an elevation of about 5,000 feet.

Nesting.—Friedmann (1929) lists 65 recorded hosts of the dwarf cowbird, mostly small flycatchers, small sparrows, vireos, warblers and other small birds, but the list includes a number of larger birds, such as the Mexican ground dove, scissor-tailed flycatcher, Rio Grande redwing, five orioles, two towhees, two cardinals, both races of blue grosbeaks, summer and Cooper's tanagers, Western mockingbird, two thrashers, and two thrushes. Bendire (1895) says: "According to my observations, the Least Vireo seems to be oftener imposed upon, in southern Arizona at least, than any other bird, the

Desert Song Sparrows, Black-throated Sparrow, and Vermilion Fly-catcher following in the order named."

Usually only one egg of this cowbird is laid in the nest of the host, but often two are laid and sometimes more. In the W. C. Hanna collection is a set of two eggs of the orchard oriole with four of the cowbird, and a set of the Arizona hooded oriole, containing four eggs of the dwarf cowbird and one of the bronzed cowbird.

W. L. Dawson (1923) tells of a pair of least vireos that "showed notable valor in driving off from time to time a snooping female who spied upon their progress. Rousing one morning to a sudden out-cry, I arrived upon the scene in time to see an irate Vireo drag a Cowbird from the nest and hold her for a dramatic moment suspended in mid-air—until the Vireo's strength gave out and both fell struggling to the ground. But in spite of this instant and summary punishment, the Cowbird had accomplished her mission."

Wilson C. Hanna writes to me that he has a set of eggs of the California black-chinned sparrow that contains an egg of the dwarf cowbird, and a set of the black-throated gray warbler with one egg of this cowbird; each nest held two eggs of the host and one of the parasite; both sets were taken in San Bernardino County, Calif.

Eggs.—Bendire (1895) says: "In general appearance and shape the eggs of the Dwarf Cowbird resemble those of the preceding species [eastern species], and the same description will answer for both; but they appear on an average to be somewhat less heavily spotted, which gives them a lighter appearance, and they are also considerably smaller.

"The average measurement of thirty-seven specimens in the United States National Museum collection is 19.30 by 14.99 millimetres, or 0.76 by 0.59 inch. The largest egg in this series measures 20.57 by 15.49 millimetres, or 0.81 by 0.61 inch; the smallest, 18.03 by 13.74 millimetres, or 0.71 by 0.54 inch."

Young.—It has been stated that the young cowbird does not push its nest mates out of the nest, as the European cuckoo does, but Daw-son (1923) says: "I once found a nest which contained only a lusty Cowbird, while three proper fledglings clung to the shrubbery below, and one lay dead upon the ground."

Behavior.—Dawson (1923) describes a habit, common to all the races of the species, as follows: "In feeding upon the ground about corrals the Cowbirds are quickly actuated by the flock impulse, rising as one bird at a fancied alarm. After alighting upon a fence or upon the unprotesting backs of cattle, they hop down again one by one as confidence becomes established. They greet each other always with quivering bodies and uplifted tails, and that upon the most trivial occasions."

DISTRIBUTION

Range.—Southwestern United States, east to southern Louisiana, south into México.

Breeding range.—The dwarf brown-headed cowbird breeds from northwestern, central, and southeastern California (Hoopa, Death Valley), the Colorado Valley in southern Nevada, extreme southwestern Utah (St. George), north-central and southeastern Arizona (northeastern slope of San Francisco Mountains, Showlow), northwestern and central-southern New Mexico (Manuelito, Grant County, Playas Valley, Las Cruces), western and southern Texas (El Paso, Houston), and southern Louisiana (Marsh Island, St. James Parish); south at least to northern Baja California (San Quintín, Colonia), southern Sonora (Guaymas, Álamos), northern Durango (Rancho Baillon), and northern Tamaulipas (Matamoros).

Winter range.—Winters from north-central California (Sacramento Valley), southern Arizona (Parker, Phoenix, Tucson), and central Texas (Fort Clark, Boerne); south to southern Baja California (San José del Cabo, Santiago), Colima (Manzanillo, Colima), Guerrero (Iguala, Rancho Correza), Oaxaca (Tehuantepec City), and western Veracruz (Orizaba).

TANGAVIUS AENEUS MILLERI van Rossem

Miller's Bronzed Cowbird

HABITS

This western race of the bronze-backed, red-eyed cowbirds is found in northwestern México, from the Territory of Tepic on the south through Sinaloa and Sonora and into southern Arizona, breeding as far north as Tucson, Sacaton and the valley of the San Pedro River, where I collected a specimen near Fairbank.

Members of this race are slightly larger than those of the eastern race, with a stouter bill. The adult male is hardly distinguishable, except that the rump is violet, like the upper tail coverts, rather than bronzy like the back. The adult female is paler, dark mouse gray above, rather than dull black, and paler mouse gray below, rather than dark sooty brownish, as in the eastern race.

Nesting.—Friedmann (1929) lists the following seven species as hosts of the bronzed cowbird; Derby flycatcher, Audubon's oriole, Arizona hooded oriole, Sclater's oriole, canyon towhee, yellow-throated sparrow and Guatemala mockingbird. In later papers (1931, 1933 and 1938) he adds western white-winged dove, western kingbird, Giraud's flycatcher, Cooper's tanager, Griscom's tanager, Xantus's becard, scarlet-headed oriole, and Durango wren.

M. French Gilman (1914) shows a photograph of a hooded oriole's nest containing two eggs of the bronzed cowbird and four eggs of the dwarf cowbird, also another that held two eggs of the bronzed cowbird and two of the oriole, both found near Sacaton, Ariz.

Herbert Brandt tells me that he examined a nest of this oriole in Sabino Canyon, Ariz., that held one egg of the oriole and six eggs of the bronzed cowbird.

George N. Lawrence (1874) quotes Col. A. J. Grayson as follows:

On the 19th of May, 1868, whilst hunting in the woods near Magatlan, I discovered a nest of the Bull-head Fly-catcher (*Pitangus derbianus*), which is a common species in this region, and builds a large nest, dome-shaped, the entrance being on the side. Whilst I was quietly looking at the nest (which was about forty feet from the ground), I observed a female Red-eyed Cowbird among the branches of the same tree looking very melancoly. Suddenly she darted towards the nest, upon the side of which she perched, and immediately attempted to enter, but the vigilance of the fly-catcher was too acute, and observing the intrusion upon her sacred domicile, quickly attacked the Cowbird and drove her instantly away. I soon after saw the same bird examining the nest of the Mazatlan Oriole (*Icterus pustulatus*), but as there had been no egg yet laid in the nest, it did not seem to suit her, and she soon disappeared in the intricacies of the forests, leaving me strongly impressed as to her intentions.

In southern Arizona, we found eggs of the bronzed cowbird in nests of Scott's oriole and the hepatic tanager.

Eggs.—The eggs of the bronzed cowbird are indistinguishable from those of the red-eyed cowbird, described under that form. The measurements given by Bendire (1895) evidently include the eggs of both forms.

Plumages.—The molts and plumages of this western race are similar to those of the eastern race, but the colors are somewhat paler and grayer in the immature plumages. The young male of this race is similar to the young female of *involucratus*, but decidedly paler. The adult male has violet rump and upper tail coverts.

Dickey and van Rossem (1938) write:

Neither sex reaches maturity until the first postnuptial molt. One-year-old males are variously intermediate in coloration between adult males and adult females, but acquire more of the male coloration at the time of the prenuptial molt in spring. One-year-old females are duller and less metallic than mature ones and are also slightly smaller. The annual (postnuptial) molt of the adults takes place in September and October. Adults as well as one-year-old birds have a spring molt, limited in extent and consisting chiefly of the replacement of a relatively few feathers about the interscapular region, breast and head.

Colors of soft parts.—Adult males in winter: iris, brownish orange to orange-brown; bill, tarsi, and feet, black. Adult males in summer: similar, but iris, scarlet to crimson. Females (adult and birds of the year alike at all seasons): iris, similar to adult males in winter, but averaging paler; bill, tarsi, and feet, brownish or plumbeous black.

Food.—The same authors found milo maize or Egyptian corn in two stomachs examined, miscellaneous seeds in nine and a caterpillar in one. Gilman (1914) says that they "stay around the barnyard where they pick up corn and other grains and scraps from the table thrown to the chickens; and they also remain around the school yard, where they eat watermelon set in the shade for birds of all kinds."

Behavior.—On this subject Dickey and van Rossem (1938) write:

Except during the breeding season, red-eyed cowbirds normally wander in large flocks. Occasionally they consort with the grackles, but as a rule each species is likely to keep pretty well to itself. The largest cowbird assemblages noted were a flock estimated at between 400 and 500, which was seen in some bare trees along the road near Santa Rosa on October 24, 1925, and another of about 100 which was beach-combing through the litter of the high-tide mark at Barra de Santiago on April 1, 1927. Groups of less than 50 were decidedly more numerous, and the average fall, winter, and early spring flocks contained 25 or 30 birds each.

The great disparity in the relative numbers of males and females is noticeable even in the winter flocks, but becomes still more apparent when the spring break-up occurs. About April 1 or even a little earlier, the flocks disintegrate into little bands consisting almost invariably of a single old male and his harem of four or five females. This small group retains its identity as a unit until the following fall. The male is in constant attendance, strutting with shoulder tufts raised and chest puffed out before first one and then another of his flock, who for the most part ignore him completely.

In southern Arizona, Harry S. Swarth (1929) observed: "Bands of six or eight attended individual horses or steers, often in company with Dwarf Cowbirds, trotting closely alongside the selected animal in order to take advantage of the small patch of shade it afforded, and showing a marked preference for feeding by the animal's head."

DISTRIBUTION

Range.—Central Arizona to western México.

Breeding range.—The bronzed cowbird breeds from central and southeastern Arizona (Wickenburg, Phoenix) and southwestern New Mexico (Guadalupe Canyon), south through central Sonora (Opodepe, Guaymas), western Chihuahya (Durazno), and Sinaloa (Labrados, Presidio) to Nayarit (Tepic) and Colima.

Winter range.—Winters throughout most of its breeding range north rarely to southern Arizona (Tucson).

Casual records.—Accidental in southeastern California (Havasu Lake).

Migration.—The data deal with the species as a whole. Early dates of spring arrival are: Veracruz—southern Veracruz, April 12. Texas—Mission, April 17. Arizona—Tucson, April 11 (median of 4 years, April 23).

Late date of spring departure: Sonora—Rancho La Arizona, May 7.

Early date of fall arrival: Sonora—Rancho La Arizona, August 18.
Late date of fall departure: Arizona—Tucson, September 22.

Egg dates.—Arizona: 9 records, May 30 to July 7; 4 records,
June 12 to June 28.

Texas: 44 records, April 1 to July 5; 22 records, May 12 to June 8.

TANGAVIUS AENEUS AENEUS (Wagler)

Red-Eyed Bronzed Cowbird

HABITS

The red-eyed cowbird is the best known and most widely distributed
of the bronze-backed cowbirds. Its breeding range extends from
eastern Texas (San Antonio) southward along the eastern coast
region of México and Central America as far as Panamá. It differs
from the type race of northwestern México in having the back and
rump entirely bronze color, thus lacking the violaceous rump of
typical *aeneus*. The common names are not helpful in distinguishing
our two subspecies, for both have red eyes and both have bronze backs.

Alexander F. Skutch contributes the following account of the
distribution and haunts of this species in Central America:

"In northern Central America the red-eyed cowbird is found from
the lowlands of both coasts far up into the mountains, breeding in the
highlands of western Guatemala at least as high as 8,500 feet above
sea-level. In southern Central America it is less widely distributed.
In Costa Rica it appears to be absent from the heavily forested
Caribbean lowlands and from the almost equally heavily forested
lowlands on the Pacific side of the country, to the southward of the
Gulf of Nicoya; but it is present in the drier lowlands around and
to the north of the Gulf of Nicoya (Guanacaste), the central highlands,
and the upper portions of the Caribbean slope. It avoids the forest,
and its local distribution is largely determined by the presence of open
country. Hence it is more abundant in the highlands, where there is
a dense human population, with many open fields and pastures,
than in the less populous and more uniformly forested lowlands.
For the same reason, it is more common in the dry and relatively
open Pacific lowlands, and in arid, mountain-rimmed valleys in the
Caribbean drainage, than in the heavily wooded coastal districts
of the Caribbean; yet in Guatemala and Honduras it has invaded
these districts where they have been extensively cleared for banana
plantations and pastures. Red-eyed cowbirds often fly in compact
flocks over some of the larger highland cities of Central America, and
I have seen many of them in the central plaza of San José, Costa Rica.

"Red-eyed cowbirds perform at least short migrations, largely altitudinal, even in the tropical portions of their range. In the mountains above Tecpán, Guatemala, I found them at between 8,000 and 9,000 feet only during the nesting season, from March until July. During this period they were a familiar sight in the pastures about the house which I occupied from early February until the end of the year. But in August they vanished, apparently having descended to lower and warmer regions, and were not seen again in this locality during the remainder of the year, although a few were found on the plateau a thousand feet lower. During the year I spent near Vara Blanca, living in a narrow clearing in the midst of the rain forest of this excessively wet region on the northern slope of the Cordillera Central of Costa Rica, at 5,500 feet above sea-level, the first red-eyed cowbird was seen on March 28, just as the nesting season was beginning for the majority of the local birds. I had been present in the same spot since the preceding July, without having seen a single individual. In this instance, I think it probable that the cowbirds had arrived from the cultivated lands of the central plateau to the south, passing over the continental divide, which here was about 6,800 feet high. To the north were scarcely broken forests leading down to the Caribbean lowlands, where the species is not known to occur."

Courtship.—A. F. Skutch (MS.) gives the following account of the courtship: "In the middle of March, I watched a flock of about 50 red-eyed cowbirds foraging in a compact group around a straw-pile beside a granary in the Guatemalan highlands, at an altitude of 7,300 feet. Apparently the birds were picking up waste grain. From time to time, from no apparent cause, they would all take wing in a body, wheel around in a close flock, then drop down again to continue their gleaning. The breeding season was approaching, and the male cowbirds were already in an amorous mood. Now and again one would rise a few feet into the air and hover prettily on beating wings above one of the females. Other males perched in the pine trees scattered about the field, where each spread and raised his cape until it surrounded his head like a black halo, and sang with low, squeaky whistles.

"One evening in July, at about sunset, I witnessed the courtship of a pair of red-eyed cowbirds beside the Ulna River in Honduras. The female was walking over a lawn, feeding, and the male followed her with his head thrown back, chest puffed out in front, and wings quivering, walking with a stiff, seemingly unnatural gait. Of a sudden he jumped into the air and remained for about a minute poised on vibrating wings, about a yard above the female. Then he dropped to the ground in front of her, and with out-fluffed plumage bobbed up and down by flexing his legs. She considered him for a moment,

but apparently was not impressed by this gallant show, for she rudely
flew away and left him to deflate himself all alone. Then he flew off
in pursuit of her."

Friedmann (1925) gives a slightly different account of the courtship:

On May 6, a pair of Red-eyes was found in a field and the male watched displaying to the female. He ruffled up the feathers of his cape or mantle first and
then all the feathers both on the upperparts and the underparts, brought his tail
stiffly forward and under, arched his wings slightly, (not more than half as far as
it was possible to arch them), and instead of bowing over forwards as does the
male of the ordinary Cowbird, merely bent his head so that his bill was touching
the feathers of his breast for its full extent. Then he suddenly bounced up and
down four times, each bounce taking him about an inch from the ground. While
bouncing up and down he gave a series of three very deep, guttural, yet bubbling
sounds, and then a set of two short and one long squeaky, thin, high notes quite
similar to the song of the ordinary Cowbird but wheezier, more throaty and
shorter. Occasionally he did bow forward a little, but nothing like the extent
to which *M. ater* does.

The sexual and territorial relations of the red-eyed cowbird are not
well known. J. C. Merrill (1877), near Brownsville, Tex., found them
scattered over the "surrounding country in little companies of one or
two females and half a dozen males." Friedmann (1929) states: "In
this species the males outnumber the females to a somewhat greater
degree than in any other Cowbird as far as I know. During my field
work in southern Texas I saw remarkably few females compared to
the number of males noticed. * * *

"That the males establish singing trees and territories is certain as
I have noted in several instances that certain males were to be found
every day in the same tree."

Nesting.—Merrill (1877) gives us our first information on the
nesting of this cowbird, of which he says:

My first egg of *M. æneus* was taken on May 14, 1876, in a Cardinal's nest.
A few days before this a soldier brought me a similar egg, saying he found it in
a Scissor-tail's (*Milvulus*) nest; not recognizing it at the time, I paid little attention to him, and did not keep the egg. I soon found several others, and have
taken in all twenty-two specimens the past season. All but two of these were
found in nests of the Bullock's, Hooded, and small Orchard Orioles (*Icterus spurius*
var. *affinis*). It is a curious fact that although Yellow-breasted Chats and Red-
winged Blackbirds breed abundantly in places most frequented by these Cowbirds, I have but once found the latter's egg in a Chat's nest, and never in a
Red-wing's, though I have looked in very many of them. * * * On six occasions
I have found an egg of both Cowbirds in the same nest; in four of these there were
eggs of the rightful owner, who was sitting; in the other two the Cowbird's eggs
were alone in the nests, which were deserted. * * * But the most remarkable
instance was a nest of the small Orchard Oriole, found June 20, containing three
eggs of *aeneus*, while just beneath it was a whole egg of this parasite, also a broken
one of this and of the Dwarf Cowbird.

Friedmann (1929) writes: "The Red-eyed Cowbird victimizes relatively few species of birds. The various species of Orioles seem to be

the chief hosts of this parasite." He says that the species, as a whole, "is definitely known to victimize eleven genera and seventeen species and subspecies, but about 75 per cent of all the eggs are laid in nests of Orioles. I have data on 76 victimized nests all in all and of these no less than 51 belong to four species of *Icterus*." In addition to those mentioned above, he includes among the victims of the eastern race the western blue grosbeak, Audubon's oriole, western mockingbird, Sennett's thrasher, and the Texas wren.

Later (1931) he adds three more victims of the eastern race: the Mexican ground dove, the Rio Grande redwing, and Scalter's towhee. And, in another paper (1933), he increases the list to 20 known forms and one hypothetical form, including such large birds as Couch's kingbird, and the Costa Rican Thrush, as well as one vireo and four finches.

Eggs.—Bendire (1895) describes the eggs of the red-eyed cowbird as "rather glossy; the shell is finely granulated and strong. Their shape varies from ovate to short and rounded ovate. They are pale bluish green in color and unspotted, resembling the eggs of the Black-throated Sparrow and Blue Grosbeak in this respect, but are much larger.

"The average measurement of thirty-eight specimens in the United States National Museum collection is 23.11 by 18.29 millimetres, or 0.91 by 0.72 inch. The largest egg of the series measures 24.64 by 18.80 millimetres, or 0.97 by 0.74 inch; the smallest, 21.84 by 16.76 millimetres, or 0.86 by 0.66 inch."

These measurements evidently include the eggs of both races of the species. Usually only one egg is laid in a nest of the host, but he mentions a nest of Audubon's oriole that held three eggs of the oriole and three of this cowbird.

Young.—Friedmann (1929) made the following observations on the growth and development of young red-eyed cowbirds:

I have not seen a newly hatched Red-eyed Cowbird but have found birds one day old. They resemble young *Molothrus ater* of corresponding age but are a little larger in size. The skin is orange-pink; the eye-skin greenish-blue; the bill and feet dusky yellowish, the claws light yellow; the gape of the bill swollen and white; the inside of the mouth reddish; the down mouse gray and present on the head, spinal, humeral, alar, and femoral tracts, longest on the head and shortest on the spinal tract; egg-tooth prominent and pyramidal in shape and about 1 mm. high.

On the second day the birds nearly double in weight; the skin becomes slightly tinged with brownish; no new neossoptiles appear and the old ones apparently do not increase in length. * * *

A four day old bird had the primaries and secondaries sprouted; the eyes were still closed; feather sheaths were present on all the tracts except the head where the greenish gray skin (formerly orange), was still covered with light mouse gray down. This bird when five days old measured 82 mm. in length, a considerable advance over the previous day when it measured only 78 mm. The eyes opened on the fifth day and the sheaths of the rectrices began to push through the skin while those of the remiges began to open. The sixth and seventh days saw little change except growth and unfurling of the feathers from their sheaths.

The young Red-eye seems to get along with the rightful young in the nest better than do the young of *M. bonariensis* or *M. ater* and it sometimes happens that some of the rightful young survive with the parasite although in most cases the legitimate young last but a few days in face of the competition of the Red-eye. However, they do seem to survive longer with the young Red-eye than do young Sparrows or Warblers with the young of *M. ater*.

This may be partly due to the fact that the victims of the Red-eye are more nearly its own size than are those of the other Cowbirds.   * * *

When about eleven days old the young Red-eyed Cowbird usually leaves the nest and as a rule stays nearby for several days. For about two weeks and possibly more it is cared for by its foster-parents and then shifts for itself. Usually by the time this happens the season is well advanced as the Red-eye is a fairly late breeder, and the young repair to the fields and marshes for the post-juvenal molt. In these places all the young congregate from all the surrounding countryside and when the molt is finished the birds come out in flocks.

Alexander F. Skutch has sent me these notes on the young of this species: "A nest of the green jay (*Xanthoura luxuosa*) found near Matías Romero on the Isthmus of Tehuantepec, on July 8, 1934, contained a single red-eyed cowbird [probably *T. aeneus assimilis*], well feathered and almost ready to fly, sitting awake and alert between two young jays, larger than itself but naked and slumbering. A nest of the yellow-green vireo (*Vireo flavoviridis*), found near Colomba, Guatemala, on July 26, 1935, contained one cowbird nestling in addition to three young vireos. * * * In the dry country about Zacapa, Guatemala, on August 15, 1935, I saw a young cowbird being fed by a female oriole—apparently either *Icterus gularis* or *I. pectoralis*, both of which were numerous in the neighborhood, and difficult to distinguish unless they are adult males. On June 14, 1933, I saw, on the edge of a maize field in the Guatemalan highlands at an elevation of 8,500 feet, a fledgling red-eyed cowbird attended by a pair of Guatemalan spotted towhees (*Pipilo maculatus repetens*). While the youngster clamored loudly for food, an adult cowbird alighted beside it, and after an interval touched its open mouth with her (?) bill, but I was too far away to see if any food was given. The adult cowbird then flew off, and the towhee came and fed the dusky fledgling." This observation does not prove that the cowbird was feeding its own young, but is significant in connection with similar observations on the eastern cowbird by Fletcher (1925).

Plumages.—The plumages and molts of the red-eyed cowbird differ from those of the eastern cowbird in that the young male does not acquire the adult plumage until after the first postnuptial molt, the second instead of the first fall; also, he has a partial spring molt.

Friedmann's (1929) descriptions of the plumages and molts of the male are as follows: The juvenal plumage is "dull sooty black or dark sooty, the feathers of the underparts of the body with more or less distinct narrow margins of paler; mandible brownish basally." The first winter plumage, acquired by a complete postjuvenal molt, is "dull black, the underparts, especially the throat sometimes dark sooty brownish; back and scapulars very faintly, the wings, upper tail coverts and tail strongly, glossed with bluish green; bill, legs and feet black. Young males in this plumage are similar to adult females.

"First Nuptial Plumage acquired by wear which shows very little, and to a slight extent by molt involving the head and neck and breast. Plumage similar to the first winter plumage but the head, neck, and breast dull bronzy."

The adult winter plumage is acquired by a complete postnuptial molt; in this the head, neck and body are "greenish bronze, rump like the back, the plumage soft and silky but not as smooth as in *T. a. aeneus*; 'presenting the appearance of having been wet and imperfectly dried' (Ridgway); tail coverts blue-black, the upper ones glossed with violet; wing coverts glossy dark greenish blue, brightest on greater coverts and tertials, less bright as well as more greenish on primaries, primary coverts and alula; lesser wing coverts dark metallic violet, the middle coverts violet-bluish; tail dark metallic bluish-green or greenish-blue; bill black; iris red; legs and feet black or blackish brown."

The adult nuptial plumage is acquired by wear and is only slightly brighter.

In the female the molts are the same as in the male. The juvenal plumage is similar to that of the male but paler and grayer, "above sepia or grayish sepia; beneath paler and grayer, with indistinct narrow paler margins to the feathers." The first winter plumage is "dull black, the underparts, especially the throat sometimes dark sooty brown; back and scapulars very faintly, the wings, upper tail converts and tail strongly, glossed with bluish-green; neck ruffs much less developed than in the male. The female of this race is much darker than that of *T. a. aeneus*." Adults and young are practically alike in winter plumage.

Food.—Friedmann (1929) writes:

As far as my observations go, the Red-eyed Cowbird is entirely graminivorous in its food habits but I feel that more extensive data would show it to feed upon insects as well. The gizzards of some twenty-odd specimens collected during May, 1924, near Brownsville, Texas, contained only weed seeds and a few oats. The oats were undoubtedly picked up in the cavalry stable yards at Fort Brown. As many as 1,500 small grass seeds were found in a single stomach and most of the birds examined had consumed large numbers of these seeds. From this it may be judged that the species is highly beneficial in its food habits although its activities at times in fields of ripening grain or rice might be less creditable, but of this we know nothing as yet.

The Red-eye associates considerably with cattle and in all probability does so for the insects it finds scared up by the grazing animals, but my stomach content examinations have failed to reveal even a trace of insect food.

In southern Vera Cruz these cowbirds evidently do considerable damage to grain crops, for Alexander Wetmore (1943) says:

Red-eyed cowbirds were found in small flocks, regularly at the village, and also around the lagoons. As the corn matured they spread out through the fields to feed on the grain in company with *Cassidix*, and at times I saw them in such localities in flocks. When the ears were ripened the natives went into the fields to bend or break the stalks at an abrupt angle below the ears, so that these instead of standing upright were turned down toward the ground and were covered by the stalks above. Whole fields treated in this way presented a curious appearance. The theory was that the ears were thus hidden so that they were protected from damage by birds. Before this, while grain was in the milky stage, men and boys went out at dawn to the fields armed with slings and slingshots, or with clods to be thrown by hand. They stood on small elevated platforms of poles that gave them clear view across the corn, where by shouting or by casting missiles they kept the birds moving and so prevented damage.

Alexander F. Skutch contributes this account of the feeding habits: "True cow-pen birds, the red-eyed cowbirds often forage close beside the heads of grazing cattle, snatching up the insects which the animals stir up from their lurking places amid the herbage. They also alight upon the backs of horned cattle and mules to vary their diet with ticks and insect pests which they pluck from the animals' skin. Often they forage in the pastures in company with the far bigger giant cowbirds (*Psomocolax oryzivorus*) and groove-billed anis.

"The red-eyed cowbirds also joined the giant cowbirds and other members of the Icteridae in another and most unexpected form of hunting. Along the Río Morjá, a small tributary of the Río Motagua in Guatemala, was a broad, bare flood plain, covered with small, water-worn stones, where I could count upon watching the cowbirds feed almost every evening, from an hour or so before sunset until the sun had sunk behind the cane brakes. The giant cowbirds formed the nucleus of these assemblages, but their party was joined by red-eyed cowbirds, a few great-tailed grackles (especially the males), and Sumichrast's blackbirds (*Dives dives*). Often a few wild Muscovy ducks would forage near these smaller birds in the shallows; at a little distance, all five species appeared sufficiently black to remind me of the truth of the old adage 'Birds of a feather flock together'. For some reason, the male giant cowbirds resented the presence of the male red-eyed cowbirds and often pursued them, although they never drove them far away.

The chief occupation of both kinds of cowbirds was stone-turning, for which their strong, black bills seemed well fitted. They moved

stone after stone, turning over the smaller ones, pushing aside those
which were somewhat larger, and merely raising slightly one side of
the biggest, to see what edible matter might lurk beneath them.
The Sumichrast's blackbirds and the great-tailed grackles joined in
this pursuit, but not so energetically as the cowbirds; for the grackles
especially preferred to hunt small creatures that lurked in the shallows,
where the other stone-turners rarely ventured.   All four of these
blackish members of the troupial family turned their stones in exactly
the same fashion: the bird's head was lowered and the tip of its bill
inserted beneath the near edge of the stone and pushed forward, in the
line of advance of the bird.   As the decisive push was delivered, the
bird's lower mandible was dropped somewhat and its bill held slightly
open.   Whatever small animals lay concealed beneath the stone were
eaten, then the bird proceeded to move another."

Behavior.—Skutch (MS.) remarks on their behavior: "As evening
fell, the red-eyed cowbirds and the Sumichrast's blackbirds finished
their supper gleaned from the stony flood plain and retired to roost in a
dense stand of young giant canes (*Gynerium sagittatum*) growing
behind the barren flats.   Until they fell asleep, the blackbirds continued to utter a delightful variety of clear and soothing whistles, but
their companions the cowbirds were rarely heard."

Voice.—Friedmann (1929) writes:

The song is confined to the male and is quite similar to that of the ordinary
Cowbird (*Molothrus ater*) but wheezier, throatier, the individual notes shorter
and the preliminary guttural notes deeper.   It may be written as follows *ugh
gub tse pss tseeee*.   Frequently the three first notes are omitted and sometimes the
last two are run together.   I have heard the song given by birds while flying
and also while on the ground.   Strangely enough I never heard it from a bird in
a tree although Visher's experience (see above) was just the reverse.

The call notes of the Red-eyed Cowbird are not yet well known and are worth
careful study.   In my experience with this species, call notes were rarely heard
and the few that were noted were all of one type, a harsh, beady, almost rasping
chuck.   This note seemed to be a feeding note and was used by both sexes.   I
never saw or heard anything to indicate that the birds have any other call notes.

Field marks.—The red-eyed cowbird is larger than the common
cowbird, nearly the size of Brewer's blackbird, with which it is often
associated, but the cowbird's eye is red, while that of the blackbird
is yellow; and the cowbird has a stouter, more conical bill than the
blackbird and its bronzy color is conspicuous, contrasting with the
violaceous-green wings and tail.   The blood-red eye of this cowbird
is distinctive when near enough; and in the mixed flocks of blackbirds
it can often be recognized by its top-heavy appearance due to the
puffing out of the feathers of the head and neck, forming a sort of ruff.

Winter.—The red-eyed cowbird is only partially migratory in the
lower Rio Grande Valley.   Merrill (1877) says; "Here they are com-

mon throughout the year, a small proportion going south in winter. Those that remain gather in large flocks, with the Long-tailed Grackles, common Cowbirds, and Brewer's, Red-winged, and Yellow-headed Blackbirds; they become very tame, and the abundance of food about the picket-lines attracts them for miles around."

DISTRIBUTION

Range.—South-central Texas to eastern and north-central México.

Breeding range.—The red-eyed bronzed cowbird breeds from south-central Texas (Eagle Pass, Lee County), and the Yucatán Peninsula (Chichén Itzá, San Felipe) south through Central America to western Panama (Calobre, Chitrá); west to Nuevo León (Galeana, Linares) and eastern San Luis Potosí (Valles, Tamazunchale).

Winter range.—Winters throughout most of breeding range except north only to southern Texas (Edinburg, Aransas Pass .

Family THRAUPIDAE: Tanagers

PIRANGA LUDOVICIANA (Wilson)

Western Tanager

Plate 30

HABITS

For many years after its discovery this brilliantly colored bird was known as the Louisiana tanager, as indicated in Wilson's scientific name, which it still bears. The name seems wholly inappropriate today, for it is only a rare migrant in what we now know as the State of Louisiana. But, at the time of its discovery, what was then known as the Louisiana Purchase, or the Territory of Louisiana, extended from the Mississippi River to the Continental Divide and northward to British Columbia. As the bird was widely distributed over much of that territory, the name seemed more suitable.

The first specimens were obtained by members of the Lewis and Clark party on their journey across the northwestern territories of this country, and the frail specimens that they obtained were figured and named by Wilson. Later, Townsend and Nuttall obtained some better specimens, from which Audubon (1841, vol. 3) drew his beautiful plate; Audubon quotes Nuttall as saying:

"We first observed this fine bird in a thick belt of wood near Lorimer's Fork of the Platte, on the 4th of June, at a considerable distance to the east of the first chain of the Rocky Mountains (or Black Hills), so that the species in all probability continues some distance down the Platte. We have also seen them very abundant in the spring, in the forests of the Columbia, below Fort Vancouver. On the Platte they appeared shy and almost silent, not having there apparently commenced breeding. About the middle of May we observed the males in small numbers scattered through the dark pine forests of the Columbia, restless, shy, and flitting when approached, but at length more sedentary when mated."

The western tanager breeds from northwestern British Columbia and southwestern Mackenzie to southern California, southern Arizona, and central-western Texas, mainly in the mountains in the southern portions of its range, and shows a decided preference for the coniferous forests of pines or firs.

In western Washington, in the vicinity of Tacoma and Seattle, we found this tanager common at lower levels, usually at an elevation of 1,500 feet or less, wherever there was a growth of tall Douglas firs, but S. F. Rathbun told me that he sometimes found it as high as 4,000 feet in the mountains.

It is a common summer resident in the western half of Montana, where Aretas A. Saunders (1921) says that it breeds "in the Transition and Canadian zones, showing a marked preference for forests of Douglas fir on the east side of the divide, and for mixed forests of Douglas fir, yellow pine and larch on the west side. Occurs in migration in cottonwood groves in the valleys."

In the Uinta Basin, Utah, Arthur C. Twomey (1942) found that "this species ranges through a number of altitudinal communities, varying from 7,500 to 10,000 feet, from the yellow pine, blue spruce, aspen, and lodgepole pine, to the alpine fir Communities.

"Western tanagers frequently were heard singing from the tops of the highest cottonwoods along the Green River during May, but at no time were they ever numerous. The last to be seen and taken (June 9) in the low country was a female with well-developed ovaries. The following day, June 10, a trip was made to the yellow-pine forests at Green Lake. Here the western tanagers were at the height of their breeding season. Males could be heard from all corners of the forest, singing their clear song."

H. W. Henshaw (1875) says: "In 1873, in Southern Colorado, the species was found in small numbers among the cottonwoods along the streams, at an elevation of about 7,500 feet. On reaching the pines, at an elevation of about 9,000 feet, they were found to be present in much greater numbers, and at 10,000 feet were still common."

In California, the western tanager occurs as a breeding bird in the coniferous forests of the mountain ranges throughout the State, more sparingly in the coastal ranges, and more abundantly in the Sierra Nevada, where it breeds from 3,000 feet up to the summits. On migrations it occurs over the entire length and breadth of the State, even in the lowlands. Referring to the Lassen Peak region, Grinnell,

Dixon, and Linsdale (1930) say: "During the migrations through the western part of the section tanagers were most often observed perched in the blue oaks, valley oaks, or digger pines. During the summer the birds were found in the pines, incense cedars, and firs in the mountainous portion of the section."

In the San Bernardino Mountains, Grinnell (1908) found it breeding at from 6,500 to 8,000 feet.

In the Guadalupe Mountains of western Texas, Burleigh and Lowery (1940) found that the "western tanager was limited in its distribution during the summer months to the thick fir woods at the tops of the ridges and was not noted at this season of the year below an altitude of 8,000 feet."

In the Huachuca Mountains of Arizona, we found this tanager in the wooded canyons at elevations around 7,000 feet. After I left, two nests were found by my companion, Frank C. Willard, one on June 7 and the other on the 14th.

Spring.—Frederick C. Lincoln (1939) outlines the spring migration of the western tanager as follows:

On the spring migration the birds enter the United States about April 20, appearing first in western Texas and the southern parts of New Mexico and Arizona. By April 30, the van has advanced evenly to an approximately east-and-west line across central New Mexico, Arizona, and southern California. But by May 10, the easternmost birds have advanced only to southern Colorado, while those in the Far West have reached northern Washington. Ten days later the northern front of the species is a great curve, extending northeastward from Vancouver Island to Central Alberta, and thence southeastward to northern Colorado. Since these Tanagers do not reach northern Colorado until May 20, it is evident that those present in Alberta on that date, instead of having traveled northward through the Rocky Mountains, which from the location of their summer and winter homes would seem to be the natural route, have reached that province by a route through the interior of California, Oregon and Washington to southern British Columbia and thence across the mountains, despite the fact that these are still partly covered with snow at that time.

Harry S. Swarth (1904) says of the migration in the vicinity of the Huachuca Mountains in southern Arizona: "They are fairly common during the spring migration, the first noted being on April 26, but are more abundant in the lower oak regions than elsewhere, going in flocks of ten or twelve, often in company with the Black-headed Grosbeaks. Such flocks were seen throughout May and early June, after which they disappeared."

W. Otto Emerson (1903) writes of a heavy spring flight in southern California:

One of the most wonderful occurrences of the movements of birds in the season of migration, which ever came under my notice, took place at Haywards during May, 1896, when countless numbers of *Piranga ludoviciana*, or Louisiana tanagers, began to make their appearance between May 12 and 14. From the 18th to the

22nd they were to be seen in endless numbers, moving off through the hills and canyons to their summer breeding range in the mountains. This continued till the 28th, and by June 1 only here and there a straggling member of the flock was to be seen.

They were first found feeding on early cherries, in an orchard situated along the steep bank of a creek, on the edge of rolling hills, well covered with a thick young growth of live oaks, which faced the orchard on the east. To this thick cover they would fly, after filling themselves with cherries, and rest till it was time to eat again. This they would keep up from daylight till dark, coming and going singly all day, without any noise whatever being heard.

Two men were kept busy shooting them as fast as they came into the trees which lay on the side next to the oak-covered hills. The tanagers at first seemed to take no notice of the gun reports, simply flying to other parts of the orchard. * * * After the first week, I found on going here (May 17), that dozens on dozens of the birds were lying about. For the first two weeks the birds so found were mostly males, but later on the greater numbers were composed of females and young of the year. * * *

Mr. H. A. Gaylord of Pasadena, Cal., in a letter under date of June 16, 1896, states that 'they were seen singly from April 23 to May 1. From this date up to May 5 their numbers were greatly increased, and by May 5 there was an unusually large number of them. Then for about ten days, until May 16, the great wave of migration was at its height." * * * He also says, "the damage done to cherries in one orchard was so great that the sales of the the fruit which was left, did not balance the bills paid out for poison and ammunition. The tanagers lay all over the orchard, and were, so to speak, 'corded up' by hundreds under the trees."

Through Oregon and Washington the western tanagers migrate in large numbers, widely spread through the valleys and open country. Gabrielson and Jewett (1940) write: "In migration, the Western Tanagers excite a great deal of comment, particularly when unusual weather conditions force them to stop over. In late May 1920 we were together in Harney County when a sudden heavy snowfall forced down a multitude of migrating birds, many of which remained for several days. It was curious to walk through the sagebrush and see the topmost stalks flame-tipped with the brilliant yellow, red, and black of these birds. Along with them were numbers of Hermit Warblers and Gray Flycatchers, certainly a combination odd enough to intrigue anyone's interest."

Referring to El Paso County, Colo., Aiken and Warren (1914) say that these tanagers "arrive from the south in small flocks of from 3 or 4 to 7 or 8, and in migration are found well out on the plains."

Nesting.—As the western tanager generally spends the breeding season among coniferous trees, the nests are usually built in pines or firs, rarely in a tamarack, and occasionally in oaks or aspens. Thomas D. Burleigh (1921) mentions four nests found near Warland, Mont., "one June 4 with five slightly incubated eggs, another June 6 with four incubated eggs, a third June 22 with four well incubated eggs, and the last July 1 with four fresh eggs. These varied from twenty-five to thirty-five feet from the ground and were all at the outer end of limbs of large Douglas firs. All were alike in construction, being compactly built of fir twigs and rootlets, lined with rootlets and a few horse hairs. The female was incubating on the first nest and would not flush and finally had to be lifted from the nest by hand."

S. F. Rathbun records in his notes a nest found near Seattle that was in an unusual situation: "It was but ten feet above the ground, being placed on one of the lower limbs of a small fir growing by the side of a well-used path leading to a house surrounded by trees. The nest was directly over the path, but three feet from a small electric light attached to the trunk of the tree. Being so well built into the small twigs growing from the place of attachment, the nest was hardly discernible from beneath and was found by seeing the female alight on the limb and fail to appear."

Near Fort Klamath, Oreg., Dr. J. C. Merrill (1888) found the nests usually in pines or firs, but one "was in a young aspen about six feet from the ground."

W. L. Dawson (1923) says that, in California, the nest is usually placed on "some horizontal branch of fir or pine, from six to fifty feet high, and from three to twenty feet out. * * *

"The nest is quite a substantial affair though rather roughly put together, of fir twigs, rootlets, and moss, with a more or less heavy lining of horse- or cow-hair, and other soft substances."

I have a set in my collection, taken by Chester Barlow in El Dorado County, Calif., from a nest 40 feet up on the limb of a black oak, and another, taken by Virgil W. Owen in Los Angeles County, from a nest 30 feet up in the top of a live oak. In some notes received a long time ago from Owen he states that, in the Chiricahua Mountains, Ariz., the nests are usually in pine trees, but that he "has examined several in sycamore trees."

One of the two Arizona nests found by Willard referred to above, was only 15 feet from the ground in a small fir tree; the other was 65 feet up, near the end of a 30-foot limb of a large pine tree (pl. 30). Claude T. Barnes (MS.) found a nest in a canyon near Salt Lake City, Utah, that was "placed on the fork of a horizontal limb of a mountain balsam (*Abies lasiocarpa*)," about 12 feet above his head.

Eggs.—The western tanager lays from three to five eggs to a set, perhaps most often three in the southern portions of its range. In 15 sets taken by Owen in southern Arizona, 10 contained three eggs and only five were sets of four. They are ovate in shape, with some tendency toward short ovate, and are moderately glossy. The following description is taken from notes sent to me by William George F. Harris. The ground color may be "pale Nile blue," "bluish glaucous," "deep bluish glaucous," or "Etain blue." The eggs are marked with irregular specks, spots or blotches of "raw umber," "mummy brown," "Prout's brown," or "Saccardo's umber." These markings are generally well distributed over the entire egg; even so, there is usually a concentration toward the large end, and often a distinct wreath is formed. On some of the finely speckled eggs the browns are so dark as to appear almost black. The undertones of "brownish drab" or "deep brownish drab" are usually more pronounced on the more heavily marked eggs, and entirely lacking on others.

The measurements of 50 eggs average 22.9 by 16.8 millimeters; the eggs showing the four extreme measure 25.9 by 16.3, 23.0 by 19.1, 20.3 by 16.3, and 23.9 by 15.2 millimeters.

Young.—Mrs. Irene G. Wheelock (1904) writes: "Incubation lasts thirteen days, and is performed by the mother bird alone, the male rarely if ever going to the nest until the brood are hatched. As soon as the nestlings are out of the shell, however, he assumes his full share of the labor of feeding them. In the case of one brood at Slippery Ford in the Sierra Nevada, the male brought fifteen large insects and countless smaller ones in the half hour between half-past four and five one June morning. During most of the day the trips to the nest with food averaged ten minutes apart. The longest period of fasting was twenty-three minutes, and the shortest one and one-half minutes." The nest was too high up for her to positively identify the food, but she saw the old birds catch insects for them in the air, and thought she recognized caterpillars and dragon-flies in the bills of the parents.

In a nest containing large young, watched for an hour by Claude T. Barnes (MS.), the male fed the young seven times and the female fed them four times. The longest interval between feedings was ten minutes and the shortest one minute. The food consisted of insects and larvae.

Plumages.—Dr. Dwight (1900) describes the juvenal plumage of the western tanager as "above, yellowish green obscurely streaked. Wings and tail dull black, edged with olive-yellow, forming on the coverts two wing bands. Below, pale yellow with dusky streaks on the breast, similar to the young of other Tanagers." A. D. Du Bois (MS.) describes a small nestling, recently out of the nest, as "conspicuously marked with buff and blackish about the head, the crown being buff, bordered by broad black stripes."

The postjuvenal molt, which occurs in July in California, involves the contour plumage and the wing coverts but not the rest of the wings nor the tail. It produces a first winter plumage which differs from the juvenal in being unstreaked and brighter colored. Dwight describes the young male as, "above, olive-yellow, brownish on the back, the wing bands strongly tinged with lemon-yellow, the one at the tips of greater coverts palest. Below, clear lemon-yellow, a slight orange tinge often on forehead and chin."

A. J. van Rossem has sent me the following notes on the molts and plumages of the males: "From the well-known postjuvenal [first winter] plumage, which is very similar to that of the adult female, they change, in the latter part of March or the first part of April, into a plumage similar to that of the adult, except that the dusky, greenish rectrices, primaries with their coverts, and secondaries are retained. The secondary coverts and tertials are renewed with the body plumage; the renewal of the tertials is usual, but not invariable. It may be unilateral or bilateral, and may involve one, two, or three pairs of feathers. The alula is occasionally renewed also. While there is considerable variation among individuals, these one-year-old males are, as a whole, somewhat less brilliant than the adults. The black of the upper part is duller and more or less intermixed with greenish; the red on the head paler, less intense, and more restricted in area; and the yellows decidedly duller and less brilliant.

"At the fall [first postnuptial] molt, the dusky flight and tail feathers are replaced with the black ones of maturity. The fall body plumage is substantially like that of the adult nuptial, save that the head is yellowish green instead of red, and most of the feathers are tipped with olive. This tipping is most pronounced dorsally, but is apparent also across the breast. The chin and throat are nearly concolor with the rest of the under parts.

"The adult spring plumage is attained in much the same manner as is the first nuptial. Whether this is a complete body molt, as is the case with the first nuptial, is uncertain. There is no question as to the entire anterior half of the body. The posterior half is molted to at least a considerable degree, but whether the spring molt of the adult includes all the posterior half, and such wing feathers as are replaced in the one-year-old, it is impossible to say at this time; the evidence is that it does not."

Dwight (1900) says of the plumages of the female: "The plumages and moults correspond to those of the male. The juvenal dress is practically indistinguishable from that of the male. The first winter plumage is rather duller, being browner above and paler below. The first nuptial plumage is acquired by a very limited prenuptial moult, such wing coverts as are acquired being duller than those of the male and the few orange-tinged feathers paler, the whole bird paler and grayish. The adult winter plumage is brighter than the first winter, and in adult nuptial plumage a few orange feathers may appear acquired by prenuptial moult."

Food.—F. E. L. Beal (1907) examined only 46 stomachs of the western tanager, taken in various parts of California from April to

September, inclusive. The food in these consisted of over 82 percent insects and nearly 18 percent fruit. Of the insect food, he says:

The largest item of the animal food is Hymenoptera, most of which are wasps, with some ants. Altogether they amount to 56 percent of the food for the six months, and in August they reach 75 percent. * * * Hemiptera stand next in importance, with 8 percent. They are mostly stink-bugs, with a few cicadas. Beetles amount to 12 percent of the food, of which less than 1 percent are useful Carabidae. The remainder are mostly click-beetles (Elateridae) and the metallic wood-borers (Buprestidae), two very harmful families. The former in the larval stage are commonly known as wireworms, and bore into and destroy or badly injure many plants. The Buprestidae, while in the larval stage, are wood-borers of the worst description. Grasshoppers were eaten to the amount of 4 percent, and caterpillars to the extent of less than 2 percent.

The greater part of the fruit eaten appeared to be the pulp of some large kind like peaches or apricots. One stomach contained seeds of elderberries; another the seeds and stems of mulberries, and two the seeds of raspberries or blackberries. Nearly all these stomachs were collected in the mountains, away from extensive orchards, but still the birds had obtained some fruit, probably cultivated.

It is to be regretted that the stomachs of those birds killed in the cherry orchards were not saved; they might have told a different story.

In Nevada, Robert Ridgway (1877) noted that in May these tanagers "were very numerous in the rich valley of the Truckee, near Pyramid Lake, where they were observed to feed chiefly on the buds of the grease-wood bushes (*Obione confertifolia*), in company with the Black-headed Grosbeak and Bullock's Oriole. * * * In September they were noticed to feed extensively on the fruit of the *Crataegus rivularis*, in company with the Red-shafted Flicker, Gairdner's Woodpecker, the Cedar-bird, and the Cross-bills (*Loxia americana* and *L. leucoptera*)."

Dawson (1923) says: "A lady in Monticito, noting the predilection of the birds for fruit, had a wheel-like arrangement placed on top of a stake driven in her lawn. Upon the end of each spoke half an orange, freshly cut, was made secure. The Tanager saw and appreciated; and the lady had the satisfaction of seeing as many as twelve Tanagers feeding on the wheel at one time."

Rathbun sent me the following note: "Aug. 15, 1922. This evening shortly before sunset, we noticed one of these birds in the garden. This individual was busy catching flying ants, termites (*Termopsis angusticollis*), its actions while so doing being identical with those of the flycatchers. We could plainly see the bird take these insects from the air, and on one occasion it ascended straight upward 40 feet and captured one of the insects with the greatest ease, then dropped almost vertically to the spot from which it flew. During intervals of watching from its perch the tanager would remain perfectly motionless with the exception of moving its head from side to side while scanning the air.   No attempt made by it was unsuccessful.''

Mrs. Wheelock (1904) tells of young tanagers fly-catching among the pines late in August: "They were following the flycatcher fashion of catching insects on the wing, beginning when the sun touched the tops of the trees and moving downward as the day advanced and the insect life nearer the ground awoke to activity.   In like manner they retreated to the tree tops as the shadows fell in the afternoon."

Economic status.—The damage done to the ripening cherries in California during the spring migration of these tanagers apparently can be quite serious at times, but this does not occur every spring and then only when the migration which often follows the foothills of the mountain ranges far away from the large fruit orchards, is unusually heavy.   This tanager destroys many injurious insects, and a careful study of its food will show that it is about 80 percent beneficial and probably not over 20 percent harmful.

Behavior.—As evidence that the western tanager is not too shy, I quote the following from some notes sent to me by A. Dawes Du Bois: "Flathead County, Mont., July 24, 1914: There was quite an assemblage of birds at the spring this morning, waiting their turns for a bath.   I stood on the brink of the spring with one foot braced up the slope.   A young Louisiana tanager bathed at my feet.   Its mother came, with food in her bill, to a branch less than 3 feet from my head, but, becoming suspicious, did not feed.   Instead, she uttered signal notes, addressed no doubt to the young one.   It paid little heed, looked me over and bathed again; then sat on a low branch close to my feet and preened its feathers.   Father tanager came—gorgeous in his coat of many colors—and, as I stood like a statue, he hopped on the ground beneath me, between my feet, but did not go in for a bath."

Barnes (MS) says: "The flight is in an unwavering line with fairly rapid wing beats."

Voice.—Aretas A. Saunders writes to me: "In my experience with this species in Montana, the song of this bird is very similar to that of the scarlet tanager.   Perhaps, if I had made records at that time, I would have found a definite difference, but the general sound of the song certainly is very similar, if not identical."

My impression of the song of the western tanager, as I heard it in British Columbia in 1911, was that it resembled the robinlike song of the scarlet tanager.   Rathbun says in his notes: "The tanager seems to chant its song.   Some of its notes are a reminder of certain of the robin's, but have roughness lacking in the former.   But its song is very pleasing, carrying a suggestion of the wildness and freedom of the woodland—not of the country that has felt the influence of mankind."

W. Leon Dawson (1923) writes: "While chiefly silent during the migrations, the arrival of the birds upon their chosen summer sites is betokened by the frequent utterance of a pettish *pit-ic* or *pit-itic*. The full-voiced song grows with the season, but at its best it is little more than an étude in R. * * * I can detect no constant difference between the song of the Western Tanager and that of the Scarlet Tanager (*P. erythromelas*), save that that of the former is oftener prefaced with the call note, thus: *Piteric whew, we soor a-ary e-eerie witooer*. This song, however, is less frequently heard than that of the Scarlet Tanager, East."

Ralph Hoffmann (1927) says of it: "A Tanager is always deliberate and often sits for a long period on one perch singing short phrases at longish intervals. The song sounds much like a Robin's; it is made up of short phrases with rising and falling inflections *pir-ri pir-ri pee-wi pir-ri pee-wi*. It is hoarser than a Robin's, lower in pitch and rarely continued for more than four or five phrases; it lacks the joyous ringing quality of the Robin's. The Tanager's call note is one of the most characteristic sounds of the mountains of California and the evergreen forests in the lowlands of Oregon and Washington. It may be written *prit-it* or *pri-titick*, followed often by a lower *chert-it*."

He writes the note of the young as *chi-wee*, "suggesting the note of the young Willow Goldfinch or the call of the Purple Finch." Du Bois (MS.) calls it "a musical *pe-o-weet*," the middle note low.

Field marks.—The adult male western tanager is unmistakable, with his brilliant yellow body, his black back, tail, and wings and with a touch of red on his head; his colors fairly gleam among the dark green of the conifers. His mate is more quietly colored, olive above and yellow below, with no black in her plumage and rarely a tinge of orange on her forehead; she might be mistaken for a female scarlet tanager, except for her two white wing bars. The white, or yellowish white, wing bars are characters of the species at all ages or seasons. The call note is more distinctive than the song.

Enemies.—The Nevada cowbird has been known, but only in a few cases, to lay its eggs in nests of the western tanager. Probably the tanager has other feathered and furred enemies, but the sharp-shinned hawk seems to be the only one recorded.

Fall.—S. F. Rathbun tells me that, in western Washington, the autumnal migration begins about the middle of August and continues up to the middle of September. Farther south, where the birds breed in the mountains, they descend to the foothills and valleys in August before they start on their southward migration to a warmer climate.

A. J. van Rossem (1936), referring to the Charleston Mountains in Nevada, says:

The first evidence of migration was noted about mid-August, when there was an obvious decrease in the number of the tanagers about our camp, and by the 28th of the month the species was so rare that perhaps not more than one a day would be seen at the spring. There seemed to be a gap in the time between the departure of the summer visitants and the arrival of extraterritorial migrants, for on September 11 and 15 we found small flocks migrating commonly at Indian Springs and on the 14th numbers were observed up to 8,500 feet in Lee Cañon. The latest date we have is October 7, 1931, at which time a single bird was taken at the lower (8,200 feet) spring in Lee Cañon. In the Sheep Mountains tanagers were migrating commonly through the yellow-pine zone from September 16 to 19, 1930."

Swarth (1904) says that in the Huachuca Mountains, in Arizona, western tanagers reappeared on the fall migration "about the third week in July, rapidly increasing in numbers from then on. Throughout August they remained in large flocks composed mostly of young birds and females, with but a sprinkling of old males, and their favorite food at this time seemed to be the wild cherries, of which there is an abundance in the mountains."

In the Uinta Basin, in Utah, according to Twomey (1942): "The main fall migration wave struck the lower river valleys of the Basin during the first two weeks of September. They never were seen in any numbers at this time. They seemed to scatter over the valleys and to move rapidly south. After the middle of September no further individuals were observed."

Winter.—The main winter range of the western tanager is from central México and Cape San Lucas southward to Costa Rica, according to the 1931 A. O. U. Check-List. Dickey and van Rossem (1938) say of its status in El Salvador:

Common, at times even abundant, winter visitant everywhere in the Arid Lower Tropical Zone and locally in adjacent parts of the oak and pine regions. Found from sea level to 3,500 feet. * * *

It was rather surprising to find western tanagers wintering so commonly nearly 200 miles south of the southernmost point from which they were previously known.

The first arrivals to be detected were two old males which were seen in the mimosa thickets at Divisadero November 12, 1925. No more were observed until collecting was started on Mt. Cacaguatique November 20, 1925, when the species was found to be extremely common everywhere through the coffee and also in the pines and oaks a few hundred feet higher. From the latter part of December until the middle of February, western tanagers were generally distributed everywhere over the lowlands, but later on were again found only above 2,000 feet. The impression in the field was that they arrived via the highlands, that part of the population spread out over the lowlands during the winter and then retired again to the hills for the short period remaining before the northward flight. At any rate, none was seen at any locality below 2,000 feet after February 19, though there were plenty of birds above that level for over two months longer.

There were no marked migrations at any time. The departure was a gradual one with ever-decreasing numbers in evidence after April 1. The last individual noted was taken at Chilata April 23, 1927.

DISTRIBUTION

Range.—Southern Alaska and central-western Canada south across western United States to Costa Rica.

Breeding range.—The westein tanager breeds from southern Alaska (lower Stikine River), northern British Columbia (Glenora, Peace River parklands), southwestern and central-southern Mackenzie (Fort Liard, Fort Smith), northeastern Alberta (Fort Chipewyan), and central Saskatchewan (Nipawin); south to northern Baja California (Sierra Juárez, Sierra San Pedro Mártir), southern Nevada (Charleston Mountains), southwestern Utah (Zion Park), central and southeastern Arizona (Bill Williams Mountain, south to Santa Catalina, Santa Rita, Huachuca, and Chiricahua Mountains), southwestern New Mexico (Black Mountains), and western Texas (Guadalupe, Davis, and Chisos Mountains); east to western South Dakota (Short Pines Hills, Black Hills), northwestern Nebraska (Black Hills), and central Colorado (Colorado Springs, Beulah). One breeding record for southern Wisconsin (Jefferson County).

Winter range.—Winters from southern Baja California (La Paz, Miraflores), Jalisco (Cruz de Vallarta), and southern Tamaulipas (Güemes, Altamira); south on the Pacific side of the Continental Divide through Guatemala and El Salvador to northwestern Costa Rica (Tempate); casually north to California (Santa Barbara, San Diego), southeastern Arizona (Tucson) and southern Texas (Brownsville).

Casual records.—Accidental in northern Alaska (Point Barrow), Yukon (Kluane), Minnesota, central Nebraska, Missouri, eastern Texas, Louisiana (New Orleans, Grand Isle), Mississippi (Gulfport), Quebec (Kamouraska), Maine (Bangor), Massachusetts (Lynn, Brookline), New York (Highland Falls), and Connecticut (New Haven).

Migration.—Early dates of spring arrival are: Tamaulipas—Gómez Farías region, March 29. Nuevo León—Monterrey, April 16. Sinaloa—Arroyo de Limones, April 16. Sonora—Alamos, March 30. Louisiana—Jefferson Parish, March 19. Minnesota—Minneapolis, May 11. Texas—Port Arthur, April 17; El Paso, April 18. Kansas—Morton County, May 2. Nebraska—North Platte, May 8. South Dakota—Custer City, May 24. Manitoba—Brandon, June 7. Saskatchewan—Big River, May 23. Mackenzie—Grand Rapids, May 21; Fort Simpson, May 31. New Mexico—Albuquerque, April 22. Arizona—Tombstone, April 8; Tucson, April 13 (median of 6 years, April 30). Colorado—Colorado Springs, May 2 (median of 26 years, May 18). Utah—Saint George, May 13. Wyoming—Wheatland, May 1; Laramie, May 13 (average of 11 years, May 23). Idaho—Grangeville, May 10; Meridian, May 13 (median of 4 years, May 16). Montana—Fortine, May 9 (median of 21 years, May 20). Alberta—Banff, April 30 (median of 7 years, May 21); Pelican Portage, Athabasca River, May 19. California—San Clemente Island, March 23; Los Angeles County, April 7 (median of 27 years, May 1). Nevada —Carson City, May 2; Colorado River, Clark County, May 7. Oregon—Josephine County, April 11. Coos County, May 1 (median of 7 years, May 5). Washington—Shelton, April 15 (median of 5 years, May 3); Spokane, April 24 (median of 6 years, May 5). British Columbia—Okanagan Landing, April 29 (median of 18 years, May 10); Comox, Vancouver Island, May 4 (median of 24 years, May 13). Alaska—Point Barrow region (only record), May 24.

Late dates of spring departure are: El Salvador—Chilata, April 23. Guatemala—La Perla, April 8. Michoacán—Tacámbaro, April 28. San Luis Potosí—Tamazunchale, April 22. Sonora—Hacienda de San Rafael, May 18. Baja California—San Antonio del Mar, April 28. Louisiana—Grand Isle, May 11. Kansas—Finney County, June 1. Nebraska—North Platte and Stapleton, May 27. Arizona—Whetstone Mountains, June 2; Casa Grande, June 1. Colorado—Denver, June 6 (median of 15 years, May 25); Yuma, June 4. California—Berkeley, June 7; Pasadena, May 26. Nevada—Humboldt National Forest, Elko County, May 31.

Early dates of fall arrival are: Washington—Yakima, July 24. Oregon—Weston, July 1. California—Death Valley, July 3; Berkeley, July 17 (median of 14 years, August 13). Colorado—Beulah, July 31 (median of 5 years, August 3). Arizona—Mount Trumbull Region, August 4; Phoenix and Whetstone Mountains, August 12. New Mexico—Cloudcroft, July 18. Manitoba—Brandon, July 19. Texas—Palo Duro State Park, August 10; El Paso, Brenkam, and Commerce, August 29. Baja California—San Lucas, September 27. Sonora—San Bernardino Ranch, August 28. Coahuila—Las Delicias, August 12. Guatemala—Panajachel, November 11. El Salvador—Divisadero, November 12. Honduras—Tegucigalpa, November 2.

Late dates of fall departure are: British Columbia—Marpole, Kootenay District, October 8; Okanagan Landing, September 16 (median of 13 years, September 6); Hazelton, August 30. Washington—Seattle, October 16 (median of 5 years, Septbemer 23); Spokane, September 29. Oregon—Klamath County, September 28; Forest Grove, September 25. Nevada—Lee Cañon, Clark County, October 7; Carson City, September 18. California—Oakland, November 2 (median of 15 years at Berkeley, September 25). Alberta—Henry

House, September 10. Montana—Fortine, September 25 (median of 18 years, August 27). Idaho—Bayview, September 24. Wyoming—Laramie, October 22 (median of 10 years, October 2). Utah—Green Lake, Uinta Basin, September 15. Colorado—Pueblo, October 14; Fort Morgan, October 11 (median of 14 years, September 28). Arizona—Tonto Basin, October 9. New Mexico—Clayton, October 3. Mackenzie—Fort Resolution, July 12. Manitoba—Treesbank, September 4. South Dakota—Rapid City, October 2. Nebraska—Sioux County, October 1. Oklahoma—Kenton, September 25. Texas—El Paso, September 25. Mississippi—Gulfport, October 25. Sonora—Caborca, October 30.

Egg dates.—Alberta: 4 records, June 7 to June 15.

California: 50 records, May 7 to July 15; 25 records, June 10 to June 21.

Colorado: 6 records, May 19 to June 28; 3 records, June 11 to June 14.

Oregon: 10 records, June 11 to July 8; 5 records, June 13 to June 21.

PIRANGA OLIVACEA (Gmelin)

Scarlet Tanager

Plates 31, 32, and 33

Contributed by WINSOR MARRETT TYLER

HABITS

The scarlet tanager is a bird of contradictions. It possesses a brilliancy of plumage almost unrivaled among North American birds, yet the tanager, even the scarlet male, is seldom conspicuous; its characteristic song is diagnostic to those of us who know it well, yet when the tanager is heard singing, it is often mistaken for a robin, a rose-breasted grosbeak, or even for a red-eyed vireo, birds which sing somewhat like it: thus, unseen and unheard (or unregarded), it is often considered a rare bird, even in localities where it breeds commonly. The behavior of the tanager largely accounts for this anomaly.

Spring.—The scarlet tanager comes back to New England from the tropics during the height of the spring migration, at a time when a multitude of birds, residents and migrants, are here in great profusion. The expanding leaves are fast shading the bare branches, shutting from our view countless perches in the treetops where a bird may be almost invisible, so that the tanager, an arboreal bird of quiet demeanor, practically disappears from sight, hidden in the labyrinth of leafy branches. Even when in full view against a bright sky, the bird often appears as a shape rather than a bit of color, although sometimes, of course, when on a prominent perch the bird seems to blaze with color—as Frank Bolles (1891) says: "his plumage seemed burning among the leaves."

At the time of his arrival the chorus of bird music is a confusing medley from which few voices stand out clear and separate, and as the tanager's song bears to many ears more than a superficial resemblance to those of some other birds, he often passes unnoticed. Yet the tanager is a fine songster and sings freely all spring and summer long.

Every few years in May comes a prolonged period of heavy rain which washes the insects from the trees and shrubs, and forces the arboreal birds to feed on the ground. During these periods the hordes of migrating warblers collect in the fields to seek insects among the grass blades. Here the tanagers gather too, making spots of glowing color in the open country.

Alexander Sprunt, Jr. (1924), calls attention to the fact that "although this species has been taken on the coast of South Carolina on a few former occasions, * * * this record [of a bird collected on one of the barrier islands of South Carolina] is of interest in that no specimen, heretofore, has been taken on any coast island, or in such close proximity to the ocean."

Audubon (1841) stresses the same point. He says: "My friend Dr. Bachman informs me that they are seldom met with in the maritime districts of South Carolina; and that there they follow the mountain range as it were for a guide."

Francis M. Weston (MS) writes: "The scarlet tanager is a regular and somewhat abundant spring migrant through the Pensacola region. Occurrence is usually restricted to the latter half of April, but, in the course of 30 years of observation, I have noted tanagers as early as April 5, 1937, and as late as May 15, 1940. As with other trans-Gulf migrants, their abundance on this coast is conditioned by the weather. In seasons when long spells of clear weather offer no obstacle to passage across this region, few tanagers are seen; in seasons when spells of rain, heavy fog, or strong northerly winds interfere with the northward flight, incoming migrants from across the Gulf stop when they reach this coast. At such times, the coastal woods swarm with tanagers until on the first favorable night, they are off again, and the woods are deserted."

Alexander F. Skutch (MS.) reports: "In Central America the scarlet tanager is known only as a passage migrant in both fall and spring. In the 15 years during which I have given attention to the spring migration, I have recorded the scarlet tanager only five times, at dates ranging from March 29 to April 30. All these records were made in Costa Rica, four of them in El General in the southern part of the country on the Pacific slope, the fifth at Vara Blanca, at an altitude of

5,500 feet in the central mountains. These migrating scarlet tanagers were all seen singly, except a male and a female which on April 21, 1940, were keeping company as though mated. While migrating through Central America, the scarlet tanagers forage high up in the forest trees, and possibly for this reason, rather than because of actual scarcity, they have been so seldom recorded.

"Early in the morning of April 29, 1942, I heard the oft-repeated song of a scarlet tanager in the forest near my home in southern Costa Rica. He sang again in the same place on the following morning—a rich, deep-toned song which brought to mind forests of oak and hickory whither he was bound, far away in the north. Later I succeeded in glimpsing him, a splendid male in full nuptial array of scarlet and black amid the golden blossoms in the top of a tall mayo tree (*Vochysia ferruginea*) at the edge of the forest. Could he expect to find, in those far northern woodlands, another tree which would provide so glorious a background for his flaming plumage and his cheerful song?"

Courtship.—The tanager presumably has no marked ritual of courtship, for the literature speaks of it seldom and sparingly. Forbush (1929) gives us a hint, saying: "In hot weather the males of this species often may be seen with the wings drooping and tail cocked up, which gives them a jaunty appearance. This posture is exaggerated during courtship by dragging the wings and fluffing up the scarlet plumage, which may add to his attractiveness in the eyes of his expectant consort." Francis H. Allen (MS.) gives a different picture. He says: "A female called *chip-err* a few times in the top of a tree and was there joined by a male, which leaned forward towards her with closely appressed plumage, giving him a very attenuated appearance, and held his wings out from the body and drooped, with a sharp bend at the wrist. That is, the primaries were not extended, but the forearm was, and was held drooping at an angle of perhaps forty-five degrees from the horizontal."

Early in June 1943, I heard a bird note which I did not recognize, repeated over and over with a slight questioning rise in pitch at the end. It was not a whistle, but a roughened note which might be spelled *kiree*; it resembled the tone which physicians term the "spoken voice," as heard through a stethoscope. I looked up, and above my head was a pair of scarlet tanagers perched close together on a branch. One of them was making the sound—perhaps both of them were. The male was beside his mate, facing the same way, almost touching her. She was crouched down on the perch, and her wings were quivering. It was evidently the moment just before the culmination of courtship.

Nesting.—The tanager builds a rather small, flat, loosely constructed nest, using as materials twigs and rootlets, lining it with weed stems and grasses. It is generally placed well out from the trunk of a tree on a horizontal branch, usually not far from 20 feet above the ground. A. C. Bent (MS.) describes a nest containing four eggs as being "15 feet up and 8 feet out, near the end of a horizontal limb of a hemlock, beside a path in woods of mixed trees; it was made of very fine twigs, coarse grass, and weed stems, and lined with fine grass."

W. G. F. Harris (MS.) found in Rehoboth, Mass., a nest "about 15 feet from the ground against the trunk of a small beech tree. A very loose and comparatively flat nest of long rootlets (many of them over 12 inches long), it was lined with finer rootlets and dry weed stems. Its outside measurements, in millimeters, were height 64, and diameter 115; the cup had a depth of 40 and a diameter of 66 millimeters."

A. D. DuBois (MS.) speaks of a nest "about 45 feet from the ground in an ash tree which stood in a pasture just outside the wood. The nest was placed well out from the trunk, at forks of a branch which extended upward at an angle of 45 degrees." F. W. Rapp (MS.) reports to A. C. Bent his discovery in Michigan of what he "considers as colony nesting." He says: "Between the latter part of May and the middle of June 1897, I found nine nests of the tanager in a area of about three acres, in the midst of a 40-acre tract of oak near Vicksburg. These nests were loosely but firmly constructed of small sticks and twigs and could be looked through from below." Eight of these nests were in oaks (white, black and scarlet) and one was in a maple tree, ranging from 25 to 32 feet above the ground. Bent tells me that he found a nest of the scarlet tanager in an apple tree in an orchard, and another on a branch of a small red cedar in an old hillside pasture.

Eggs.—The scarlet tanager lays from three to five eggs to a set, usually four. They are usually ovate in shape, sometimes tending to short ovate, and are only moderately glossy. The ground color is "bluish glaucous," "deep bluish glaucous," "light Niagara green," "pale Niagara green," or "etain blue." The irregular spots or blotches are of "auburn," "chestnut," "bay," and "argus brown." There is considerable variation in the amount of the markings; the eggs may be minutely speckled or boldly spotted. These spots may be evenly distributed over the entire egg, but there is usually a concentration toward the large end, where often a wreath is formed; and occasionally they are confluent, forming a solid cap at the large end. The undertones of "deep brownish drab" are not visible on all eggs, but are best seen on the more boldly marked types. In most cases the markings are distinct, but occasionally they are somewhat clouded and portions of the ground color are concealed by a suffusion of light brown. Rarely a set of eggs may be found with one or more eggs entirely lacking the blue ground color, which is replaced by a creamy white color, upon which are the usual brown spots.

The measurements of 50 eggs, according to Harris, average 23.3 by 16.5 millimeters; the eggs showing the four extremes measure 26.9 by 16.8, 23.9 by 17.8, 19.8 by 16.0 ,and 21.3 by 15.2 millimeters.

Young.—Louis S. Kohler (1915a) reports on the study of three nests of the scarlet tanager found in New Jersey. The incubation period was 13 days in one nest, and 14 days in the other two nests, the incubation in all three nests being performed wholly by the female parent. In the first nest both parents fed the young, which were hatched on July 5, until they were fledged, a period of 15 days, when "the male disappeared from the vicinity and the young were seen daily with the adult female until August 1st when they all disappeared." In the second nest the young were fed by both parents for 2 days. After this "the male discontinued his efforts and only visited the nest at intervals of perhaps 30 minutes bringing no food, and finally "left the vicinity." The female, unaided, raised three of the four young. At the third nest the young were fed for 2 days by the female, the male "never approaching the nest closer than five or six feet. However, at the beginning of the third day the male began bringing food to the youngsters and continued to do so for five days thereafter. At this time, for some inconceivable reason, he took a great dislike to his mate and their offspring and began administering vicious pecks and jabs with his beak at her and the young. She quickly took on a defensive mood and after several hours of conflict drove him off and kept him away. * * * The young of this brood progressed with equal regularity with Number One and about August 1st moved from the vicinity of the nest about two hundrd feet down the valley and here were seen with the mother bird until the 15th when they also disappeared."

Plumages.—[AUTHOR'S NOTE: Dwight (1900) describes the juvenal plumage as "above, olive-yellow, including sides of head and neck, the back greener with dusky edgings. Wings and tail dull brownish black, the secondaries, wing coverts, tertiaries, and rectrices edged with olive-yellow, whitish on the tertiaries and primaries. Below, dull white, sulphur-yellow on the abdomen and crissum, broadly streaked on the breast and sides with grayish olive-brown." The sexes are alike in this plumage.

A partial postjuvenal molt occurs in late July and August, involving the contour plumage and the wing coverts but not the rest of the wings nor the tail. I have seen a young bird in full juvenal plumage as late as August 11, and have seen the beginning of the molt as early as July 24.

Dwight (1900) describes the first winter plumage of the young male as "above, including sides of head deep olive-yellow or pale olive-green. Below, citron-yellow. The wing coverts are jet-black edged with olive-yellow, but frequently only a part of them are renewed."

He says of the first nuptial plumage:

Acquired by a partial prenuptial moult probably in March and April which involves the body plumage, wing coverts, tertiaries and the tail but not the primaries, their coverts, the secondaries and usually not the alulae. The body plumage becomes scarlet vermilion varying in intensity sometimes pale or mixed with orange, usually paler but often indistinguishable from the adult. The tibiae become black and red often retaining a few old greenish feathers. Black tertiaries and black wing coverts without edgings are assumed in sharp contrast to the worn brown flight feathers which mark adults in nuptial dress. It is not unusual for only a part of the wing coverts or tertiaries to be renewed and as a freak, scarlet coverts are occasionally assumed. Greenish feathers of the first winter dress left over are comparatively infrequent on the body, the moult usually being quite complete.

I have seen young males in this first nuptial plumage in which the body color is decidedly yellowish, varying from "cadmium orange" to "cadmium yellow" or "light cadmium," often more tinged with orange above and yellower below. Year-old birds and adults have a complete postnuptial molt, beginning sometimes as early as July 17, and sometimes not completed before September 21. At this molt the wings and tail become entirely jet-black, and the yellow-green of the upper parts is deeper than in first winter plumage.

Of the female, Dwight (1900) says: "In first winter plumage the female is greener with less yellow and duller than the male and without black wing coverts. The first nuptial plumage is yellowish and so fresh that a prenuptial moult is indicated, probably more limited than that of the male. At the postnuptial moult an orange tinged adult winter plumage is acquired and sometimes black wing coverts appear, seen in the adult nuptial plumage in which only the body feathers are renewed by a partial prenuptial moult."]

Food.—Edward H. Forbush (1907) gives an admirable account of the tanager's food. He says:

In its food preferences the Tanager is the appointed guardian of the oaks. It is drawn to these trees as if they were magnets, but the chief attraction seems to be the vast number of insects that feed upon them. It is safe to say that of all the many hundreds of insects that feed upon the oaks few escape paying tribute to the Tanager at some period of their existence. We are much indebted to this beautiful bird for its share in the preservation of these noble and valuable trees. It is not particularly active, but, like the Vireos, it is remarkably observant, and slowly moves about among the branches, continually finding and persistently destroying those concealed insects which so well escape all but the sharpest eyes. Nocturnal moths, such as the *Catocalas*, which remain motionless on the tree trunks by day, almost invisible because of their protective coloring, are captured by the Tanager. Even the largest moths, like *cecropia* and *luna*, are killed and eaten by this indefatigable insect hunter. * * * I once saw a male Tanager swallow what appeared to be a hellgramite or dobson (*Corydalis cornuta*) head first and apparently entire, though not without much effort. * * * As a caterpillar hunter the bird has few superiors. It is often very destructive to the gipsy moth, taking all stages but the eggs, and undoubtedly will prove equally useful against the brown-tail moth. Leaf-rolling caterpillars it skillfully takes from the rolled leaves, and it also digs out the larvae of gall insects from their hiding places. Many other injurious larvae are taken. Wood-boring beetles, bark-boring beetles, and weevils form a considerable portion of its food during the months when these insects can be found. Click beetles, leaf-eating beetles, and crane flies are greedily eaten. These beneficial habits are not only of service in woodlands, but they are exercised in orchards, which are often frequented by Tanagers. Nor is this bird confined to trees, for during the cooler weather of early spring it goes to the ground, and on plowed lands follows the plow like the Blackbird or Robin, picking up earthworms, grubs, ants, and ground beetles. Grasshoppers, locusts, and a few bugs are taken, largely from the ground, grass, or shrubbery.

Forbush enlarges on the value of the tanager to apple orchards, saying: "Two Scarlet Tanagers were seen eating very small caterpillars of the gipsy moth for eighteen minutes, at the rate of thirty-five a minute. These birds spent much time in that way. If we assume that they ate caterpillars at this rate for only an hour each day, they must have consumed daily twenty-one hundred caterpillars, or fourteen thousand seven hundred in a week. Such a number of caterpillars would be sufficient to defoliate two average apple trees, and so prevent fruitage. The removal of these caterpillars might enable the trees to bear a full crop."

Waldo L. McAtee (1926) does not give unqualified praise to the tanager's feeding habits: "In its choice of animal food the Scarlet Tanager must be criticized for preying more extensively upon useful Hymenoptera than upon any other group of insects. We do not imagine that the bird makes a special search for these insects, but believe that it merely happens to encounter them frequently in the particular places where it habitually feeds. After Hymenoptera the important insect groups on the Tanager's bill-of-fare are beetles, lepidoptera, and bugs. * * *

"While the Scarlet Tanager feeds on the useful parasitic wasps and their allies to a greater extent than would seem desirable, it does enough good so that judgment from an economic point of view must be rendered in its favor."

McAtee, summarizing, says: "One-eighth of the food of this species is derived from the plant, and seven-eighths from the animal world. Wild fruits are the chief vegetable food, those of juneberry, huckle-

berry, bayberry, sumac, blackberry, elderberry, and blueberry being
most frequently taken."

Francis H. Allen (MS.) writes: "I saw a pair of scarlet tanagers
in a rum cherry, feeding on the ripe fruit and catching flies on the
wing." Francis M. Weston (MS.), speaking of Pensacola, Fla., says:
"During their brief, enforced stays in this region, the favorite food of
the scarlet tanager is the ripe berries of the red mulberry (*Morus rubra*).
Because of the dense foliage of these trees, it is not possible to make
an actual count of the birds in their branches, but an observer at some
point of vantage, watching the tanagers come and go to and from a
large mulberry in full fruit, would not hesitate to guess the presence
of 40 or 50 birds at one time."

Walter B. Barrows (1912) states that Professor Aughley records
the capture of a scarlet tanager "which had 37 locusts in its craw and
nothing else that I could identify."

Behavior.—As we watch a scarlet tanager at close range, we note
its quiet, unhurried manner as it moves leisurely about its favorite
woodland of oak trees. But that it can move rapidly on occasion is
shown by E. H. Forbush (1929) who remarks: "Mr. A. C. Bagg says
that he saw one drop a red berry from its bill and recover it before it
had fallen eight inches."

As a rule, however, the bird gives us the impression of a placid,
indolent, somewhat self-conscious personality, almost lethargic, paying
little attention to the life about it. The tanager, a bird seemingly of
neutral qualities, compares unfavorably in the popular mind with the
more striking, buoyant species in its neighborhood, and this prejudice
perhaps explains why so little has been written about it in the literature. Indeed, Frank Bolles (1894) speaks disparagingly of the bird,
saying: "Mr. and Mrs. Tanager, he in scarlet coat and she in yellow
satin, are best measured by contrast with the refined warblers. Their
voices are loud, their manners brusque, their house without taste or
real comfort. They have no associates, no friends. They never seem
at ease, or interested in the misfortunes or joys of those beneath
them."

Lyle Miller, of Youngstown, Ohio, writes to us: "June 6, 1926—As
I was hurrying through a woods to my car, my attention was directed
towards a male tanager. He was sitting in a low shrub, quietly
watching me. I stopped and eyed him. It was then I noticed his
mate a short distance away, also in a frozen attitude. I stood quietly
watching the birds for two or three minutes. Neither one moved
although both were quite close to me. Glancing cautiously around,
I spied the nest. It was only inches from my head, 5 feet up in a
dogwood tree. The strange behaviour of the birds was explained.

Not till I walked towards them did they move and give their customary call note of alarm. The nest held four fresh eggs."

Voice.—The tanager's song is rather pleasing, although the bird is by no means a great artist. The song resembles the robin's in form, that is it consists of short phrases alternating in pitch, continuing on indefinitely, usually with no concluding phrase, as in the rosebreasted grosbeak's song, which satisfies the ear in a musical sense. However, in rare instances the tanager does introduce a final phrase, rounding out the song into a finished sentence. Such a song, from my notes of 1913, might be written, *querit, queer, queery, querit, queer.*

The quality of the tanager's voice, with a hoarse burr running through it, gives individuality to the song, making it stand out distinctively among the songs of North American birds. The syllables *Weer weera*, pronounced with a faint hum to suggest the huskiness of the tanager's voice, call the song to mind. The phrases, repeated half a dozen times or more with little range in pitch, are spoken or hummed rather than whistled; although they carry well, they are not overloud and at a little distance might not be noticed if one were unfamiliar with the song.

Aretas A. Saunders (MS.) sends to A. C. Bent this analysis: "The song of the scarlet tanager consists of a series of from 3 to 9 notes and slurs, with short pauses between them. The notes and slurs are usually of equal length, and also the pauses between them, so that the song has an éven rhythm, and one that is quite similar to that of the robin. The quality is best described as a harsh whistle. The pitch varies from C''' to D'''', and songs range from 1½ to 3½ tones each. Their length varies from 1⅘ to 3 seconds." The frequently uttered note commonly written *Chip-churr* (*Chick-kurr*, I think, is better) may consist of two or three *chips* before the *churr*, according to Saunders, and Eugene P. Bicknell (1884) says: "Speaking of this well know [sic] *chip-chir*, Mr. Fred T. Jenks, of Providence, R. I., has called my attention to what is undoubtedly a clear instance of geographical variation in utterance. Mr. Jenks writes that he has observed that in 'Illinois and Indiana it has three notes, *chip-chirree*'." William Brewster (1886a), speaking of North Carolina, says: "The song is normal, the call note *chip-churr*, as in New England, not *chip-prairie*, as in Southern Illinois."

A. A. Saunders (MS.) says that during courtship "the female has a call in a somewhat husky whistle. It is a single note, sounding like *whee* or an upward slur, like *puwee*. Young birds that have recently left the nest sometimes go astray or get temporarily separated from the parents. In such cases, when they get hungry, they call, over and over, a note that is distinctive of the species, but not, so far as I am aware, ever used by adult birds. This call is a husky whistle of three connected notes with an upward inflection and may be written *taylilee*."

Albert R. Brand (1938) gives the approximate mean vibration frequency of the tanager's song as 2925, a little higher than that of the robin's song.

Enemies.—Herbert Friedmann (1929) reports that the scarlet tanager is "a fairly common victim" of the cowbird, and that it "is parasitized throughout most of its range." Edward H. Forbush (1929) says of the tanager: "His concealment among the leaves, together with his ventriloquial powers, must serve him well, for I have seldom found remains indicating the demise of one of these male birds by the talons of a hawk."

Fall.—Francis M. Weston (MS.) states: "Scarlet tanagers, usually single birds, pass through the Pensacola region in small numbers in October on their southward migration. It is very certain that the fall migration route of the species as a whole is not a reversal of the spring route for, in 30 years of observation, I have never seen a concentration of tanagers in fall under any conditions of favorable or adverse flying weather." Alexander F. Skutch (MS.), speaking of Central America, says: "In the autumn I have met this tanager only twice. A female or young male was seen near Tela, Honduras, on October 7, 1930, and the same or another individual in the same locality on the following day."

DISTRIBUTION

Range.—Southern Canada from Manitoba eastward through Quebec south across central and eastern United States to Perú and Bolivia.

Breeding range.—The scarlet tanager breeds from central Nebraska (North Platte, Neligh), eastern North Dakota (Fargo, Grafton), southeastern Manitoba (Winnipeg, Indian Bay), central-western Ontario (Lac Seul, Port Arthur), northeastern Minnesota (Duluth), northern Michigan, southern Ontario (Liard, Lake Nipissing), southern Quebec (Montreal, Hatley), New Brunswick (Beechmount), and central and central-southern Maine (Kineo, Hancock County); south to central-northern and southeastern Oklahoma (Pushmataha County, McCurtain County), central Arkansas (Rich Mountain, Hot Springs National Park), west-central Tennessee (Wildersville), northwestern and central Alabama (Florence, Talladega Mountains), northern Georgia (East Point), northwestern South Carolina (Walhalla, Spartansburg), western North Carolina (Statesville), central Virginia (Naruna, Petersburg), and Maryland. Reported breeding, but unconfirmed, in southeastern Manitoba (Brandon), and northeastern Texas (Tyler, Harrison County).

Winter range.—Winters from northwestern and central Colombia (Remedios, Bogotá Plateau) south through Ecuador to central Perú (Monterico, Chanchamayo) and central-western Bolivia (Yungas); casually north to El Salvador (Mount Cacaguatique, Pierto del Triunfo); accidental in North Carolina (Mount Olive).

Casual records.—Casual in summer in Saskatchewan (Indian Head), Wyoming (Cheyenne), Colorado (Palmer Lake, Pueblo) and Texas (Pease River Valley, McLennan County. During migrations reported from Nova Scotia (Wolfville, Seal Island, Halifax), Bermuda, Bahama Islands (Andros, New Providence, Cay Lobos), Cuba, Jamaica, Mona Island, the Lesser Antilles (St. Croix, Barbuda, Antigua, Santa Lucia, Mustique, Barbados), and eastern Central America.

Accidental in Alaska (Point Barrow), British Columbia (Comox), California (San Nicolas Island), Arizona (Tucson), and Colorado (Grand Junction, New Castle, Pueblo, Fort Morgan); sight records from Ontario (Pancake Bay, Timmins), Western North Dakota (Charlson), and central South Dakota (Faulkton, Rosebud Indian Reservation).

Migration.—Early dates of spring arrival are: Costa Rica—El General, March 29. Quintana Roo—Chetumal, May 5. Yucatán—Dzidzantún, April 26. Leeward Islands—Barbuda, April 29. Puerto Rico—Mona Island, May 3. Cuba—Havana, April 9. Bahama Islands—Cay Lobos, April 15. Bermuda—April 18. Florida—Princeton, March 25; Tortugas, March 29; Pensacola, April 1 (median of 18 years, April 14). Alabama—Autaugaville, April 2; Decatur, April 11. Georgia—Savannah, April 1; Atlanta, April 4. South Carolina—Greenville, April 15; Spartanburg, April 20 (median of 17 years, April 26). North Carolina—Piney Creek, April 8; Raleigh, April 15 (average of 20 years, April 30). Virginia—Lynchburg, April 10; Naruna, April 19. West Virginia—Bluefield and Wheeling, April 15. District of Columbia, April 17 (average of 42 years, April 29). Maryland—Laurel, April 20 (median of 8 years, April 28). Pennsylvania—York, April 10; Warren, April 14. New Jersey—Morristown, April 12; Vineland, April 23. New York—New York City, April 19; Nyack, April 28. Connecticut—Old Saybrook, April 27. Rhode Island—Apponaug, April 4. Massachusetts—Hyannis, April 17; Vineyard Haven, April 18; Orange, April 22. Vermont—Rutland, May 4. New Hampshire—Hanover, May 4. Maine—Mount Desert Island—April 17; Yarmouth, April 19. Quebec—Hatley, May 15. New Brunswick—Bathurst, May 12. Nova Scotia—Westport, April 15; Bridgetown, May 1. Louisiana—Grand Isle, April 3. Mississippi—Rosedale, March 27; Deer Island, April 3. Arkansas—Rogers, April 4; Tillar, April 5. Tennessee—Nashville, April 6 (median of 12 years, April 17); Elizabethton, April 10. Kentucky—Lexington, April 14. Missouri—St. Louis, April 21. Illinois—Chicago region, April 1 (average, May 6); Murphysboro, April 10. Indiana—Richmond, April 15. Ohio—central Ohio, April 17 (average, April 30). Michigan—southern Michigan, April 6; Vicksburg, April 26; Saulte Ste. Marie, May 9. Ontario—St. Thomas, April 28; Ottawa, May 13 (average of 25 years, May 20). Iowa—Burlington and Keokuk, April 27. Wisconsin—Milwaukee, April 23. Minnesota—Hutchinson, April 26 (average of 34 years in southern Minnesota, May 11); Stearns County, May 9 (average of 23 years in northern Minnesota, May 18). Texas—Kemah, April 6; San Antonio and Austin, April 12. Kansas—Fort Leavenworth, April 16. Nebraska—Omaha, April 14; Stapleton, April 18. South Dakota—Yankton, May 4; Faulkton, May 7. North Dakota—Fargo, May 16. Manitoba—Treesbank and Reaburn, May 3. Saskatchewan—Indian Head, May 22. Colorado—Boulder, May 8.

Late dates of spring departure are: Ecuador—below San José, March 13. Panamá—Chiriqui, March 26. Costa Rica—El General, April 30. Windward Islands—St. Lucia, May 19. Bermuda—Ireland Island, May 6. Florida—Daytona Beach, May 12. Alabama—Long Island, May 6. Georgia—Atlanta, June 4. South Carolina—Charleston, May 22. North Carolina—Chapel Hill, May 20; Raleigh, May 14 (average 6 years, May 11). Virginia—Naruna, June 11. Maryland—Laurel, May 18 (median of 3 years, May 16). Massachusetts—Boston, May 27. Louisiana—Monroe, May 31. Mississippi—Oxford, May 24. Kentucky—Bardstown, May 24. Illinois—Chicago, June 1 (average of 16 years, May 26). Ohio—Buckeye Lake, median, May 23. Texas—Brownsville, May 22; Dallas County, May 21. Oklahoma—Tulsa County, May 22. Nebraska—Plattsmouth, June 2; Red Cloud, May 29. North Dakota—Jamestown, June 4. Colorado—Fort Morgan, June 10.

Early dates of fall arrival are: South Dakota—Faulkton, July 6. Nebraska—Hastings, July 14. Texas—Port Arthur, July 20. Illinois—Chicago, September 9 (average of 13 years, September 18). Kentucky—Bowling Green, September 6. Mississippi—Bay St. Louis, August 24. Louisiana—Thibodaux, August 25. Massachusetts—Nantucket, August 29. Maryland—Laurel, August 17. North Carolina—Raleigh, September 8 (average of 3 years, September 18). Georgia—Young Harris, August 12; Savannah, August 20. Alabama—Greensboro, September 18. Florida—Pensacola, September 9 (median of 15 years, September 25). Honduras—near Tela, October 7. El Salvador—Monte Mayor, October 6. Nicaragua—Río Escondido, September 27. Costa Rica—Bonilla, October 3.

Late dates of fall departure are: British Columbia—Comox, November 17. Manitoba—Treesbank, August 27. North Dakota—Fargo, October 4. South Dakota—Yankton, September 29. Nebraska—Cedar Creek, October 7. Oklahoma—Tulsa County, October 1. Texas—Cove, October 19. Minnesota—Minneapolis, October 26 (average of 7 years, September 19); Isanti County, September 30 (average of 7 years in northern Minnesota, September 16). Wisconsin—Milwaukee, October 13; Mazomanie, October 7. Iowa—Marble Rock, September 30. Ontario—Point Pelee, October 14; Ottawa, October 8 (average of 11 years, September 14). Michigan—Sault Ste. Marie, October 21; Detroit, October 6. Ohio—Hillsboro, November 20; New Bremen, October 22; central Ohio, October 11 (average, September 23). Indiana—Richmond, October 15. Illinois, Rantoul, November 2; Chicago, October 6 (average of 13 years, September 28). Missouri—Bolivar, November 7. Kentucky—Danville, October 7. Tennessee—Elizabethton, October 13. Arkansas—Monticello, October 20. Mississippi—Saucier, November 13; Oriel, October 20. Quebec—Montreal, October 31; Hatley, September 14. Maine—Castine, November 6; Mount Desert Island, October 14. New Hampshire—Sandwich, October 15; Jefferson region, October 12. Vermont—Bennington, October 7. Massachusetts—Brockton, November 11; Concord, November 8. Rhode Island—Kingston, October 24; Providence, October 6. Connecticut—West Hartford, October 21; Stamford, October 15. New York—Watertown, November 3; Orient, October 19. New Jersey—Englewood, October 22. Pennsylvania—Germantown, November 3; Renovo, October 13; Berks County, October 3 (average of 14 years, September 21). Maryland—Baltimore, October 23. District of Columbia, November 13 (average of 21 years, October 4). West Virginia—Bluefield, November 15. Virginia—Lexington, November 17. North Carolina—Weaverville, October 20; Raleigh, October 17 (average of 3 years, October 8). Georgia—Atlanta, October 29. Alabama—Greensboro, October 16. Florida—St. Marks, October 25; Pensacola, October 24 (median of 16 years, October 16). Bermuda—October 17.

Egg dates.—Connecticut: 16 records, June 1 to June 22; 10 records, June 3 to June 8.

Illinois: 11 records, May 28 to Aug. 2; 6 records, June 9 to June 26.

Massachusetts: 42 records, May 24 to June 25; 23 records, May 28 to June 6.

Pennsylvania: 7 records, May 27 to June 11.

PIRANGA FLAVA DEXTRA Bangs

Eastern Hepatic Tanager

HABITS

Outram Bangs (1907) described this eastern race as: "Similar to true *P. hepatica*, but smaller; the adult ♂ much more richly colored; back much redder, less grayish; pileum darker, more intense red—dull scarlet—vermilion; under parts, darker, deeper red—deep orange—vermilion (flame-scarlet in true *P. hepatica*), Adult ♀ darker in color throughout with the back decidedly less grayish."

The range of this race extends from southwestern New Mexico and western Texas through eastern México to Guatemala.

Nothing seems to have been published on the habits of this race, but we have no reason to suppose that they differ materially from those of the better known western form.

DISTRIBUTION

The eastern hepatic tanager breeds from the mountains east of the Continental Divide from north-central New Mexico (Willis, Mesa Yegua), through western Texas (Guadalupe, Davis, and Chisos Mountains), Nuevo León (Cerro de la Silla), Tamaulipas (Realito), and Puebla (Huauchinango) to central Veracruz (Las Vigas, Jalapa, Jico) eastern Oaxaca, and Chiapas (San Cristóbal 28 miles east-southeast of Comitán). It winters from central Nuevo León (Mesa del Chipinque) and northern Tamaulipas (Matamoros) south to western Guatemala (Chanquejelve, Momostenango, Chichicastenango), and is casual in southern Texas (Flour Bluff).

PIRANGA FLAVA HEPATICA Swainson

Western Hepatic Tanager

Plate 34

HABITS

Two races of this species are now currently recognized as occurring north of México, and at least two others have been described and named. For a full discussion of the claims of these races for recognition and their ranges, the reader is referred to a paper on the subject by Sutton and Phillips (1942), based on the study of a large series of specimens. The subject is too complicated to be discussed here. This western race is the form that breeds in Arizona, in parts of north central New Mexico, and in western México.

Dr. Coues (1878) gives the following account of the introduction of this species to our fauna:

During Capt. L. Sitgreaves's expedition down the Zuñi and Colorado Rivers * * * Dr. S. W. Woodhouse observed this beautiful Tanager in the San Francisco Mountains, and secured a full-plumaged male, adding to the then recognized fauna of the United States a species long before described by Mr. Swainson as a bird of Mexico. In 1858, Baird recorded a second specimen from Fort Thorn, New Mexico; and, in 1866, I wrote of the bird as a summer resident in the vicinity of Fort Whipple, Arizona, where it arrives during the latter part of April. * * *

Meantime, however, in 1873, Mr. Henshaw had been busy with birds in Arizona, and had taken a female specimen at Camp Apache, Arizona. * * * There this Tanager was not rare; perhaps half a dozen individuals were seen in the course of one afternoon, in a grove of oaks that skirted some pine woods.

In 1922, we found hepatic tanagers fairly common in the Huachuca Mountains, in Arizona, nesting in the tall yellow pines in the upper parts of the canyons, above 5,000 feet and near the lower limit of the heavy pine timber.

In the Chiricahua Mountains, Ariz., they were seen mostly in the pines, but sometimes in neighboring oaks.

Harry S. Swarth (1904) says of its status in the Huachuca Mountains: "A fairly common summer resident, generally distributed over the mountains during migration, but in the breeding season restricted more to the canyons between 5,000 and 7,500 feet. In 1902 the first arrival was noted on April 11th, and the following year on April 16th; about the middle of May they were quite abundant in the higher pine regions, going in flocks of eight or ten, feeding in the tree tops and but seldom descending to the ground."

Nesting.—On May 26, 1922, we collected a set of three eggs and a pretty nest of the hepatic tanager in Stoddard Canyon, a branch of Ramsey Canyon at about 7,000 feet, in the Huachuca Mountains. The nest was about 50 feet from the ground and about 12 feet out from the trunk, in a fork near the end of a horizontal branch of a tall yellow pine. It was suspended by its edges between the two prongs of the fork. It was made of green grass, green and gray weed stems, flower stalks and blossoms, and was neatly lined with finer dry and green grasses. My companion Frank C. Willard made a difficult climb with the use of ropes to secure it; he cut off the branch, near the nest, which was photographed near the ground. In my collection is another set of four eggs, taken by Virgil W. Owen in the same region on May 14, 1907; the nest was 18 feet up and 12 feet out near the end of a pine limb.

After I left Arizona, Frank C. Willard collected a nest and two eggs of the hepatic tanager in Miller Canyon, on June 15, 1922. The nest was placed at the tip of a branch of a large sycamore, 25 feet above ground.

In the Thayer collection is a set, taken by O. W. Howard, that was found 19 feet from the ground in an upright fork of a madroña.

Eggs.—Four eggs usually make up a full set for the hepatic tanager, but often only three are laid and rarely as many as five. These are usually ovate in shape and are moderately glossy. William George F. Harris has contributed the following description of the colors: The ground color may be "pale Nile blue," "Etain blue," "pale Niagara green," or "bluish glaucous." The eggs are speckled or spotted with "bay," "chestnut," "chestnut-brown," "carob brown," or "liver brown." The markings, usually in the form of fine speckling or small spots, are generally well distributed over the entire egg, but with concentration toward the large end, where frequently they form a wreath. Some eggs show undertones of "pale neutral gray," but these markings are seldom prominent and often indistinct. In general, the eggs of this species seem to show a tendency to be less heavily or boldly marked than those of either the western tanager or the summer tanager.

The measurements of 50 eggs average 24.5 by 17.7 millimeters; the eggs showing the four extremes measure 26.8 by 18.5, 25.3 by 18.9, 21.5 by 16.9, and 22.0 by 16.6 millimeters.

Young.—Nothing seems to be known about the period of incubation nor about the development and care of the young.

Plumages.—Ridgway (1902) describes the nestling (juvenal) plumage as: "Conspicuously streaked beneath with dusky on a pale buffy ground, more indistinctly streaked above on a grayish olive ground; middle and greater wing-coverts margined terminally with buff; otherwise like adult female."

The sexes are alike in juvenal plumage and in the first winter plumage.

The first winter plumage is acquired by a partial postnuptial molt, in July and August, which involves the contour feathers and the wing coverts, but not the rest of the wings nor the tail. After this molt, young birds of both sexes are essentially like the adult female, light olive green above, more yellowish on the crown, with grayish cheeks, and yellow beneath. This plumage is apparently worn through the first breeding season with little change, except that some young males may acquire a few red or orange feathers on the head and throat.

The fully adult plumage is acquired in late summer at the first postnuptial molt, which is complete and is practically the only seasonal molt of any consequence. In the fall and winter male the back and scapulars are more strongly tinged with brownish gray than in spring birds, but wear produces a clearer color effect before the nuptial season. There is much individual variation in the seasonal changes of the males, some retaining a few vestiges of immaturity during their second winter. After the postjuvenal molt the female produces no color changes of consequence, except that some old females acquire orange feathers on the throat and forehead.

Food.—In a grove of oaks, near Camp Apache, according to Henshaw (1875), "they appeared to be feeding upon insects, which they gleaned from among the foliage and smaller branches of the oaks." He also saw them moving "slowly about in the tops of the pines searching for insects. At this season [July], they capture these generally while at rest, but occasionally sally forth and take them in mid-air."

Mrs. Bailey (1928) says: "Those seen by Major Goldman in the Burro Mountains the middle of September were feeding on wild grapes and wild cherries in a northeast slope canyon at 6,500 feet."

Behavior.—Like all tanagers, the hepatic tanagers are rather slow and deliberate in their movements. We did not find them particularly shy and were able to observe them at their nest building. Henshaw (1875), however, found them so "excessively shy" that he had difficulty in getting within gunshot of them. On July 21, "young, just from the nest, were taken. The old birds manifested much affection and solicitude for their progeny, flying down on the low branches, and, after venting their anger in harsh notes, returned to the side of their young and led them away to a place of safety."

Voice.—Henshaw (1875) says: "With the exception of the call notes, used by both sexes, and which resemble the syllables *chuck*, *chuck*, several times repeated, they were perfectly silent, and neither here nor elsewhere did I ever hear any song." But Frank Stephens wrote William Brewster (1881) that "the song is loud and clear, but short."

Enemies.—Frank C. Willard told me that he found an egg of the bronzed cowbird in a nest of an hepatic tanager. This is probably a rare occurrence, for I can find no such report by anyone else.

Winter.—The winter range of this form of the hepatic tanager seems to lie in western México, from Sonora southward. Col. A. J. Grayson wrote to George N. Lawrence (1874): "I discovered this species to be quite frequent in the Sierra Madre Mountains, on their western slope between Mazatlán and Durango in December, but I have never met with it in the *tierras calientes* proper. It seems to be a mountain species."

DISTRIBUTION

Range.—Arizona to western México.

Breeding range.—The Western hepatic tanager breeds from northwestern and central Arizona (Hualpai Mountains, Bill Williams

Mountain, Flagstaff) and southwestern New Mexico (Silver City, Head of Rio Mimbres); south through the highlands of México west of the Sierra Madre del Oriente at least to Michoacán (Cerro del Estribo) and Guerrero (Omilteme); west to western Chihuahua (Jesús María, Pinos Altos); eastern Sonora, and Oaxaca (25 miles northeast of Oaxaca).

Winter range.—Winters from southeastern Arizona (Patagonia) and southern coastal Sonora (San José de Guaymas) south to limits of breeding range and into coastal and lowland areas.

Migration.—The data deal with the species as a whole. Early dates of spring arrival are: Texas—Brewster County, April 28; Rockport, May 5. New Mexico—Silver City, May 10. Arizona— Tucson area, April 4 (median of 6 years, April 27); Beaverdam, May 6.

Late dates of fall departure are: Arizona—Huachuca Mountains, October 25; Prescott, October 2. New Mexico—Burro Mountains, Grant County, September 16. Texas—Davis Mountains, October 6.

Egg dates.—Arizona: 16 records, May 21 to July 10; 8 records, June 1 to June 19.

New Mexico: 1 record, June 29.

PIRANGA RUBRA RUBRA (Linnaeus)

Eastern Summer Tanager

Plates 35, 36, and 37

HABITS

This wholly red tanager is the southern representative of the family, breeding throughout the central United States east of the Prairies and southward to Florida, the gulf coast, and northeastern México. It occurs as a straggler only in New England and on our northern borders.

The favorite haunts of the summer redbird, as it is often called, are open dry, upland woods, among oaks, hickories, and other hardwood trees. In North Carolina, according to Pearson and the Brimleys (1919), it is "equally at home in pine forests, mixed woods, groves of shade trees near houses, or mulberry orchards." In South Carolina, Arthur T. Wayne (1910) says: "This species prefers open pine woods with an undergrowth of scrubby oaks and small hickory trees in which to breed."

Spring.—Alexander F. Skutch writes to me: "As it arrives late in Central America, so the summer tanager leaves early. By the beginning of April the species is already becoming rare in Costa Rica; my latest record for this country is April 12. In Guatemala, farther north, it has been recorded as late as April 25 by myself and April 27 by Griscom; but these were stragglers that had lingered behind the main migration."

The summer tanager is one of many species of small birds that evidently migrate directly across the Gulf of Mexico from Central America to the Gulf States. M. A. Frazar (1881) reported that a few of this species were observed while he was cruising from the coast of Texas to Mobile, Ala., and when his small schooner was about 30 miles south of the mouths of the Mississippi. Further evidence of trans-Gulf migration is given in the following contribution from Francis M. Weston regarding the spring migration of the summer tanager, as observed near Pensacola, Fla.: "Abundance of the summer tanager at this season depends upon weather conditions, as is the case with most of the trans-Gulf migrant species that make their first landfall on this part of the coast. A season of long periods of good weather brings us no more tanagers than would be needed to provide our rather sparse breeding population, and it is presumed that at such times great numbers must pass over unseen on their way to more northerly nesting grounds. In bad weather, however, when an incoming flight meets rain, heavy fog, or strong north winds, and halts on the coast instead of continuing on its way inland, tanagers in uncountable abundance swarm in city gardens and parks and in coastwise patches of woods. I recall my amusement, one spring, at the confusion of a visiting ornithologist who, delighted at the sight of several tanagers, set out to count the number he could find in a single vacant, wooded city block. All went well, the birds flitting along before him as he slowly traversed the block. Then, looking back, he saw that more new birds had come into the area behind him than he had already chased out and counted. He finally gave up the project as hopeless and contented himself with noting the species as 'very abundant'. A swarm of tanagers like that can be expected during any or every spell of bad weather from the last week of March through all of April. This spring influx persists even into May, for, on May 8, 1945, a mixed flight of incoming migrants, halted by bad weather, included a fair sprinkling of summer tanagers."

The tanagers of this species that breed in Florida, and perhaps some of those that nest farther north, evidently pass over Cuba and the Florida Keys to reach the mainland of Florida. A heavy storm, resulting in many casualties to this and other species, at Key West and the Tortugas, is described by Commander F. M. Bennett (1909). Migration through Texas serves to bring the birds to the more western portions of their breeding range, though some of these apparently cross a portion of the Gulf of Mexico. The earliest birds reach Florida before the end of March, but the main northward migration is accomplished through the month of April, reaching the northern limits of the breeding range early in May.

Around the middle of April, 1929, a remarkable visitation of summer tanagers reached New England, blown northward by a severe storm, of which John B. May (Forbush, 1929) says: "This storm first appeared in Texas April 13, travelled east rather slowly to South Carolina, then swung northeast along the coast, reaching its greatest intensity between New Jersey and Massachusetts on April 16."

Audubon (1841) says: "Whilst migrating, they rise high above the trees, and pursue their journeys only during the day, diving towards dusk into the thickest parts of the foliage of tall trees, from which their usual unmusical but well-known notes of *chicky-chucky-chuck* are heard, after the light of day has disappeared."

Nesting.—F. M. Weston contributes the following account of the nesting habits of this species in northwestern Florida, near Pensacola: "During the nesting season, the summer tanager deserts the city and the coastal strip of woods and retires to the pine areas a few miles inland. Here they select as nest sites the dogwoods (*Cornus florida*) and the scrub oaks scattered through the pine lands. I suspect, too, that they nest in the pine trees that are still young enough to bear branches within 10 or 15 feet of the ground. The nests are hard to find and I have but little data to offer. This, first, because the birds are far from common and, secondly, because the flimsy, inconspicuous nests can be concealed by a single leaf of such large-foliaged scrub oaks as *Quercus catesbaei* and *Q. marilandica*, two species especially favored as nest sites. Nests containing eggs have been found from the last week of May until the middle of June, and I have no data that would indicate a second nesting. Both sexes feed the young birds, but I do not know if the male parent assists with incubation or nest building."

Farther south in Florida, according to Arthur H. Howell (1932): "Summer Tanagers live in open woodland, preferring the pines, but are found to some extent in oak hammocks. Their nests are placed usually on a horizontal limb of a pine or oak, 12 to 35 feet above the ground, and are very loosely constructed of weed stems and Spanish moss, and lined with fine grasses." Charles R. Stockard (1905) says that, in Mississippi: "These birds seem to have a foolish fancy for building their nests on horizontal branches that overhang roadways. * * * They build a nest home of smooth contour and always lined with a golden yellow grass straw or a similar greenish straw giving to the concavity of the nest a very characteristic appearance; the common 'pepper grass' stems make a favorite material for the outer layer."

A nest in the U. S. National Museum collection measures 4 inches in diameter and 2 inches in height; it has an internal depth of barely half an inch.

Eggs.—Four eggs usually constitute the set for summer tanager, but often only three are laid and rarely as many as five. They are ovate in shape, with some variations toward elongate or short ovate. The shell is moderately glossy. William George F. Harris has given me the following description of the eggs: The ground color may be "pale Nile blue," "Etain blue," "pale Niagara green," or "pale glaucous green." This is speckled, spotted, blotched, and occasionally clouded, with "Argus brown," "Brussels brown," "raw umber," "chestnut brown," "mummy brown," or "Prout's brown," with undertones of "light mouse gray" or "Quaker drab." There is great variation in size and arrangement of the markings; in general, they are well distributed over the entire egg, but there is a tendency to concentrate toward the large end, where sometimes they are confluent and form a solid wreath or cap. The gray or drab undertones are as a rule not particularly prominent. The markings are generally bolder than on the eggs of the scarlet or the hepatic tanagers. Rarely, a set may partially or entirely lack the blue coloring and have instead a creamy white ground color with the usual brown markings.

The measurements of 50 eggs average 23.1 by 17.1 millimeters; the eggs showing the four extremes measure 25.4 by 17.8, 24.9 by 18.3, 21.1 by 16.3, and 23.5 by 16.1 millimeters.

Young.—The period of incubation for the summer tanager is said to be 12 days. Information on the development and care of the young seems to be lacking, beyond the fact, mentioned by Weston (MS.), that both sexes are known to feed the young.

Plumages.—Dwight (1900) describes the juvenal plumage of the summer tanager as: "Above, ruddy or yellow tinged sepia-brown with darker edgings and feather centres producing a faintly streaked appearance. Wings deep olive-brown with olive-yellow or greenish edgings, usually reddish tinged on the outer primaries, the coverts duller, the tertiaries paler. Tail bright olive-green or olive-yellow often reddish tinged basally, the shafts sepia-brown. Below, dull white tinged with sulphur-yellow on abdomen and crissum, distinctly and broadly streaked on the throat, breast and sides with deep olive-brown."

The sexes are alike in the juvenal plumage and nearly alike in the first winter plumage. The postjuvenal molt in July and August involves the contour plumage and the wing coverts, but not the rest of the wings nor the tail. This produces the first winter plumage, which Dwight describes as: "Above pale olive-green with a strong orange tinge, reddish in many specimens. Below chrome-yellow often strongly tinged with orange especially on the crissum and 'edge of the wings.' The wing coverts are edged with olive-green strongly tinged with yellow or orange according to individual vitality. The orbital ring is usually chrome-yellow or paler."

The first nuptial plumage in the young male is, he says—

Acquired by a partial prenuptial moult which involves portions of the body plumage, wing coverts, tertiaries and the tail. There is an unusual amount of individual variation in the extent of this moult accentuated by the contrast of the new vermilion or poppy-red feathers among the old greenish or yellow ones. Some birds become entirely red except for the old greenish primaries, their coverts and the secondaries and there are all sorts of intermediates ranging down to those with a mere sprinkling of red feathers. The central quills only of the tail may be renewed, sometimes only part of the tertiaries and wing coverts, but in every case it is easy to see that the process of moult has stopped at points where the checking of its normal advance would produce the varied plumages found.

The prenuptial molt takes place in winter or early spring, beginning in February or earlier, while the birds are in their winter quarters.

The first and subsequent postnuptial molts occur in August and are complete; at this molt young males assume the fully red plumage, which is never again replaced by an olive-green body plumage, as in the scarlet tanager.

The adult nuptial plumage is the result of wear, which is very slight. Adults have only one complete annual molt in August. The molts of the female are similar to those of the male, but young females are often yellower than young males in their first winter, and old females sometimes show a mixture of red feathers in the body plumage, or tinges of red in the wings.

A. F. Skutch tells me that he has seen young males with red in their plumage as early as December in Costa Rica.

Food.—Arthur H. Howell (1932) writes: "The food habits of this bird have not been thoroughly studied. Many observers have reported its habit of visiting beehives and destroying the bees. It is known to feed also on beetles, wasps, tomato worms, and spiders, and on certain small wild fruits, such as blackberries and whortleberries. Examination in the Biological Survey of the stomachs of 6 birds taken in Alabama and of 2 taken in Florida showed that the bird has a decided preference for Hymenoptera (bees, wasps, etc.), these insects being present in 7 of the 8 stomachs in proportions varying from 30 to 98 per cent of the total contents. Other insects taken were dragon flies and click beetles."

A. F. Skutch writes to me: "Summer tanagers are expert flycatchers and capture many insects on the wing. As in the United States, so in Central America, they sometimes arouse the ire of apiculturalists by catching bees, and are shot for this offense; but careful study might reveal that it is only the drones that they attack. They are also fond of the soft, white larvae of wasps, and in the Tropics find an immense variety of these insects, with nests of the most diverse forms. The outer walls and rafters of my house are a veritable museum of wasps' nests, and the wintering summer tanagers often come to feast upon the young brood. But they are excessively shy while close to the house, and it is difficult to watch them at this activity. Sometimes, while occupied indoors, I have heard a scratching on the outer walls, and gone to the window only to see a summer tanager fly away from a wasps' nest. I actually watched the tanagers plunder nests of three different kinds, on the house or in the surrounding trees; but more often I have found evidence of their visits in the form of nests torn open. * * *

"In addition to insect food of varied kinds, the summer tanagers eat a certain amount of fruit. They come to my feeding table to share the bananas and plantains with eight non-migratory species of the family."

Paul H. Oehser has sent me the following extract from a paper by Phil Rau (1941), of Kirkwood, Mo.

Some years ago I recorded that birds sometimes pierce the paper nests of *Polistes pallipes* and feed on the larvae; on one particular occasion fifty per cent of the small newly-founded nests of this species were destroyed by an unknown species of bird (Can. Ent., 62: 144, 1930). More direct evidence, however, was obtained by Dr. E. S. Anderson who informed me that for two weeks during 1939, he observed a male Summer Tanager (*Piranga rubra rubra*) remove the larvae of *Polistes pallipes* and *P. variatus* from time to time from nests under the eaves of his barn at Gray's Summit, Missouri. These the bird carried to its young in a nest in a nearby tree.

One sometimes finds Polistes' nests with whole series of cells destroyed, and at first we thought that this was done by the wasps themselves, who removed the cells to obtain building material for new nests, or, if it was a 'live' nest, for cells on other parts of the same nest. In this I find I was mistaken; the damaged condition, when it appears, is quite certainly done by birds when removing the larvae from the nests. Of the twelve species of Polistes wasps studied by me in Missouri, Panama, and Mexico, I have never found any evidence of wasps obtaining building material from old nests, or from portions of 'live' nests.

J. I. Hamaher (1936) thus describes the tanager's method of attack on a wasps' nest:

The nest of this common black and white paper nest wasp was in a pine tree near the kitchen window from which I watched the performance for about half an hour. When I first noted some unusual activity the bird was pecking at something which he held. Then perching on a twig about three feet from the wasp nest, he sat for a moment facing the nest. I noted then that about a dozen wasps were flying about the nest in an excited manner. The bird then made a dive toward the swarm, seized a wasp and flew off to a resting place nearby. I was at first in doubt whether he was eating the wasps or merely killing them. I afterward found several dead wasps beneath the tree on the ground. After several times repeating the attack the wasps all suddenly disappeared whereupon the Tanager alighted on the nest and rapidly tore the upper protecting layers away and attacked the comb.

Several observers have referred to the summer tanager's habit of catching bees, but Floyd Bralliar (1922) tells the best story. A friend of his, a beekeeper, complained that his bees were not doing well, though there were plenty of flowers in the vicinity and no disease was apparent in the hives. They sat down to watch, and saw one tanager catch 15 or 20 bees within a few minutes. Then another tanager came and satisfied its appetite.

"We did not know," he says, "how many birds were feeding there, but it was evident that there were more than two, for no two birds could possible eat so many bees as we saw caught that day. After watching them for a week, my friend, himself something of a naturalist and a great lover of birds, decided he would have to do a distasteful thing in self protection, so he took his gun and began shooting tanagers. The first day, he killed eight of these birds feeding on his bees. Within a few days the bees began to grow strong, showing that this had been their only trouble; and as he had killed all the summer tanagers near by, he had no more trouble."

He says further: "The summer tanager feeds largely on beetles, caught on the wing or in trees. * * * It eats beetles so large that it seems impossible for it to swallow them. After these insects are digested the indigestible feet, legs, and shells are rolled into a ball by the bird's stomach and disgorged."

Behavior.—The summer tanager does not differ materially in its mannerisms from the other tanagers. It is very deliberate in its movements and rather solitary in its habits, spending much of its time in the concealing foliage of the woodland trees, where it is surprisingly inconspicuous in spite of the brilliant plumage of the male. Were it not for its loud voice, the bird might easily be overlooked.

Voice.—Aretas A. Saunders writes to me: "In the spring of 1908, in Alabama, I became very familiar with the song of the summer tanager. According to my note-books, I heard the song practically daily from the bird's arrival in April to the end of my stay in early June. My recollections of the song are that it is not harsh, as is the scarlet tanager's, but musical, with more liquid consonent sounds between the notes."

Ridgway (1889) says that its notes are much louder than those of the scarlet tanager: "The ordinary one sounds like *pa-chip-it-tut-tut-tut*, or, as Wilson expresses it, *chicky-chucky-chuck*. The song resembles in its general character, that of the Scarlet Tanager, but is far louder, better sustained, and more musical. It equals in strength that of Robin, but is uttered more hurriedly, is more 'wiry' and much more continued."

Mrs. Nice (1931) says: "The song is rich, musical and varied, from 3 to 6 seconds in length, from 4 to 6 given a minute. One typical song went as follows: *hée para vée-er chewit terwee hée para vée-er.*"

A. F. Skutch tells me that the summer tanager "is by no means a silent bird during its sojourn between the Tropics. It often utters its somewhat rattling call-note, *chicky-tucky-tuck*. Often I have heard the familiar voice floating down from the tops of the forest trees, where intervening masses of foliage hid from my view the brilliant red form. But the summer tanager sings far less in its winter home than many another bird. Early one October, in southern Costa Rica, I found a newly arrived male who sang in a loud voice for several minutes." On April 24 and 25, 1932, he heard one sing a sweet song; this was the only one he ever heard sing in the Tropics in the spring.

Enemies.— Friedmann (1929) lists the summer tanager as "an uncommon host" of the cowbirds; he found five records of such parasitism, involving two races of *Molothrus ater* and both races of *Piranga rubra*.

Harold S. Peters (1936) records one louse and one mite as external parasites on this tanager.

Fall.—Francis M. Weston writes to me of the migration near Pensacola, Fla.: "Migration commences early and is, I believe, a reversal of the spring route. As early as the last week of August, tanagers appear commonly in the coastwise woods, several miles south of the nearest known nesting areas, and from then until mid-October a few birds can always be found. Stormy spells in September halt flights of southbound migrants of many species that, in good weather, presumably pass overhead undetected, and summer tanagers are always present in these gatherings. Sometimes, particularly when a succession of bad days, such as was experienced here from September 18 to 20, 1937, dams up several days flights, tanagers are as abundant as in spring. In the fall of 1925, the only year I was able to get satisfactory returns from the Pensacola Lighthouse (a first-class light almost on the Gulf beach) a single tanager was picked up among a host of casualties of the night of October 26–27."

The migrating tanagers referred to above would probably cross the Gulf of Mexico to Central America. Others that breed in Florida apparently migrate southward through Cuba to Yucatán, and then on to their winter homes in Central and South America.

Winter.—Alexander F. Skutch contributes the following account: "My earliest record of the arrival of the summer tanager in Central America was made in the Coast Rican highlands on September 18, 1935, but this is an exceptionally early date, and the species is rarely met before October, when it begins to become abundant. Rapidly spreading over most of Central America, it is one of the common and widespread winter visitants from Guatemala to Panamá, on both sides of the Cordillera. During the winter months it resides at altitudes ranging from sea-level up to 8,500 feet, but is most abundant in the lower and warmer regions. It frequents both the treetops of the heavy forest and the scattered trees of shady pastures, plantations, and orchards. The spreading willows that grow along the water-courses in the Caribbean lowlands of Guatemala and Honduras, seeming so exotic amid the heavy foliage of the majority of the trees, are very attractive to the summer tanagers, which dart actively through their open crowns.

"Throughout the six months of their sojourn in Central America, the summer tanagers are solitary and unsociable. They never form flocks; and when two are close together, attentive watching will usually reveal that they are quarreling, probably over territorial rights; for it seems that these tanagers, like some of the warblers, claim exclusive feeding territories while in their winter homes. Early in the afternoon of October 13, 1944, soon after the arrival of the species in this locality of southern Costa Rica, I found two individuals in a guava tree behind the house. Both wore the yellowish plumage of the female and the young male; on neither could I detect any red. One sang sweetly in a low voice, repeating its lilting melody over and over as it flitted about the second, who moved less frequently and from time to time uttered a low, liquid monosyllable. They continued this for many minutes. Then they flew into the tall hedge at the back of the yard, thence through the yard from tree to tree, then down into the lower pasture, and across it to the bank of the creek. I lost sight of them for a few minutes, but soon the song led me to them once more, in some trees in the midst of the pasture. I watched them for the better part of an hour; and all this time they continued to behave as already described: one (I believe always the same), singing in a sweet, low voice; the other uttering the liquid note; the two flitting around each other. Rarely one would fly at the other and make it retreat. Finally they flew up into the forest and I lost sight of them. Were these summer tanagers, a young male and a female, contesting the same winter territory? This seems the most probable explanation of the episode I witnessed."

DISTRIBUTION

Range.—Central and east-central United States south through México and Central America to Perú, Bolivia, and Brazil.

Breeding range.—The summer tanager breeds from central Texas (San Angelo), central Oklahoma (Fort Cobb, Ponca City), eastern Kansas (Geary region), southeastern Nebraska (Falls City), northwestern Missouri (Albany), southeastern Iowa (Keokuk), central Illinois (Camp Point, Philo), southern Indiana (Silverwood, Greensburg), southwestern, central and central-eastern Ohio (Cincinnati, Columbiana County), throughout West Virginia (except in high mountains), northeastern Tennessee (Johnson City), western North Carolina (Morgantown), central Virginia (Lexington), eastern Maryland and southern Delaware; south to southern Texas (Lomitas, Houston), the Gulf coast and southern Florida (Fort Myers, Fort Lauderdale). Formerly bred north to central Iowa (Des Moines). northern Illinois (Lacon, Chicago region), southern Wisconsin (Albion, Milwaukee), central Indiana (Kokomo), and southern New Jersey (Cape May).

Winter range.—Winters from Michoacán and Puebla (Metlaltoyuca), Veracruz (Motzorongo, Jaltipan), Campeche (Pacaytun, Matamoros), Yucatán (Chichén Itzá) and Quintana Roo (Palmul and Xcopén); south throughout Central America and in South America to south-central Perú, western Bolivia, western Brazil (Rio Uapés), and southeastern Venezuela (Mount Roraima); casually north to southern Texas (Brownsville) and western Cuba (Santiago de las Vegas).

Casual records.—Casual in California (Los Angeles, Wilmington, San Diego), Baja California (Laguna Salada, Guadalupe Island, La Jolla), Arizona (Tucson), Colorado (Boulder, Denver), Minnesota (Pipestone), Michigan (Pinckney), Ontario (Point Pelee, Rondeau Park, Penetanguishene, Scarboro Heights), and New York (Cincinnatus), along the Atlantic Coast north to Maine (Wiscasset), New Brunswick (Grand Manan), and Nova Scotia (Seal Island, Halifax); also casual or accidental in Sonora (Rancho la Arizona), Nayarit (Río las Canas), Bermuda, Bahama Islands (New Providence, Andros), Jamaica, Swan Island, and Trinidad.

Migration.—The data deal with the species as a whole. Early dates of spring arrival are: Sinaloa—Escuinapa, March 25. Sonora—Magdalena, April 19. Bahama Islands—Nassau, April 5. Bermuda, April 9. Florida—Winter Park, March 2; Pensacola, March 8 (median of 39 years, March 31). Alabama—Coosada, March 31. Georgia—Savannah, March 18. South Carolina—Charleston, March 25; Frogmore, April 1. North Carolina—Raleigh, April 2 (average of 29 years, April 19). Virginia—New Market, April 14. West Virginia—Charleston, April 27. District of Columbia, April 18 (average of 23 years, May 1). Maryland—Montgomery County, April 21. Pennsylvania—Sharon, May 6. New Jersey—Leonia, May 5. New York—Brooklyn, April 6. Connecticut—New Haven. April 8; Wallingford, April 16. Rhode Island—Block Island, April 7.

Massachusetts—Boston, April 11. Vermont—Putney, May 23. Maine—Sebasco, April 7. Nova Scotia—Wolfville, April 17. Louisiana—Bains, March 18; Grand Isle, March 31. Mississippi—Biloxi and Bay St. Louis, March 31. Arkansas—Glenwood, March 29. Tennessee—Maryville, April 2; Nashville April 9 (median of 12 years, April 15). Kentucky—Versailles and Bowling Green, April 2; Lexington, April 7. Missouri—Bolivar, March 30; Marionville, April 2. Illinois—Olney, April 18. Indiana—Bloomington, April 1; Richmond, April 14. Ohio—Circleville, April 17. Ontario—Toronto, April 13; Point Pelee, May 7. Iowa—Burlington, April 20. Wisconsin—Milwaukee, April 30. Minnesota—Frontenac, May 14. Texas—Kerrville, March 5; Dallas, April 4. Oklahoma—Oklahoma City, April 1. Kansas—Elmdale, April 23. Nebraska—Red Cloud, April 30. New Mexico—State College, May 2. Arizona—San Xavier Mission, April 14. Colorado—Boulder, May 1. California—Piacacho, April 20.

Late dates of spring departure are: Ecuadór—Quito, April 7. Colombia—Quimari, April 14. Panamá—Loma del León, March 29. Costa Rica—El General valley, April 17 (median of 9 years, April 9). Nicaragua—Río Escondido, April 13. Guatemala—near Quiriguá, April 25. British Honduras—Mountain Cow, April 18. Veracruz—Volcán San Martín, April 22. Tamaulipas—Galindo, April 20. San Luis Potosí—Tamazunchale, April 14. Cuba—Havana, May 30. Bahama Islands—Andros Island, April 19. Bermuda—April 29. Florida—Tortugas, May 14. Maryland—Laurel, May 29. New York—Speonk, May 25. Mississippi—Cat Island, May 10. Texas—Cove, May 18.

Early dates of fall arrival are: Texas—El Paso, August 27. Mississippi—Deer Island, August 9. Louisiana—Thibodaux, August 20. Florida—Paradise Key, August 6. Cuba—Havana, September 15. Baja California—Guadelupe Island, October 12. Tamaulipas—Matamoros, August 26. Yucatán—Cayos Arcas, August 29. Honduras—near Tela, October 3. Costa Rica—San Miguel de Desamparados, September 18; El General valley, September 29 (median of 4 years, October 5). Panamá—Río Caimitillo valley, October 21. Colombia—Santa Marta region, October 19. Brazil—Madeira River, November 22. Ecuador—Pastaza Valley, October 17. Perú—Huachipa, October 5.

Late dates of fall departure are: Arizona—San Francisco River, October 10. New Mexico—Mesilla, October 1. Kansas—Lake Quivira, October 4. Oklahoma—Tulsa, October 5. Texas—Anahuac and Cove, October 18. Iowa—Wall Lake, September 26. Michigan—Pinckney, November 6 (only record). Ohio—central Ohio, October 7.

Indiana—Carlisle, October 15. Illinois—Odin, October 1. Missouri—Bolivar, October 16. Kentucky—Eubank, October 10. Tennessee—Elizabethton, October 20. Arkansas—Delight, October 13. Mississippi—Gulfport, October 25. Louisiana—Monroe, October 29. Novia Scotia—Annapolis Royal, October 20. Maine—Monhegan Island, October 21; Winthrop, September 23. Massachusetts—Middleboro, October 26; Nantucket, October 9. Rhode Island—Kingston, September 29. Connecticut—Hartford, September 28. New York—Ward's Island, September 19. New Jersey—Long Beach, September 29. Pennsylvania—Philadelphia, Octover 23. Maryland—Baltimore County, September 29. District of Columbia, September 17 (average of 7 years, September 14). West Virginia—Bluefield, October 10. Virginia—Lynchburg, October 17. North Carolina—Raleigh, October 30 (average of 10 years, September 7). South Carolina—Charleston, October 14. Georgia—Atlanta, October 26. Alabama—Piedmont, October 20. Florida—Miami, November 12; Pensacola, November 8 (median of 21 years, October 20). Sonora—San Pedro River, October 5.

Egg dates.—Arizona: 9 records, May 27 to August 5.

Florida: 5 records, May 9 to June 2.

Georgia: 11 records, May 10 to June 17; 6 records, May 23 to May 31.

South Carolina: 25 records, May 11 to June 2; 13 records, May 19 to May 26.

Texas: 4 records, March 24 to June 5.

PIRANGA RUBRA COOPERI Ridgway

Cooper's Summer Tanager

HABITS

This western race of our well-known summer tanager is decidedly larger than its eastern representative, with paler coloration. Its range includes the southwestern corner of the United States, from middle Texas to New Mexico, Arizona, and the lower Colorado Valley in extreme southeastern California, and extends southward through western México to Colima. The report of its accidental occurrence in Colorado has been shown by Gordon Alexander (1936) to be an error.

In Arizona, we found Cooper's tanager common only in the lower valleys, particularly along the San Pedro River, where my companion Frank C. Willard took a set of four eggs on June 9, 1922. Swarth (1904) reports it to be "of very rare occurrence in the mountains, during migration." We did not see it in the mountains at all.

In the Valley of the lower Colorado River, Grinnell (1914b) found this tanager "strictly confined to the willow association. Not one bird was seen even so far from this association as the mesquite belt." And in New Mexico, it is reported by Mrs. Bailey (1928) as frequenting the cottonwoods along the rivers and in canyons. Referring to Brewster County, Tex., Van Tyne and Sutton (1937) record Cooper's tanager as a common nesting species "where there are cottonwood, mesquite, or willow trees. It is apparently not so fond of oaks, although singing males were noted more than once in oak woods in lower parts of the Chisos Mountains." * * *

"On May 13 * * * Sutton observed an adult male that was singing and displaying before a parti-colored young male which also was singing fervently. In display, the adult male spread its wings and tail and stuck its bill straight up."

Nesting.—Van Tyne and Sutton (1937) found two nests of Cooper's tanager in Brewster County: "On May 11 a nest with four fresh eggs and the female parent were collected at Castalon. The nest was about fifteen feet from the ground close against the trunk of a slender willow that stood not far from the Rio Grande. On May 29 a nest and four eggs were found on the Combs ranch, thirteen miles south of Marathon. This nest was built on a horizontal willow bough, about twenty feet above a stagnant pool along the Maravillas."

A New Mexico nest, reported by Mrs. Bailey (1928), "was found by Mr. Ligon in the top branches of a walnut tree growing in a canyon bed. Its one egg was eaten and the nest destroyed by a Woodhouse Jay."

The Arizona nest, referred to above, was taken by Frank C. Willard on June 9, 1922, near Fairbank in the San Pedro Valley. It was placed in the extreme top of a large willow, 35 feet from the ground. The nest was made of grass and green weed stems, with a lining of fine grass.

Eggs.—The four eggs in the normal set for Cooper's tanager are apparently indistinguishable from the eggs of the summer tanager. The measurements of 38 eggs average 23.3 by 17.4 millimeters; the eggs showing the four extremes measure 25.4 by 18.3, 22.9 by 18.8, 21.8 by 17.4, and 22.4 by 16.3 millimeters.

Food.—Evidently, Cooper's tanager is quite as fond of honey bees as is its eastern relative. In a letter to Herbert Brandt, H. E. Weisner, who operates a large apiary near Tucson, Ariz., complains of the damage done to his bees by this and the western tanager. He writes as follows: "It was several years before I realized the fact that their food in the areas about my apiaries consisted almost entirely of bees, and worker bees at that. Or, I had better say parts of bees, for they skillfully avoided contact with the stinger end of their victims by breaking off that end. They accomplished this by catching the bees across the middle of the body and, upon alighting on a branch or other perch, breaking off the protruding end of the abdomen by giving it a 'swipe' on the perch. * * * All summer long, the top of nearly every hive was sprinkled with the abdomens of bees, and since just as many fall upon each equal area of uncovered ground throughout the apiary, it is very evident that the toll was very great."

DISTRIBUTION

Range.—Southwestern United States to central México.

Breeding range.—Cooper's summer tanager breeds from southeastern California (Colorado River Valley from Needles to Potholes), southern Nevada (Colorado River opposite Fort Mojave), central-western, central, and southeastern Arizona (Fort Mohave, Aquarius and Juniper Mountains, the Tonto Basin, Clifton), southwestern, central, and southeastern New Mexico (Cooney, Los Pinos; probably Carlsbad), western Texas (Frijole, Davis Mountains, Brewster County), and northeastern Coahuila (Sabinas); south to northeastern Baja California (Cerro Prieto), central-northern and southeastern Sonora (Rancho la Arizona, Magdalena, Opodepe, Guirocoba), northern Durango (Río Sestín), southeastern Coahuila (Sierra de Guadalupe), and central Nuevo León (Cerro de la Silla, Allende, Montemorelos).

Winter range.—Winters in southern Baja California (San José del Cabo), southern Sinaloa (Mazatlán), Michoacán (Los Reyes, Mount Tancítaro), Morelos (Morelos) and central Guerrero (Chilpancingo).

Casual records.—Casual in southwestern California (Santa Barbara, Hueneme, Pasadena, San Clemente Island).

Literature Cited

AIKEN, CHARLES EDWARD HOWARD, and WARREN, EDWARD ROYAL
  1914. Birds of El Paso County, Colorado. Colorado College Publ., gen.
      ser. No. 74 (sci. ser., vol. 12, No. 13, pt. 2), pp. 497–603.
ALEXANDER, GORDON
  1936. Eastern summer tanager in Colorado. Auk, vol. 53, p. 452.
ALLEN, ARTHUR AUGUSTUS
  1914. The red-winged blackbird: A study in the ecology of a cat-tail marsh.
      Abstr. Proc. Linn. Soc., New York, Nos. 24–25, 1911–1913, pp.
      43–128.
  1944. Touring for birds with microphone and color camera. Nat. Geogr.
      Mag., vol. 85, pp. 689–696.
ALLEN, CHARLES N.
  1881. Songs of the western meadowlark (*Sturnella neglecta*). Bull. Nuttall
      Ornith. Club, vol. 6, pp. 145–150.
ALLEN, CHARLES SLOVER
  1892. Breeding habits of the fish hawk on Plum Island, New York. Auk.
      vol. 9, pp. 313–321.
ALLEN, FRANCIS HENRY
  1922. Some little known songs of common birds. Natural History, vol. 22,
      pp. 235–242.
ALLEN, GLOVER MORRILL
  1925. Birds and their attributes.
ALLEN, JOEL ASAPH
  1868. Notes on birds observed in western Iowa, in the months of July,
      August and September; also on birds observed in northern Illinois
      in May and June, and at Richmond, Wayne Co., Indiana, between
      June third and tenth. Mem. Boston Soc. Nat. Hist., vol. 1, pp.
      488–526.
ALLIN, ALBERT ELLIS, and DEAR, LIONEL SEXTUS
  1947. Brewer's blackbird breeding in Ontario. Wilson Bull., vol. 59, pp.
      175–176.
AMERICAN ORNITHOLOGISTS' UNION
  1910. Check-List of North American birds, ed. 3.
  1931. Check-List of North American birds, ed. 4.
  1944. Nineteenth supplement to the American Ornithologists' Union Check-
      List of North American birds. Auk, vol. 61, pp. 441–464.
  1957. Check-List of North American birds, ed. 5.
ANDERSON, RUDOLPH MARTIN
  1907. The birds of Iowa. Proc. Davenport Acad. Sci., vol. 40, pp. 125–417.
ANTHONY, ALFRED WEBSTER
  1894. *Icterus parisorum* in western San Diego County, Calif. Auk, vol. 11,
      pp. 327–328.
  1921. Strange behavior of a Bullock's oriole. Auk, vol. 38, p. 277.
ATTWATER, HENRY PHILEMON
  1892. List of birds observed in the vicinity of San Antonio, Bexar County,
      Texas. Auk, vol. 9, pp. 229–238.

AUDUBON, JOHN JAMES
    1834. Ornithological biography, vol. 2.
    1839. Ornithological biography, vol. 5.
    1841. The birds of America, vols. 2 and 3.
    1842. The birds of America, vol. 4.
    1844. The birds of America, vol. 7.
BAGG, AARON CLARK, and ELIOT, SAMUEL ATKINS, Jr.
    1937. Birds of the Connecticut Valley in Massachusetts.
BAILEY, FLORENCE MERRIAM
    1902. Handbook of birds of the western United States.
    1910. The palm-leaf oriole. Auk, vol. 27, pp. 33–35.
    1928. Birds of New Mexico.
BAILEY, HAROLD HARRIS
    1913. The birds of Virginia.
BAILLAIRGE, WILSON.
    1930. A bronzed grackle foster parent. Canadian Field-Nat., vol. 44, pp.
        166–167.
BAIRD, SPENCER FULLERTON; BREWER, THOMAS MAYO; and RIDGWAY, ROBERT
    1874. A history of North American birds: Land birds, vol. 2.
BANGS, OUTRAM
    1899. The Florida meadowlark. Proc. New England Zool. Club, vol. 1,
        pp. 19–21.
    1907. A new race of the hepatic tanager. Proc. Biol. Soc. Washington,
        vol. 20, pp. 29–30.
BARBOUR, THOMAS
    1923. The birds of Cuba. Mem. Nuttall Ornith. Club, No. 6.
BARROWS, WALTER BRADFORD
    1889. The English sparrow (*Passer domesticus*) in North America. U. S.
        Dep. Agric. Bull. 1.
    1912. Michigan bird life.
BASSETT, FRANK N.
    1931. Brewer blackbirds roosting in duck blinds. Condor, vol. 31, p. 171.
BAUMGARTNER, FREDERICK M.
    1934. Bird mortality on the highways. Auk, vol. 51, pp. 537–538.
BEAL, FOSTER ELLENBOROUGH LASCELLES
    1897. Some common birds in their relation to agriculture. U. S. Dep.
        Agric. Farmers' Bull. 54.
    1900. Food of the bobolink, blackbirds, and grackles. U. S. Dep. Agric.
        Biol. Surv. Bull. 13.
    1907. Birds of California in relation to the fruit industry, pt. 1. U. S. Dep.
        Agric. Biol. Surv. Bull. 30.
    1910. Birds of California in relation to the fruit industry, pt. 2. U. S. Dep.
        Agric. Biol. Surv. Bull. 34.
    1926. Some common birds useful to the farmer. U. S. Dep. Agric. Farmers'
        Bull. 630 (1915), rev.
    1948. Some common birds useful to the farmer. U. S. Dep. Interior,
        Conserv. Bull. 18.
BEAL, F. E. L.; McATEE, WALDO LEE; and KALMBACH, EDWIN RICHARD
    1927. Common birds of southeastern United States in relation to agriculture.
        U. S. Dep. Agric. Farmers' Bull. 755.
BENDIRE, CHARLES EMIL
    1895. Life histories of North American birds. U. S. Nat. Mus. Spec. Bull. 3.

BENNETT, F. M.
    1909. A tragedy of migration. Bird-Lore, vol. 11, pp. 110–113.
BENT, ARTHUR CLEVELAND
    1903. A North Dakota slough. Bird-Lore, vol. 5, pp. 146–151.
    1919. Life histories of North American diving birds. Order Pygopodes.
        U. S. Nat. Mus. Bull. 107.
BERGTOLD, WILLIAM HARRY
    1913. A study of the house finch. Auk, vol. 30, pp. 40–73.
    1921. The English sparrow (*Passer domesticus*) and the motor vehicle.
        Auk, vol. 38, pp. 244–250.
BETTS, NORMAN DE WITT
    1913. Birds of Boulder County, Colorado. Univ. Colorado Studies, vol. 10,
        No. 4.
BICKNELL, EUGENE PINTARD
    1884. A study of the singing of our birds. Auk, vol. 1, pp. 126–140.
BOLLES, FRANK
    1891. Land of the lingering snow.
    1894. From Blomidon to Smoky and other papers.
BOND, RICHARD MARSHALL
    1939. Observations on raptorial birds in the lava beds, Tule Lake region of
        northern California. Condor, vol. 41, pp. 54–61.
    1947. Food items from red-tailed hawk and marsh hawk nests. Condor,
        vol. 49, p. 84.
BONWELL, J. R.
    1895. A strange freak of a cowbird. Nidiologist, vol. 2, p. 153.
BRALLIAR, FLOYD
    1922. Knowing birds through stories.
BRAND, ALBERT RICH
    1938. Vibration frequencies of passerine bird song. Auk, vol. 55, pp.
        263–268.
BREWSTER, WILLIAM
    1881. Notes on some birds from Arizona and New Mexico, with a description
        of a supposed new whip-poor-will. Bull. Nuttall Ornith. Club,
        vol. 6, pp. 65–73.
    1886a. An ornithological reconnaissance in western North Carolina. Auk,
        vol. 3, pp. 94–112.
    1886b. Bird migration. Mem. Nuttall Ornith. Club, No. 1.
    1902. Birds of the Cape region of Lower California. Bull. Mus. Comp.
        Zool., vol. 41, pp. 1–241.
    1906. The birds of the Cambridge region of Massachusetts. Mem. Nuttall
        Ornith. Club, No. 4.
    1936. October Farm. Compiled from the Concord journals and diaries of
        William Brewster.
    1937. The birds of the Lake Umbagog region of Maine, pt. 3. Bull Mus.
        Comp. Zool., vol. 66, pp. 408–521.
BROOKS, ALLAN
    1928. Are the boat-tailed and great-tailed grackles specifically distinct?
        Auk, vol. 45, pp. 506–507.
    1932. The iris of the Florida boat-tailed grackle. Auk, vol. 49, pp. 94–95.
BRUNER, LAWRENCE
    1896. Some notes on Nebraska birds. Rep. Nebraska State Hort. Soc.
        1896, pp. 48–178.

BRYANT, HAROLD CHILD
1911. The relation of birds to an insect outbreak in northern California during the spring and summer of 1911. Condor, vol. 13, pp. 195–208.
1912. Birds in relation to a grasshopper outbreak in California. Univ. California Publ. Zool., vol. 11, pp. 1–20.
1914. A determination of the economic status of the western meadowlark (*Sturnella neglecta*) in California. Univ. California Publ. Zool., vol. 11, pp. 377–510.

BRYANT, WALTER [Pierc]E
1890. A catalogue of the birds of Lower California, Mexico. Proc. California Acad. Sci., ser. 2, vol. 2, pp. 237–320.

BUCKALEW, HERBERT
1934. Nesting of boat-tailed grackle and blue-winged teal in Delaware. Auk, vol. 51, p. 384.

BURLEIGH, THOMAS DEARBORN
1921. Breeding birds of Warland, Lincoln Co., Montana. Auk, vol. 38, pp. 552–565.
1925. Notes on the breeding habits of some Georgia birds. Auk, vol. 42, pp. 396–401.
1931. Notes on the breeding birds of State College, Centre County, Pennsylvania. Wilson Bull., vol. 43, pp. 37–54.
1936. Egg laying by the cowbird during migration. Wilson Bull., vol. 48, pp. 13–16.
1939. Alta Mira oriole in Texas—an addition to the A. O. U. "Check-list." Auk, vol. 56, pp. 87–88.
1944. The bird life of the Gulf coast region of Mississippi. Occas. Pap. Mus. Zool. Louisiana State Univ., No. 20, pp. 329–490.

BURLEIGH, THOMAS DEARBORN, and LOWERY, GEORGE HINES, Jr.
1940. Birds of the Guadalupe Mountain region of western Texas. Occas. Pap. Mus. Zool. Louisiana State Univ., No. 8.

BURLEIGH, THOMAS DEARBORN, and PETERS, HAROLD SEYMOUR
1948. Geographic variation in Newfoundland birds. Proc. Biol. Soc. Washington, vol. 61, pp. 111–124.

BURNS, FRANKLIN LORENZO
1915. Comparative periods of deposition and incubation of some North American birds. Wilson Bull., vol. 27, pp. 275–286.

BURROUGHS, JOHN
1879. Locusts and wild honey.

BUTLER, AMOS WILLIAM
1898. The birds of Indiana. Indiana Dep. Geol. and Nat. Resources, 22nd. Ann. Rep. (1897), pp 515–1197.

BUTTRICK, P. L.
1909. Observations on the life history of the bobolink. Bird-Lore, vol. 11, pp. 125–126.

CAMERON, EWAN SOMERLED
1907. The birds of Custer and Dawson Counties, Montana. Auk, vol. 24, pp. 389–406.

CAMPBELL, H. C.
1891. Orchard orioles nesting near kingbirds. Ornithologist and Oologist, vol. 16, p. 88.

CAMPBELL, LOUIS WALTER
1936. The subspecies of red-winged blackbirds wintering near Toledo. Wilson Bull., vol. 48, pp. 311–312.

CARRIKER, MELBOURNE ARMSTRONG, JR.
  1910.   An annotated list of the birds of Costa Rica, including Cocos Island.
          Ann. Carnegie Mus., vol. 6, pp. 314–915.
CASSIN, JOHN
  1862.   Illustrations of the birds of California, Texas, Oregon, British and
          Russian America.
CATESBY, MARK
  1731.   The natural history of Carolina, Florida and the Bahama Islands.
CHAPMAN, FRANK MICHLER
  1890.   On the winter distribution of the bobolink (*Dolichonyx oryzivorous*)
          with remarks on its routes of migration.   Auk, vol. 7, pp. 39–45.
  1892.   A preliminary study of the grackles of the subgenus *Quiscalus*.
          Bull. Amer. Mus. Nat. Hist., vol. 4, pp. 1–20.
  1912.   Handbook of the birds of Eastern North America, rev. ed.
  1921a. Notes on the plumage of North American birds, fifty-ninth paper.
          Bird-Lore, vol. 23, pp. 83–84.
  1921b. Notes on the plumage of North American birds, sixtieth paper.   Bird-
          Lore, vol. 23, pp. 195–196.
  1922.   Notes on the plumage of North American birds, sixty-fourth paper.
          Bird-Lore, vol. 24, p. 204.
  1923a. Notes on the plumage of North American birds, sixty-sixth paper.
          Bird-Lore, vol. 25, p. 128.
  1923b. Notes on the plumage of North American birds, sixty-eighth paper.
          Bird-Lore, vol. 25, p. 389.
  1935a. Further remarks on the relationships of the grackles of the subgenus
          Quiscalus.   Auk, vol. 52, pp. 21–29.
  1935b. *Quiscalus quiscula* in Louisiana.   Auk, vol. 52, pp. 418–420.
  1936.   Further remarks on *Quiscalus* with a report on additional specimens
          from Louisiana.   Auk, vol. 53, pp. 405–417.
  1939a. *Quiscalus* in Mississippi.   Auk, vol. 56, pp. 28–31.
  1939b. Nomenclature in the genus *Quiscalus*.   Auk, vol. 56, pp. 364–365.
  1940.   Further studies of the genus *Quiscalus*.   Auk, vol. 57, pp. 225–233.
CHRISTOFFERSON, KARL
  1927.   The bronzed grackle as a bird of prey.   Bird-Lore, vol. 29, p. 119.
COATNEY, G. ROBERT, and WEST, EVALINE.
  1938.   Some blood parasites from Nebraska birds.   II.   Amer. Midland Nat.,
          vol. 19, pp. 601–612.
COMPTON, LAWRENCE VERLYN
  1947.   The great-tailed grackle in the upper Rio Grande Valley.   Condor,
          vol. 49, pp. 35–36.
COOKE, MAY THACHER
  1937.   Some longevity records of wild birds.   Bird-Banding, vol. 8, pp. 52–65.
  1943.   Returns from banded birds: Some miscellaneous recoveries of interest.
          Bird-Banding, vol. 14, pp. 67–74.
COTTAM, CLARENCE
  1929.   The fecundity of the English sparrow in Utah.   Wilson Bull., vol. 43,
          No. 3, pp. 193–194.
  1943.   Unusual feeding habit of grackles and crows.   Auk, vol. 60, pp.
          594–595.
COTTAM, CLARENCE, and UHLER, FRANCIS MOREY
  1935.   Bird records new or uncommon to Maryland.   Auk, vol. 52, pp.
          460–461.

COUES, ELLIOTT
    1874. Birds of the Northwest.
    1878. Birds of the Colorado Valley. U. S. Geol. Surv. Terr. Misc. Publ.
        No. 11.
COUES, ELLIOTT, and PRENTISS, DANIEL WEBSTER
    1883. Avifauna Columbiana: Being a list of birds ascertained to inhabit the
        District of Columbia, with the time of arrival, departure of such
        as are non-resident, and brief notices of habits, etc. U. S. Nat.
        Mus. Bull. 26, ed. 2.
COWAN, IAN McTAGGART
    1939. The vertebrate fauna of the Peace River District of British Columbia.
        Occas. Pap. British Columbia Prov. Mus., No. 1.
    1942. Termite-eating birds in British Columbia. Auk, vol. 59, p. 451.
CURRIER, EDMONDE SAMUEL
    1904. Summer birds of the Leach Lake region, Minnesota. Auk, vol. 21,
        pp. 29–44.
DAVIS, MALCOLM
    1944. Purple grackle kills English sparrow. Auk, vol. 61, pp. 139–140.
DAWSON, WILLIAM LEON
    1903. The birds of Ohio.
    1923. The birds of California.
    1927. The sociable redwing. Nat. Mag., Oct. 1927, pp. 227–229.
DAWSON, WILLIAM LEON, and BOWLES, JOHN HOOPER
    1909. The birds of Washington, vol. 1.
DEANE, RUTHVEN
    1895. Strange habits of the rusty and crow blackbirds. Auk, vol. 12, pp.
        303–304.
    1908. Destruction of English Sparrows. Auk, vol. 25, pp. 477–478.
DEMERITT, WILLIAM W.
    1936. *Agelaius humeralis* a new bird for North America. Auk, vol. 53, p.
        453.
DENIG, ROBERT L.
    1913. A meadowlark's unusual nest-site. Bird-Lore, vol. 15, pp. 113–114.
DICKEY, DONALD RYDER, and VAN ROSSEM, ADRIAAN JOSEPH
    1938. The birds of El Salvador. Publ. Field Mus. Nat. Hist., zool. ser,
        vol. 23.
DINGLE, EDWARD VON SIEBOLD
    1932. The color of the iris in the boat-tailed grackle. Auk, vol. 49, pp.
        356–357.
DU MONT, PHILIP ATKINSON
    1931. Summary of bird notes from Pinellas County, Florida. Auk, vol.
        48, pp. 246–255.
DWIGHT, JONATHAN, JR.
    1900. The sequence of plumages and moults of the passerine birds of New
        York. Ann. New York Acad. Sci., vol. 13, pp. 73–360.
EATON, ELON HOWARD
    1914. Birds of New York. New York State Mus. Mem. 12, pt. 2.
EATON, WARREN FRANK
    1924. Decrease of the English sparrow in eastern Massachusetts. Auk, vol.
        41, pp. 604–606.

EIFRIG, CHARLES WILLIAM GUSTAVE
   1915.   Field notes from the Chicago area. Wilson Bull., vol. 27, pp. 417–419.
   1919.   Notes on birds of the Chicago area and its immediate vicinity. Auk, vol. 36, pp. 513–524.
EMERSON, WILLIAM OTTO
   1903.   A remarkable flight of Louisiana tanagers. Condor, vol. 5, pp. 64–66.
   1904.   *Icterus bullocki* as a honey-eater. Condor, vol. 6, p. 78.
EMLEN, JOHN THOMPSON, JR.
   1937.   Bird damage to almonds in California. Condor, vol. 39, pp. 192–197.
ERNST, STANTON GRANT
   1944.   Observations on the food of the bronzed grackle. Auk, vol. 61, pp. 644–645.
EWING, HENRY ELLSWORTH
   1911.   The English sparrow as an agent in the dissemination of chicken and bird mites. Auk, vol. 28, pp. 335–340.
FARLEY, FRANK LEGRANGE
   1932.   Birds of the Battle River region of Central Alberta. Institute of Applied Art, Ltd.
   1938.   Churchill, Manitoba, and its bird life. Canadian Field-Nat., vol. 52, pp. 118–119.
FAUTIN, REED WINGATE
   1941a.   Development of nestling yellow-headed blackbirds. Auk, vol. 58, pp. 215–232.
   1941b.   Incubation studies of the yellow-headed blackbird. Wilson Bull., vol. 53, pp. 107–122.
FINLEY, WILLIAM LOVELL
   1907.   American birds.
FISHER, ALBERT KENRICK
   1893.   The hawks and owls of the United States, in their relation to agriculture. U. S. Dep. Agric., Div. Ornith. and Mamm., Bull. 3.
   1896.   Summer roosts of swallows and red-winged blackbirds. Observer, vol. 7, pp. 382–384.
FLETCHER, LAURENCE B.
   1925.   A cowbird's maternal instinct. Bull. Northeastern Bird-Banding Assoc., vol. 1, pp. 22–24.
FLOYD, CHARLES BENTON
   1926.   Bronzed grackle recoveries. Bull. New England Bird-Banding Assoc., vol. 2, p. 13.
FORBES, STEPHEN A.
   1907.   An ornithological cross-section of Illinois in autumn. Bull. Illinois State Lab. Nat. Hist., vol. 7, pp. 305–335.
   1908.   The mid-summer bird life of Illinois: A statistical study. Amer. Naturalist, vol. 42, pp. 505–519.
FORBES, STEPHEN ALFRED, and GROSS, ALFRED OTTO
   1923.   On the numbers and local distribution of Illinois land birds of the open country in winter, spring and fall. Bull. Illinois Nat. Hist. Surv., vol. 15, pp. 397–519.
FORBUSH, EDWARD HOWE
   1907.   Useful birds and their protection.
   1927.   Birds of Massachusetts and other New England States, pt. 2.
   1929.   Birds of Massachusetts and other New England States, pt. 3.
FOSTER, FRANK B.
   1927.   Grackles killing young pheasants. Auk, vol. 44, p. 106.

FRAZAR, MARSTON ABBOTT
1881. Destruction of birds by a storm while migrating. Bull. Nuttall Ornith. Club, vol. 6, pp. 250–252.

FRIEDMANN, HERBERT
1925. Notes on the birds observed in the lower Rio Grande Valley of Texas during May, 1924. Auk, vol. 42, pp. 537–554.
1929. The cowbirds. A study in the biology of social parasitism.
1931. Addition to the list of birds known to be parasitized by the cowbirds. Auk, vol. 48, pp. 52–65.
1933. Further notes on the birds parasitized by the red-eyed cowbird. Condor, vol. 35, pp. 189–191.
1934. Further additions to the list of birds victimized by the cowbird. Wilson Bull., vol. 46, pp. 25–36 and 104–114.
1938. Additional hosts of the parasitic cowbirds. Auk, vol. 55, pp. 41–50.
1943. Further additions to the list of birds known to be parasitized by the cowbirds. Auk, vol. 60, pp. 350–356.
1949. Additional data on victims of parasitic cowbirds. Auk, vol. 66, pp. 154–163.

GABRIELSON, IRA NOEL
1914. Ten days' bird study in a Nebraska swamp. Wilson Bull., vol. 26, pp. 51–68.
1915. Notes on the red-winged blackbird. Wilson Bull., vol. 27, pp. 293–302.
1922. Short notes on the life histories of various species of birds. Wilson Bull., vol. 34, pp. 201–204.

GABRIELSON, IRA NOEL, and JEWETT, STANLEY GORDON
1940. Birds of Oregon.

GANDER, FRANK FORREST
1927. Swimming ability of fledgling birds. Auk, vol. 44, pp. 574–575.

GILL, GEOFFREY
1946. Age records of banded birds. EBBA News, vol. 9, No. 9.

GILLESPIE, MABEL
1930. Grackle recoveries. Bird-Banding, vol. 1, pp. 45–46.

GILLESPIE, MABEL and JOHN ARTHUR
1932. Color of the iris in grackles. Auk, vol. 49, p. 96.

GILMAN, MARSHALL FRENCH
1914. Breeding of the bronzed cowbird in Arizona. Condor, vol. 16, pp. 255–259.

GOELITZ, WALTER ADOLPH
1916. The lazy bird. Oologist, vol. 33, pp. 146–147.

GOETZ, CHRISTIAN JOHN
1938. Some bronzed grackle and blue jay age records. Bird-Banding, vol. 9, pp. 199–200.

GOSS, NATHANIEL STICKNEY
1891. History of the birds of Kansas.

GOSSE, PHILIP HENRY
1847. The birds of Jamaica.

GOWANLOCK, J. NELSON
1914. The grackle as a nest-robber. Bird-Lore, vol. 16, pp. 187–188.

GREENE, EARLE ROSENBURY
1946. Birds of the lower Florida Keys. Quart. Journ. Florida Acad. Sci., vol. 8, No. 3, pp. 199–265, 1945; and Florida Audubon Soc. Spec. Bull., 1946.

GRIMES, SAMUEL ANDREW
    1931. Notes on the orchard oriole. Florida Naturalist, vol. 5, pp. 1–7.
GRINNELL, JOSEPH
    1898. Birds of the Pacific slope of Los Angeles County. Pasadena Acad.
        Sci. Publ. No. 2.
    1908. The biota of the San Bernardino Mountains. Univ. California Publ.
        Zool., vol. 5, pp. 1–170.
    1909. A new cowbird of the genus *Molothrus*, with a note on the probable
        genetic relationships of the North American forms. Univ. Cali-
        fornia Publ. Zool., vol. 5, pp. 275–281.
    1910. The Scott oriole in Los Angeles County. Condor, vol. 12, p. 46.
    1914a. A new red-winged blackbird from the Great Basin. Proc. Biol. Soc.
        Washington, vol. 27, pp. 107–108.
    1914b. An account of the mammals and birds of the lower Colorado Valley,
        with especial reference to the distributional problems presented.
        Univ. California Publ. Zool., vol. 12, pp. 51–294.
    1915. A distributional list of the birds of California. Pacific Coast Avi-
        fauna, No. 11 (Publ. of Cooper Ornith. Club).
    1927. Six new subspecies of birds from Lower California. Auk, vol. 44,
        pp. 67–72.
GRINNELL, JOSEPH; DIXON, JOSEPH; and LINSDALE, JEAN MYRON
    1930. Vertebrate natural history of a section of northern California through
        the Lassen Peak region. Univ. California Publ. Zool., vol. 35,
        pp. 1–594.
GRINNELL, JOSEPH, and LINSDALE, JEAN MYRON
    1936. Vertebrate animals of Point Lobos Reserve. 1934–35. Carnegie
        Inst. Washington Publ. No. 481.
GRINNELL, JOSEPH, and MILLER, ALDEN HOLMES
    1944. The distribution of the birds of California. Pacific Coast Avifauna
        No. 27 (Publ. of Cooper Ornith. Club).
GRINNELL, JOSEPH, and STORER, TRACY IRWIN
    1924. Animal life in the Yosemite. Contr. Mus. Vert. Zool., Univ. of
        California.
GRISCOM, LUDLOW
    1923. Birds of the New York City region. Amer. Mus. Nat. Hist. Handb.
        Ser. No. 9.
    1932. The distribution of bird-life in Guatemala. Bull. Amer. Mus.
        Nat. Hist., vol. 64, pp. 439
    1934. The ornithology of Guerrero, México. Bull. Mus. Comp. Zool.,
        vol. 75, pp. 367–422.
GROFF, MARY EMMA and BRACKBILL, HERVEY
    1946. Purple grackles "anting" with walnut juice. Auk, vol. 63, pp. 246–
        247.
HAMAHER, J. I.
    1936. Summer tanager (*Piranga rubra*) eating wasps. Auk, vol. 53, pp.
        220–221.
HANN, HARRY WILBUR
    1937. Life history of the oven-bird in southern Michigan. Wilson Bull.,
        vol. 49, pp. 145–237.
    1941. The cowbird at the nest. Wilson Bull., vol. 53, pp. 211–221.
HARPER, FRANCIS
    1920. The song of the boat-tailed grackle. Auk, vol. 37, pp. 295–297.

HARRIS, HARRY
1919. Birds of the Kansas City region. Trans. Acad. Sci. St. Louis, vol. 23, pp. 213–371.

HENSHAW, HENRY WETHERBEE
1875. Report on the ornithological collections made in portions of Nevada, Utah, California, Colorado, New Mexico and Arizona during the years 1871, 1872, 1873 and 1874. Wheeler's Rep. Expl. Survey West of the 100th Meridian.

HERRICK, FRANCIS HOBART
1901. The home life of wild birds.
1935. Wild birds at home.

HESS, ISAAC ELMORE
1910. One hundred breeding birds of an Illinois ten-mile radius. Auk, vol. 27, pp. 19–32.

HOFFMANN, RALPH
1904. A guide to the birds of New England and eastern New York.
1927. Birds of the Pacific states.

HOLLAND, HAROLD MAY
1923. Cowbird-like behavior of red-winged blackbird. Auk, vol. 40, pp. 127–128.

HOLT, ERNEST GOLSAN, and SUTTON, GEORGE MIKSCH
1926. Notes on birds observed in southern Florida. Ann. Carnegie Mus., vol. 16, pp. 409–439.

HOWARD, WILLIAM JOHNSTON
1937. Bird behavior as a result of emergence of seventeen year locusts. Wilson Bull., vol. 49, pp. 43–44.

HOWELL, ARTHUR HOLMES
1906. Birds that eat the cotton boll weevil, a report of progress. U. S. Dep. Agric. Biol. Surv. Bull. 25.
1907. The relation of birds to the cotton boll weevil. U. S. Dep. Agric. Biol. Surv. Bull. 29.
1924. Birds of Alabama.
1932. Florida bird life.

HOWELL, ARTHUR HOLMES, and VAN ROSSEM, ADRIAAN JOSEPH
1928. A study of the red-winged blackbirds of southeastern United States. Auk, vol. 45, pp. 155–163.

HUEY, LAURENCE MARKHAM
1926. Notes from northwestern Lower California, with description of an apparently new race of the screech owl. Auk, vol. 43, pp. 347–362.
1931. *Icterus pustulatus*, a new bird to the A. O. U. Check-list. Auk, vol. 48, pp. 606–607.
1942. A vertebrate faunal survey of the Organ Pipe Cactus National Monument, Arizona. Trans. San Diego Soc. Nat. Hist., vol. 9, pp. 353–376.

HUNN, JOHN TOWNSEND SHARPLESS
1926. An oriole tragedy. Bird-Lore, vol. 28, p. 335.

HURD, THEODORE D.
1890. Nesting of the Arizona hooded Oriole at Riverside, Cal. Ornithologist and Oologist, vol. 15, p. 13.

HYDE, ARTHUR SIDNEY
1939. The ecology and economics of the birds along the northern boundary of New York State. Roosevelt Wildlife Bull., vol. 12, No. 2.

ILLINGWORTH, J. F.
    1901. The Bullock's and Arizona hooded orioles. Condor, vol. 3, pp.
        98–100.
IVOR, HANCE ROY
    1941. Observations on "anting" by birds. Auk, vol. 58, pp. 415–416.
JACKSON, HARTLEY HARRAD THOMPSON
    1923. Notes on summer birds of the Mamie Lake region, Wisconsin. Auk,
        vol. 40, pp. 478–489.
JONES, LYNDS
    1897. The Oberlin summer grackle roost. Wilson Bull., old ser. vol. 9,
        new ser. vol. 4, No. 15, pp. 39–56.
JUDD, SYLVESTER DWIGHT
    1896. Feeding habits of the English sparrow and crow. Auk, vol. 13, pp.
        285–289.
    1901. The relation of sparrows to agriculture. U. S. Dep. Agric. Biol.
        Surv. Bull. 15.
    1902. Birds of a Maryland farm. U. S. Dep. Agric. Biol. Surv. Bull. 17.
KALMBACH, EDWIN RICHARD
    1914. Birds in relation to the alfalfa weevil. U. S. Dep. Agric. Bull. 107.
    1930. English sparrow control. U. S. Dep. Agric. Leaflet 61.
    1940. Economic status of the English sparrow in the United States. U. S.
        Dep. Agric. Techn. Bull. 711.
KALTER, LOUIS B.
    1932. Influence of an osprey on bronzed grackles and pied-billed grebes.
        Wilson Bull., vol. 44, p. 42.
KENDEIGH, SAMUEL CHARLES
    1941. Birds of a prairie community. Condor, vol. 43, pp. 165–174.
KENNARD, FREDERIC HEDGE
    1920. Notes on the breeding habits of the rusty blackbird in northern New
        England. Auk, vol. 37, pp. 412–422.
KENNEDY, CLARENCE HAMILTON
    1914. The effects of irrigation on the bird life of the Yakima Valley, Wash-
        ington. Condor, vol. 16, p. 254.
    1915. Adaptability in the choice of nesting sites of some widely spread
        birds. Condor, vol. 17, p. 66.
KENYON, KARL W., and UTTAL, LEONARD J.
    1941. The strange death of a young grackle. Wilson Bull. vol. 53, p. 197.
KEYES, CHARLES R.
    1884. Notes from central Iowa. Ornithologist and Oologist, vol. 9, p. 34.
    1888. Blackbird flights at Burlington, Iowa. Auk, vol. 5, pp. 207–208.
KNIGHT, ORA WILLIS
    1908. The birds of Maine.
KNOWLTON, GEORGE F.
    1944. Red–wings eat pea aphids. Auk, vol. 61, p. 138.
KNOWLTON, GEORGE F., and HARMSTON, FRED CARL
    1943. Grasshoppers and crickets eaten by Utah birds. Auk, vol. 60, p.
        589–591.
KNOWLTON, GEORGE F., and TELFORD, P. E.
    1946. Insects eaten by Brewer's blackbirds. Auk, vol. 63, p. 589.
KOHLER, LOUIS S.
    1915a. Home life of the scarlet tanager. Oriole, vol. 3, pp. 4–8.
    1915b. An albino meadowlark. Oologist, vol. 32, p. 119.

KOPMAN, HENRY HAZLITT
1915. List of the birds of Louisiana: Part VI. Auk, vol. 32, pp. 15–29.
LACK, DAVID, and EMLEN, JOHN THOMPSON, Jr.
1939. Observations on breeding behavior in tricolored red-wings. Condor, vol. 41, pp. 225–230.
LACKEY, J. B.
1913. Notes from Mississippi. Oologist, vol. 30, pp. 257–258.
LAMB, CHESTER C.
1944. Grackle kills warbler. Condor, vol. 46, p. 245.
LANGILLE, JAMES HIBBERT
1884. Our birds in their haunts.
LA RIVERS, IRA
1941. The Mormon cricket as food for birds. Condor, vol. 43, pp. 65–69.
1944. Observations on the nesting mortality of the Brewer's blackbird, Euphagus cyanocephalus. Amer. Midland Nat., vol. 32, pp. 417–437.
LASKEY, AMELIA RUDOLPH
1937. Notes on the song of immature birds. Migrant, vol. 8, pp. 67–68.
1940. Bertram, a blackbird with personality. Bird-Lore, vol. 42, pp. 25–30.
LAWRENCE, GEORGE NEWBOLD
1874. Birds of western and northwestern Mexico, based upon collections made by Col. A. J. Grayson, Capt. J. Xantus and Ferd. Bischoff, now in the museum of the Smithsonian Institution at Washington, D. C., Mem. Boston Soc. Nat. Hist., vol. 2, pp. 265–319.
LEWIS, JOHN BARZILLAI
1931. Behavior of rusty blackbird. Auk, vol. 48, pp. 125–126.
LEWIS, WALTER E.
1925. Large flocks of wintering blackbirds at Gate, Oklahoma. Wilson Bull., vol. 37, p. 91.
LIGON, JAMES STOKLEY
1926. Nesting of the great-tailed grackle in New Mexico. Condor, vol. 28, pp. 93–94.
LINCOLN, FREDERICK CHARLES
1925. Notes on the bird life of North Dakota with particular reference to the summer waterfowl. Auk, vol. 42, pp. 50–64.
1931. Some causes of mortality among birds. Auk, vol. 48, pp. 538–546.
1935. The migration of North American birds. U. S. Dep. Agric. Circ. 363.
1939. The migration of American birds.
1940. The Arizona hooded oriole in Kansas. Auk, vol. 57, p. 420.
LINSDALE, JEAN MYRON
1931. Facts concerning the use of thallium in California to poison rodents— its destructiveness to game birds, song birds and other valuable wildlife. Condor, vol. 33, pp. 95–102.
1932. Further facts concerning losses to wild animal life through pest control in California. Condor, vol. 34, p. 134.
1936a. The birds of Nevada. Pacific Coast Avifauna, No. 23 (Publ. of Cooper Ornith. Club).
1936b. Coloration of downy young birds. Condor, vol. 38, p. 112.
1938. Environmental responses of vertebrates in the Great Basin. Amer. Midland Nat., vol. 19, pp. 1–206.

LOCKWOOD, SAMUEL
　　1872. The Baltimore oriole and carpenter bee. Amer. Naturalist, vol. 6,
　　　　pp. 721–724.
LOFBERG, LILA MCKINLEY
　　1928. Bird banding at Florence Lake, 7300 feet altitude. Condor, vol. 30,
　　　　p. 312.
　　1933. Notes on the eggs of a few Florence Lake birds. Condor, vol. 35,
　　　　p. 243.
LOWERY, GEORGE HINES., JR.
　　1938. A new grackle of the *Cassidix mexicanus* group. Occas. Pap. Mus.
　　　　Zool. Louisiana State Univ., No. 1.
　　1946. Evidence of trans-Gulf migration. Auk, vol. 63, pp. 175–211.
MACOUN, JOHN, and MACOUN, JAMES M.
　　1909. Catalogue of Canadian birds., ed. 2.
MAILLIARD, JOSEPH
　　1900. Breeding of *Agelaius tricolor* in Madera County, California. Condor,
　　　　vol. 2, pp. 122–124.
　　1910. The status of the California bi-colored blackbird. Condor, vol. 12,
　　　　pp. 63–70.
　　1914. Notes on a colony of tri-colored redwings. Condor, vol. 16, pp.
　　　　204–207.
　　1915a. The Kern redwing—*Agelaius phoeniceus aciculatus*. Condor, vol.
　　　　17, pp. 12–15.
　　1915b. Further remarks upon the Kern redwing. Condor, vol. 17, pp.
　　　　228–230.
MANWELL, REGINALD D.
　　1941. Homing instinct of the red-winged blackbird. Auk, vol. 58, pp.
　　　　184–187.
MATHEWS, FERDINAND SCHUYLER
　　1921. Field book of wild birds and their music, rev. ed.
MAYNARD, CHARLES JOHNSON
　　1883. The naturalist's guide.
　　1896. The birds of eastern North America, ed. 2.
MAYR, ERNST
　　1941. Red-wing observations in 1940. Proc. Linn. Soc. New York, 1940–
　　　　1941, pp. 75–83.
MCATEE, WALDO LEE
　　1919. Observations on the shifting range, migration and economic value
　　　　of the bobolink. Auk, vol. 36, pp. 430–431.
　　1922. Local suppression of agricultural pests by birds. Ann. Rep. Smith-
　　　　sonian Inst., 1920, pp. 411–438.
　　1926. The relation of birds to woodlots in New York State. Roosevelt
　　　　Wild Life Bull., vol. 4, pp. 7–152.
MCCABE, THOMAS TONKIN
　　1932. Wholesale poison for the red-wings. Condor, vol. 34, pp. 49–50.
MCCANN, HORACE DOLBEY
　　1931. Recoveries of purple grackles banded at Paoli, Pennsylvania, 1923–
　　　　1931. Bird-Banding, vol. 2, pp. 174–178.
MCCLURE, H. ELLIOTT
　　1945. Effects of a tornado on bird life. Auk, vol. 62, pp. 414–418.

McILHENNY, EDWARD AVERY

1936. Purple gallinules (*Ionornis martinica*) are predatory. Auk, vol. 53, pp. 327–328.

1937. Life history of the boat-tailed grackle in Louisiana. Auk, vol. 54, pp. 274–295.

1940. Sex ratio in wild birds. Auk, vol. 57, pp. 85–93.

McILWRAITH, THOMAS

1894. The birds of Ontario.

MERRILL, JAMES CUSHING

1877. Notes on *Molothrus aeneus*, Wagl. Bull. Nuttall Ornith. Club, vol. 2, pp. 85–87.

1878. Notes on the ornithology of southern Texas, being a list of birds observed in the vicinity of Fort Brown, Texas, from February, 1876, to June, 1878. Proc. U. S. Nat. Mus., vol. 1, pp. 118–173.

1888. Notes on the birds of Fort Klamath, Oregon. Auk, vol. 5, pp. 357–366.

MILLIKEN, A.

1932. A return-2 Baltimore oriole. Bird-Banding, vol. 3, p. 32.

MILLER, ALDEN HOLMES

1931. Notes on the song and territorial habits of Bullock's oriole. Wilson Bull., vol. 43, pp. 102–108.

MILNE, LORUS J.

1928. Further notes on the bronzed grackle as a fisherman. Canadian Field-Nat., vol. 42, p. 177.

MINOT, HENRY DAVIS

1877. The land-birds and game-birds of New England. Revised and edited by William Brewster.

MOUSLEY, HENRY

1916. Five years personal notes and observations on the birds of Hatley, Stanstead County, Quebec—1911–1915. Auk, vol. 33, pp. 168–186.

MULFORD, ALICE S.

1936. Ecological relations of the Brewer blackbird. (Thesis for a degree.)

MUNRO, JAMES ALEXANDER

1929. Blackbirds feeding on the forest tent caterpillar. Condor, vol. 31, p. 80.

1947. Observations of birds and mammals in central British Columbia. Occas. Pap. British Columbia Prov. Mus., No. 6.

MURIE, ADOLPH, and BRUCE, H. D.

1935. Some feeding habits of the western sandpiper. Condor, vol. 37, pp. 258–259.

MURPHEY, EUGENE EDMUND

1937. Observations on the bird life of the middle Savanna Valley, 1890–1937. Contr. Charleston Museum, No. 9.

NEFF, JOHNSON ANDREW

1937. Nesting distribution of the tri-colored red-wing.

NEFF, JOHNSON ANDREW, and MEANLEY, BROOKE

1957. Status of Brewer's blackbird on the Grand Prairie of eastern Arkansas. Wilson Bull., vol. 69, pp. 102–105.

NELSON, EDWARD WILLIAM
1897. Preliminary descriptions of new birds from Mexico and Guatemala in the collection of the United States Department of Agriculture. Auk, vol. 14, pp. 42–76.
1900. Descriptions of thirty new North American birds, in the Biological Survey collection. Auk, vol. 17, pp. 253–270.

NERO, ROBERT W.
1956. A behavior study of the red-wing blackbird. Wilson Bull., vol. 68, pp. 5–37, 129–150.

NICE, MARGARET MORSE
1931. The birds of Oklahoma. Publ. Univ. Oklahoma, vol. 3, No. 1, pp. 1–224, rev. ed.
1935. Some observations on the behavior of starlings and grackles in relation to light. Auk, vol. 52, pp. 91–92.
1937. Studies in the life history of the song sparrow. I. A population study of the song sparrow. Trans. Linn. Soc. New York, vol. 4, pp. 1–247.
1939. The watcher at the nest.
1949. The laying rhythm of cowbirds. Wilson Bull., vol. 61, pp. 231–234.

NICHOLSON, DONALD JOHN
1929. The peculiar suspiciousness of nesting southern meadowlarks. Wilson Bull., vol. 41, p. 104.

NIETHAMMER, GÜNTHER
1937. Handbuch der deutschen Vogelkunde. Band 1: Passeres.

NORRIS, JOSEPH PARKER
1890. A series of eggs of the European tree sparrow. Ornithologist and Oologist, vol. 15, pp. 24–25.
1891. English sparrows by the thousand. Ornithologist and Oologist, vol. 16, pp. 13–14.

NORRIS, RUSSELL TAPLIN
1944. Notes on a cowbird parasitizing a song sparrow. Wilson Bull., vol. 56, pp. 129–132.
1947. The cowbirds of Preston Frith. Wilson Bull., vol. 59, pp. 83–103.

NUTTALL, THOMAS
1832. A manual of the ornithology of the United States and of Canada.

OBERHOLSER, HARRY CHURCH
1907. A new *Agelaius* from Canada. Auk, vol. 24, pp. 332–336.
1919a. Description of a new red-winged blackbird from Texas. Wilson Bull., vol. 31, pp. 20–23.
1919b. Notes on the races of *Quiscalus quiscula* (Linnaeus). Auk, vol. 36, pp. 549–555.
1930. Notes on a collection of birds from Arizona and New Mexico. Sci. Publ. Cleveland Mus. Nat. Hist., vol. 1, pp. 83–124.
1938. The bird life of Louisiana. Louisiana Dep. Conserv. Bull. 28.

ODUM, EUGENE PLEASANTS, and PITELKA, FRANK ALOIS
1939. Storm mortality in a winter starling roost. Auk, vol. 56, pp. 451–455.

PACKARD, FRED MALLERY
1936. An analysis of some banding records of the eastern red-wing. Bird-Banding, vol. 7, pp. 28–37.
1937. A further analysis of some banding records of the eastern red-wing. Bird-Banding, vol. 8, pp. 139–144.
1946. Midsummer wandering of certain Rocky Mountain birds. Auk, vol. 63, pp. 152–158.

PARKS, GEORGE HAPGOOD
  1945. Strange behavior of a bronzed grackle. Bird-Banding, vol. 18, p. 144.
PEARSON, THOMAS GILBERT
  1921. Notes on the bird-life of southeastern Texas. Auk, vol. 38, pp.
          513–523.
  1925. The red-winged blackbird, *in* Portraits and habits of our birds, vol. 1,
          p. 99.
PEARSON, T. G.; BRIMLEY, CLEMENT SAMUEL; and BRIMLEY, HERBERT HUTCHIN-
  SON.
  1919. Birds of North Carolina. North Carolina Geol. and Econ. Surv.,
          vol. 4.
  1942. Birds of North Carolina, rev. ed.
PECK, CLARK J.
  1905. The Overbrook grackle roost. Cassinia, No. 9, pp. 36–39.
PELLET, FRANK C.
  1926. The fishing habit of the bronzed grackle. Wilson Bull., vol. 38,
          p. 235.
PENNOCK, CHARLES JOHN
  1931. On the color of the iris and other characteristics of the boat-tailed
          grackle. Auk, vol. 48, pp. 607–609.
PERKINS, SAMUEL E., III
  1932. Indiana bronzed grackle migration. Bird-Banding, vol. 3, pp. 85–94.
PETERS, HAROLD SEYMOUR
  1936. A list of external parasites from birds of the eastern part of the
          United States. Bird-Banding, vol. 7, pp. 9–27.
PETERS, HAROLD SEYMOUR, and BURLEIGH, THOMAS DEARBORN
  1951. The birds of Newfoundland.
PETERSON, ROGER TORY
  1939. Great-tailed grackle breeding in New Mexico. Condor, vol. 41,
          p. 217.
  1941. A field guide to western birds.
PHILLIPS, ALLAN ROBERT
  1940. Two new breeding birds for the United States. Auk, vol. 57, pp.
          117–118.
  1950. The Great-tailed Crackles of the Southwest. Condor, vol. 52, No. 2
          (March), pp. 78–81.
PHILLIPS, JOHN CHARLES
  1912. Rusty blackbirds (*Euphagus carolinus*) wintering in Essex Co., Mass.
          Auk, vol. 29, p. 395.
  1915. Notes on American and Old World English sparrows. Auk, vol. 32,
          pp. 51–59.
PHILLIPS, W. J., and KING, KENNETH M.
  1923. The corn earworm: Its ravages on field corn and suggestions for control.
          U. S. Dep. Agric. Farmers' Bull., No. 1310, p. 12.
PIERCE, FRED J.
  1921. The meadowlark as a conversationalist. Wilson Bull., vol. 33, pp.
          154–155.
PROCTOR, THOMAS
  1897. An unusual song of the red-winged blackbird. Auk, vol. 14, pp.
          319–320.

RATHBUN, SAMUEL FREDERICK
    1917.   Description of a new subspecies of the western meadowlark. Auk, vol. 34, pp. 68–70.
    1934.   Notes on the speed of birds in flight. Murrelet, vol. 15, pp. 23–24.
RAU, PHIL
    1941.   Birds as enemies of Polistes wasps. Canadian Ent., vol. 73, p. 196.
RAY, MILTON SMITH
    1909.   Some Sierran nests of the Brewer blackbird. Condor, vol. 11, pp. 194–196.
REED, CHESTER ALBERT
    1904.   North American birds' eggs.
RICHARDSON, FRANK
    1947.   Water surface feeding of blackbirds. Condor, vol. 49, p. 212.
RIDGWAY, ROBERT
    1877.   United States geological exploration of the fortieth parallel, pt. 3: Ornithology.
    1889.   The ornithology of Illinois.
    1902.   The birds of North and Middle America. U. S. Nat. Mus. Bull. 50, pt. 2.
RIPLEY, LEWIS W.
    1914.   A successful campaign against grackles and starlings in Hartford, Conn. Bird-Lore, vol. 16, pp. 362–364.
ROBERTS, THOMAS SADLER
    1909.   A study of a breeding colony of yellow-headed blackbirds; including an account of the destruction of the entire progeny of the colony by some unknown natural agency. Auk, vol. 26, pp. 371–389.
    1932.   The Birds of Minnesota, ed. 1, vol. 2.
    1936.   The Birds of Minnesota, ed. 2, vol. 2.
ROBERTSON, JOHN McB.
    1930.   Roads and birds. Condor, vol. 32, pp. 142–146.
ROWLEY, JOHN STUART
    1939.   Breeding birds of Mono County, California. Condor, vol. 41, pp. 247–254.
RUST, HENRY J.
    1917.   An annotated list of the birds of Fremont County, Idaho, as observed during the summer of 1916. Condor, vol. 19, pp. 29–43.
SAMUELS, EDWARD AUGUSTUS
    1883.   Our northern and eastern birds.
SANBORN, COLIN CAMPBELL, and GOELITZ, WALTER ADOLPH
    1915.   A two year nesting record in Lake County, Ill. Wilson Bull., vol. 27, pp. 434–448.
SAUNDERS ARETAS ANDREWS
    1914.   Birds of Teton and Lewis and Clark counties, Montana. Condor, vol. 16, pp. 124–143.
    1921.   A distributional list of the birds of Montana. Pacific Coast Avifauna, No. 14 (Publ. of Cooper Ornith. Club).
    1935.   A guide to bird songs.
SAUNDERS, WILLIAM EDWIN
    1920.   Additional notes on the birds of Reed Deer, Alberta. Auk, vol. 37, pp. 304–306.
SAVAGE, JAMES
    1895.   Some rare birds of recent occurrence near Buffalo, N. Y. Auk, vol. 12, pp. 312–314.

SCHORGER, ARLIE WILLIAM

1927. Notes on the distribution of some Wisconsin birds, I. Auk, vol. 44, p. 235–240.

1934. Notes on the distribution of some Wisconsin birds, II. Auk, vol. 51, pp. 482–486.

1941. The bronzed grackle's method of opening acorns. Wilson Bull. vol. 53, pp. 238–240.

SCOTT, WILLIAM EARL DODGE

1885. On the breeding habits of some Arizona birds. Auk, vol. 2, pp. 1–7, 159–165.

SEMPLE, JOHN BONNER

1932. Nest material of the Baltimore oriole, *Icterus galbula* (Linnaeus). Cardinal, vol. 3, pp. 69–70.

SENNETT, GEORGE BURRITT

1878. Notes on the ornithology of the lower Rio Grande of Texas, from observations made during the season of 1877. Bull. U. S. Geol. and Geogr. Surv., vol. 4, pp. 1–66.

1879. Further notes on the ornithology of the lower Rio Grande of Texas, from observations made during the spring of 1878. Bull. U. S. Geol. and Geogr. Surv., vol. 5, pp. 371–440.

SETON, ERNEST THOMPSON (Ernest Evan Seton-Thompson)

1891. The birds of Manitoba. Proc. U. S. Nat. Mus., vol. 13, pp. 457–643.

1901. Lives of the Hunted.

SHAININ, VINCENT EVERETT

1940. Bobolink rises from ocean surface. Auk, vol. 57, pp. 256–257.

SHARP, CLARENCE SAUGER

1903. Some unusual nests of the Bullock oriole. Condor, vol. 5, pp. 38–42.

SHERMAN, ALTHEA ROSINA

1932. Red-winged blackbirds nesting in treetops near top of hill. Auk, vol. 49, p. 358.

SIMMONS, GEORGE FINLAY

1925. Birds of the Austin region.

SKINNER, MILTON PHILO

1928. A guide to the winter birds of the North Carolina sandhills.

SLADE, ELISHA

1881a. Peculiar nidification of the bobolink. Bull. Nuttal Orinth. Club. vol. 6, pp. 117–118.

1881b. *Icterus baltimorei* and *Populus tremuloides*. Bull. Nuttall Ornith. Club, vol. 6, pp. 181–182.

SMITH, LOREN B., and HADLEY, CHARLES H.

1926. The Japanese beetle. U. S. Dep. Agric., Dep. Circ. 363.

SMITH, WINNIFRED

1947. Orchard orioles. Migrant, vol. 9, pp. 8–16.

SMYTH, ELLISON A., JR.

1912. Birds observed in Montgomery County, Virginia. Auk, vol. 29, pp. 508–530.

SNYDER, LESTER LYNNE

1928. On the bronzed grackle. Canadian Field-Nat., vol. 42, p. 44.

1937. Some measurements and observations from bronzed grackles. Canadian Field-Nat., vol. 51, pp. 37–39.

SORIANO, PABLO S.
  1931. Food habits and economic status of Brewer and red-winged black-
          birds. California Fish and Game, vol. 17, pp. 361–395.
SPRUNT, ALEXANDER, JR.
  1924. The scarlet tanager (*Piranga erythromelas*) on the coast of South
          Carolina. Auk, vol. 41, pp. 484–485.
  1931. Observations on the color of the iris in the boat-tailed grackle (*Mega-
          quiscalus major*). Auk, vol. 48, pp. 431–432.
  1932. Distribution of yellow and brown-eyed males of boat-tailed grackle
          in Florida. Auk, vol. 49, p. 357.
  1934. A new grackle from Florida. Charleston Mus. Leaflet, No. 6.
  1941. Predation of boat-tailed grackles on feeding glossy ibises. Auk, vol.
          58, pp. 587–588.
STAEBLER, ARTHUR E.
  1941. Number of contour feathers in the English Sparrow. Wilson Bull.,
          vol. 53, pp. 126–127.
STEPHENS, FRANK
  1903. Bird notes from eastern California and western Arizona. Condor,
          vol. 5, pp. 100–105.
STEPHENS, KATE
  1906. Scott orioles at San Diego. Condor, vol. 8, p. 130.
STEPHENS, THOMAS CALDERWOOD
  1917. A study of a red-eyed vireo's nest which contained a cowbird's egg.
          Bull. Lab. Nat. Hist., State Univ. Iowa, vol. 7, No. 4, pp. 25–38.
STEVENS, ORIN ALVA
  1925. A redwing census. Bird-Lore, vol. 27, p. 250.
STOCKARD, CHARLES RUPERT
  1905. Nesting habits of birds in Mississippi. Auk, vol. 22, pp. 273–285.
STONE, WITMER
  1897. The genus *Sturnella*. Proc. Acad. Nat. Sci. Philadelphia, pp. 146–
          152.
  1937. Bird studies at old Cape May, vol. 2.
STONER, DAYTON
  1942. Longevity and other data on a captive English Sparrow. Auk, vol.
          59, pp. 440–442.
STURGIS, BERTHA BEMENT
  1928. Field book of birds of the Panama Canal Zone.
SUMNER, EUSTACE LOWELL, JR.
  1928. Notes on the development of young screech owls. Condor, vol. 30,
          pp. 333–338.
SUTTON, GEORGE MIKSCH
  1928. The birds of Pymatuning Swamp and Conneaut Lake, Crawford
          County, Pennsylvania. Ann. Carnegie Mus., vol. 18, pp. 19–239.
  1942. A pensile nest of the red-wing. Wilson Bull., vol. 54, pp. 255–256.
SUTTON, GEORGE MIKSCH, and BURLEIGH, THOMAS DEARBORN
  1940. Birds of Tamazunchale, San Louis Potosi. Wilson Bull., vol. 52.
          pp. 221–233.
SUTTON, GEORGE MIKSCH, and PETTINGILL, OLIN SEWALL, JR.
  1943. The Alta Mira oriole and its nest. Condor, vol. 45, pp. 125–132.
SUTTON, GEORGE MISKCH, and PHILLIPS, ALLAN ROBERT
  1942. The northern races of *Piranga flava*. Condor, vol. 44, pp. 277–279.

SWARTH, HARRY SCHELWALDT
  1904. Birds of the Huachuca Mountains, Arizona. Pacif. Coast Avifauna, No. 4 (Publ. of Cooper Ornith. Club).
  1929. The faunal areas of southern Arizona: A study in animal distribution. Proc. California Acad. Sci., vol. 18, pp. 267–383.
TATE, RALPH C.
  1926. Some materials used in nest construction by certain birds of the Oklahoma panhandle. Univ. Oklahoma Bull., vol. 5, pp. 103–104.
TATUM, GEORGE F.
  1915. A reconstructed Baltimore oriole's nest. Bird-Lore, vol. 17, p. 291.
TAVERNER, PERCY ALGERNON
  1928. On the bronzed grackle. Canadian Field-Nat., vol. 42, pp. 44–45.
  1934. Birds of Canada.
  1939. The red-winged blackbirds of the Canadian prairie provinces. Condor, vol. 41, pp. 244–246.
TAYLOR, WALTER PENN
  1912. Field notes on amphibians, reptiles and birds of northern Humboldt County, Nevada, with a discussion of some of the faunal features of the region. Univ. California Publ. Zool., vol. 7, pp. 319–436.
TIPPENS, JAMIE ROSS
  1936. Some hot weather observations. Migrant, vol. 7, pp. 56–57.
TODD, WALTER EDMOND CLYDE
  1940. Birds of western Pennsylvania.
TODD, WALTER EDMOND CLYDE, and CARRIKER, MELBOURNE ARMSTRONG
  1922. The birds of the Santa Marta region of Colombia, a study of altitudinal distribution. Ann. Carnegie Mus., vol. 14, viii–611 pp.
TORREY, BRADFORD
  1894. A Florida sketch-book.
TOWNSEND, CHARLES HASKINS
  1883. Some albinos in the museum of the Philadelphia Academy. Bull. Nuttall Ornith. Club, vol. 8, p. 126.
TOWNSEND, CHARLES WENDELL
  1909. Some habits of the English sparrow (Passer domesticus). Auk, vol. 26, pp. 13–22.
  1920. Supplement to the birds of Essex County, Massachusetts. Mem. Nuttall Ornith. Club, No. 5.
  1927. Notes on the courtship of the lesser scaup, everglade kite, crow, and boat-tailed and great-tailed grackle. Auk, vol. 44, pp. 549–554.
TRAUTMAN, MILTON BERNARD
  1940. The birds of Buckeye Lake, Ohio. Misc. Publ. Mus. Zool. Univ. Michigan, No. 44.
TWOMEY, ARTHUR CORNELIUS
  1942. The birds of the Uinta Basin, Utah. Ann. Carnegie Mus., vol. 28, pp. 341–490.
TYLER, JOHN GRIPPER
  1907. A colony of tri-colored blackbirds. Condor, vol. 9, pp. 177–178.
  1913. Some birds of the Fresno district, California. Pacific Coast Avifauna, No. 9 (Publ. of Cooper Ornith. Club).
TYLER, WINSOR MARRETT
  1923. Courting orioles and blackbirds from the female bird's eye-view. Auk, vol. 40, pp. 696–697.

VAN ROSSEM, ADRIAAN JOSEPH

   1926. The California races of *Agelaius phoeniceus* (Linnaeus).   Condor, vol.
          28, pp. 215–230.

   1936. Birds of the Charleston Mountains, Nevada.   Pacific Coast Avi-
          fauna, No. 24 (Publ. of Cooper Ornith. Club).

   1945. A distributional survey of the birds of Sonora, Mexico.   Occas. Pap.
          Mus. Zool. Louisiana State Univ., No. 21.

VAN TYNE, JOSSELYN, and SUTTON, GEORGE MIKSCH

   1937. The birds of Brewster County, Texas.   Misc. Publ. Mus. Zool. Univ.
          Michigan, No. 37.

WALKER, FRED S.

   1910. The wintering of meadowlarks at Pine Point, Maine.   Auk, vol. 27,
          p. 341.

WALKINSHAW, LAWRENCE H.

   1949. Twenty-five eggs apparently laid by a cowbird.   Wilson Bull., vol. 61,
          pp. 82–85.

WARREN, BENJAMIN HARRY

   1890. Report on the birds of Pennsylvania, ed. 2.

WAYNE, ARTHUR TREZEVANT

   1910. Birds of South Carolina.   Contr. Charleston Mus., No. 1.

   1918. Some additions and other records new to the ornithology of South
          Carolina.   Auk, vol. 35, pp. 437–442.

WEAVER, RICHARD LEE

   1939. Winter observations and a study of the nesting of English sparrows.
          Bird-Banding, vol. 10, pp. 73–79.

   1942. Growth and development of English sparrows.   Wilson Bull., vol.
          54, pp. 183–191.

   1943. Reproduction in English sparrows.   Auk, vol. 60, pp. 62–74.

WEBER, JAY ANTHONY

   1912. A case of cannibalism among blackbirds.   Auk, vol. 29, pp. 394–395.

WELLMAN, GORDON BOIT

   1928. Baltimore oriole feeding on larvae of needle miner.   Auk, vol. 45,
          p. 507.

WETMORE, ALEXANDER

   1920. Observations on the habits of birds at Lake Burford, New Mexico.
          Auk, vol. 37, pp. 393–412.

   1926. The migration of birds.

   1936. The number of contour feathers in passeriform and related birds.
          Auk, vol. 53, pp. 159–169.

   1937. Observations on the birds of West Virginia.   Proc. U. S. Nat. Mus.,
          vol. 84, pp. 401–441.

   1939. Observations on the birds of northern Venezuela.   Proc. U. S. Nat.
          Mus., vol. 87, pp. 173–260.

   1943. The birds of southern Veracruz, Mexico.   Proc. U. S. Nat. Mus.,
          Vol. 93, pp. 215–340.

WETMORE, ALEXANDER, and SWALES, BRADSHAW HALL

   1931. The birds of Haiti and the Dominican Republic.   U. S. Nat. Mus.
          Bull. 155.

WHEATON, JOHN MAYNARD

   1882. Report on the birds of Ohio.   Rep. Geol. Surv. Ohio, vol. 4, pt. 1,
          Zool., pp. 187–628.

WHEELOCK, IRENE GROSVENOR
  1904. Birds of California.
  1905. Regurgitative feeding of nestlings. Auk, vol. 22, pp. 54–70.
WHITE, FRANCIS BEACH
  1937. Local notes on the birds at Concord, New Hampshire.
WIDMANN, OTTO
  1907. A preliminary catalog of the birds of Missouri. Trans. Acad. Sci.
    St. Louis, vol. 17, pp. 1–288.
WILLIAMS, J. FRED
  1940. The sex ratio in nestling eastern red-wings. Wilson Bull., vol. 52,
    pp. 267–277.
WILLIAMS, LAIDLAW
  1952. Breeding behavior of the Brewer blackbird. Condor, vol. 54, pp. 3–47.
WILSON, ALEXANDER
  1832. American ornithology, vols. 1 and 2.
WING, LEONARD
  1943. Spread of the starling and English sparrow. Auk, vol. 60, pp. 74–87.
WITHERBY, H. F.
  1919. A practical handbook of British birds, pt. 2, pp. 104–106.
WOLCOTT, ROBERT HUGH
  1899. Red-winged blackbird and cowbird. Bull. Michigan Ornith. Club,
    vol. 3, p. 18.
WOOD, HAROLD BACON
  1938. Nesting of red-winged blackbirds. Wilson Bull., vol. 50, pp. 143–144.
  1945. The sequence of molt in purple grackles. Auk, vol. 62, pp. 455–456.
WORTHINGTON, WILLIS WOODFORD, and TODD, WALTER EDMOND CLYDE
  1926. The birds of the Choctawhatchee Bay region of Florida. Wilson
    Bull., vol. 38, pp. 204–229.
WRIGHT, MABEL OSGOOD
  1907. The red-winged blackbird. Bird-Lore, vol. 9, pp. 93–96.
YARRELL, WILLIAM
  1876. A history of British birds, vol. 2, pp. 82–88.
YOUNGWORTH, WILLIAM
  1931. Unusual food of the Baltimore oriole. Wilson Bull., vol. 43, p. 58.
  1932. Notes on the nesting of the bronzed grackle and Say's phoebe. Wil-
    son Bull., vol. 44, p. 41.

Index

(Page numbers of principal entries in *italics*.)

PLATES

PLATE 1

Kent County, Mich., June 14, 1944   B. W. Baker

MALE BOBOLINK

PLATE 2

Eliot Porter

FEMALE BOBOLINK

Illinois, June 21, 1942

PLATE 3

Butler County, Pa., June 1947   H. H. Harrison

NEST OF BOBOLINK

Urbana, Ill., May 20, 1907   A. O. Gross

NEST OF EASTERN MEADOWLARK

PLATE 4

Near Toronto, Ontario   H. M. Halliday

EASTERN MEADOWLARK

PLATE 5

Duval County, Fla., May 27, 1931   S. A. Grimes

FLEDGLING SOUTHERN MEADOWLARK

PLATE 6

Mud Lake, Minn., June 1943.   R. T. Peterson
FEMALE YELLOW-HEADED BLACKBIRD AT NEST

PLATE 7

Siskiyou County, Calif., June 7, 1944   J. E. Patterson

G. A. Ammann

ADULT MALE YELLOW-HEADED BLACKBIRD AND NEST

PLATE 8

G. A. Ammann

Fifteen days old

Hennepin County, Minn., June 1929　　　S. A. Grimes

YOUNG YELLOW-HEADED BLACKBIRDS

PLATE 9

Jacksonville Beach, Fla., June 1939   S. A. Grimes   Near Toronto, Ontario   W. V. Crich

MALE (LEFT) AND FEMALE RED-WINGED BLACKBIRD

PLATE 10

Chatham, Mass., May 29, 1916 · A. C. Bent

Buffalo, N. Y., May 1927 · S. A. Grimes

NESTS OF RED-WINGED BLACKBIRD

PLATE 11

Crawford County, Pa., June 1949

H. H. Harrison

MALE ORCHARD ORIOLE AND NEST

PLATE 12

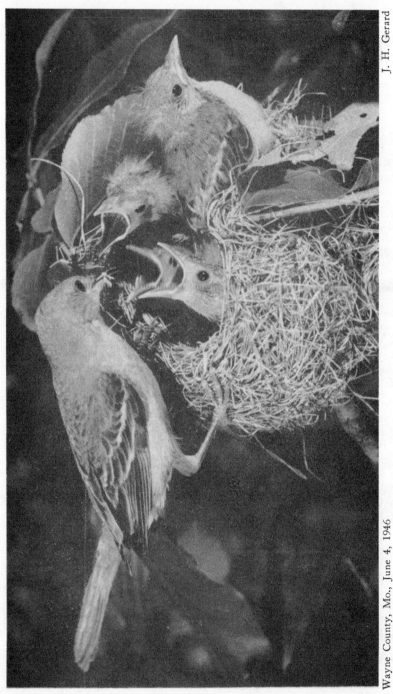

J. H. Gerard

FEMALE ORCHARD ORIOLE AND YOUNG

Wayne County, Mo., June 4, 1946

PLATE 14

Arizona   Eliot Porter

MALE ARIZONA HOODED ORIOLE AT NEST

PLATE 13

H. O. Todd, Jr.

May 26, 1947

S. A. Grimes

Duval County, Fla., May 1933

NESTS OF ORCHARD ORIOLE IN SLASH PINE (LEFT) AND 7 FEET UP IN PLUM TREE

PLATE 15

Arizona   Eliot Porter

FEMALE ARIZONA HOODED ORIOLE AT NEST

PLATE 16

Los Angeles County, Calif.,
May 11, 1946   E. M. Hall

Cochise County, Ariz., June 1, 1922
A. C. Bent

E. N. Harrison

NESTING SITES AND NEST OF SCOTT'S ORIOLE

PLATE 17

Mojave Desert, Calif., May 1922   W. M. Pierce

NEST OF SCOTT'S ORIOLE 6 FEET UP IN JOSHUA TREE

PLATE 18

Near Toronto, Ontario   H. M. Halliday

MALE BALTIMORE ORIOLE AT NEST

PLATE 19

Butler County Pa., June 1947

H. H. Harrison

FEMALE BALTIMORE ORIOLE AT NEST

PLATE 20

Cameron County, Tex., June 1940   S. A. Grimes

MALE BULLOCK'S ORIOLE AT NEST IN MESQUITE

PLATE 21

Penobscot County, Maine, June 1919
F. H. Kennard

Penobscot County, Maine, 1919
F. H. Kennard

Grand Lake, Newfoundland, June 17, 1912   A. C. Bent

NESTING SITE AND NESTS OF RUSTY BLACKBIRD

PLATE 22

Shasta County, Calif., June 15, 1944   J. E. Patterson

NEST OF BREWER'S BLACKBIRD

PLATE 23

J. H. Gerard

BREWER'S BLACKBIRD, WINTER ADULT

PLATE 24

Butler County, Pa., May 1947   H. H. Harrison

NEST AND EGGS OF BRONZED GRACKLE

PLATE 25

Butler County, Pa.   H. H. Harrison

EASTERN COWBIRD LAYING EGGS

Top: in nest of red-eyed vireo at 4:43 a.m. (E.S.T.), June 11, 1945.
Bottom: in nest of song sparrows at 4:40 a.m., May 29, 1944.

PLATE 26

Butler County, Pa., May 25, 1946   H. H. Harrison   Tompkins County, N. Y., June 9, 1911   A. D. Du Bois

EASTERN COWBIRD REMOVING EGG FROM NEST OF CHESTNUT-SIDED WARBLER
(LEFT), AND NESTLING (RIGHT)

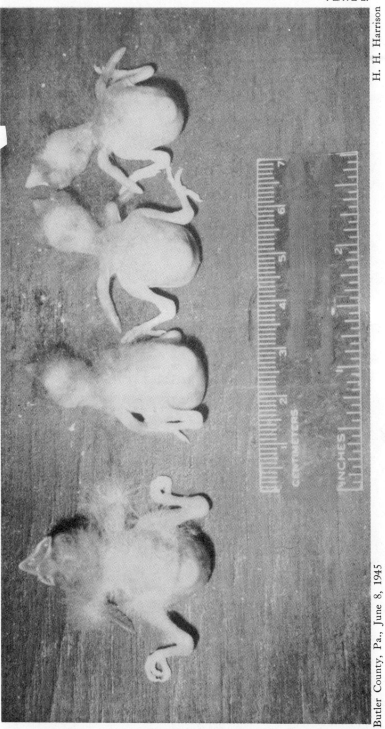

PLATE 27

H. H. Harrison

NESTLING EASTERN COWBIRD (LEFT) AND VEERIES, 1 DAY OLD

Butler County, Pa., June 8, 1945

PLATE 28

Herbert Friedmann

EASTERN COWBIRD IN COURTSHIP DISPLAY

Ithaca, N. Y.

PLATE 29

Tarentum, Pa., October 1949

H. H. Harrison

CAPTIVE EASTERN COWBIRDS, FEMALE (LEFT) AND MALE (RIGHT)

PLATE 30

Arizona   F. C. Willard

J. E. Galley

NEST OF WESTERN TANAGER AND ADULT MALE

PLATE 31

K. H. Maslowski

FEMALE SCARLET TANAGER

PLATE 32

June 1940   K. H. Maslowski

Erie County, N. Y., June 10, 1927   S. A. Grimes

FEMALE ON NEST (TOP) AND EGGS OF SCARLET TANAGER

PLATE 33

H. H. Harrison

Crawford County, Pa., June 1947

SCARLET TANAGER FAMILY, MALE (LEFT) AND FEMALE (RIGHT)

PLATE 34

Huachuca Mts., Ariz., May 26, 1922   A. C. Bent

HABITAT AND NEST OF HEPATIC TANAGER

PLATE 35

S. A. Grimes

Duval County, Fla., Apr. 30, 1926

NEST OF SUMMER TANAGER

PLATE 36

J. H. Gerard

YOUNG SUMMER TANAGERS

Madison County, Ill., June 22, 1940

PLATE 37

Near Martinsburg, W. Va., June 1949   H. H. Harrison

FEMALE SUMMER TANAGER BROODING

A CATALOGUE OF SELECTED DOVER BOOKS
IN ALL FIELDS OF INTEREST

A CATALOG OF SELECTED DOVER
BOOKS IN ALL FIELDS OF INTEREST

LASERS AND HOLOGRAPHY, Winston E. Kock. Sound introduction to burgeoning field, expanded (1981) for second edition. 84 illustrations. 160pp. 5⅜ × 8¼. (EUK) 24041-X Pa. $3.50

FLORAL STAINED GLASS PATTERN BOOK, Ed Sibbett, Jr. 96 exquisite floral patterns—irises, poppie, lilies, tulips, geometrics, abstracts, etc.—adaptable to innumerable stained glass projects. 64pp. 8¼ × 11. 24259-5 Pa. $3.50

THE HISTORY OF THE LEWIS AND CLARK EXPEDITION, Meriwether Lewis and William Clark. Edited by Eliott Coues. Great classic edition of Lewis and Clark's day-by-day journals. Complete 1893 edition, edited by Eliott Coues from Biddle's authorized 1814 history. 1508pp. 5⅜ × 8½.
21268-8, 21269-6, 21270-X Pa. Three-vol. set $22.50

ORLEY FARM, Anthony Trollope. Three-dimensional tale of great criminal case. Original Millais illustrations illuminate marvelous panorama of Victorian society. Plot was author's favorite. 736pp. 5⅜ × 8½. 24181-5 Pa. $8.95

THE CLAVERINGS, Anthony Trollope. Major novel, chronicling aspects of British Victorian society, personalities. 16 plates by M. Edwards; first reprint of full text. 412pp. 5⅜ × 8½. 23464-9 Pa. $6.00

EINSTEIN'S THEORY OF RELATIVITY, Max Born. Finest semi-technical account; much explanation of ideas and math not readily available elsewhere on this level. 376pp. 5⅜ × 8½. 60769-0 Pa. $5.00

COMPUTABILITY AND UNSOLVABILITY, Martin Davis. Classic graduate-level introduction th theory of computability, usually referred to as theory of recurrent functions. New preface and appendix. 288pp. 5⅜ × 8½. 61471-9 Pa. $6.50

THE GODS OF THE EGYPTIANS, E.A. Wallis Budge. Never excelled for richness, fullness: all gods, goddesses, demons, mythical figures of Ancient Egypt; their legends, rites, incarnations, etc. Over 225 illustrations, plus 6 color plates. 988pp. 6⅛ × 9¼. (EBE) 22055-9, 22056-7 Pa., Two-vol. set $20.00

THE I CHING (THE BOOK OF CHANGES), translated by James Legge. Most penetrating divination manual ever prepared. Indispensable to study of early Oriental civilizations, to modern inquiring reader. 448pp. 5⅜ × 8½.
21062-6 Pa. $6.50

THE CRAFTSMAN'S HANDBOOK, Cennino Cennini. 15th-century handbook, school of Giotto, explains applying gold, silver leaf; gesso; fresco painting, grinding pigments, etc. 142pp. 6⅛ × 9¼. 20054-X Pa. $3.50

AN ATLAS OF ANATOMY FOR ARTISTS, Fritz Schider. Finest text, working book. Full text, plus anatomical illustrations; plates by great artists showing anatomy. 593 illustrations. 192pp. 7⅛ × 10¼. 20241-0 Pa. $6.00

EASY-TO-MAKE STAINED GLASS LIGHTCATCHERS, Ed Sibbett, Jr. 67 designs for most enjoyable ornaments: fruits, birds, teddy bears, trumpet, etc. Full size templates. 64pp. 8¼ × 11. 24081-9 Pa. $3.95

TRIAD OPTICAL ILLUSIONS AND HOW TO DESIGN THEM, Harry Turner. Triad explained in 32 pages of text, with 32 pages of Escher-like patterns on coloring stock. 92 figures. 32 plates. 64pp. 8¼ × 11. 23549-1 Pa. $2.50

KEYBOARD WORKS FOR SOLO INSTRUMENTS, G.F. Handel. 35 neglected works from Handel's vast oeuvre, originally jotted down as improvisations. Includes Eight Great Suites, others. New sequence. 174pp. 9⅜ × 12¼.

24338-9 Pa. $7.50

AMERICAN LEAGUE BASEBALL CARD CLASSICS, Bert Randolph Sugar. 82 stars from 1900s to 60s on facsimile cards. Ruth, Cobb, Mantle, Williams, plus advertising, info, no duplications. Perforated, detachable. 16pp. 8¼ × 11.

24286-2 Pa. $2.95

A TREASURY OF CHARTED DESIGNS FOR NEEDLEWORKERS, Georgia Gorham and Jeanne Warth. 141 charted designs: owl, cat with yarn, tulips, piano, spinning wheel, covered bridge, Victorian house and many others. 48pp. 8¼ × 11.

23558-0 Pa. $1.95

DANISH FLORAL CHARTED DESIGNS, Gerda Bengtsson. Exquisite collection of over 40 different florals: anemone, Iceland poppy, wild fruit, pansies, many others. 45 illustrations. 48pp. 8¼ × 11.

23957-8 Pa. $1.75

OLD PHILADELPHIA IN EARLY PHOTOGRAPHS 1839-1914, Robert F. Looney. 215 photographs: panoramas, street scenes, landmarks, President-elect Lincoln's visit, 1876 Centennial Exposition, much more. 230pp. 8⅞ × 11¾.

23345-6 Pa. $9.95

PRELUDE TO MATHEMATICS, W.W. Sawyer. Noted mathematician's lively, stimulating account of non-Euclidean geometry, matrices, determinants, group theory, other topics. Emphasis on novel, striking aspects. 224pp. 5⅜ × 8½.

24401-6 Pa. $4.50

ADVENTURES WITH A MICROSCOPE, Richard Headstrom. 59 adventures with clothing fibers, protozoa, ferns and lichens, roots and leaves, much more. 142 illustrations. 232pp. 5⅜ × 8½. 23471-1 Pa. $3.50

IDENTIFYING ANIMAL TRACKS: MAMMALS, BIRDS, AND OTHER ANIMALS OF THE EASTERN UNITED STATES, Richard Headstrom. For hunters, naturalists, scouts, nature-lovers. Diagrams of tracks, tips on identification. 128pp. 5⅜ × 8. 24442-3 Pa. $3.50

VICTORIAN FASHIONS AND COSTUMES FROM HARPER'S BAZAR, 1867-1898, edited by Stella Blum. Day costumes, evening wear, sports clothes, shoes, hats, other accessories in over 1,000 detailed engravings. 320pp. 9⅜ × 12¼.

22990-4 Pa. $9.95

EVERYDAY FASHIONS OF THE TWENTIES AS PICTURED IN SEARS AND OTHER CATALOGS, edited by Stella Blum. Actual dress of the Roaring Twenties, with text by Stella Blum. Over 750 illustrations, captions. 156pp. 9 × 12.

24134-3 Pa. $7.95

HALL OF FAME BASEBALL CARDS, edited by Bert Randolph Sugar. Cy Young, Ted Williams, Lou Gehrig, and many other Hall of Fame greats on 92 full-color, detachable reprints of early baseball cards. No duplication of cards with *Classic Baseball Cards*. 16pp. 8¼ × 11. 23624-2 Pa. $2.95

THE ART OF HAND LETTERING, Helm Wotzkow. Course in hand lettering, Roman, Gothic, Italic, Block, Script. Tools, proportions, optical aspects, individual variation. Very quality conscious. Hundreds of specimens. 320pp. 5⅜ × 8½.

21797-3 Pa. $4.95

YUCATAN BEFORE AND AFTER THE CONQUEST, Diego de Landa. Only significant account of Yucatan written in the early post-Conquest era. Translated by William Gates. Over 120 illustrations. 162pp. 5⅜ × 8½. 23622-6 Pa. $3.50

ORNATE PICTORIAL CALLIGRAPHY, E.A. Lupfer. Complete instructions, over 150 examples help you create magnificent "flourishes" from which beautiful animals and objects gracefully emerge. 8⅛ × 11. 21957-7 Pa. $2.95

DOLLY DINGLE PAPER DOLLS, Grace Drayton. Cute chubby children by same artist who did Campbell Kids. Rare plates from 1910s. 30 paper dolls and over 100 outfits reproduced in full color. 32pp. 9¼ × 12¼. 23711-7 Pa. $2.95

CURIOUS GEORGE PAPER DOLLS IN FULL COLOR, H. A. Rey, Kathy Allert. Naughty little monkey-hero of children's books in two doll figures, plus 48 full-color costumes: pirate, Indian chief, fireman, more. 32pp. 9¼ × 12¼.
24386-9 Pa. $3.50

GERMAN: HOW TO SPEAK AND WRITE IT, Joseph Rosenberg. Like *French, How to Speak and Write It.* Very rich modern course, with a wealth of pictorial material. 330 illustrations. 384pp. 5⅜ × 8½. (USUKO) 20271-2 Pa. $4.75

CATS AND KITTENS: 24 Ready-to-Mail Color Photo Postcards, D. Holby. Handsome collection, feline in a variety of adorable poses. Identifications. 12pp. on postcard stock. 8¼ × 11. 24469-5 Pa. $2.95

MARILYN MONROE PAPER DOLLS, Tom Tierney. 31 full-color designs on heavy stock, from *The Asphalt Jungle, Gentlemen Prefer Blondes,* 22 others. 1 doll. 16 plates. 32pp. 9⅜ × 12¼. 23769-9 Pa. $3.50

FUNDAMENTALS OF LAYOUT, F.H. Wills. All phases of layout design discussed and illustrated in 121 illustrations. Indispensable as student's text or handbook for professional. 124pp. 8⅛ × 11. 21279-3 Pa. $4.50

FANTASTIC SUPER STICKERS, Ed Sibbett, Jr. 75 colorful pressure-sensitive stickers. Peel off and place for a touch of pizzazz: clowns, penguins, teddy bears, etc. Full color. 16pp. 8¼ × 11. 24471-7 Pa. $2.95

LABELS FOR ALL OCCASIONS, Ed Sibbett, Jr. 6 labels each of 16 different designs—baroque, art nouveau, art deco, Pennsylvania Dutch, etc.—in full color. 24pp. 8¼ × 11. 23688-9 Pa. $2.95

HOW TO CALCULATE QUICKLY: RAPID METHODS IN BASIC MATHE-MATICS, Henry Sticker. Addition, subtraction, multiplication, division, checks, etc. More than 8000 problems, solutions. 185pp. 5 × 7¼. 20295-X Pa. $2.95

THE CAT COLORING BOOK, Karen Baldauski. Handsome, realistic renderings of 40 splendid felines, from American shorthair to exotic types. 44 plates. Captions. 48pp. 8¼ × 11. 24011-8 Pa. $2.25

THE TALE OF PETER RABBIT, Beatrix Potter. The inimitable Peter's terrifying adventure in Mr. McGregor's garden, with all 27 wonderful, full-color Potter illustrations. 55pp. 4¼ × 5½. (Available in U.S. only) 22827-4 Pa. $1.50

BASIC ELECTRICITY, U.S. Bureau of Naval Personnel. Batteries, circuits, conductors, AC and DC, inductance and capacitance, generators, motors, trans-formers, amplifiers, etc. 349 illustrations. 448pp. 6½ × 9¼. 20973-3 Pa. $7.95

THE PRINCIPLE OF RELATIVITY, Albert Einstein et al. Eleven most important original papers on special and general theories. Seven by Einstein, two by Lorentz, one each by Minkowski and Weyl. 216pp. 5⅜ × 8½.   60081-5 Pa. $3.50

PINEAPPLE CROCHET DESIGNS, edited by Rita Weiss. The most popular crochet design. Choose from doilies, luncheon sets, bedspreads, apron—34 in all. 32 photographs. 48pp. 8¼ × 11.   23939-X Pa. $2.00

REPEATS AND BORDERS IRON-ON TRANSFER PATTERNS, edited by Rita Weiss. Lovely florals, geometrics, fruits, animals, Art Nouveau, Art Deco and more. 48pp. 8¼ × 11.   23428-2 Pa. $1.95

SCIENCE-FICTION AND HORROR MOVIE POSTERS IN FULL COLOR, edited by Alan Adler. Large, full-color posters for 46 films including King Kong, Godzilla, The Illustrated Man, and more. A bug-eyed bonanza of scantily clad women, monsters and assorted other creatures. 48pp. 10¼ × 14¼. 23452-5 Pa. $8.95

TECHNICAL MANUAL AND DICTIONARY OF CLASSICAL BALLET, Gail Grant. Defines, explains, comments on steps, movements, poses and concepts. 15-page pictorial section. Basic book for student, viewer. 127pp. 5⅜ × 8½.
21843-0 Pa. $2.95

STORYBOOK MAZES, Dave Phillips. 23 stories and mazes on two-page spreads: Wizard of Oz, Treasure Island, Robin Hood, etc. Solutions. 64pp. 8¼ × 11.
23628-5 Pa. $2.25

PUNCH-OUT PUZZLE KIT, K. Fulves. Engaging, self-contained space age entertainments. Ready-to-use pieces, diagrams, detailed solutions. Challenge a robot, split the atom, more. 40pp. 8¼ × 11.   24307-9 Pa. $3.50

THE HUMAN FIGURE IN MOTION, Eadweard Muybridge. Over 4500 19th-century photos showing stopped-action sequences of undraped men, women, children jumping, running, sitting, other actions. Monumental collection. 390pp. 7⅞ × 10⅝.   20204-6 Clothbd. $18.95

PHOTOGRAPHIC SKETCHBOOK OF THE CIVIL WAR, Alexander Gardner. Reproduction of 1866 volume with 100 on-the-field photographs: Manassas, Lincoln on battlefield, slave pens, etc. 224pp. 10⅝ × 8¼.   22731-6 Pa. $6.95

FLORAL IRON-ON TRANSFER PATTERNS, edited by Rita Weiss. 55 floral designs, large and small, realistic, stylized; poppies, iris, roses, etc. Victorian, modern. Instructions. 48pp. 8¼ × 11.   23248-4 Pa. $1.95

AUTOBIOGRAPHY: The Story of My Experiments with Truth, Mohandas K. Gandhi. Boyhood, legal studies, purification, the growth of the Satyagraha (nonviolent protest) movement. Critical, inspiring work of the man who freed India. 480pp. 5⅜ × 8½.   24593-4 Pa. $6.95

ON THE IMPROVEMENT OF THE UNDERSTANDING, Benedict Spinoza. Also contains Ethics, Correspondence, all in excellent R Elwes translation. Basic works on entry to philosophy, pantheism, exchange of ideas with great contemporaries. 420pp. 5⅜ × 8½.   20250-X Pa. $5.95

Prices subject to change without notice.

Available at your book dealer or write for free catalog to Dept. GI, Dover Publications, Inc., 31 East 2nd St. Mineola, N.Y. 11501. Dover publishes more than 175 books each year on science, elementary and advanced mathematics, biology, music, art, literary history, social sciences and other areas.